Springer Protocols Handbooks

For further volumes:
http://www.springer.com/series/8623

Springer Protocols Handbooks collects a diverse range of step-by-step laboratory methods and protocols from across the life and biomedical sciences. Each protocol is provided in the Springer Protocol format: readily-reproducible in a step-by-step fashion. Each protocol opens with an introductory overview, a list of the materials and reagents needed to complete the experiment, and is followed by a detailed procedure supported by a helpful notes section offering tips and tricks of the trade as well as troubleshooting advice. With a focus on large comprehensive protocol collections and an international authorship, Springer Protocols Handbooks are a valuable addition to the laboratory.

Characterization of Plant Viruses

Methods and Protocols

Alangar Ishwara Bhat

Division of Crop Protection, ICAR-Indian Institute of Spices Research, Kozhikode, Kerala, India

Govind Pratap Rao

Division of Plant Pathology, ICAR-Indian Agricultural Research Institute, New Delhi, India

Humana Press

Alangar Ishwara Bhat
Division of Crop Protection
ICAR-Indian Institute of Spices Research
Kozhikode, Kerala, India

Govind Pratap Rao
Division of Plant Pathology
ICAR-Indian Agricultural Research Institute
New Delhi, India

ISSN 1949-2448 ISSN 1949-2456 (electronic)
Springer Protocols Handbooks
ISBN 978-1-0716-0333-8 ISBN 978-1-0716-0334-5 (eBook)
https://doi.org/10.1007/978-1-0716-0334-5

This Humana imprint is published by the registered company Springer Science+Business Media, LLC, part of Springer Nature.
The registered company address is: 233 Spring Street, New York, NY 10013, U.S.A.

Foreword

Characterization of Plant Viruses: Methods and Protocols written by Drs. Bhat and Rao provides a timely overview of all major protocols currently used in the discipline of plant virology. These include classic methods of indicator host inoculation, insect transmission, and virion isolation. Also addressed are serological and nucleic acid-based methods in widespread use. Inclusion of recently developed biotechnological approaches for plant virus diagnosis, characterization, and management brings the reader to the cutting edge of our field. Thus, this new book will be a valuable reference for students and practitioners of all levels with an interest in plant virus characterization.

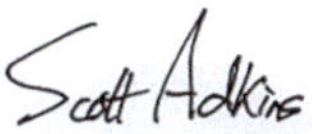

USDA, ARS, USHRL, Fort Pierce, FL, USA **_Scott Adkins_**

Preface

Viruses caused diseases in plants for many centuries before they were described and shown as the causal agents. After the first identification of the mosaic disease in tobacco caused by a virus in the early 1900s, many viral diseases on different crops are described. Intensive cultivation of crops coupled with changing climate scenario has made virus disease as one of the major production constraints in cereals, vegetables, fruits, and other crops. At present, more than 1300 viruses infecting different crops worldwide have been characterized. Enormous developments have been made in the diagnosis of plant viruses, and efficient detection tools are available for large number of viruses. Biological, physicochemical, protein, and nucleic acid-based methods are the broad methods used for diagnosis of viruses. The biological diagnosis methods such as symptoms, isolation, host range, and transmission play an important role in the preliminary identification of the viral pathogen. Serological (protein) and nucleic acid-based methods offer more reliable and sensitive methods for detection. Among serological methods, various forms of enzyme-linked immunosorbent assays (ELISA) have become very popular and are largely used for routine detection of viruses. Lateral flow assay is an onsite detection method that can be used by the cultivator himself in the field without the aid of either a laboratory or technical knowledge. Among the different nucleic acid-based methods, polymerase chain reaction (PCR) and real-time PCR are the most sensitive methods (10^2–10^5 times more than ELISA) available for detection and is important when viruses occur at low concentrations. It also has the potential to detect more than one pathogen in one reaction, and diagnosis is amenable to automation. Isothermal amplification methods such as loop-mediated isothermal amplification (LAMP), rolling circle amplification (RCA), and recombinase polymerase amplification (RPA) are recently added assay methods that can be performed in laboratory with minimum facilities and equally or even more sensitive than PCR. When sufficient sequence information of pathogens is available, microarray technology can be used to detect large number of pathogens in a single reaction. Next-generation sequencing (NGS) is an unbiased approach for the identification of viruses when all other methods fail. There is a quite good advancement in the management of viruses especially production of virus-resistant transgenic crops using various approaches including the latest CRISPR-Cas system. Elimination of viruses and production of virus-free plants are possible using approaches such as somatic embryogenesis and meristem-tip culture.

The objective of this book on *Characterization of Plant Viruses: Methods and Protocols* is an attempt to update and describe the protocols for the biological, serological, and nucleic acid-based assays for detection, diagnosis, and management of plant viruses. The book contains 50 chapters, Appendix and Glosarry. Chapters 1 to 19 of the book deals with techniques used in the biological characterization of plant viruses such as symptoms, host range, transmission by mechanical, graft, and different vectors including insects, fungi, mites, and nematodes. Chapters 20 to 26 of the book deals with the protocols for purification of viruses, electron microscopy, coat protein molecular weight determination, and nature of viral nucleic acids. Chapters 27 to 30 consists of *in vitro* expression of coat protein, production of antibodies, and various serological assays. Isolation of nucleic acids, PCR assays, isothermal amplification assays, and next-generation sequencing approaches are dealt in Chaps. 31 to 41 of the book. Chapters 42 to 46 of the book deals with characterization of the virus through cloning, sequencing, sequence analysis, and production of infectious clones while Chaps. 47 to 50 of the book provides biotechnological approaches for management of plant viruses. There are numerous colleagues to whom we are indebted for their valuable support in correction of the text and providing images used in the book. We most sincerely acknowledge and thank the publisher, Springer Nature for their help in various ways to publish this book in time. We strongly hope that this book will be useful to every laboratory, student, teacher, and researcher in plant virology, plant pathology, plant biology, and molecular biology and serve as a practical manual on identification, characterization, and management of plant viruses.

Kozhikode, India ***Alangar Ishwara Bhat***
New Delhi, India ***Govind Pratap Rao***

About the Book

The book on *Characterization of Plant Viruses: Methods and Protocols* gives detailed methodology used in the biological, serological, and nucleic acid-based assays for the plant virus detection, diagnosis, and management. The book contains 50 chapters, appendix, and glossary. Chapters 1 to 19 of the book deals with techniques used in the biological characterization of viruses such as symptoms, host range, transmission of viruses by mechanical, graft, and different vectors including insects, fungi, mites, and nematodes. Chapters 20 to 26 of the book deals with protocols for purification of viruses representing different species, properties of purified virus, and techniques for physicochemical properties such as determination of molecular weight of coat protein, isolation, and determination of nature of virus nucleic acid. Chapters 27 to 29 consists of *in vitro* expression of coat protein of viruses and production of polyclonal and monoclonal antibodies. Various serological assays from precipitin tests to ELISA, dot-blot, tissue blot, and electro-blot immunoassay, immunosorbent electron microscopy (ISEM), immunoflourescence, and lateral flow immuno assay (LFIA) are also dealt in Chap. 30 of the book. Isolation of DNA and RNA from virus-infected plants and nucleic acid-based assays such as dot-blot, polymerase chain reaction (PCR), real-time PCR, loop-mediated isothermal amplification (LAMP), rolling circle amplification (RCA), recombinase polymerase amplification (RPA), and next-generation sequencing approaches are dealt in Chaps. 31 to 41 of the book. Chapters 42 to 46 of the book deals with characterization of the virus through cloning, sequencing, sequence analysis, and production of infectious clones while Chaps. 47 to 50 of the book provides biotechnological approaches for management of viruses such as production of transgenic plants and use of CRISPR-Cas system for virus resistance; somatic embryogenesis and meristem-tip culture for production of virus-free plants. Appendix on DNA and protein data, online resources, and glossary are provided at the end. This book will be useful to every laboratory, student, teacher, and everyone interested in plant virology, plant pathology, plant biology, and molecular biology and serve as a practical manual on various aspects of plant viruses.

Contents

About the Authors

ALANGAR ISHWARA BHAT obtained his M.Sc. (Agri.) with specialization in Plant Pathology from University of Agricultural Sciences, Bangalore, and Ph.D. degree in Plant Pathology with specialization in Plant Virology from the ICAR-Indian Agricultural Research Institute, New Delhi. He joined Agricultural Research Service (ARS) in 1993; from 1994 to 2001 he worked as Scientist at the ICAR-Indian Agricultural Research Institute, New Delhi; from 2002 to 2008 as Senior Scientist and from 2009 onwards as Principal Scientist at the ICAR-Indian Institute of Spices Research, Kozhikode. His current area of research includes identification, characterization, development of diagnostics, and application of biotechnological approaches for management of viruses and phytoplasma infecting important spice crops such as black pepper, cardamom, ginger, and vanilla. His major contributions are on etiology, characterization, development of diagnostics, and application of biotechnological approaches for the management of plant viruses belonging to *Ampelovirus, Badnavirus, Cucumovirus, Ilarvirus, Nucleorhabdovirus, Panicovirus, Potyvirus,* and *Tospovirus* genera infecting different crop plants. Etiology of diseases such as necrosis disease in sunflower caused by *Tobacco streak virus* (TSV), bud blight of soybean and other legumes by *Groundnut bud necrosis virus* (GBNV), stunted disease in black pepper caused by *Piper yellow mottle virus* (PYMoV), chlorotic streak in cardamom caused by *Banana bract mosaic virus* (BBrMV), vein clearing disease in cardamom caused by *Cardamom vein clearing virus* (CdVCV), chlorotic fleck in ginger caused by *Ginger chlorotic fleck associated virus 1* (GCFaV-1), and GCFaV-2; mosaic disease in vanilla caused by *Cucumber mosaic virus* (CMV), *Cymbidium mosaic virus* (CymMV), *Bean yellow mosaic virus* (BYMV), and *Bean common mosaic virus* (BCMV) were identified for the first time. Cloning, molecular characterization, and sequencing of viruses infecting several crops were done to understand the possible origin and evolutionary relationship with other viruses. Complete genome sequencing of PYMoV and CMV infecting black pepper, BBrMV and CdVCV infecting cardamom, and GcFV-1 infecting ginger were done. Serological and nucleic acid-based diagnostics were developed for many viruses infecting different crops, such as multiplex PCR methodology for simultaneous detection of viruses in black pepper, cardamom, ginger, and vanilla; ELISA, RT-PCR/PCR/real-time PCR, and loop-mediated isothermal amplification (LAMP) methodology for the detection of viruses in black pepper, cardamom, ginger, and vanilla, TSV in sunflower and other crops; GBNV in legumes and tomato. Virus elimination through somatic embryogenesis and meristem culture in black pepper and vanilla were developed. Protocols were developed for genetic transformation of black pepper and vanilla using viral sequences as transgenes to get virus-resistant plants.

During April 1998 to March 1999, Dr. Bhat was a visiting Scientist under the Department of Science and Technology, Govt. of India sponsored BOYSCAST Fellowship at University of Georgia, USA, and he was on deputation under NAIP sponsored training at the Food and Environment Research Agency, UK, during April to June 2011. He has guided nine Ph.D., one M.Phil., and eight M.Sc. students. He has handled ten externally funded (DBT, SERB, and ICAR) projects. Dr. Bhat has a total of 89 publications in reputed refereed journals. He has edited two books and contributed 28 book chapters, three technical bulletins, and was awarded with several national awards.

Govind Pratap Rao is working as a Principal Scientist (Plant Pathology) at Indian Agricultural Research Institute, New Delhi. He did his M.Sc. (Botany) in 1981 and Ph.D. in Plant Virology from Gorakhpur University in 1986. He did postdoc at the University of Urbana–Champaign, Illinois, USA, with Prof. R.E. Ford on characterization of sugarcane mosaic and maize dwarf mosaic viruses in 1994 and on sugarcane yellow leaf virus at Cedex, Montpellier, France, in 1998. Dr. Rao has 32 years of research experience on plant pathology especially on plant virology and phytoplasmas. He did significant contributions in characterization of viruses and phytoplasmas infecting sugarcane, vegetables, legumes, cucurbits, ornamentals, wheat, rice, maize, cucurbits, maize, and sorghum. He is the authority of phytoplasma research in India and characterized so far more than 50 new phytoplasma diseases on different crops and weeds in India and identified several insect vectors for the identified phytoplasma strains. He has published over 150 research publications and authored and edited 25 books to his credit. He has been awarded several prestigious awards to his credit. The most important ones are: National Biotechnology Associateship Award (1991–1992), DBT, Govt. of India; Young Scientist Award (1994–1995) from DST, Govt. of India; Overseas BOYSCAST Award (1996) from DST, Govt. of India; President Award, Society for General Microbiology, UK, 1998; Best U.P. Agriculture Scientist Award (UPCAR), Govt. of Uttar Pradesh in 2002; Vigyan Ratna Award by CST, Govt. of UP for the year 2003–2004; Jin Xiu Qiu Award in 2006 by People's Govt. of Guangxi Province, Nanning, China; Global Award of Excellence, IS 2008, Al-Arish, Egypt; Dr. Ram Badan Singh Vishisht Krishi Vaigyanik Puraskar–2014 by UPCAR, Lucknow, India, and Leadership Excellence Award in Sugarcane Crop Protection by Thailand Society of Sugar Cane Technologists, Bangkok. Dr. Rao is Editor-in-Chief of *Sugar Tech*

(an international journal of sugar crops and related industries) and *Phytopathogenic Mollicutes* (an international journal of phloem-limited microorganisms). Dr. Rao is also Secretary General of Indian Virological Society, New Delhi, and member of several prestigious scientific societies and organizations like APS, USA; ASM, USA; ISSCT, Mauritius; IAPSIT, China; IPWG, Italy: SSRP, New Delhi, and IPS, New Delhi. Dr. Rao has guided 5 M.Sc. and 15 Ph.D. students on different aspects of characterization, epidemiology, and management of plant viruses and phytoplasmas. He has handled ten externally funded (DST, DBT, SERB, and ICAR) research projects from Govt. of India. Besides, Dr. Rao has visited over 30 countries as visiting scientists, expert, for invited talk, postdoc fellow, research training, panel discussion, and for attending workshop and conferences. At present, Dr. Rao is working on characterization, epidemiology, and management of viruses infected cereal crops, millets, and maize and phytoplasmas infecting important agriculture and horticultural crops in India.

Abbreviations

A	Absorption
AAP	Acquisition access period
Ab	Antibody
AFP	Acquisition feeding period
AGPC	Acid guanidinium thiocyanate phenol chloroform
APS	Ammonium persulfate
BA	Benzyl adenine
BCIP	5-Bromo-4-chloro-3-indolyl-phosphate
BIP	Backward inner primer
BL	Backward loop
BLAST	Basic local alignment search tool
bp	Base pair
BPB	Bromo phenol blue
BSA	Bovine serum albumin
Bst	*Bacillus stearothermophilus*
CBB	Coomassie brilliant blue
cDNA	Complementary DNA
CDS	Coding regions
CFA	Complete Freund's adjuvant
Ci	Curie
cm	Centimeter
CP	Coat protein
CPMR	Coat protein-mediated resistance
CRISPR	Clustered regularly interspaced short palindromic repeats
crRNA	CRISPR RNA
Ct	Cycle threshold
CTB	Cetyl trimethyl ammonium bromide
cv	Cultivar
Da	Dalton
DAC-ELISA	Direct antigen coating enzyme-linked immunosorbent assay
DAS-ELISA	Double antibody sandwich enzyme-linked immunosorbent assay
DBBJ	DNA databank of Japan
ddNTP	dideoxy nucleoside triphosphate
DEP	Dilution end point
DIBA	Dot immunobinding assay
DIECA	Diethyl dithio carbamate
DIG	Digoxigenin
DMF	Dimethyl formamide
DMSO	Dimethyl sulfoxide
DNA	Deoxyribo nucleic acid
DNase	Deoxyribonuclease
dNTP	Deoxy nucleoside triphosphate

dpm	Disintegrations per minute
DSB	Double-strand break
dsDNA	Double-stranded DNA
dsRNA	Double-stranded RNA
DTT	Dithiothreitol
E. coli	*Escherichia coli*
EBIA	Electro-blot immunoassay
EDTA	Ethylene diamine tetra methyl acid
eIF4G	Eukaryotic translation initiator factor 4G
ELISA	Enzyme-linked immunosorbent assay
EM	Electron microscopy
EMBL	European Molecular Biology Laboratory
EtBr	Ethidium bromide
$F(ab')_2$-ELISA	$F(ab')_2$-enzyme-linked immunosorbent assay
FCA	Freund's complete adjuvant
FIA	Freund's incomplete adjuvant
FIP	Forward inner primer
FL	Forward loop
g	Gram or Gravity
Gb	Giga base pair
gRNA	Guide RNA
GUS	Glucoronidase
h	Hour
HCl	Hydrogen chloride
HDR	Homology-directed repair
HEPES	*N*-2-hydroxylthylpiperazine *N′*-2-ethane sulfonic acid
His	Histidine
HP construct	Hairpin construct
HRP	Horseradish peroxidase
IAP	Inoculation access period
IC-PCR	Immunocapture polymerase chain reaction
IC-RT-PCR	Immunocapture reverse transcription polymerase chain reaction
IFA	Incomplete Freund's adjuvant
IFP	Inoculation feeding period
IgG	Immunoglobulin G
in	Inch
IP	Incubation period
IPTG	Isopropyl β-D-1-thiogalactopyranoside
ISEM	Immunosorbent electron microscopy
Kb	Kilo base pair
KDa	Kilodalton
kg	Kilogram
KN	Kinetin
L	Liter
LAMP	Loop-mediated isothermal amplification
LFA	Lateral flow assay
LFIA	Lateral flow immunoassay
LiCl	Lithium chloride

LIV	Longevity *in vitro*
LP	Latent period
m	Meter
M	Molar
mA	Milli ampere
mAb	Monoclonal antibody
Mb	Mega base pair
mCi	Milli Curie
mg	Milligram
min	Minute
mL	Milli liter
mm	Milli meter
mM	Milli molar
mPCR	Multiplex polymerase chain reaction
Mr	Relative molecular mass
mRNA	Messenger RNA
mRT-PCR	Multiplex reverse transcription polymerase chain reaction
MS medium	Murashige and Skoog medium
MSA	Multiple sequence alignment
MW	Molecular weight
NAA	Naphthalene acetic acid
NaCl	Sodium chloride
NaOH	Sodium hydroxide
NBT	Nitro blue tetrazolium
NCBI	National Center for Biotechnology Information
NCM	Nitrocellulose membrane
ng	Nanogram
NGS	Next-generation sequencing
NHEJ	Non-homologous end joining
nm	Nanometer
nmole	Nanomole
NPT	Neomycin phosphotransferase
nt	Nucleotide
O.D.	Optical density
ORF	Open reading frame
pAb	Polyclonal antibody
PAGE	Polyacrylamide gel electrophoresis
PAM	Protospacer adjacent motif
PBS	Phosphate-buffered saline
PCR	Polymerase chain reaction
PDR	Pathogen-derived resistance
PEG	Polyethylene glycol
Pg	Picogram
pmole	Picomole
pNPP	*Para* nitro phenyl phosphate
ppm	Parts per million
PTGS	Post-transcriptional gene silencing
PVDF	Polyvinylidenedifluoride

PVP	Polyvinylpyrrolidone
qPCR	Quantitative polymerase chain reaction
qRT-PCR	Quantitative reverse transcription polymerase chain reaction
RCA	Rolling circle amplification
rDNA	Recombinant DNA
Rf value	Relative distance of migration
RH	Relative humidity
RIA	Radioimmunoassay
RNA	Ribonucleic acid
RNAi	RNA interference
RPA	Recombinase polymerase amplification
rpm	Revolutions per minute
rRNA	Ribosomal RNA
RT-LAMP	Reverse transcription loop-mediated isothermal amplification
RT-PCR	Reverse transcription PCR
s	Sedimentation coefficient (expressed in Svedburg units)
SDS	Sodium dodecyl sulfate
SDS-PAGE	Sodium dodecyl sulfate-polyacrylamide gel electrophoresis
SE	Somatic embryo
s	Seconds
sgRNA	Single guide RNA
SH medium	Schenk and Hildebrandt medium
siRNA	Small interfering RNA
sRNA	Small RNA
SSB	Single stranded DNA binding protein
SSC	Saline sodium citrate
ssDNA	Single-stranded DNA
ssRNA	Single-stranded RNA
TAE	Tris-acetate-EDTA
Taq	*Thermus aquaticus*
TAS-ELISA	Triple antibody sandwich-ELISA
TBIA	Tissue blot immunoassay
TBS	Tris-buffered saline
T-DNA	Transfer DNA
TE	Tris EDTA
TEM	Transmission electron microscope
TEMED	*N*,*N*,*N*′,*N*′-tetramethylethylene diamine
Tm	Melting temperature
Tris	Tris (hydroxymethyl) amino ethane
UA	Uranyl acetate
UV	Ultraviolet
v/v	volume/volume
V	volt
var	Variety
VIGS	Virus-induced gene silencing

VSR	Viral suppressors of RNA silencing
w/v	weight/volume
w/w	weight/weight
WPM	Woody plant medium
x-gal	5-Bromo-4-chloro-3-indolyl β-D-galactopyranoside
μL	Microliter
μm	Micrometer
μM	Micromolar
μg	Microgram
°C	Degree centigrade
%	Percent
<	Less than
>	More than
=	Equal

Chapter 1

Glasshouse for Maintenance of Virus and Insect Culture

Abstract

Glasshouse is an essential requirement for plant virology works. The purpose of the glasshouse is to provide a regulated and controlled environment for maintenance of virus-free and virus-infected plants and propagation of virus in susceptible hosts and to carry out transmission experiments under insect-proof conditions. This chapter provides information on some basic requirements and precautions essential for every plant virus worker to raise and maintain healthy nursery and virus cultures to carry out experiments under regulated environment and to prevent contamination.

Key words Greenhouse, Controlled environment, Growing plants, Insect rearing, Prevention of contamination, Maintenance of virus culture

1.1 Introduction

In the glasshouse, test plants are grown under a favourable artificially controlled environment conditions, viz. temperature, humidity, light intensity, photoperiod, ventilation, soil media, disease control, irrigation, fertigation and other agronomical practices throughout the season irrespective of the natural conditions outside (Hull 2002; Dijkstra and de Jager 1998).

1.2 Construction Principles of a Glasshouse

The planning of a glasshouse starts with the acquirement of a suitable area. They should be designed to withstand wind currents of 100 km/h. The method of fixing roof to the frame should be strong enough to withstand similar velocities. In glasshouses the light, ventilation, watering, temperature and hygiene conditions are important for the good growth of plants. The amount of available light and the possibilities for an adequate temperature control are dependent on the shape of the glasshouse. The best results may be expected in glasshouses with the maximum light transmission. Plain glass sheets (Fig. 1.1a) or polyhouse with polyethylene sheet (Fig. 1.1b) are fitted in iron frames. Thermostat systems are

Alangar Ishwara Bhat and Govind Pratap Rao, *Characterization of Plant Viruses*, Springer Protocols Handbooks, https://doi.org/10.1007/978-1-0716-0334-5_1, © Springer Science+Business Media, LLC, part of Springer Nature 2020

Fig. 1.1 View of a glass/polyhouse. (**a**) A glasshouse with a double-door system; (**b**) Polyhouse with different chambers; (**c**) Polyhouse with double-door entrance; (**d**) Maintenance of the virus culture in a chamber of the glasshouse; (**e**) Maintenance of virus culture on plants with proper labelling inside glasshouse

generally installed for temperature control with proper light and water arrangements. The irrigation facility may be provided with mechanization and automation to get the irrigation accomplished at the right time and with enough precision. Misting unit (humidification system) to create mist inside the chamber to increase relative humidity (RH) need may be installed if necessary. Glasshouse should be made insect-proof. All around the glasshouse, a suitable drain channel should be provided and connected to soil sump made at least 10 m away from the glasshouse (Fig. 1.1c).

At the entrance, a double-door system (Fig. 1.1b, c) is provided to avoid inflow of air currents from outside and also to prevent direct entry of insects. The inner chamber is provided with a long corridor with separate chambers or benches on either side of the corridor (Fig. 1.1c). One side of the chamber may be used for raising the healthy nursery of host plants. Several small inoculation chambers on either side of the corridor can be made for different viruses. Each chamber is provided with cemented bench (Fig. 1.1d) for keeping pots with suitable light, temperature, watering and drainage arrangements (Maramorosch and Mahmood 2014).

1.3 Preventing Contamination in the Glasshouse

Since insects are the potential virus vectors, the glasshouse should be made with every precaution to avoid them entering into the glasshouse. Insects may enter the glasshouse through netting in glasshouse, through doors and ventilators and through clothing of the personnel entering into the glasshouse. Hence, insecticidal sprays are required at regular intervals in the glasshouse. Insect traps such as yellow sticky traps may be installed inside the glasshouse to monitor the presence of insects.

Pots should be cleaned regularly, and weeds growing in pots and other areas of the glasshouse should be removed regularly. The prophylactic control programme is again the major requirement in the glasshouse. All infected plants should be regularly removed. The soil and potting mixture used should be steam sterilized for at least 2 h at 121 °C, and the glasshouse should be fumigated with 2% formalin at regular intervals to avoid unwanted microorganisms.

Some viruses are stable for long periods outside the host and may be transmitted by only slight contact, so extreme care needs to be taken to avoid contamination with such viruses. Tobacco mosaic virus is the most infectious virus and remains infectious in dried state for a very long period. Hence, in order to avoid contamination, smoking may be forbidden in glasshouses. Hands should be thoroughly washed before and after inoculation with soap and then rectified spirit. Avoid touching anything not known to be virus-free after washing hands. Cotton cheese cloth should be discarded after use. Contaminated equipment should not be left on benches after use. After handling infected plants or making inoculation, hands should be thoroughly washed and wiped with rectified spirit.

1.4 Cleaning of Equipment

1.4.1 Glass Wares, Pestle and Mortar

Brush with detergent and water, wash thoroughly and after drying wrap them with brown paper and sterilize either by autoclaving at 15 lbs. psi for 20 min or by keeping in an incubator at 180 °C for 30 min.

1.4.2 Pipettes

Rinse them immediately after use in tap water in a pipette washer and keep them overnight in a tray with potassium dichromate and concentrated sulphuric acid solution. Transfer them to a washing cylinder for half a day. Put in the pipette washer for 1 h and place in

Fig. 1.2 Cages used for rearing insect vectors

an oven at 180 °C for 30 min or after drying put little non-absorbent cotton at the back of each pipette, wrap each one separately with brown paper and tie with thread. Sterilize in the autoclave at 15 lbs. psi for 20 min.

1.5 Glasshouse for Insect Rearing (Maramorosch and Mahmood 2014)

It is a temperature-controlled room in which insects are maintained on suitable hosts in small cages, made of fine wire nets and wood (Fig. 1.2). The insects (aphids, whiteflies, hopper, etc.) collected from the field are reared on healthy plants of the same species with regular weekly transfers. Such host plants after each transfer should be kept separately for observation to see if they produce any symptoms. The presence of virus in the insect should be checked through a series of hosts, so that it should not be the carrier of any virus. After ensuring that insects are free from viruses, they are allowed to multiply on healthy hosts. The brushes used for transmission studies should be thoroughly washed and steam sterilized after use. Whenever cages are used, they should be dry and should not be kept in very hot or cold places.

1.6 Notes

1. Always maintain the glasshouse completely insect-proof. Yellow sticky traps can be kept in every chamber of the glasshouse to monitor the presence of insects. Whenever insects are seen, take appropriate measures such as spraying with recommended insecticides preferably green-labelled ones (Maramorosch and Mahmood 2014).

2. There should be a time gap of entry to the glasshouse when a worker is coming from the field to avoid entry of insects along with the worker.
3. Test or culture plants if grown outside should be thoroughly sprayed to make them vector-free before taking inside the glasshouse.
4. Worker while entering the glasshouse should not open both the doors at the same time. It is advisable that one should open the first door, enter inside and close the first door. Then open the second door, enter inside the glasshouse and close the second door. This will help in preventing entry of any outside insects into the glasshouse (Maramorosch and Mahmood 2014).
5. The worker should be careful not to touch more than one virus-infected plant at a time, and while handling a virus, one should not touch the bench or any other implements that are kept in the glasshouse to avoid contamination through touch (Hull 2002).
6. The cultural operation should be done rapidly and carefully, and the implements used should be washed and sterilized before and after use.
7. The operator should be very careful in each step for maintenance of proper sanitation.
8. Prophylactic spraying of insecticides for the control of insect vectors from time to time is necessary.
9. Mortar, pestle, glassware and other implements used in the glasshouse should be steam sterilized in the autoclave.
10. Hands should be washed with tri-sodium phosphate soap.
11. Glasshouse and its surroundings should always be kept clean.

References

Dijkstra J, de Jager CP (1998) Practical plant virology: protocols and exercises. Springer, Amsterdam

Hull R (2002) Matthews' plant virology, 4th edn. Matthews Academic, San Diego

Maramorosch K, Mahmood F (2014) Rearing animal and plant pathogen vectors. CRC Press, Boca Raton

Chapter 2

Symptoms of Virus-Infected Plants

Abstract

Symptoms are outward expression of plants indicating deviation from the normal physiological, biochemical and biological function. Symptoms may be due to biotic or abiotic factors. Viruses are one among several biotic factors that can cause different disease symptoms in plants. All viruses cause systemic infections in susceptible plants. Symptoms caused by viruses fall into two categories, namely internal and external. In the present chapter, both external and internal symptoms caused by viruses are explained.

Key words External symptoms, Internal symptoms, Leaf curl, Local lesion, Mosaic mottle, Chlorosis, Necrosis, Ringspot, Stunting, Pin wheel inclusions

2.1 Introduction

Symptoms are the visible manifestation in plants that result due to the interaction between the virus and plant under the influence of environment. In earlier days, symptoms were the most important criteria for the characterization of viruses. Names of the most viruses have been derived based on the specific type of symptoms produced by the host. Though symptoms on the diseased plants are an easy criterion to identify the viral diseases, it is not a reliable criterion as symptoms caused by virus infection sometimes resemble those caused by other pathogens, mutations, nutrient deficiencies, physiological disorders, herbicide injury or toxicity (Gibbs and Harrison 1980). However, the distribution or sequence of symptoms in virus-infected plants varies because of the typical way in which the virus infects, spreads and multiplies in the plant. It is also very difficult to differentiate symptoms caused by one virus from the other. Symptoms caused by a particular virus in a host are also variable and may even remain latent depending on environmental conditions and seasons. This is particularly more common in perennial plants infected with the virus. Use of symptoms alone for identification and classification of viruses lead to confusion as: (1) single virus may cause variable symptoms depending on the strain; (2) mixture of viruses may affect disease expression; (3) different cultivars may react differently to a virus and (4) soil and

Alangar Ishwara Bhat and Govind Pratap Rao, *Characterization of Plant Viruses*, Springer Protocols Handbooks, https://doi.org/10.1007/978-1-0716-0334-5_2,

weather conditions may alter disease expression. To overcome these situations, experimental host plants should be raised under standard conditions in the glasshouse, which will give consistent and distinguishing disease symptoms with a particular virus. Such plants are known as indicator hosts (Matthews 1991).

Though viruses cause a wide range of symptoms, the most noticeable ones are changes in colour, necrosis and malformation of tissues. Symptoms are variable and depend on the host plant, how long it has been infected, the virus strain and environmental conditions. Proper understanding of symptoms is the basic criteria for identifying and differentiating a viral disease from disease due to fungi, bacteria, phytoplasmas and other nutritional deficiencies. Some characteristic symptoms are often associated with a particular virus or a virus group, e.g. yellow mosaics and leaf curls are mainly associated with begomoviruses, necrosis with tospo- and ilar viruses, shoe string symptoms with papaya ringspot virus and cucumber mosaic virus. Thus, proper identification of symptoms is the first step in identification which can narrow down the further steps in identification (Smith 1977; Matthews 1991; Hull 2002, 2009). Symptoms can be grouped into two major categories—external and internal.

In this chapter we have described different external and internal symptoms caused by plant viruses.

2.1.1 External Symptoms

2.1.1.1 Local Lesion

Local lesion develops near the site of entry of viruses on the leaves. They are important for biological assay. Infected cells may lose chlorophyll giving rise to chlorotic local lesions (Fig. 2.1a) or cause the death of cells, giving rise to necrotic lesions (Fig. 2.1b). Hosts producing local lesions are not common natural hosts of the viruses. Some of the experimentally used local lesion hosts include *Nicotiana glutinosa*, *Chenopodium amaranticolor*, *C. quinoa*, *Phaseolus vulgaris* and *Vigna unguiculata* (Fig. 2.1c).

Fig. 2.1 (**a**) Chlorotic local lesions on *Chenopodium amaranticolor*; (**b**) necrotic local lesions on *Nictotiana glutinosa*; (**c**) local lesions on *Vigna unguiculata*

Fig. 2.2 Stunting of banana due to banana bunchy top virus (**a**); wheat due to wheat dwarf virus (**b**)

Fig. 2.3 Mosaic mottling symptom on a dicotyledonous plant, *Nicotiana glutinosa* (**a**); Mosaic symptoms on a monocotyledonous plant, cardamom (**b**)

2.1.1.2 Systemic Symptoms

Reduction in Growth (Stunting)

Stunting of the plant growth is the most common symptom induced by virus infection. Stunting (reduction in growth) may affect all parts of the plant such as leaves, flowers, fruits, roots, petioles and internodes (Fig. 2.2a, b).

Chlorosis of Plant Parts/ Mosaic Patterns

One of the most common effects of virus infection is the development of a pattern of light green and dark green areas giving a mosaic pattern in infected leaves. In most of the dicotyledons, the mosaic is composed of irregular outlines of dark green and light green or pale or yellow-green patches (Fig. 2.3a). In monocotyledons, mosaic is composed of stripes or streaks of tissue of lighter and darker colour on the entire length of leaf lamina. The shades of colour vary from pale green to yellow, and the streaks or stripes run parallel to the length of the leaf (Fig. 2.3b).

Mosaics caused by the viruses are of different kinds:

(a) *Mottling*: If leaves display an irregular diffused chlorosis pattern of light and dark green areas, it is referred to as mottling.

Fig. 2.4 Yellow mottling symptoms on black pepper leaf (**a**); chlorotic flecks on ginger (**b**)

Fig. 2.5 Symptoms of green mosaic (**a**); elongated chlorotic streaks on maize (**b**); yellow mosaic (**c**) and golden mosaic (**d**) on cowpea

Mottling patches do not have distinct boundaries but are diffused all over the leaf surface (Fig. 2.4a).

(b) *Flecking*: If the discoloured parts are sharply bordered, it is called flecking or spotting (Fig. 2.4b).

(c) *Mosaic*: Dark and light green colour intermingled on the leaf gives a mosaic pattern. It may be green mosaic (Fig. 2.5a),

Fig. 2.6 Vein yellowing (a); vein banding (b); vein clearing (c)

yellow mosaic or golden mosaic (Fig. 2.5c, d). In monocots, the mosaic generally consists of elongated chlorotic stripes or steaks in part or throughout the length of leaf lamina (Fig. 2.5b).

(d) *Vein clearing*: Main vein and side veins of a leaf become translucent and more distinct (Fig. 2.6c).

(e) *Vein yellowing*: In some viral diseases, veins become chlorotic or yellow, and a netting appearance of yellow vein symptoms is clearly observed (Fig. 2.6a).

(f) *Vein banding*: Veins sometimes resist the yellowing process resulting in veins having a dark green appearance (Fig. 2.6b).

Ringspots

Chlorotic or necrotic rings on leaves, stems and fruits are the major symptoms in some of the virus-infected plants, such as the one caused by cucumber mosaic virus, tospovirus and potyviruses (Fig. 2.7a–d).

Fig. 2.7 Ringspot on papaya fruit skin caused by papaya ringspot virus (**a**); Rings on tomato fruit caused by tomato spotted wilt virus (**b**); Chlorotic rings on tobacco leaves caused by ringspot virus (**c**); Chlorotic ringspot on citrus leaves due to Indian citrus ringspot virus (**d**)

Fig. 2.8 Necrotic symptoms caused by groundnut bud necrosis virus on mungbean (**a**) and stem and petiole and veinal necrosis of sunflower due tobacco streak virus (**b**)

Necrosis

In necrosis, death of tissues, organs or the whole plants are seen (Fig. 2.8a, b). Necrotic patterns follow the veins as virus moves into the leaf. In some diseases, the complete leaf becomes necrotic. Necrosis spreads rapidly to the growing point, causing killing of the growing buds (bud necrosis), e.g. tospovirus in mungbean (Fig. 2.8a), and subsequently leaves may collapse and die. Necrosis may also occur on stem, petiole (Fig. 2.7b), fruits, seeds and tubers.

Symptoms on Stem

A few of the virus-infected plants cause symptoms on the stem that include bark scaling symptoms on citrus due to critrus psorosis virus (Fig. 2.9a) and stem pitting by citrus tristeza virus (Fig. 2.9b).

Fig. 2.9 Bark scaling symptoms on citrus due to citrus psorosis virus (**a**) and stem pitting on citrus due to citrus tristeza virus (**b**)

Fig. 2.10 Colour breaks in tulips (**a**) and gladiolus (**b**) caused by a potyvirus

Symptoms on Flower

Many of the mosaic-causing viruses affect the pigmentation of flowers where petals become flecked or longitudinally streaked (Fig. 2.10a, b). Colour breaking of tulip and gladiolus is caused by a *Potyvirus*.

Symptoms on Fruit

Affected fruit's skin shows mottled rings and mosaic symptoms (Fig. 2.11a, b). In some cases, seed coats of infected seeds may also be mottled. Fruit distortion in eggplant is caused by tomato bushy stunt virus. Ringspot and mosaic on the skin of cucurbits is caused by watermelon mosaic virus and ringspot on tomato by

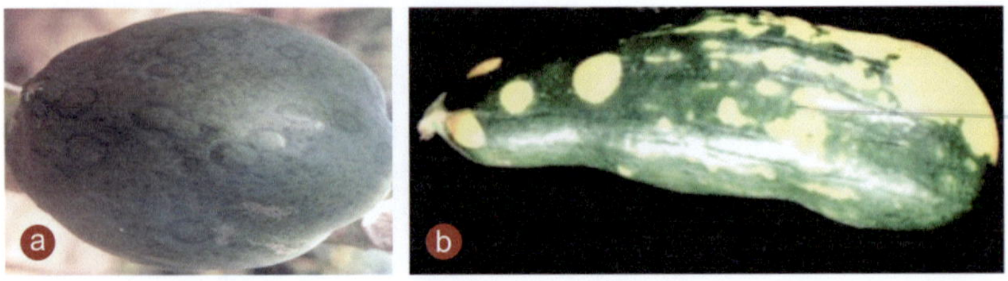

Fig. 2.11 Rings and spots on fruit skin of papaya (**a**) and pumpkin (**b**) caused by virus

Fig. 2.12 Plum pox virus symptoms on infected (**a**) plum fruits, (**b**) peach fruit, (**c**) apricot and (**d**) stone seed apricot

tomato spotted wilt virus and in papaya by papaya ringspot virus. Ringspot and mosaic mottling symptoms are caused by plum pox virus on different stone fruits (Fig. 2.12a–d).

Symptoms in Seed/Tuber

Tuber cracking and ringspot in potato are caused by potato mop top virus and seed mottling in soybean due to a potyvirus (Fig. 2.13).

Malformation

Virus-infected plants show a wide range of abnormalities leading to malformation of plants or organs. Reduction in growth leads to stunting and dwarfing, uneven growth of leaf lamina; dark green areas raised to give a blistering effect, and margin of leaves are irregular and twisted (Fig. 2.14a); leaf rolling (upward or downward) and curling are also seen (Fig. 2.14b). Leaf narrowing (restricted expansion of laminar tissue) and rugosity (retarded growth of veinal tissue) or leaves become thick and brittle. The veins on the lower surface of leaves may become thickened and twisted. In some cases, flowers and fruits or seeds fail to form or abort. New tissues or organs may also arise after virus infection.

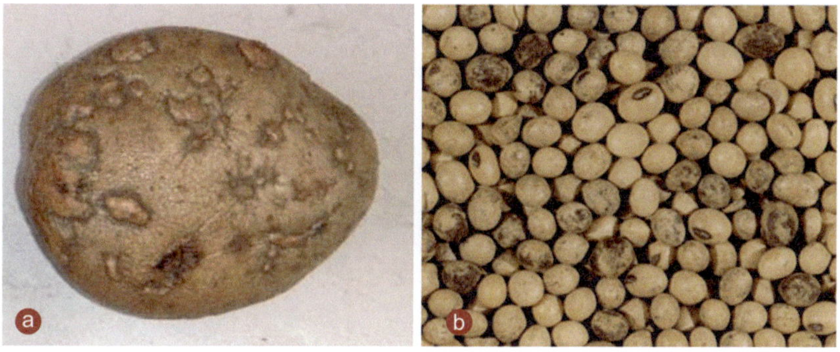

Fig. 2.13 Tuber ringspot in potato (**a**); seed mottling in soybean (**b**)

Fig. 2.14 Malformation caused by plant viruses. Shoe string on papaya caused by papaya ringspot virus (**a**); leaf curl caused by papaya leaf curl virus (**b**); puckering and malformation on cucurbits caused by a begomovirus (**c**); leaf deformation in cassava due to cassava mosaic virus (**d**)

Small outgrowths on leaves, especially veins or stems are called enations (Fig. 2.14).

2.1.2 Internal Symptoms (Hull 2002)

Internal symptoms are mainly due to the changes in the internal organs of the plant. Anatomical and histological changes in the form of necrosis, hypoplasia or hyperplasia or combination of them are seen with some virus-infected plants. In mosaic symptoms, low chlorophyll contents give translucence appearance to the

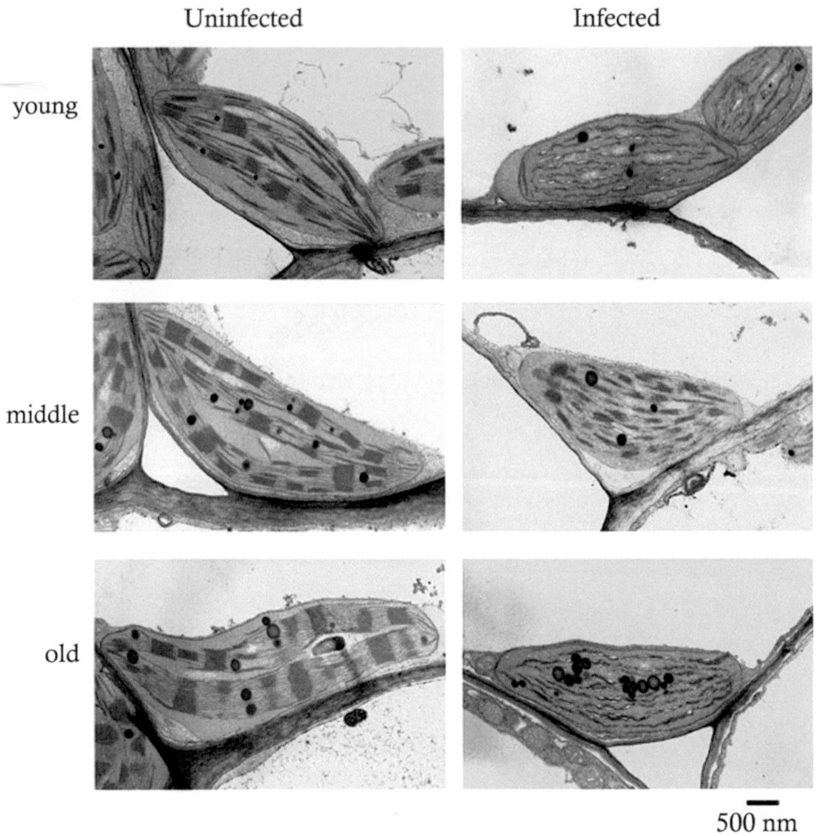

Fig. 2.15 Effect of *Eupatorium* yellow vein virus infection on chloroplast structure in leaves of *Eupatorium makinoi* bands. (Reproduced from Funayama-Noguchi and Terashima 2006 with permission from CSIRO, Australia)

leaves. Mesophyll cells become smaller and less differentiated with fewer or no intercellular spaces. Hyperplasia in the form of enlargement of vascular bundles is seen in plants showing vein clearing symptoms. Necrosis of phloem and epidermal cells is seen in potato with leaf roll disease. Virus infection also alters organelle like nucleus, chloroplast and mitochondria. In general, although nuclei are less affected, and in a few cases nuclei may harbour intranuclear inclusions or virus particles which may affect size and shape of the nucleus. Chloroplast in a virus-affected plant may be swollen, clumped or rounded (Fig. 2.15); the colour may vary from green to colourless (Funayama-Noguchi and Terashima 2006). The cell wall of virus-infected cell may sometimes protrude or thicken.

One of the important effects of the virus infection is the development of inclusion bodies. Virus infection of cells leads to the development of microscopic bodies differing from other cellular structures called as inclusion bodies. They are located either in the cytoplasm or in the nucleus. Their presence in a diseased plant is a

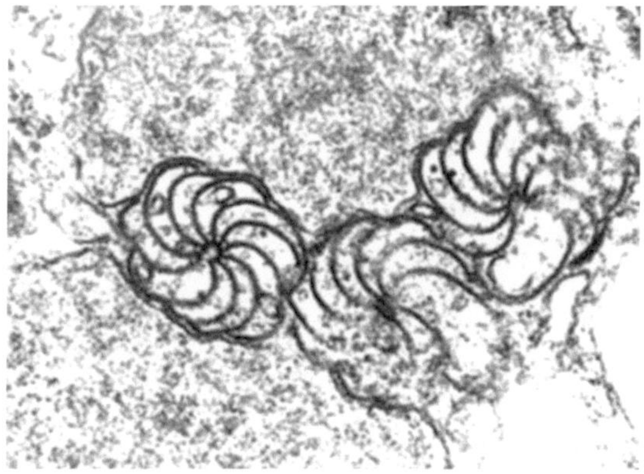

Fig. 2.16 Pinwheel inclusions caused by potyvirus in cytoplasm of infected plant tissue

diagnostic value as they are characteristic of virus groups, and the type of inclusion depends solely on the virus and not on the host plant. Some inclusion bodies are large enough to be seen in the ordinary microscope. Inclusion bodies may contain cellular material or mixtures of these constituents in various proportions. They may occur in the nucleus (nuclear inclusion) or the cytoplasm (cytoplasmic inclusion). They may be amorphous, crystalline or pin wheel shaped (Figs. 2.16 and 2.17). In general, amorphous inclusion bodies are round or oval, frequently vacuolated and in the size of 1–30 microns. They are located in the cytoplasm and rarely in the nucleus. They give the usual reactions for proteins and stain well with acid dyes. Their structure and mode of formation vary. They are classified into different groups based on their structure. Crystalline inclusions are found in both cytoplasm and nucleus. They may be numerous crystals, large plates, sometimes hexagonal and para-crystalline needles. Crystalline inclusions in the nucleus may be thin rectangular or isometric crystals.

By cytochemical tests and specific stains for light microscopy, inclusion bodies have been found to contain nucleoproteins. Some amorphous inclusions also contain lipids and cellular organelles such as mitochondria, plastids and chloroplasts. Different methods have been employed for the examination of tissues for the presence of inclusion bodies. A simple and rapid method involves direct microscopic observation of epidermal strips mounted in tap water. Other methods use stains such as trypan blue, phloxine, pyronin-methyl green, calcomine orange brilliant green and congo rubin-methyl green. Protocols for microscopic examination of inclusion bodies in virus infected tissues are discussed herein.

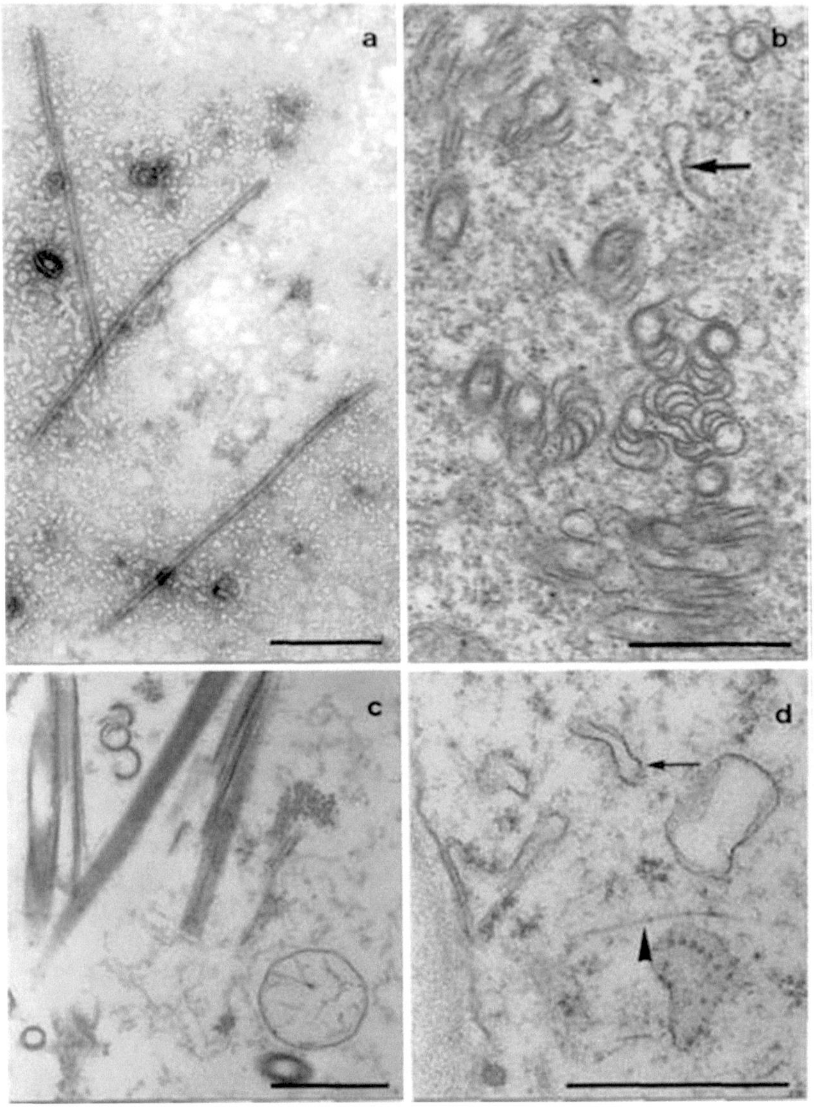

Fig. 2.17 Transmission electron micrographs. Bars: **a** = 200 nm; **b–d** = 500 nm. (**a**) Zucchini yellow mosaic virus (ZYMV) particles detected in the sap of infected leaf-material by negative staining with phosphotungstate; (**b**) Infected older leaf cell showing cylindrical inclusions (CI) and proliferated endoplasmaic reticulum (arrow); (**c**, **d**) Sieve elements containing CI, proliferated endoplasmaic reticulum (arrow) and virus particles (arrowhead). (Reproduced from Zechmann et al. 2003 with permission from Springer)

2.1.2.1 Direct Microscopic Examination (Christie and Edwardson 1987)

Materials

1. Cover slips.
2. Infected and healthy leaves.
3. Light microscope.
4. Razor blade.
5. Slide.
6. Watch glass.

Method (Direct Microscopic Examination)

1. Cut very thin cross section of healthy and infected leaves, mount on a slide in water and observe under the microscope.
2. Cytoplasmic streaming, movements and formation of inclusion are seen in large trichomes. Besides, other anatomical changes in infected leaves may also be visible such as in mosaic-affected leaves, and chlorotic areas are thinner because of the deficiency of palisade cells which instead of tubular appearance become round with fewer and smaller chloroplasts than usual. Chloroplasts are granular and less deeply coloured. Large intercellular spaces and degeneration of some spongy cells produce large spaces. In yellow and leaf roll-infected leaves, accumulation of starch, the formation of tylosis in xylem elements and degeneration of phloem cells are also visible.

2.1.2.2 Examination of Inclusion Bodies Through Staining with Trypan Blue

Materials

1. Same as in Subheading 2.1.2.1, Materials.
2. Trypan blue stain (0.5%): Make stock solution of 0.5% trypan blue by weighing 5 mg of trypan blue powder and dissolving it in 1 mL of water or 1× PBS (phosphate-buffered saline). To make trypan blue stain, mix 1:1 the trypan blue solution with 1× PBS.
3. Phosphate-buffered saline (PBS) (10×): Dissolve 8.00 g NaCl, 0.20 g KH_2PO_4, 2.90 g Na_2HPO_4 (or 1.15 g anhydrous), 0.20 g KCl, 0.20 g NaN_3 in distilled water and make up the volume to 1000 mL. For working solution of 1× PBS, mix 100 mL of 10× PBS with 900 mL of distilled water.

Method (Examination of Inclusion Bodies Through Staining with Trypan Blue)

1. Take out epidermal peels of infected leaf from the ventral surface (as it contains a lesser number of the stoma).
2. Keep the peels in 0.5% trypan blue stain in a watch glass for 1 min.
3. Wash with water to remove excess stain and mount in water.
4. Observe under microscope. Nuclei turn blue and amorphous inclusion stain strongly, so that it is distinguishable from nuclei.

2.1.2.3 Examination of Inclusion Bodies Through Staining with Phloxine

Materials

1. Same as in Subheading 2.1.2.1, Materials.
2. Glycerine.
3. Phloxine 1%: Dissolve 100 mg phloxine in 10 mL of water.
4. Trypan blue 0.5% (*see* Subheading 2.1.2.2, Materials above for the preparation of this stain).

Method (Examination of Inclusion Bodies Through Staining with Phloxine)

1. Take out epidermal peels of infected leaf from the ventral surface as it contains the lesser number of the stoma.
2. Keep the peels in phloxine + trypan blue (1:1) in a watch glass for 5 min.
3. Wash with water to remove excess stain.
4. Mount in water/glycerine and observe under microscope. Nucleus takes purple stain, and inclusion bodies appear pink in colour. Tobacco infected with tobacco mosaic virus (TMV) produces needle-shaped and hexagonal crystalline inclusions in epidermal cells, trichomes and subsidiary cells of the stoma in profuse numbers. Cowpea infected with cowpea mosaic virus (CPMV) shows amorphous mass in the cytoplasm of epidermal cells. Healthy peels will have clear cytoplasm and distinct nucleus.

2.1.2.4 Examination of Inclusion Bodies by Staining with Pyronin-Methyl Green

Materials

1. Same as in Subheading 2.1.2.1, Materials.
2. Methyl-green-pyronin stain: Weigh 200 mg of methyl green and dissolve in 100 mL of 0.2 M acetate buffer, pH 5.3. This solution is extracted repeatedly with chloroform to remove residual methyl violet, and then 200 mg of pyronin B is dissolved in it, store the stain in brown-coloured bottle.
3. Acetate buffer (0.2 M, pH 5.3): Dissolve 1.32 g of sodium acetate (anhydrous) and 0.230 g of acetic acid in about 80 mL of water, adjust pH of the solution to 5.3 using either NaOH or HCl and make up the volume to 100 mL.

Method (Examination of Inclusion Bodies by Staining with Pyronin-Methyl Green)

1. Take out epidermal peels of infected leaf from the ventral surface as it contains lesser number of stoma.
2. Keep the peels in solution containing methyl-green-pyronin stain in a watch glass for 5–10 min.
3. Wash with water to remove excess stain and mount in water.
4. Observe under microscope. Nuclei stain blue while inclusions red that also indicates the chemical nature of the inclusion. The blue colour indicates the presence of DNA and the red indicates the presence of RNA.

2.1.2.5 Examination by Staining with Toluidine Blue

Materials

1. Same as in Subheading 2.1.2.1, Materials.
2. Potassium phosphate buffer (0.05 M, pH 7.0): Dissolve 8.7 g of dipotassium hydrogen orthophosphate (K_2HPO_4) in 1 L of distilled water (0.05 M solution), and to this add slowly 0.05 M solution of potassium dihydrogen orthophosphate (KH_2PO_4) prepared by dissolving 1.7 g of K_2HPO_4 in 250 mL of distilled water till to desired pH of 7.0 is obtained.
3. Toluidine blue O stain: Prepare 0.05% solution by dissolving 0.05 g of toluidine blue O in 100 mL of 0.05 M potassium phosphate buffer, pH 7.0.
4. Triton X 100.

Method (Examination of Inclusion Bodies by Staining with Toluidine Blue)

1. Wash the leaves and blot them dry.
2. Peel off the epidermal layer with a fine point forceps.
3. Float the epidermal strip on 2.5% aqueous Triton X 100 solution for 2–3 min to remove plastids.
4. Transfer the strips to distilled water for 1 min to wash off the Triton X 100.
5. Mount the strips directly in toluidine blue O stain on a glass slide.
6. Cover the strips with a cover slip, leave for 1–2 min, examine under a light microscope. Nuclei will stain blue and nucleoli will stain purple. Inclusion body will appear in light blue. Both diseased and healthy leaves should be used for comparison.

References

Christie RG, Edwardson JR (1987) Light and electron microscopy of plant virus inclusions. The University of Florida, Florida Agricultural Experiment Station, Monograph series

Funayama-Noguchi S, Terashima I (2006) Effects of *Eupatorium* yellow vein virus infection on photosynthetic rate, chlorophyll content and chloroplast structure in leaves of *Eupatorium makinoi* during leaf development. Funct Plant Biol 33:165–175

Gibbs A, Harrison BD (1980) Plant virology—the principles. Edward Arnald, London

Hull R (2002) Matthews' plant virology, 4th edn. Academic, San Diego

Hull R (2009) Comparative plant virology, 2nd edn. Academic, San Diego

Matthews REF (1991) Plant virology, 2nd edn. Academic, London

Smith KM (1977) Plant viruses. Chapman and Hall, London

Zechmann B, Muller M, Zellnig G (2003) Cytological modifications in zucchini yellow mosaic virus (ZYMV) infected Syrian pumpkin plants. Arch Virol 148:1119–1133

Chapter 3

Isolation and Diagnosis of Virus Through Indicator Hosts

Abstract

Isolation followed by propagation of viruses is the first step in the characterization of a virus. Under natural field conditions, plants may be infected with more than one virus and other pathogens. Hence, it is important to isolate the virus in question under in vitro or glasshouse condition. This is usually done by inoculating the field-collected samples onto a host that produce local lesion symptoms. Inoculation may be done by mechanical means using sap extracted from infected plants or through vectors (if virus in question is not mechanically transmitted). Pure culture of the virus isolate is then made by inoculating single local lesion on to hosts that produce either local or systemic symptoms. Pure culture of the virus is inoculated (either mechanically or through vectors) onto a set of diagnostic hosts that are known to produce characteristic symptoms that would help in the identification of the causal virus.

Key words Diagnosis, Indicator host, Local lesion host, Mechanical transmission, Propagation host, Pure culture, Systemic host, Vector transmission

3.1 Introduction

Many a times' disease symptoms on plants in the field are inadequate for the identification as (1) different viruses may cause similar symptoms in the same crop; (2) single virus may cause variable symptoms, depending on the strain; (3) mixture of viruses or presence of satellite viruses may greatly affect disease expression; (4) different cultivars may react differently to a virus and (5) soil and weather conditions may alter disease expression (Hull 2014). Hence, the first and foremost thing before characterizing any virus is the isolation and establishment of pure culture of the virus isolate in a suitable propagation host. In the case of mechanically transmitted viruses, this is done by inoculating the symptomatic leaves/ tissues (collected from the field) onto a local lesion host. A pure culture of the virus is established on the suitable propagation host by mechanical inoculation using the single local lesion. Pure culture thus established can then be used for further studies to characterize the virus isolate. For viruses that cannot be transmitted by mechanical means, the pure culture may be established through vector transmission using a single vector. For viruses that do not have

Alangar Ishwara Bhat and Govind Pratap Rao, *Characterization of Plant Viruses*, Springer Protocols Handbooks,
https://doi.org/10.1007/978-1-0716-0334-5_3,

any local lesion host, the pure culture may be established on the systemic host raised under insect-proof conditions (Hill 1984; Dijkstra and de Jager 1998). The procedures of mechanical and vector transmissions are provided in the subsequent chapters (Chapters 6, 11–19).

In order to diagnose the virus in question, pure culture of the virus isolate is inoculated either mechanically (through sap inoculation) or through vectors onto a set of indicator host plants (at right growth stage) raised under identical conditions in an insect-proof glasshouse. The inoculated plants are kept under insect-proof conditions for symptom development that may take up to 2 weeks from the date of inoculation. Symptoms developed on each of the indicator hosts are recorded, which would aid in the diagnosis of viruses. Some of the commonly used indicator hosts include *Chenopodium amaranticolor*, *Cucumis sativus*, *Nicotiana glutinosa*, *N. tabacum*, *N. benthamiana*, *Datura stramonium*, *Physalis vulgaris*, *Vigna unguiculata*, *Gomphrena globosa* and *Physalis floridana* (Hill 1984) (Fig. 3.1).

3.2 Materials

1. Earthen pots of desired size/pro-trays.
2. Glasshouse.
3. Infected plants.
4. Potting mixture: prepared by mixing soil : sand : farm yard manure in the ratio of 1:1:1.
5. Seeds/propagative propagules of test plants.

3.3 Method

1. Fill earthen pots or pro-trays of required size with soil : sand : farm yard manure (1:1:1 ratio) and keep in an insect-proof glasshouse.
2. Seeds or propagative propagules of test plants are sown or planted in pots.
3. Pots are labelled with date of sowing or planting and name of the test plant.
4. If seeds cannot be directly used for sowing, in such cases seedlings are initially raised in an insect-proof glasshouse and transplanted in pots of desired size.
5. Pots are watered regularly with nutrients as required.
6. When test plants attain two-leaf stage, inoculate them with sap extracted from field-infected sample either mechanically (*see*

Fig. 3.1 Commonly used local lesion and indicator hosts. (**a**) *Nicotiana benthamiana*, (**b**) *N. glutinosa*, (**c**) *Cucumis sativus*, (**d**) *Datura stramonium*, (**e**) *Vigna unguiculata*, (**f**) *N. tabacum*, (**g**) *Chenopodium amaranticolor*

Chapter 6 for details of mechanical inoculation) or through vector transmission (*see* Chapters 11–19 for details of vector transmission). Keep inoculated plants in insect-proof glasshouse and observe for symptom development. In many cases,

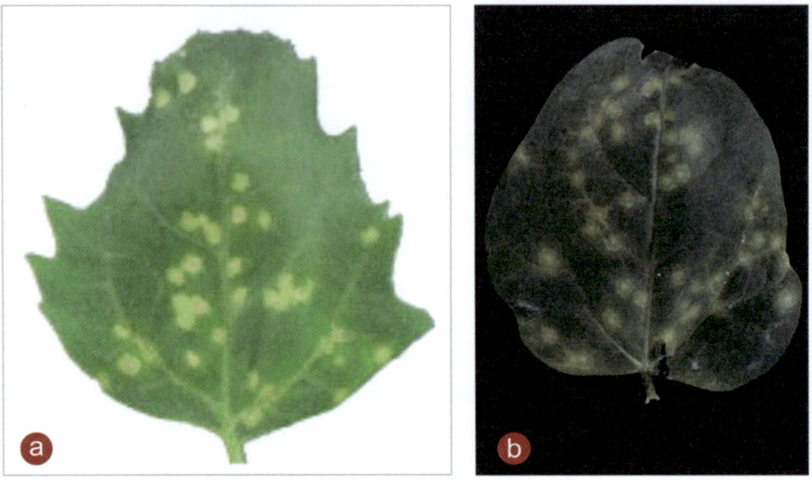

Fig. 3.2 Production of local lesions on (**a**) *Chenopodium amaranticolor* upon inoculation with cucumber mosaic virus, (**b**) *Vigna unguiculata* upon infection with groundnut bud necrosis virus

local lesions are produced within 48–72 h after inoculation (Fig. 3.2).

7. Establish pure culture of the virus isolate by inoculating a single local lesion onto either a local lesion host or a systemic host of the virus raised under insect-proof condition.
8. Inoculate pure culture of the virus isolate on to a range of diagnostic or indicator hosts and study the type of symptoms produced on each of the hosts. This will aid to some extent in the initial identification of the causal virus associated with the disease. Confirm the presence of the virus in indicator hosts by back inoculation onto local lesion host of the virus isolate.

3.4 Notes

1. The experimental plants can be raised through seeds or through propagative propagules such as rhizomes, suckers, cuttings and tubers. Seeds are the most commonly used source of healthy test plants. But at the same time, many viruses are seed borne or carried through propagative propagules, and hence, precautions should be taken to test the experimental plants for freedom from viruses before being used for inoculation (Hull 2014).
2. If isolation is by vectors, ensure that vectors used for acquisition are virus-free.
3. The test plants that are susceptible to virus infection are commonly used as propagative host. The nature of propagative host should be that it grows fast, is less fibrous and have less

phenolics and secondary metabolites, which help in fast multiplication of virus inside the propagative host. The propagative hosts should be suitable for maintaining virus culture and also for purification of viruses (Dijkstra and de Jager 1998).

4. Use young growing tissue of infected plant as a source for inoculation onto local lesion hosts.
5. If isolating a new or unknown virus, use all possible local lesion hosts.
6. In the case of vector transmission, use right stage (instar) of the vector for transmission.
7. For viruses that infect only one particular species, pure culture may be established by inoculating (either mechanically or through vectors) healthy plants of the same species raised under insect-proof conditions (Hull 2014).

References

Dijkstra J, de Jager CP (1998) Practical plant virology: protocols and exercises. Springer, Amsterdam

Hill SA (1984) Methods in plant virology. Blackwell, Oxford

Hull R (2014) Matthews' plant virology, 5th edn. Academic, San Diego

Chapter 4

Host Range of Viruses

Abstract

Host range of a virus is one of the biological properties of the virus. A host is a plant that a virus can infect and within which that virus can replicate. A few of the viruses have the ability to infect only a limited number of hosts while a few other viruses have the ability to infect large number of hosts belonging to several plant genera and families. Plant hosts play an important role in biological characterization of a plant virus.

Key words Broad host range, Narrow host range, Mechanical inoculation, Vector transmission

4.1 Introduction

Host range is an important criterion in the biological characterization of plant viruses. Plant viruses are obligate symbionts that depend on their host plants for reproduction. To colonize new hosts, most plant viruses also require vectors, which may include insects, mites, nematodes, or fungi. Although some viruses can be transmitted intergenerationally from parent to offspring in the seed, transmission by vectors clearly plays an important role in determining fitness for most of the viruses, and it can result in extremely rapid spread of a virus. Transmission within a host generation, such as that carried out by vectors, is known as horizontal transmission, whereas transmission between host generations is known as vertical transmission. Host range data is useful for also developing strategies for the virus disease management. Viruses vary widely in their host range. There are viruses whose natural host range is highly restricted (such as *Badnavirus*, *Babuvirus*, *Fijivirus*, *Tobamovirus*) while others have a very broad host range (such as cucumber mosaic virus, bean common mosaic virus, tobacco streak virus, tomato spotted wilt virus). The most suitable diagnostic hosts for the propagation of plant viruses are *Nicotiana benthamiana*, *Chenopodium amaranticolor*, cucumbers, cowpea, datura and French beans (Hull 2002; Nayudu 2008).

To determine the host range of the virus under study, the virus is inoculated to a range of plant species (either mechanically or through the viruliferous vectors of the virus). The inoculated plants

Alangar Ishwara Bhat and Govind Pratap Rao, *Characterization of Plant Viruses*, Springer Protocols Handbooks, https://doi.org/10.1007/978-1-0716-0334-5_4, © Springer Science+Business Media, LLC, part of Springer Nature 2020

are then kept under insect-proof glasshouse conditions and observed for the development of disease symptoms. Back inoculation to the propagation host or original host is used to check for symptomless infections (details of mechanical and vector transmission of viruses are discussed in next chapters).

4.2 Materials

1. Earthen pots of desired size/pro-trays.
2. Glasshouse.
3. Pure culture of the virus isolate.
4. Potting mixture: prepared by mixing soil : sand : farm yard manure in the ratio of 1:1:1.
5. Seeds/propagative propagules of test plants.
6. Test virus.

4.3 Method

1. Earthen pots or pro-trays of required size are filled with soil : sand : farm yard manure (1:1:1 ratio) and kept in an insect-proof glass house.
2. Seeds or propagative propagules of test plants are sown or planted in pots.
3. Pots are labelled with date of sowing or planting and name of the test plant.
4. If seeds cannot be directly used for sowing, in such cases seedlings are initially raised in an insect-proof glass house conditions and transplanted in pots of desired size as per requirement.
5. Pots are watered regularly with nutrients as required.
6. When test plants attain two leaf stage, inoculate them with pure culture of the virus isolate either mechanically (*see* Chapter 6 for details of mechanical inoculation) or through vector transmission and keep inoculated plants in insect-proof glass house and observe for symptom development.
7. Observe and record symptoms produced on different host-type of symptoms and days taken for symptom development.
8. Confirm the presence of virus by back inoculation onto local lesion host of the virus isolate.

4.4 Notes

1. The absence of symptoms following inoculation of a test plant needs to be always checked by back inoculation to an indicator or propagation host.
2. Certain plant species might contain inhibitors of virus infection that prevent mechanical inoculation. In such cases use of additives in the extraction buffer is important.
3. It is always a good idea to keep at least five plants of each host for inoculation. An equal number of plants of each host should also be kept as control by mock inoculation.
4. Certain viruses are known to be seed transmitted in some host plant species. Extra care should be taken while using such host plants for testing. The plants raised from these hosts may be confirmed for absence of test virus through ELISA or PCR tests (Hull 2002).
5. It is advisable to make tests only under one set of conditions especially when large number of species are to be tested.
6. A given species may vary widely, in susceptibility to a virus depending on the growth condition. Abiotic factors such as temperature, humidity, nutrient status of medium used for host range studies will have a great impact on the results. Poor soil nutrition may sometimes aggravate the disease symptoms. Hence, it is very important to apply nutrients as per the recommended dose including the micronutrients (Hull 2002).
7. Even within a species, cultivars/varieties may differ in their reaction to a virus.
8. Even closely related strains of a virus may differ in the range of plants they will infect.
9. Transmission though vectors may have to be employed for determining host range of viruses that cannot be mechanically transmitted.
10. Test plants have to be inoculated at the right growth stage (cotyledonary/primary leaf).
11. Mock inoculation of healthy test plants is important to compare the results obtained by inoculation with infected plants.

References

Hull R (2002) Matthew's plant virology, 4th edn. Academic, San Diego

Nayudu MV (2008) Plant viruses. Tat McGraw-Hill Publishing Company Limited, New Delhi

Chapter 5

Physico-chemical Properties of Virus in Crude Sap

Abstract

The physico-chemical properties of the virus in crude sap such as dilution end point, thermal inactivation point and longevity in vitro would provide useful data on the concentration of the virus in the living host and its stability that would help in the transmission and virus purification experiments. Although, these properties are not being used at present to identify and characterize a plant virus, but still they are quite useful in getting idea about concentration and storage duration of a virus in active state in crude sap for future course of study.

Key words Dilution end point, Thermal inactivation point, Longevity in vitro

5.1 Introduction

Physico-chemical properties of virus in crude sap, such as its concentration and resistance to inactivation by heating, dilution and ageing (dilution end point, thermal inactivation point and longevity in vitro), have been considered to be of diagnostic value (Smith 1972; Francki 1980; Hill 1984; Kurstak 1981). Recently, it has been shown that these properties are very variable and hardly reflect the intrinsic properties of the virus. However, information about the concentration of virus in a plant and its stability outside the plant is still useful for optimal success in mechanical inoculation, selection of propagation host and purification of viruses. In this chapter, we have discussed the protocols for determining the dilution end point, thermal inactivation point and longevity in vitro for the plant viruses in crude sap.

5.1.1 Determination of the Dilution End Point (DEP)

The dilution end point is the highest dilution of sap from a virus-infected plant which is still infectious when mechanically inoculated to test plants but is quoted as the dilution limits between this dilution and the next one at which the infectivity is lost (Fig. 5.1). This will obviously depend on virus concentration, diluent used and the condition and species of source and assay plants (Francki 1980). Dilutions are usually on a logarithmic scale, and each dilution is inoculated on the assay plants, preferably a local lesion host.

Alangar Ishwara Bhat and Govind Pratap Rao, *Characterization of Plant Viruses*, Springer Protocols Handbooks,
https://doi.org/10.1007/978-1-0716-0334-5_5, © Springer Science+Business Media, LLC, part of Springer Nature 2020

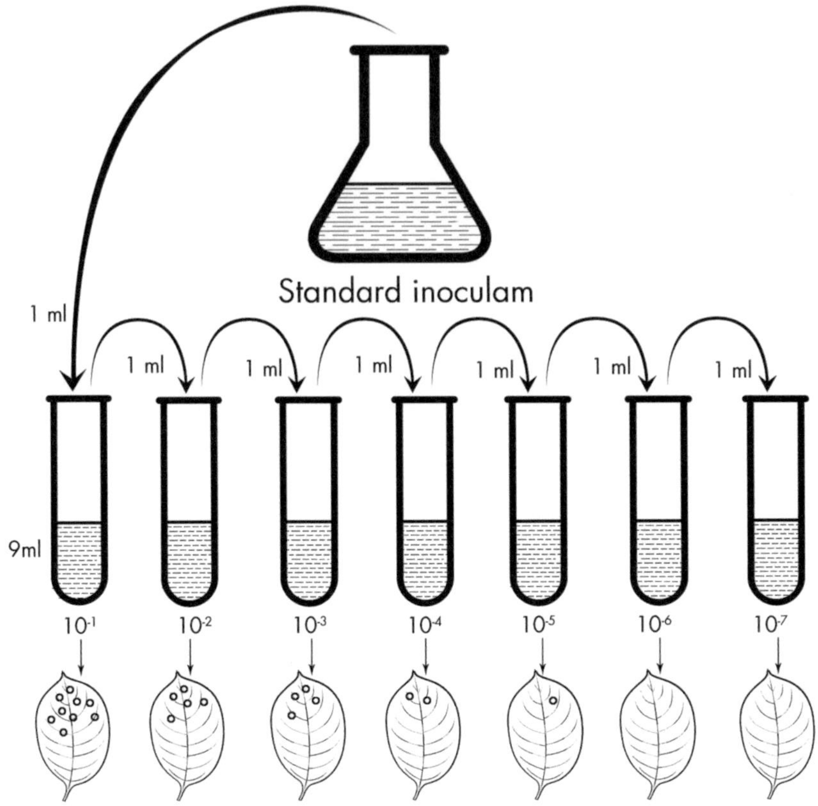

Fig. 5.1 Schematic diagram showing determination of dilution end point

5.1.2 Determination of the Thermal Inactivation Point (TIP)

Thermal inactivation point (TIP) is the maximum temperature required for the complete inactivation of the virus in untreated crude sap during a 10 min exposure. In practice, it is expressed as one temperature as the inactivation temperature, or to mention twin temperatures in between which the virus is inactivated completely (Smith 1972; Francki 1980). Experiments are started by heating the crude sap for 10 min to different temperatures such as 40, 50, 60, 70, 80, 90 and 100 °C and inoculating them onto the assay plants (Fig. 5.2). In the range of inactivation, the experiment is repeated at 5 °C intervals and sometimes at 2 °C intervals. The experiment is extended here by counting the local lesions that result from the inoculation of each sample. The test is affected by the factors that influence dilution end point.

5.1.3 Longevity In Vitro (LIV)

Longevity in vitro (LIV) is defined as the time (days, weeks or hours) that virus in crude sap kept at room temperature or freeze temperature (4 °C) remains infectious. To determine the LIV of a virus, the samples of crude sap are removed from storage at different intervals and tested on assay plants. In the absence of information on the stability of a virus in sap, the first series of intervals should be at a geometric progression, e.g. 1, 2, 4, 8, 16,

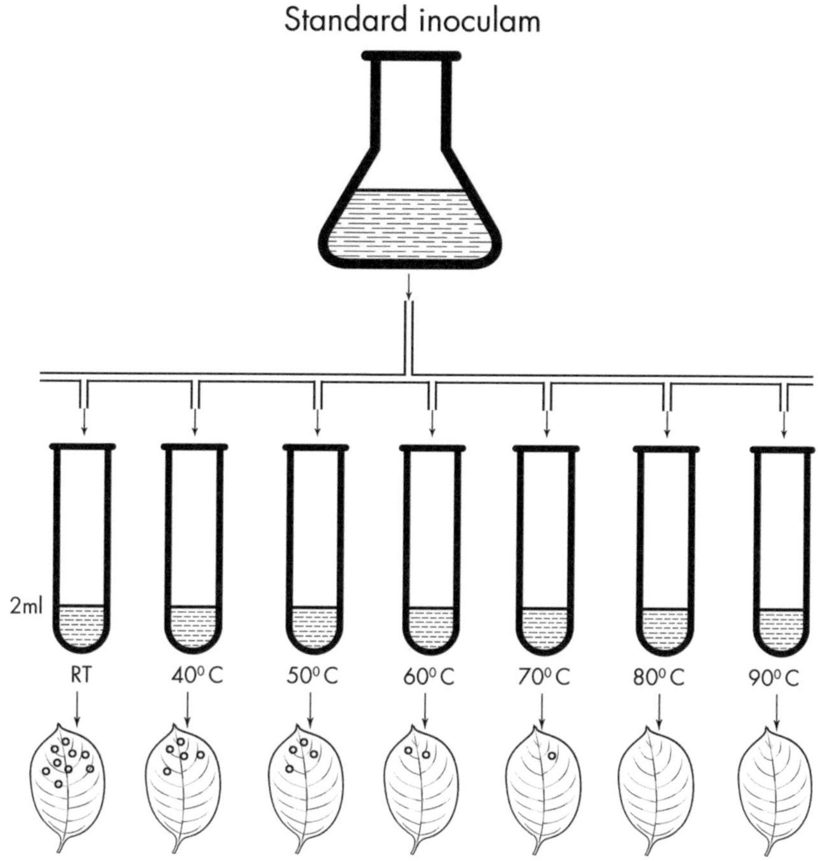

Fig. 5.2 Schematic diagram showing determination of thermal inactivation point

32,. . . days, until infectivity is lost. Once LIV is roughly established, the test can be repeated over a narrow range of shorter intervals. LIV depends on the original concentration of the virus in the sap and on properties of the sap.

5.2 Materials

1. Abrasive (carborundum, 600 mesh or celite powder).
2. Glasshouse.
3. Incubator (20–22 °C).
4. Labels.
5. Micropipette (1 and 5 mL).
6. Muslin cloth.
7. Pestle and mortar.
8. Pipette.
9. Phosphate buffer (0.1 M, pH 7.0): Dissolve 17.4 g of di-potassium hydrogen orthophosphate (K_2HPO_4) in 1 L of

distilled water (0.1 M solution), and to this, add slowly 0.1 M solution of potassium dihydrogen orthophosphate (KH_2PO_4) prepared by dissolving 3.4 g of KH_2PO_4 in 250 mL of distilled water till the desired pH of 7.0 is obtained.

10. Test tubes with stand.
11. Thermometers.
12. Thin-walled test tubes.
13. Virus-infected source plants.
14. Virus-free assay plants.
15. Wash bottle.
16. Water bath with test tube racks.

5.3 Methods

5.3.1 Determination of the Dilution End Point (DEP)

1. Place the required number of test tubes in a row in a test tube rack and number them serially.
2. Grind the leaves of the infected source plant in a mortar at 1 g/mL of the diluent (water or phosphate buffer), press the pulp through a muslin cloth and collect the sap.
3. Transfer the sap to test tube 1 (undiluted) and prepare dilutions 10^{-1} to 10^{-7} (Fig. 5.1).
4. Fill the test tube no. 2–8 with 9 mL of diluent.
5. From the test tube no. 1, pipette 1 mL sap to test tube 2 and mix thoroughly (dilution 10^{-1}).
6. Transfer with a clean pipette 1 mL of the 10^{-1} diluted sap in test tube no. 2 to test tube no. 3 and mix thoroughly (dilution 10^{-2}).
7. Repeat this procedure with test tubes 4–8, up to dilution 10^{-7} (Fig. 5.1).
8. Inoculate the eight suspensions (10^{-1} to 10^{-7}) obtained to assay plants, as described in chapter on mechanical inoculation (*see* Chapter 6).
9. Label and keep the plants under insect-proof glasshouse for observation. Record the number of local lesions produced in each of the assay plants and determine the dilution end point for the test virus.

5.3.2 Determination of Thermal Inactivation Point (TEP)

1. Place the required number of test tubes (eight) in a test tube rack and number them serially (1–8) (Fig. 5.2).
2. Grind the leaves of the infected source plants in a mortar, press the pulp through a muslin cloth and collect the sap (standard inoculum).

3. Pipette 2 mL sap in each of the eight test tubes (Fig. 5.2).
4. Fill the water bath(s) until the water level is at least 3 cm above the level of sap in the test tubes when in the water bath.
5. Preheat the water bath(s) to the required temperature(s).
6. When the required temperature has reached, put one test tube in the water bath rack and place the thermometer close to the test tube and at the same level. Keep the temperature constant for exactly 10 min.
7. Remove the test tube from the water bath after 10 min and cool it immediately in running tap water.
8. Inoculate the heat-treated sap without delay to assay plants as described in the experiment on mechanical inoculation (*see* Chapter 6).
9. Local lesions are seen on the assay hosts by 2–4 days post-inoculation (Fig. 5.2). Count the number of local lesions obtained on assay plant for each of the temperature treatments. Plot the sums on a logarithmic scale on the ordinate against the temperature on an arbitrary linear scale on the abscissa. If the infectivity is found lost between temperature 50 and 60 °C, the experiment is again repeated at 50, 52, 54, 56, 58 and 60 °C temperatures as described above to note the TIP of the test virus, and the TIP is recorded as between two temperatures.

5.3.3 Determination of Longevity In Vitro (LIV)

1. Place the required number (seven) of test tubes in a test tube rack and number them serially.
2. Grind the leaves of the infected source plants in a mortar, press the pulp through a muslin cloth and collect the sap (standard inoculum).
3. Pipette 2 mL of sap in each of the test tubes, add a pinch of sodium azide and mix well. Close the test tubes with a stopper or aluminium foil.
4. Place all test tubes in two batches (except for tube no. 1, time zero) in an incubator (20–22 °C) as well as in a refrigerator at 4 °C.
5. Inoculate the sap sample from no. 1 to assay plants as described in mechanical inoculation experiment (*see* Chapter 6).
6. Repeat **step 5** for the tube samples from no. 2 to 7 at time intervals chosen 1, 2, 4, 8, 16 and 32 days.
7. Local lesions appear in the assay plants within 2–3 days after inoculation. Count the number of local lesions and calculate their average for each of the treatments. Plot them on a logarithmic scale on the ordinate against time on the arbitrary scale on the abscissa and record the LIV of the virus under study. For example, if the test virus lost its activity at 16 days, the experiment is repeated again in a narrow range of 2 days between

8 and 16 days like 8, 10, 12, 14 and 16 days and after having an idea of LIV in this range, the next test duration range will be 1 day to confirm the exact LIV value for a test virus.

5.4 Notes

1. Dilution end point is dependent on the concentration of the virus in the tissue used for extraction. Concentration of the virus may vary for the same virus depending on the host, environmental conditions and different tissue within the same host.
2. Extraction can be done in distilled water for stable viruses such as tobacco mosaic virus. It is always advisable to use buffer for extraction to prevent inactivation of the virus due to release of inhibitors during extraction. Addition of reducing agents such as sodium sulphite or mercaptoethanol will further help in preventing oxidation of the sap constituents (Hill 1984).
3. Test plants should be inoculated at the primary leaf stage.
4. Provide proper environmental conditions for the growth of plants after inoculation.
5. Dilution end point may provide good data on virus concentration in the host which may be useful for planning harvesting time for purification of the virus.

References

Francki RIB (1980) Limited value of thermal inactivation point, longevity *in vitro* and dilution end point as criteria for the characterization, identification and classification of plant viruses. Intervirology 13:91–98

Hill SA (1984) Methods in plant virology. Blackwell, Oxford

Kurstak E (1981) Hand book of plant virus infections. Comparative diagnosis. Elsevier, Amsterdam

Smith KM (1972) A textbook of plant virus diseases. Longman, London

Chapter 6

Mechanical Sap Transmission

Abstract

The transmission of a virus from infected to healthy plant is an important procedure to study the biological properties of the virus. Mechanical transmission is an in vitro method of virus transmission used for a variety of assays such as isolation and propagation of virus isolates, establishing pure culture and sub-culturing of virus isolates, assay for virus infectivity, diagnosis of the virus through indicator hosts, determination of host range and properties of the virus in crude sap and so on. In the laboratory, this is usually accomplished by grinding the leaf of a diseased plant, and rubbing the infectious sap on to the leaf of a healthy plant. The procedure is referred to as mechanical or sap transmission. The different steps and requirements for the process of mechanical transmission are discussed in the chapter.

Key words Virus-infected plant, Extraction, Sap inoculation, Spray inoculation

6.1 Introduction

Plant viruses are transmitted from plant to plant by various means. For successful transmission, penetration to the living cells is required and essential, which takes place either through mechanical injury followed by the contact of a virus or through insects that carry the virus. Viruses in the genera *Cucumovirus*, *Potexvirus*, *Potyvirus* and *Tobamovirus* are contagious and easily transmitted by mechanical means in nature. Mechanical transmission of such viruses occurs when plants come in contact or by the action of animals and human beings. Also, many viruses can be mechanically transmitted in the laboratory and/or glasshouse to study the properties of viruses. In this process, infectious virus is manually introduced into a living plant tissue by mechanical trauma to facilitate the initiation of virus infection in the host plant cells. In theory, all viruses can be introduced into any living cells by standard methods of mechanical inoculation, but whether the virus becomes functional depends on a variety of conditions such as the degree of susceptibility of the recipient cell, functioning of the viral genome within the cell and conditions necessary for virus replication. Wounding or abrasion of cells during inoculation is a prerequisite for the success in establishing infection. The main steps involved in mechanical transmission are the extraction of the virus and

Alangar Ishwara Bhat and Govind Pratap Rao, *Characterization of Plant Viruses*, Springer Protocols Handbooks,
https://doi.org/10.1007/978-1-0716-0334-5_6,

Fig. 6.1 Inoculation process of infected virus sap through (**a**) pestle and (**b**) spraying and (**c**) by hand on test plants

administration of extracted virus sap to the infectible sites by mechanical means (Kado 1972; Hill 1984; Walkey 1991). The success of transmission depends on the source of inoculum, method of preparing inoculum, constituents in the inoculum and stability of the extracted virus.

Young actively growing leaves are the best source of inoculum as they contain a good concentration of virus. The concentration of the virus varies with the host and kind of the test virus. Within the host again concentration and distribution of virus would vary depending on season, growth stage and tissues. Various substances such as host metabolites and cellular debris may possess inhibitory properties. The removal of contaminating host components increases the effectiveness of the inoculum. Further, the concentration of virus in the extract will also influence the extent of transmission. Compared to stable viruses, mechanical transmission of unstable viruses are difficult. In general, phloem-borne viruses are difficult to be transmitted through mechanical sap inoculation (Hull 2002; 2014).

Extraction of the virus is done by homogenizing the infected tissue in a suitable buffer (containing a reducing agent if needed). The buffer, its molarity and pH would have a great impact on the success of transmission. Extraction process consists of ruptures of the infected cells and release of virus particles into the buffer solution. The tissue debris is strained through absorbent cotton or muslin cloth. This solution is diluted appropriately with buffer 1:1 or 1:10 and used for inoculation immediately.

6.1.1 Methods of Mechanical Transmission

Mechanical inoculation can be done by two different methods, viz. hand inoculation or spray inoculation, depending on the numbers of plants to be inoculated (Fig. 6.1). The hand inoculation is done when the number of test plants to be inoculated are limited but when large number of test plants are to be inoculated, spray inoculation practise is preferred (Walkey 1991).

6.2 Materials

1. Cotton absorbent.
2. Extraction buffer (0.1 M phosphate buffer pH 7.0): Dissolve 17.4 g of di-potassium hydrogen orthophosphate (K_2HPO_4) in 1 L of distilled water (0.1 M solution). To this, slowly add 0.1 M solution of potassium dihydrogen orthophosphate (KH_2PO_4) prepared by dissolving 3.4 g of KH_2PO_4 in 250 mL of distilled water till the desired pH of 7.0 is obtained.
3. Healthy test plants.
4. Infected plants.
5. Abrasive (carborundum/celite powder).
6. Insect-proof glasshouse.
7. Mortar with pestle.
8. Muslin cloth.
9. Pipette.
10. Wash bottle.
11. Air brush.
12. Hand sprayer.
13. Pump and pressure gauge.
14. Spray gun.

6.3 Methods

6.3.1 Hand Inoculation (Kado 1972; Walkey 1991)

1. Grind the leaves of the infected source plant in a mortor at 1 g/ 10 mL of 0.1 M phosphate buffer, pH 7.0.
2. Pass the homogenate through a muslin cloth/absorbent cotton and collect the filtrate.
3. Pre-dust celite powder (0.2%) or carborundum powder 400 or 600 mesh (2%) on the leaf surface of the test plants or add them in the inoculum before inoculation.
4. Place the palm under the leaf to be inoculated and apply inoculum with rubbing by fingers or using brush, cotton swab, gauze pad or pestle, which are the commonly used for inoculation (Fig. 6.2).
5. After about a minute, wash the leaf surface with water with the help of a wash bottle.
6. Label the plants and keep them for symptom development under insect-proof conditions.

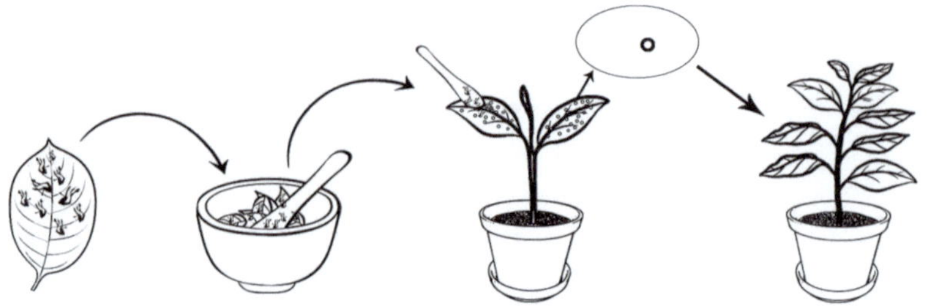

Fig. 6.2 Schematic diagram showing steps involved in mechanical sap inoculation of viruses

Fig. 6.3 Local lesions produced on *Chenopodium amaranticolor* upon mechanical inoculation with cucumber mosaic virus

6.3.2 Spray Inoculation (Mandal et al. 2008)

1. Prepare inoculum as mentioned above (**steps 1–3**).
2. Fill the inoculum containing carborundum/celite powder into the sprayer and shake properly before use.
3. Set the pressure at 50–70 psi depending on the type of the test plants used.
4. Place the air brush about 6 in. above the plant and apply the inoculum by spraying (Fig. 6.1).
5. Shake the inoculum bottle intermittently to prevent any clogging of the nozzle by the abrasive used.
6. Allow the plant to grow further for symptom expression under insect-proof conditions.
7. Local lesions (chlorotic/necrotic) may appear in 2–3 days while systemic infection will start appearing 5–7 days after inoculation (Fig. 6.3).

6.4 Notes

1. Use young symptomatic tissues having good virus concentration as a source of infected material.
2. The grinding (extraction) buffer may be changed depending on the virus and host. In some cases, addition of reducing agents like sodium sulphite or 2 mercaptoethanol may be required to prevent inactivation of the virus particles due to oxidation of sap constituents (Walkey 1991; Hull 2014).
3. For unstable viruses, carry out extraction under ice and hold the inoculum at 4 °C during inoculation process.
4. Inoculation should be done in evening hours as it helps in establishing of the virus and helps better recovery of the plant from injury made during inoculation.
5. After extraction, carry out inoculation immediately without any delay.
6. Test plants should be inoculated at the primary leaf stage.
7. Provide proper environmental conditions for the growth of plants after inoculation.
8. To test infectivity of purified virus preparation, inoculate the purified preparation on test plants dusted with abrasives such as celite or carborundum.

References

Hill SA (1984) Methods in plant virology. Blackwell Scientific Publications, Oxford

Hull R (2002) Matthews' plant virology, 4th edn. Academic, San Diego

Hull R (2014) Matthews' plant virology, 5th edn. Academic, San Diego

Kado CI (1972) Mechanical and biological inoculation principles. In: Kado CI, Agrawal HO (eds) Principles and techniques in plant virology. Van Nostrand Reinhold Company, New York

Mandal B, Csinos AS, Martinez-Ochoa N, Pappu HR (2008) A rapid and efficient inoculation method for tomato spotted wilt tospovirus. J Virol Methods 149:195–198

Walkey DGA (1991) Applied plant virology. Springer, Dordrecht

Chapter 7

Transmission Through Grafting and Budding

Abstract

Grafting is a horticultural practice used for many purposes. In this, a shoot (scion) or bud isolated from one plant is joined to the rooted part (stock) of another plant, the union of two produces one plant. If either scion or stock contains the virus, the whole plant established through grafting also contains the virus. The success of grafting depends on the compatibility and union of stock and scion. The time required for transmission of virus through grafting may vary from several days to months. Graft inoculation is one of the important and widely used experimental means of transmitting viruses. It is also used as a diagnostic tool for detection of viruses, especially of woody plants. Grafting may succeed in transmitting a virus where other methods fail, thus grafting can be used to differentiate plants infected with virus from plants showing virus-like symptoms due to nutritional deficiencies.

Key words Approach grafting, Bud grafting, Leaf patch grafting, Wedge grafting, Tongue grafting, Stock plant, Scion, Double bud grafting, Petiole grafting, Chip bud grafting

7.1 Introduction

Grafting technique (placing the cut end of one plant onto immediate contact of tissues of other plants to establish a union product in one plant) has been well practised since time immemorial. There is a wide variety of grafting techniques such as stem grafting or wedge grafting, tuber grafting (in potatoes) and root grafting. Grafting is widely used commercially for propagation of plants. Graft inoculation was the first widely used technique for the transmission and detection of viruses, especially of woody plants. In this, tissue from plant to be tested is grafted to a plant species sensitive to infection with the virus in question. Though many of the viruses can now be detected by rapid methods such as enzyme-linked immunosorbent assay (ELISA) or polymerase chain reaction (PCR), for some viruses especially in woody plants, graft inoculation to indicator plants is still the most sensitive detection method.

Grafting is an ancient horticultural practice to establish the organic union between the cut surfaces of tissues of two different plants. Various parts of the plant including roots, tubers, corms, etc. may be used for grafting. The commonly practised grafting is the transfer of the detached shoot portion of one plant referred to

Alangar Ishwara Bhat and Govind Pratap Rao, *Characterization of Plant Viruses*, Springer Protocols Handbooks, https://doi.org/10.1007/978-1-0716-0334-5_7, © Springer Science+Business Media, LLC, part of Springer Nature 2020

as scion, onto the root bearing portion of another, called stock. If either the scion or stock is infected, the virus usually moves to the healthy partner who may express visible symptoms (Matthews 1991; Nayudu 2008). A pre-requisite for successful graft transmission is the perfect union of the cambial layers of the stock and scion. While interspecific and intergenic grafts are often possible, grafts between taxonomically distant species are much less likely to be successful. The success of the graft or bud depends upon the union of stock and scion that takes place as a result of formation of callus by both stock and scion. The cambium layer produces this callus. Hence, when scion is placed over stock, effort should be made to have two cambium layers coincide or match. Thus, in all kinds of grafting, the partners must be firmly held together by grafting tape or other binding material till the union is complete. The callus formation is influenced by the vigour and species of a plant, by the presence of inhibitors, such as gums, resins, latex and environmental factors of temperature and humidity (Corbett and Sisleer 1956; Mahlstede and Haber 1957; Gibbs and Harrison 1980; Adriance and Brison 2006). The grafted plants have to be kept under humid conditions to prevent water loss. The time required for the virus to get established in the healthy partner may vary from several days to few weeks in herbaceous plants while woody plants usually take several months to express symptoms. Graft inoculation of monocotyledons plants is difficult due to the absence of the ring of cambial tissue. Thus, in general, success or failure of the graft is based on compatibility of stock and scion, the closeness of fit and cambial contact.

Grafting can transmit all viruses that cause systemic infection on hosts. Graft inoculation is the widely used technique for the transmission and detection of viruses that cannot be transmitted mechanically. Grafting is successful when carried out between within or closely related species. The speed of movement of the virus across the graft junction can vary enormously depending on the virus, the plant partners involved, the grafting technique used and the environmental condition. Viruses that reach high concentration and are well distributed in plant tissues tend to move very quickly, in actively growing tissues within 2–4 days. Viruses that reach only low concentration and have a restricted distribution in plants take longer (7–30) days (Hill 1984; Basu and Giri 1993; Nayudu 2008).

There are many different ways of grafting. The two commonest ones are called detached-scion grafting and approach grafting. With detached-scion grafting, it is better to transmit the virus from scion to stock rather than from stock to scion because this will distinguish virus diseases from physiological disorders, such as nutrient deficiencies, which might be induced in a normal scion grafted onto an affected stock. Some of the commonly used methods of grafting and budding are explained below.

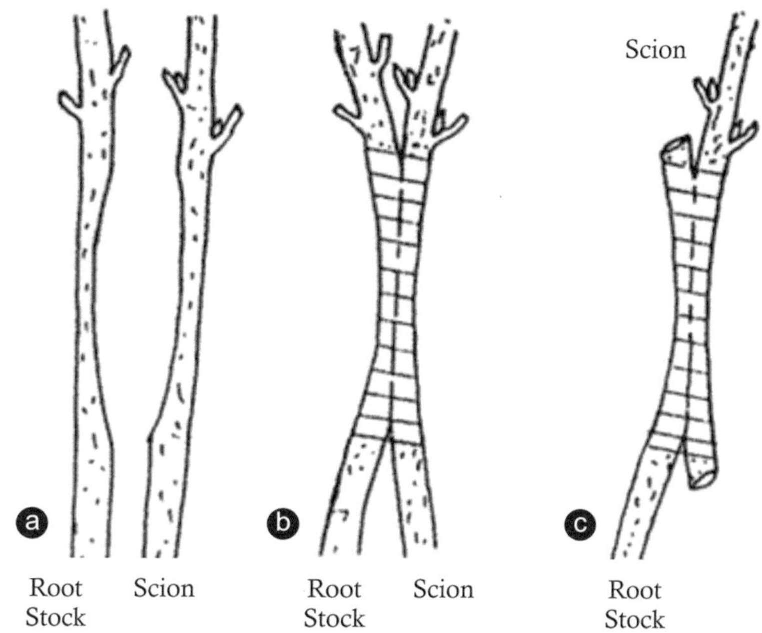

Fig. 7.1 Steps in approach grafting: (**a**) Preparation of root stock and scion; (**b**) binding of root stock and scion; (**c**) tiding up of root stock and scion with cello-tape or rubber ribbon

7.1.1 Approach Grafting

In this, two intact rooted plants are brought together and grafted, thereby minimizing graft failure due to dehydration of tissues (Fig. 7.1).

7.1.1.1 Materials

1. Polythene bags and sheets.
2. Two rooted plants (one infected and one healthy).

7.1.1.2 Method

1. Take diseased and healthy plants in two separate pots.
2. Make a long cut through the cambium and slightly into the wood on both plants at the same length of the stem.
3. Hold together the cut regions of both plants and tie it with a polythene.
4. Cover the plants if needed, with polythene bags till the establishment of the graft.
5. Observe the plants for symptom development. After successful graft union, new shoots arising out of healthy plant would show typical symptoms of virus infection.

7.1.2 Wedge or Top Cleft Grafting

In this method, a slit is made completely through the stem of the stock plant and into this is placed a piece of similar sized stem or petiole tissue, tapered at both ends ('V' shape) from the virus donor

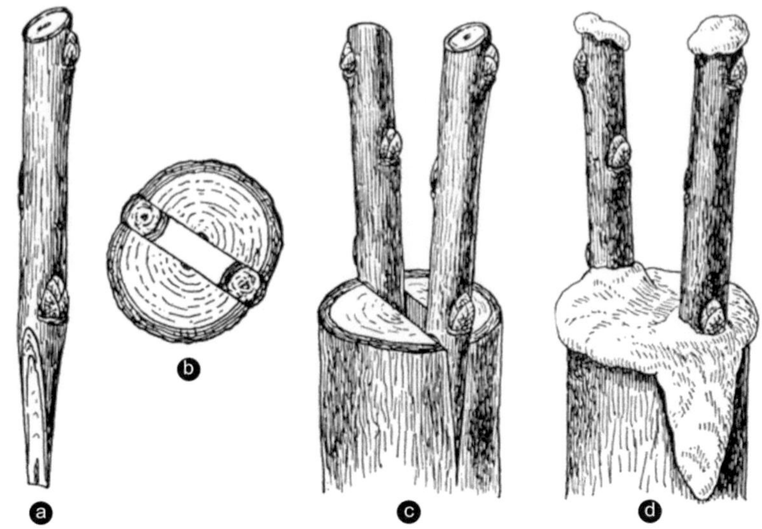

Fig. 7.2 Steps in wedge grafting: (**a**) scion; (**b**) stock; (**c**) the scion is gently tapped into V-shaped cut of stock and scions should be inserted at an angle so that the cambium layer of stock and scion are closely matched, barely crossing each other; (**d**) after scion is in place, all cut surfaces are covered with grafting wax

plant (scion). Grafting tape is then used to cover the grafted area entirely to prevent dehydration of tissues (Fig. 7.2).

7.1.2.1 Materials

1. Blade.
2. Cotton.
3. Polythene bags and sheets.
4. Scion-infected plant.
5. Stock plant (healthy plant).
6. Grafting wax.

7.1.2.2 Method

1. Cut upper portion of the stock plant with a blade, and a piece of moist cotton is put at the top immediately (Fig. 7.2).
2. Give a longitudinal incision at the top of cut region.
3. Cut a shoot from the infected plant of thickness equal to that of stock and immediately dip it in water to serve as the scion.
4. Give two deep incisions to the scion shoot, one from either side at the base to expose the cambium. This forms the wedge. Remove all leaves except two or three from the scion.

5. Fix the scion between the longitudinally incised arms of the stock.
6. Wrap the joint with a moist cotton swab and tie with a polythene ribbon or cover with grafting wax for retention of moisture.
7. Cover the graft with polythene bag and place a moistened absorbent cotton inside polythene bag so that graft can get sufficient moisture for survival.
8. Remove the polythene bag everyday for some time so that graft plant gets air.
9. Put appropriate label providing details of the graft and keep under observation.
10. The union will take place within 10–12 days and the virus will pass from the scion to the stock. New branches on the stock will produce typical symptoms.

7.1.2.3 Notes (Adriance and Brison 2006; Nayudu 2008)

1. Healthy and diseased plant should be of the same species.
2. Scion and stock should be of equal thickness.
3. Cotton swab wrapped around the graft should be moistened from time to time.
4. Inner walls of the polythene bag used to tie grafted plants should be kept moist.
5. Graft should be wrapped tightly with polythene ribbon or with grafting wax.
6. Only two or three leaves should be allowed to remain on scion to avoid excess transpiration.

7.1.3 Tongue Grafting

When handling plants with soft stem or stolon tissue, the tongue graft may be used. A downward slit of about 15–20 mm extending only part way into the centre of the stem of the stock plant is made with a razor or sharp scalpel blade. A corresponding, similar sized upward slit is made into the stem of the scion. The epidermis of each 'tongue' produced by the incisions in the plant partners is sometimes removed before the tongue of the scion is fitted into the incision of the stock plant. The grafted tissues are bound tightly together to cover tissues (Fig. 7.3).

7.1.3.1 Materials

Same as in Subheading 7.1.2.1.

7.1.3.2 Method

1. Cut upper portion of the stock plant with a blade, and a piece of moist cotton is put at the top immediately (Fig. 7.3).
2. Give a longitudinal incision at the top.
3. Cut a shoot from the infected plant of thickness equal to that of stock and dip it in water that will serve as the scion.

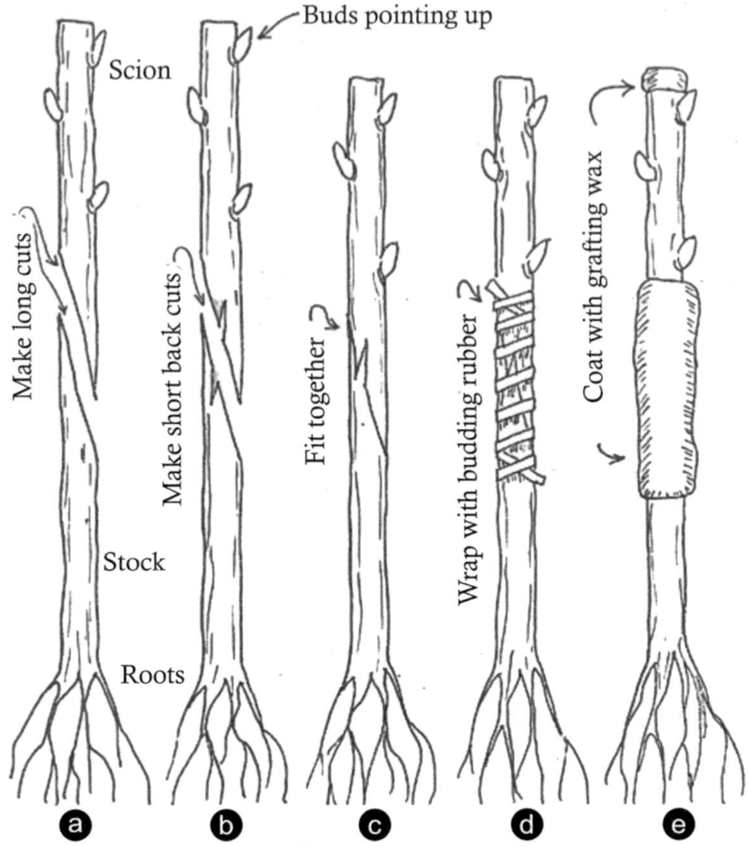

Fig. 7.3 Steps in tongue grafting: (**a**) Make a sloping cut in root stock with a tongue pointing up and in the scion with a tongue pointing up; (**b**) Match the tongues of root stock and scion end to end; (**c**) Fit together the tongues ensuring maximum contact of cambium layers; (**d**) bind with budding rubber; (**e**) coat and seal with grafting wax

4. Make a downward slit of about 15–20 mm extending only part way into the centre of the stem of the stock plant with a razor.
5. Make a corresponding similar sized upward slit into the stem of the scion. Remove all leaves except two or three from the scion.
6. Remove the epidermis of each 'tongue' produced by the incisions in both plant partners (scion and stock).
7. Fix the scion between the incised arms of the stock.
8. Wrap the joint with a moist cotton swab and tie with a polythene ribbon and seal with grafting wax for retention of moisture.

9. Cover the graft with polythene bag after placing a moistened cotton inside to provide the graft sufficient moisture for survival.
10. Open the polythene bag everyday for some time so that graft plant gets air.
11. Put appropriate label providing details of the graft.
12. The union will take place within 10–12 days and the virus will pass from the scion to the stock. New branches on the stock will produce typical symptoms.

7.1.4 Leaf Patch Grafting

Leaf patch grafting is practised in woody plants. Leaves should contain chitin so that it can withstand the stress and strain due to this process. In this, cells unite easily and successfully while in other leaves due to the absence of chitin cells will shrink and die.

7.1.4.1 Materials

1. Blade.
2. Cardboard piece.
3. Cello-tape.
4. Diseased and healthy plants.

7.1.4.2 Method

1. Hold two leaves, one from healthy and another from the infected plant, at the same level in a juxtaposed position.
2. Cut a square piece from both the leaves giving support with a cardboard underneath.
3. Cut leaf pieces are mutually exchanged and fixed in the position, taking care that veins should match, with cello-tape pasted from both the sides.
4. After 15 days to 1 month, the healthy plant with diseased leaf patch will show symptoms of the virus.

7.1.4.3 Notes

1. The leaves should be of equal size and at the same height of both stock and scion plants.
2. Leaves should be held in correct position properly while cutting the patch.
3. While fixing the patch, veins should match and there should not be any gap between the patch and cut surface of the leaf.

7.1.5 Bud (Shield) Grafting

Bud grafting is a form of lateral grafting that makes use of a scion having only a single bud. The bud is removed with a few surrounding tissues and inserted into the bark of stock plant between nodes. In this technique, a 'T' shaped cut is made in the bark of the stock plant and the two bark flaps are lifted up to place the bud (Fig. 7.4). The bud is sliced from a twig so that the cut surface beneath the bud is wider than the bud but smaller than 'T' shaped opening in

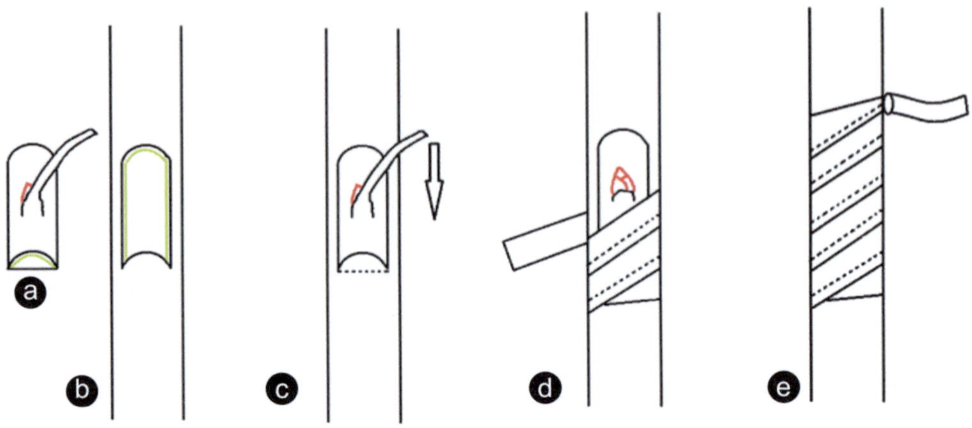

Fig. 7.4 Steps in bud (shield) grafting: (**a**) bud is removed with a few surrounding tissues; (**b**) a 'T' shaped cut is made in the bark of the stock plant and the two bark flaps are lifted up to place the bud between nodes; (**c**) the removed bud is inserted into opening of the bark of stock plant; (**d** and **e**) the bark flaps folded over to hold it in place, and the grafted area either sealed with wax or bound tightly to hold the tissues together, but leaving the bud exposed

the bark of the stock plant. The bud is inserted in the opening; the bark flaps folded over to hold it in place, and the grafted area either sealed with wax or bound tightly to hold the tissues together, but leaving the bud exposed. In addition to buds, wood chips and leaves have been used successfully as inoculation sources in bud grafts.

7.1.5.1 Materials

1. Beaker with water.
2. Budding knife.
3. Healthy and infected plants.
4. Polythene ribbon.
5. Knife.

7.1.5.2 Method

1. Select an infected plant bearing mature buds.
2. Cut the buds with a knife.
3. Remove the wood portion without damaging the bud and put it in a beaker containing water.
4. Make a superficial 'T' shaped cut on the stock healthy plant.
5. Loosen the bark slowly with the knife without damaging the wood.
6. Put the bud inside the bark flap of the plant in such a way that the only bud remains emerged in a vertical position and bark portion of the bud remains covered with the bark flap of the stem (Fig. 7.4).

7. Tie with a polythene ribbon to keep the bud in position and bud union with the stem. See that the bud remains exposed.

7.1.5.3 Notes

1. Removal of the flap from stem should be done carefully so that wood is not injured.
2. Incision of the stem of the healthy plant should be a superficial one and should not touch the wood.
3. Bud should not get injured by any chance.
4. The material to be used for grafting should not be too far from apical tissue and rather woody than too soft, to guarantee virus presence and to support growing on of scion to the rootstock.

7.1.6 Double Budding

An extension of bud grafting technique is referred to as double budding. It is used for virus indexing of fruit trees. In this, two buds are grafted on a rootstock plant, one directly above the other, about 40–60 mm apart. The upper bud is from a healthy virus indicator plant and the lower one from a virus source plant or a plant under test. Virus (if present) moves across the grafted surfaces from the lower bud to the upper virus indicator bud during the dormant season. When the upper bud grows out in the spring, symptoms appear on the young growth.

7.1.7 Petiole Grafting

In this method, infected leaflet tissue is used as scion for grafting. The terminal end of the stock plant is excised and an incision of approximately 5–10 mm down the centre of the stem is made. The scion (leaflet from infected plant) is collected and its petiole is trimmed into a wedge shape and inserted into the stem slit of the stock plant (Reddy et al. 2002). The graft insertion area is sealed or tied tightly to hold the tissues together leaving the petiole exposed. This method of grafting was successfully used for screening pigeon-pea accessions against pigeonpea sterility mosaic virus (PPSMV) (Reddy et al. 2002).

7.1.7.1 Materials

Same as in Subheading 7.1.5.1.

7.1.7.2 Method

1. Collect leaflets from virus-infected plant.
2. Trim the petiole of the collected leaflet into a wedge shape and place it in a beaker containing water.
3. Excise terminal end of the stock (healthy) plant using a knife and make an incision of approximately 5–10 mm deep in the centre of the stem.
4. Insert the trimmed petiole (from **step 2**) into the cut made in the stem of stock plant (**step 3**).
5. Tie with polythene ribbon to keep the petiole in position for proper union of tissues.

6. Cover the grafted plants with polythene bag till the establishment of the graft.
7. Observe the plants for symptom development. After successful graft union, new shoots arising out of healthy (stock) plant would show typical symptoms of virus infection.

7.1.7.3 Notes

1. Petiole grafting is more efficient than chip or cleft grafting in the transmission of virus in pigeonpea.
2. It is convenient and simple to perform and allows testing of plants at the seedling stage as well as the detection of the virus within 2 weeks of grafting.

7.1.8 Chip Bud Grafting (Wagaba et al. 2013)

In this method of grafting, axillary buds from virus-infected plants are used as scion. It is a modified method of top and side grafting methods. The procedure involves removal of stem portion of healthy (stock) plant to expose the cambium tissue. The axillary bud excised from non-lignified portion of the virus-infected plant with petiole attached (scion) is inserted into the root stock plant and bud graft is secured with a parafilm. Callus formation is seen within 1 week and virus transmission takes place within 2–6 weeks after graft inoculation (Fig. 7.5) (Wagaba et al. 2013). This method is successfully used in the screening of cassava plants against cassava brown streak virus.

7.1.8.1 Materials

1. Same as in Subheading 7.1.5.1.
2. Parafilm.
3. Double edged knife.

7.1.8.2 Method

1. Collect axillary buds of 3–6 mm in thickness from non-lignified stems of virus-infected plants to be used as scion (Fig. 7.5).
2. Excise buds to a depth of 2 mm (to expose the cambium) with the petiole and leaf attached 4–12 nodes below the apical point by making a triangular cut with a double edged razor blade.
3. Excise axillary buds of equivalent size from the root stock plant six to eight nodes above soil level.
4. Insert the axillary bud with petiole attached excised above from **step 2** into the root stock plant.
5. Secure the bud graft with a parafilm.
6. Retain the petiole and remove the leaf blade from the scion bud.
7. Keep grafted plants in the greenhouse. Remove parafilm wrapping 1 week after grafting. Retention of green colour of scion indicates success of grafting. Successful grafting can also be seen with the formation of callus tissues within a week after

Fig. 7.5 Steps in chip bud grafting of cassava: (**a**) axillary bud is removed from stem portion of 6- to 8-week-old rootstock test plant to expose cambium tissue, (**b**) axillary bud excised from non-lignified portion of virus-infected plants with petiole attached is inserted into rootstock test plant, (**c**) bud graft secured with parafilm, (**d**) successful bud graft 1 week after graft initiation with bud remaining healthy and attached to rootstock with visible callus formation and (**e**) whole plant after completion of chip bud grafting (Reproduced from Wagaba et al. 2013)

grafting. Transmission and expression of symptoms in the grafted plants appear 2–6 weeks after grafting.

7.1.8.3 Notes

1. Chip bud grafting is an improvised version of top and side grafting system which is more efficient in transmitting viruses.
2. It is used for screening cassava plants for viruses.
3. The same root stock can be used for the second grafting using scion from another plant 10–14 days after the first grafting.

References

Adriance GW, Brison FR (2006) Propagation of horticultural plants. Biotech Books, Delhi

Basu AN, Giri BK (1993) Essentials of viruses, vectors and plant diseases. Wiley Eastern Limited, New Delhi

Corbett MK, Sisleer HD (1956) Plant virology. University Press of Florida, Gainesville

Gibbs A, Harrison BD (1980) Plant virology—the principles. Edward Arnald, London

Hill SA (1984) Methods in plant virology. Blackwell Scientific Publication, New Jersey

Mahlstede JP, Haber ES (1957) Plant propagation, vol 83. Wiley, New Jersey, p 413

Matthews REF (1991) Plant virology. Academic, San Diego

Nayudu MV (2008) Plant viruses. Tata McGraw-Hill Publishing Company Limited, New Delhi

Reddy AS, Kulkarni NK, Kumar PL, Jones AT, Muniyappa V, Reddy DVR (2002) A new graft inoculation method for screening for resistance to pigeonpea sterility mosaic virus. International Chickpea and Pigeonpea Newsletter 9:44–46

Wagaba H, Beyene G, Trembley C, Alicai T, Fauquet CM, Taylor NJ (2013) Efficient transmission of Cassava brown streak disease viral pathogens by chip bud grafting. BMC Res Notes 6:516

Chapter 8

Transmission Through Dodder

Abstract

Dodder plants are leafless parasitic vines, members of the family Convolvulaceae, which grows as climbing tendrils and forming bridge between plants. The stems of these plants wind around other plants with haustoria which connect with vascular tissue of the host and facilitate transmission of the virus from the infected to healthy plants. The dodder can be used for experimental transmission of viruses under controlled conditions.

Key words *Cuscuta* spp., Phanerogamic parasite, Virus transmission

8.1 Introduction

Dodder (*Cuscuta* spp.) is an obligate phanerogamic parasite on higher plants. Dodder plants are characterized by rootless green or yellow stems with scale-like leaflets and very low level of chlorophyll (Fig. 8.1a) (Machado and Zetsche 1990). The plants have close physical linkages with their hosts and are completely dependent on the host for the supply of assimilates and water. At points of contact with the host, the dodder tendrils produce a haustorium, a specialized organ that penetrates host tissues. Dodder forms direct contact with host xylem. A well-known aspect of host dodder connections is the transmission of plant viruses: a single dodder plant simultaneously parasitizing two hosts may transmit plant viruses from one host to the other (Hosford 1967; Hull 2002; Walkey 2012). Dodders wind around the host and penetrate its haustoria into host tissue sending up to vascular tissue and draw their food. If the host is infected with a virus, the virus particles are also taken up by the haustoria of the parasite and get transmitted to another host through the dodder (Fig. 8.2). Dodder is a good virus transmitter because it parasitizes on a large number of plants belonging to different families that cannot be grafted. Dodder acts as a bridge through which the virus particles pass from the diseased into the healthy plant. Out of about 20 species tested for experimental transmission, *Cuscuta campestris, C. europea, C. reflexa* and *C. subinclosa* have been commonly used to transmit

Alangar Ishwara Bhat and Govind Pratap Rao, *Characterization of Plant Viruses*, Springer Protocols Handbooks, https://doi.org/10.1007/978-1-0716-0334-5_8,

Fig. 8.1 (**a**) Dodder plant; (**b**) Transmission of virus using dodder

Fig. 8.2 Wrapping of dodder tendril on the host during the process of transmission procedure

different viruses and phytoplasmas. They have wide host range. Transmission through dodder may be just a passive transport of the virus since its multiplication in dodder is not essential for transmission.

Haustoria acquire virus from the infected plants that eventually transmit to the new hosts. Dodder-transmitted viruses are sugar-beet curly top virus, tomato bushy stunt virus, tobacco rattle virus, etc.

8.2 Materials

1. Dodder.
2. Seed trays.
3. Pots with compost.
4. Healthy plants (test plants).

5. Virus-infected plants (source plants).
6. Access to controlled insect-proof glasshouse.

8.3 Method

1. Collect healthy dodder, *Cuscuta reflexa* Roxb maintained on a healthy plant grown from the seed.
2. Place the dodder by twinning anticlockwise on young growing shoots of infected source plant.
3. Once the dodder is established, one of the growing ends is twinned onto the young shoot of healthy test plant.
4. Keep the pots at 25 °C, 70–90% humidity along with supplementary intensive light.
5. Allow the dodder to grow as a bridge between donor and receptor plants for 30–45 days.
6. Remove the dodder and keep the test plants in the insect-proof glasshouse for observation.
7. Observe the symptoms on test plants and confirm the presence of virus transmission in recipient host through ELISA or PCR assays.
8. After transmission has been confirmed, remove the dodder strands from the test plants and keep in an insect-proof glasshouse under controlled conditions for further study.

8.4 Notes (Hull 2002; Walkey 2012)

1. Use of suitable dodder species for transmission is crucial for successful transmission.
2. A dodder seedling must form connections with a host within several days after germination.
3. Place one end of the strand of dodder around the infected plant's stem and place the other end around a healthy plant's stem. The dodder must parasitize both the donor (infected) plant and the recipient (healthy) plant.
4. Transmission duration may differ in herbaceous and woody plants. Therefore, sufficient time should be provided for virus transmission through dodder for different plant species.

References

Hosford RM (1967) Transmission of plant viruses by dodder. Bot Rev 33:387–406

Hull R (2002) Matthews' plant virology, 4th edn. Academic, San Diego, CA

Machado MA, Zetsche K (1990) A structural, functional and molecular analysis of plastids of the holoparasite *Cuscuta reflexa* and *C. europea*. Planta 181:91–96

Walkey DGA (2012) Applied plant virology. Springer, Heidelberg

Chapter 9

Virus Transmission Through Pollen

Abstract

Pollens from some of the virus-infected plants can serve as source of natural infection and transmission of the virus to other plants. When pollens consisting of viruses fall on stigma of female plants, they germinate and eventually facilitate the virus to infect the ovules of plants. Such viruses are called pollen-borne viruses. Virus-infected pollens may be carried by humans, wind or insects. Pollen-mediated virus transmission occurs vertically and/or horizontally. Members of the genera, *Ilarvirus*, *Nepovirus*, *Sobemovirus*, *Idaeovirus*, *Potyvirus* and many viroids are transmitted through pollens. At least 45 viruses, belonging to 16 genera and five viroids, have been described as pollen-transmitted. The process of virus transmission through pollen is discussed in the chapter.

Key words Idaeovirus, Ilarvirus, Nepovirus, Sobemovirus, Transmission

9.1 Introduction

Pollination is a vital process in the reproduction of most angiosperms and requires the transfer of pollen (male gamete) to ovule (female gamete) where fertilization occurs. The pollen grains escape when the pollen is mature. The entrance of pollen-borne viruses into ovules occurs through the pollen tube as it grows into the embryo sac. The virus may accompany the male gametes, which unites with the egg cell during fertilization. When the stigma of mother plants was pollinated with pollen infected with raspberry bushy dwarf virus, the virus infects stigma by penetration of pollen tubes followed by subsequent systemic infection (Isogai et al. 2014). Many plant species require a vector support (such as humans, insects, birds or bats) for successful pollination, whereas some plant species rely on abiotic factors such as wind and, in a few instances, water for pollination. Pollination process is very important for horticulture crops, as the fruits will not develop unless the ovules are fertilized. Some plant viruses have also evolved mechanisms that exploit the plants' own reproductive processes and can be transmitted by seed and/or pollen (Mink 1993, 1998; Johansen et al. 1994; Hull 2002). Many insect groups are potential vectors for plant viruses and some of them may contaminate pollen with viruses when feeding to healthy plants, e.g. thrips, pollen beetles

Alangar Ishwara Bhat and Govind Pratap Rao, *Characterization of Plant Viruses*, Springer Protocols Handbooks, https://doi.org/10.1007/978-1-0716-0334-5_9,

and honey bees (Brunt et al. 1996). Pollen and seed transmission are closely related (Hull 2002; Mink 1993). Pollen-transmitted viruses are generally transmitted by seeds but not necessarily *vice versa* (Brunt et al. 1996).

Example of pollen-transmitted viruses are: barely stripe virus, tobacco ring spot virus, bean common mosaic virus, fruit ring spot virus, Datura mosaic virus, blueberry leaf mottle virus and raspberry bushy dwarf virus (RBDV) (Murant et al. 1974; Childress and Ramsdell 1986; Boylan-Pett et al. 1991).

Pollen-mediated virus transmission occurs vertically and horizontally (Card et al. 2007). In horizontal pollen-mediated transmission, viruses infect new host plants from pollen grains carrying virus to maternal tissue after pollination. In vertical pollen-mediated transmission, the virus infects mother host plants and invades the ovule. The horizontal pollen-mediated transmission is of great significance especially for perennial plants, where the viral infection persists in the following year. The pollen transmission of plant viruses is more important in cross-pollinated woody perennials in comparison to annual crops (Hull 2002).

The efficacy with which viruses are transmitted by pollen is determined by a range of factors including the species or strain of the virus, species or cultivar of the host, growth stage at time of infection and, to a lesser extent, environmental factors such as temperature and moisture (Bos 1983; Mink 1993, 1998).

Members of the genera, *Ilarvirus* (*Prunus necrotic ringspot virus, Tobacco streak virus, Blueberry shock virus*), *Nepovirus* (*Artichoke yellow ringspot virus, Blueberry leaf mottle virus, Cherry leaf roll virus*), *Sobemovirus* (*Sonchus mottle virus*), *Idaeovirus* (*Raspberry bushy dwarf virus*) and many viroids are transmitted through pollens (Card et al. 2007). Pollen transmission has been successfully demonstrated when thrips carrying pollen infected with *Prunus necrotic ringspot virus* (family *Bromoviridae*) are placed on test plants (Milne and Walter 2008). At least 45 viruses belonging to 16 genera have been described as pollen-transmitted (Card et al. 2007).

The method of plant virus transmission through pollen is discussed herein:

9.2 Materials

1. Pollens.
2. Seeds (Healthy).
3. Indicator host.
4. Phosphate buffer for virus inoculation.
5. Petri plate or glass vial.
6. Brush.
7. Cages.

9.3 Method (Liu et al. 2014; Atsumi et al. 2015)

1. Establish two separate plant populations from uninfected seeds in cages.
2. The first population is mechanically inoculated with virus at the 3-leaf stage (cage 1) (details of mechanical inoculation explained in Chapter 6).
3. The second population (cage 2) is kept virus-free until artificial pollination using pollen from the flowers is produced in cage 1.
4. Before pollination, ELISA and PCR are used to confirm the presence or absence of virus in both populations (cage 1 and cage 2).
5. Female flowers are plucked from cage 1, the male flowers in cage 2 are removed before opening to prevent self-pollination.
6. The open male flowers in cage 1 are picked and their petals are removed to expose the stamens and pollen, which are then rubbed against the stigma of the female flowers in cage 2.
7. Or the pollen grains are attached to a stigma of female flower in cage 2 using a small ink brush (at **step 6**).
8. Plants in cage 2 are pollinated several times on different days.
9. Plants grown without inoculation or artificial pollination are used as the controls in cage 3.
10. Two months later, the mature seeds are collected from both cages.
11. The RNA/DNA was extracted from these seeds and tested for the presence of virus tested by PCR assays (for detailed process of PCR, please *see* Chapter 35).
12. The presence of virus in cage 2 confirms the pollen transmission.

9.4 Notes (Hull 2002; Card et al. 2007; Atsumi et al. 2015)

1. Pollens should be collected fresh for pollination.
2. Presence of the test virus in the pollens should be checked through ELISA or PCR.
3. Male flowers in test cages where pollination has to be done are removed just to avoid the self-pollination possibility.
4. Pollination is done properly on the female flowers and seeds are collected for the proof of virus transmission.

References

Atsumi GO, Tomita R, Yamashita T, Sekine K-T (2015) A novel virus transmitted through pollination causes ring-spot disease on gentian (*Gentiana triflora*) ovaries. J Gen Virol 96:431–439

Bos L (1983) Introduction to plant virology. Longman Group Ltd., Essex, UK

Boylan-Pett W, Ramsdell DC, Hoopingarner RA, Hancock JF (1991) Honeybee foraging behaviour, in-hive survival of infectious, pollen-borne blueberry leaf mottle virus and transmission of the virus in highbush blueberry. Phytopathology 81:1407–1412

Brunt A, Crabtree K, Dallwitz M, Gibbs A, Watson L (1996) Viruses of plants: descriptions and lists from the VIDE database. Wallingford, CABI International

Card SD, Pearson MN, Clover GRG (2007) Plant pathogens transmitted by pollen. Australas Plant Pathol 36:455–461

Childress AM, Ramsdell DC (1986) Detection of blueberry leaf mottle virus in highbush blueberry pollen and seed. Phytopathology 76:1333–1337

Hull R (2002) Transmission 2: mechanical, seed, pollen and epidemiology. In: Matthews' plant virology. Elsevier, Academic, Amsterdam

Isogai M, Kamata Y, Ando S, Kamata M, Shirakawa A, Sekine KT, Yoshikawa N (2014) Horizontal pollen transmission of gentian ovary ring-spot virus is initiated during penetration of the stigma and style by infected pollen tubes. Virology 503:6–11

Johansen E, Edwards MC, Hampton RO (1994) Seed transmission of viruses: current perspectives. Annu Rev Phytopathol 32:363–386

Liu HW, Luo LX, Li JQ, Liu PF, Chen XY, Hao JJ (2014) Pollen and seed transmission of cucumber green mottle mosaic virus in cucumber. Plant Pathol 63:72–77

Milne RG, Walter GH (2008) The potential for pollen borne virus transfer in a plum orchard infected with Prunus necrotic ringspot virus. J Phytopathol 145:105–111

Mink GI (1993) Pollen and seed transmitted viruses and viroids. Annu Rev Phytopathol 31:375–402

Mink GI (1998) Viruses spread in pollen. In: Ogawa JM, Zehr EI, Bird GW, Ritchie DF, Uriu K, Uyemoto JK (eds) Compendium of stone fruit diseases. APS Press, St Paul, MN

Murant AF, Chambers J, Jones AT (1974) Spread of raspberry bushy dwarf virus by pollination, its association with crumbly fruit, and problems of control. Ann Appl Biol 77:271–281

Chapter 10

Transmission Through Seeds

Abstract

Seeds provide an efficient means in disseminating plant virus diseases. Transmission of viruses through seeds plays an important role in the long distance spread of viruses. Unlike vectors, infected seeds can be carried away to distant places where they act as primary source of inoculum for spread of a virus. A few of the seed-borne viruses may be carried as surface contaminants while others are carried through embryo considered as true seed-borne viruses.

Key words *Hordeivirus*, *Ilarvirus*, *Nepovirus*, *Potyvirus*, *Tobamovirus*, *Tobravirus*, Virus transmission, Surface contamination, Embryo-borne virus

10.1 Introduction

About one-seventh of known plant viruses are transmitted through seed. As viruses persist in seed for an extended period, distribution of such seeds may result in widespread and long distance spread of virus infections. Transmission through seed plays a significant role in the ecology and epidemiology of many plant viruses. They act as a primary source of inoculum for the secondary spread within an area through vectors. Members of the genera, *Hordeivirus, Ilarvirus, Nepovirus, Potyvirus, Cucumovirus, Tobamovirus, Tobravirus* and viroids are seed borne in nature (Hull 2014). Members of the virus genera that are confined to the phloem tissue of their hosts are usually not seed borne probably due to the lack of direct vascular connections with the embryo. Some viruses are carried on the seed coat as a surface contaminant (like tobacco mosaic virus in tomato seed). Certain treatments can readily inactivate such external virus associations. A large majority of seed-borne viruses are carried in the embryo, endosperm or seed coat. In majority of the cases, the seed offers a highly effective barrier to transmit most of the viruses from one generation to the next generation. The efficiency of virus transmission through the embryo may be influenced by the ability of virus to infect male and female gametophytes (Bennett 1969). Virus testing of seed is mostly performed for quarantine testing.

Alangar Ishwara Bhat and Govind Pratap Rao, *Characterization of Plant Viruses*, Springer Protocols Handbooks, https://doi.org/10.1007/978-1-0716-0334-5_10,

Early infection of the mother plant allows sufficient time for the virus to invade the embryonic tissue before its cytoplasmic separation from the mother plant. However, seed transmissibility of any particular virus is often host specific, sometimes particular strains of the virus can be temperature dependent. Different isolates of a virus may have different seed transmission rates on the same host. Some viruses are seed transmitted by a wide range of host species although at different rates. Different varieties of the same host species vary in the rate of seed transmission. Further, variation in the rate of transmission can even occur between plants of a given variety. The time of infection is critical in determining whether seed transmission occurs or not and the extent of transmission to progeny seedlings. Within seeds, distribution of the virus varies in different parts. The virus may be found in various parts such as embryo, endosperm and seed coat. A few seed-borne viruses are present in immature seeds or seeds nearing maturity but are eliminated from mature ripened seeds by viral inactivation during seed maturation. In the case of embryo infection, developing embryo can become infected either before fertilization by infection of gametes or by direct invasion after fertilization. Some of the infected seeds show symptoms on seeds. Percentage of seeds getting infected in a plant varies widely from 0 to 100% (Shepherd 1972; Hull 2014).

10.2 Materials

1. Healthy seeds.
2. Infected seeds.
3. Insect-proof glasshouse.
4. Earthen pots.
5. Sterile potting mixture.
6. Hydrochloric acid (4%): 4 mL of concentrated hydrochloric acid dissolved in water and made up to 100 mL.
7. Bleach solution: Commercially available bleach solution may be used.

10.3 Method

1. Collect seeds from symptomatic plants (seeds collected from non-symptomatic plants may be used as the control).
2. Extract the seeds in 4% hydrochloric acid and wash in a 10% bleach solution to ensure that any viral infection that occurred is not simply the result of virus on the seed coat, but rather the result of embryonic infection.

3. Sow them in earthen pots containing sterile soil, sand and manure mix under insect-proof glasshouse.
4. Provide optimum conditions for germination and growth.
5. Observe for the appearance of any typical symptoms on the seedlings.
6. Further, identification of the virus can be done by back inoculation onto indicator hosts, electron microscopy, serological or nucleic acid-based diagnostic assays.

10.4 Notes (Shepherd 1972; Hull 2014)

1. Seeds used for transmission test should be grown in an insect-proof glasshouse to avoid secondary spread of any virus by insects or other contaminants.
2. Irrespective of the number of seeds collected for each suspected virus isolate, testing is done on a minimum of 10 randomly selected seeds.
3. The level of seed transmission of each virus varies (0.1–50%) depending on the virus and host.
4. The seeds should be extracted in 4% hydrochloric acid and washed in a 10% bleach solution to remove surface contaminants and viruses borne on the seed coat.

References

Bennett CW (1969) Seed transmission of plant viruses. Adv Virus Res 14:221–261

Hull R (2014) Matthews' plant virology, 5th edn. Academic, San Diego, CA

Shepherd RJ (1972) Transmission of viruses by seed and pollen. In: Kado CI, Agrawal HO (eds) Principles and techniques in plant virology. Van Nostrand Reinhold Company, New York

Chapter 11

Transmission of Viruses by Aphids

Abstract

The majority of plant-infecting viruses are transmitted to their host plants by different species of aphid vectors. Aphids are efficient virus vectors and transmit a large number of plant viruses. Piercing and sucking type of mouth parts make them as efficient vectors. Aphids can transmit viruses in different modes: non-persistent (stylet-borne), semi-persistent, persistent and persistent propagative. Majority of aphid species are involved in the transmission of non-persistent or stylet-borne viruses.

Key words Acquisition access period, Culturing of aphids, Inoculation access period, Non-persistent transmission, Persistent transmission, Persistent propagative transmission, Semi-persistent transmission

11.1 Introduction

Aphids are the most notorious, important and largest group of virus vectors, as they can transmit a large number of different viruses. About 300 species of aphids are known as vectors. Their biology, feeding behaviour and worldwide distribution make them ideally suited for transmission of plant viruses (Dijkstra and de Jager 1998; Gray 2008; Nayudu 2008). Most aphids produce many generations in a year and reproduce both parthenogenetically and sexually. Plants on which sexual forms of aphids mate and lay eggs are called primary hosts; secondary hosts are those where asexual generations reproduce with great efficiency by parthenogenesis, and young ones are born viviparously. The secondary hosts are often agricultural crops. Some aphids may pass through their life cycle on one host species or several species within one genus. Among all, *Myzus persicae* is an important aphid vector (known to transmit more than 50 different viruses). Some of the polyphagous aphid vectors of viruses include: *Aphis craccivora* (usually black in colour, feeds on legumes), *A. gossypii* (variable body colour from cream to greenish, feeds on many hosts including cotton, cucurbits and solanaceous hosts), *Myzus persicae* (pale green to pinkish body colour, feeds on many hosts), *A. rhamni* (potato aphid), *Rhopalosiphum maidis* (feeds on graminaceous host), *Toxoptera citricidus*

Alangar Ishwara Bhat and Govind Pratap Rao, *Characterization of Plant Viruses*, Springer Protocols Handbooks, https://doi.org/10.1007/978-1-0716-0334-5_11,

Fig. 11.1 Mustard aphid (*Lipaphis erysimi*) (**a**), Cabbage aphid (*Brevicoryne brassicae)* (**b**), Cardamom aphid (*Pentalonia caladii*) (**c**)

(affect rutaceous crops), *Pentalonia nigronervosa* (banana aphid) and *Lipaphis erysimi* (affect cruciferous crops) (Fig. 11.1) (Harris and Maramorosch 1980; Pirone and Harris 1977; Nayudu 2008). Examples of viruses transmitted by different species of aphids are: barely yellow dwarf virus, potato virus S, alfalfa mosaic virus, cucumber mosaic virus, potato leaf roll virus, lettuce necrotic yellow virus, bean mosaic virus, barely mosaic virus, pea mosaic virus, chilli mosaic virus, etc.

Aphids probe plants by inserting their stylets into host epidermis to search for preferred hosts. The aphids can transmit plant viruses in different ways: (a) viruses transmitted by probing aphids are acquired and transmitted within a few minutes are called as non-persistent or stylet-borne viruses (Pirone and Harris 1977), (b) aphids transmitting semi-persistent viruses frequently move after colonizing a plant, (c) aphids transmitting persistent viruses require more hours to acquire the virus and once acquired can retain the virus for their entire life (Sylvester 1980). Persistent viruses may be carried to great distances by aphids. Aphids lose their ability to transmit non-persistent viruses within an hour after acquisition while semi-persistent and persistent viruses are retained for a longer duration (up to 2 days in case of semi-persistent and for several days to life in case of the persistent virus). Non-persistent and semi-persistent viruses have no latent period. Only persistent viruses survive through a moult. Non-persistent viruses can easily be transmitted by mechanical inoculation while semi-persistent and persistent viruses are difficult to transmit mechanically.

Aphid transmission may be non-persistent, semi-persistent or persistent. Of about 300 aphid-borne viruses are known, most of them are non-persistent type. The genera of viruses transmitted in a non-persistent manner include viruses with helical and isometric particles, and with DNA and RNA, mono-, bi- and tripartite genomes such as *Alfamovirus*, *Caulimovirus*, *Cucumovirus*, *Fabavirus*, *Macluravirus* and *Potyvirus*. Members of caulimoviruses and closteroviruses are reported to be transmitted by semi-persistent transmission while luteoviruses and nanoviruses get transmitted in

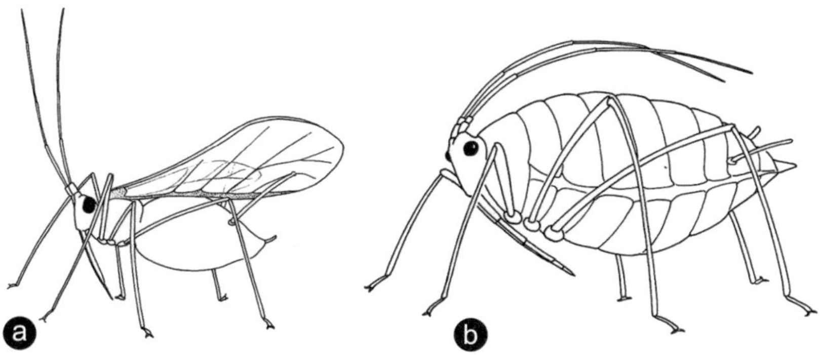

Fig. 11.2 Morphology of winged (alate) (**a**) and wingless (apterous) (**b**) aphid

persistent circulative manner. Several members of *Rhabdoviridae* (like *Lettuce necrotic yellows virus*, *Sowthistle yellow vein virus*) are transmitted in a persistent propagative manner.

The mouth parts of aphids consist of two outer mandibular stylets and two inner maxillary stylets. When an aphid starts feeding on a leaf, it first produces a drop of gelling saliva then its stylets penetrate the epidermis in a brief probe for few seconds, thus favouring non-persistent transmission of virus present in the epidermis. Subsequently, when the host is accepted, the aphid proceeds to deeper cell layers until it finally reaches the phloem sieve tubes from which it derives the nutrient. As most of the circulative and propagative viruses are restricted to phloem tissue, these viruses are usually acquired only during longer feeding periods.

11.1.1 Rearing of Virus-Free Aphids

Aphids grow well at a temperature between 15 and 20 °C and under illumination by a single fluorescent lighting tube. These conditions help to increase their multiplication rate and also suppress the development of alate forms (Fig. 11.2), which are difficult to handle. Plants used for rearing of aphids need to be changed at weekly intervals. As far as possible, use plants that are non-hosts of the viruses to be tested. Turnip, Brassica and Chinese cabbage are better hosts for culturing aphids (Harris and Maramorosch 1977; Watson 1972).

As most virus-carrying aphids do not transmit the virus to their offspring, virus-free aphids can be reared from newly born ones. To ensure regular supply of wingless (apterous) aphids of similar age, they are subcultured (Fig. 11.2).

Approximately 10–15 apterous adult aphids from a stock culture are placed in a cage on *Brassica* plant and transferred daily to another plant, leaving behind the nymphs that are less than 1 day old. After 3 days, the old adults are killed, and new adults from a previous series are used to start the next 3-day cycle.

11.2 Materials

1. Glasshouse compartment at a 16-h photoperiod at 22 ± 3 °C.
2. Large aphid cage (for instance 40 × 40 × 55 cm) for virus-free stock colony.
3. Cylindrical Perspex cages of different sizes (ranging from 5 to 8 cm diameter and 10 to 14 cm height) with its top and side holes covered with muslin cloth.
4. Healthy test plants.
5. Infected plant.
6. Insect-proof glasshouse.
7. Magnifying glass.
8. Painting brush.
9. Petri dish.
10. Aluminium foil or parafilm membrane.
11. Sprayer.
12. Systemic insecticide.
13. Virus-free aphid colony.

11.3 Methods

11.3.1 Culturing of Aphids

1. Grow the plants through seeds in 4-in. diameter pots that are preferred for feeding and breeding by the aphids.
2. Keep the plants under insect-proof cage and release the aphids onto these plants at 4–6 leaf stage on each grown seedlings.
3. Maintain the inoculated plants at 25 °C with 16 h duration light.

11.3.2 Transmission of Non-persistent Viruses by Aphids

1. Disturb the colony of aphids so that they withdraw their stylets. This is done by carefully tapping the abdomen with the painting brush, or by breathing on them.
2. Carefully pick up individual wingless aphids, using the tip of the moistened paint brush and transfer to the Petri dish lined with moist (not wet) filter paper.
3. Cover the Petri dish with the aluminium foil.
4. Store aphids in a cool, shaded place for about 1 h to starve them.
5. Open the aluminium foil.
6. Transfer aphids to infected plant material (detached leaves in a separate Petri dish) to feed for about 2 min. Watch if they feed with the help of a magnifying glass.

7. Transfer at least five aphids to each test (healthy) plant.
8. Cover to contain aphids with the help of cylindrical Perspex cages and allow a transmission feed of 1 h.
9. Kill the aphids by spraying insecticide (dimethoate @ 0.1% or 0.05% imidacloprid).
10. Keep the plants in insect-proof glasshouse for symptom development.
11. Symptoms on test plants, consisting of leaf mottle, mosaic veinal chlorosis and malformation of younger leaves, can usually be observed after 7–10 days.
12. Count the number of plants with symptoms and calculate the percentage of infected plants.

11.3.3 Transmission of Persistent Viruses

1. Aphids are handled in the same way as explained under non-persistent virus transfer, except that no starvation period is required before the acquisition.
2. Transfer the aphids to infected plant material in a Petri dish and allow them to feed for 24 h.
3. Carefully transfer the aphids to test (healthy) plants and leave for a further 24 h (duration can be increased if the virus has a latent period). Use five non-inoculated plants as a control.
4. Kill the aphids by spraying insecticide and keep the plants under insect-proof conditions for symptom development.
5. Symptoms on the test plants can be observed in 12–14 days.
6. Count the number of plants with symptoms and calculate the percentage of infected plants.

11.4 Notes (Harris and Maramorosch 1980; Dijkstra and de Jager 1998; Nayudu 2008)

1. *Acquisition access period (AAP):* It is the total time for which vector has been kept on the infected plant to acquire the virus.
2. *Acquisition feeding time (AFP):* It is the actual feeding time by the vector to acquire the virus from the infected plant.
3. *Latent period:* It is the time after an acquisition feed for which the vector is unable to transmit a virus.
4. *Inoculation access period (IAP):* Duration for which a viruliferous vector is kept in contact with the healthy plant.
5. *Inoculation feeding time (IFP):* It is actual feeding time required by a viruliferous vector to deliver the virus into a healthy plant.
6. *Retention time:* It is the time for which a vector can retain the viruses in them.

7. *Characteristics of non-persistent/non-circulative/stylet-borne viruses*:
   (a) No latent period.
   (b) Time required for acquisition and inoculation is only a few seconds.
   (c) Retention of inoculativity (without further access to virus) is only for a few minutes.
   (d) Starving vectors before acquisition feeding increases inoculativity by several folds.
   (e) Virus is restricted to foregut only. No evidence of virus in haemocoel or salivary system of the aphid.
   (f) Loss of inoculativity following a moult.
8. *Characteristics of semi-persistent viruses*:
   (a) No detectable latent period.
   (b) Loss of inoculativity following moult.
   (c) Starving before AAP has no effect on inoculativity.
   (d) Time required for acquisition and inoculation is several minutes.
   (e) Retention of inoculativity is few hours.
   (f) Continued AFP increases inoculativity for several hours.
9. *Characteristics of persistent/circulative viruses.*
   (a) Virus is acquired, translocated and circulated in the vector body.
   (b) Latent period follows AFP.
   (c) Inoculation feeding involves ejection of virus in saliva from maxillary saliva canal.
   (d) Virus does not multiply in the vector body.
   (e) In the case of circulative propagative viruses, virus multiplies in the vector body and can transmit the virus throughout its life.
10. Transovarial transmission refers to the progeny of the infected vector remain viruliferous, while transstadial blockage refers to the progeny of the infected vector remain non-viruliferous.
11. The cage used for culturing should be made using wooden or wire frame covered with a muslin cloth or nylon material (about 20 mesh) with a door facility.
12. The covers used in cages should be washed and dried every time cultures are changed, and cages washed and sprayed with fungicide.
13. As most virus-carrying aphids do not transmit the virus to their offspring, virus-free aphids can be reared from newly born ones.

14. To ensure regular supply of wingless (apterous) aphids of similar age, they should be subcultured regularly.

References

Dijkstra J, de Jager CP (1998) Virus transmission by aphids. In: Practical plant virology. Springer, Berlin, Heidelberg

Gray SM (2008) Aphid transmission of plant viruses. Curr Protoc Microbiol 10 (1):16B.1.1–16B.1.10

Harris KF, Maramorosch K (1977) Aphids as virus vectors. Academic, New York

Harris KF, Maramorosch K (1980) Vectors of plant pathogens. Academic, New York

Nayudu MV (2008) Plant viruses. Tata McGraw-Hill Publishing Company Limited, New Delhi

Pirone TP, Harris KF (1977) Nonpersistent transmission of plant viruses by aphids. Annu Rev Phytopathol 15:55–73

Sylvester ES (1980) Circulative and propagative virus transmission by aphids. Ann Rev Entmol 25:257–286

Watson MA (1972) Transmission of plant viruses by aphids. In: Kado CI, Agrawal HO (eds) Principles and techniques in plant virology. Van Nostrand Reinhold Company, New York

Chapter 12

Transmission of Viruses by Leafhoppers

Abstract

Leafhoppers are one of the most abundant groups of plant feeding insects in the world of family Cicadellidae of the order Homoptera. Nearly 25 genera with more than 50 species of different leafhoppers are reported to be vector of different plant viruses. All the leafhopper-borne viruses are transmitted in a semi-persistent or persistent circulative or persistent propagative manner. This chapter highlights materials and methods for raising leafhopper colonies and their utilization for virus transmission experiments. Specific methods for rearing of the *Nephotettix virescens* for the transmission of rice tungro spherical virus are discussed and explained in the chapter.

Key words Leafhopper, Semi-persistent viruses, Persistent viruses, Persistent propagative viruses

12.1 Introduction

Leafhoppers are small, slender, sap-sucking insects of the large insect family Cicadellidae of the order Homoptera. Leafhoppers are reported as efficient vectors of some viruses which are responsible for huge economic losses of major crops (Nault and Ammar 1989). The life cycle of leafhoppers is an incomplete metamorphosis as they hatch from eggs and mature through multiple nymphal stages (between 4 and 5 times over the course of 2–7 weeks) before reaching adulthood (Fig. 12.1). Many species of leafhoppers complete two or more generations each year. About 50 species of leafhoppers in 25 genera have been reported as virus vectors (Sastry 2013). Leafhopper transmission may be semi-persistent, persistent or persistent propagative (Carter 1974).

Members of *Rhabdovirus, Phytoreovirus, Tenuivirus* and *Marafivirus* have a persistent propagative relationship with their leafhopper vectors. Some of the important leafhopper vectors include: *Circulifer tenellus* (transmits sugarbeet curly top virus), *Nephotettix impicticeps* (transmits rice tungrovirus), *Agallia constricta* (transmits potato yellow dwarf virus and wound tumour virus), *Cicadulina mobile* (transmits maize streak virus) and *Psammotettix alienus* (transmits nucleorhabdovirus infecting wheat) (Liu et al. 2018).

Alangar Ishwara Bhat and Govind Pratap Rao, *Characterization of Plant Viruses*, Springer Protocols Handbooks,
https://doi.org/10.1007/978-1-0716-0334-5_12,

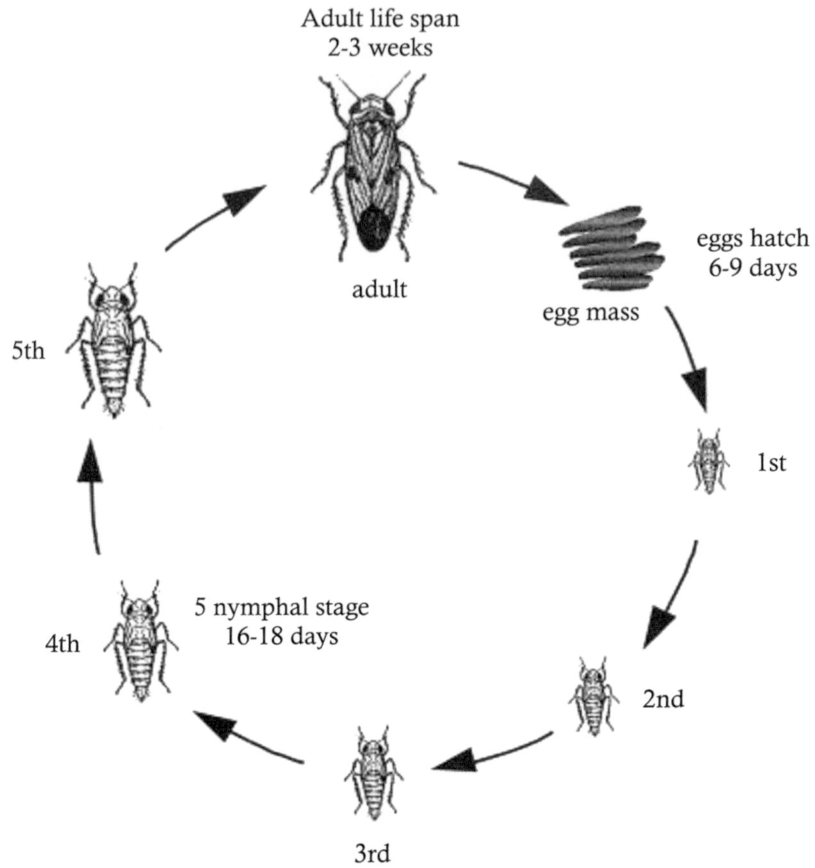

Fig. 12.1 Life cycle of a leafhopper

For transmission of rice tungro spherical virus through leafhopper, *Nephotettix virescens*, following protocol may be used.

12.1.1 Raising Insect Colonies (Maramorosch 1999)

Raising *Nephotettix* leafhopper colonies is essential for the study of virus transmission. Maintaining leafhopper colonies allows to test transmission efficiency of a vector. The leafhopper colonies also facilitate the study of the vector interactions with their plant hosts. Leafhopper colonies can be raised from field collected adults or from eggs-laden plant material collected from distant fields (Ling and Tiongco 1979; Maramorosch 1999).

12.2 Materials

1. Growth chamber with sealed access, filtered ventilation, temperature control, water drainage and humidity monitoring.
2. Glasshouse suitable for growing plants for maintenance of insect colonies.
3. 40–45-day-old virus-free rice plant (for rearing vector).

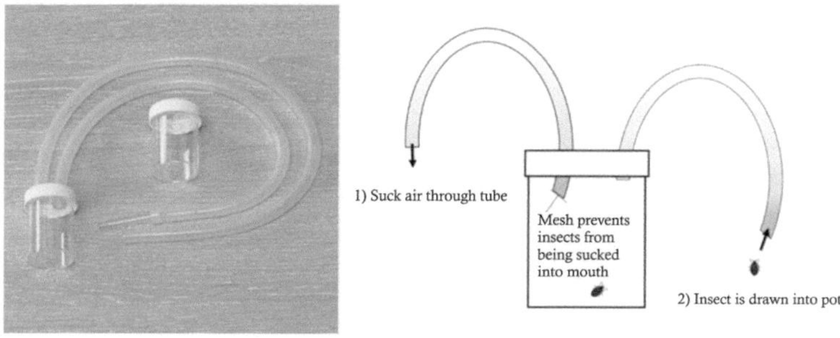

Fig. 12.2 Aspirator used to collect insect vectors

4. Seven-day-old virus-free TN1 rice seedlings (test plants).
5. Aspirator: It is a glass tube of 0.7–0.8 cm diameter and about 12 cm long, one end of the tube is covered by muslin cloth and inserted into a plastic tubing of slightly bigger diameter and about 75 cm long (Fig. 12.2). The insects are sucked from the open end of the plastic tube. The insects are released on the encaged plants by tapping the glass tube; blowing is avoided because it may cause injury to insects.
6. Black cloth.
7. Diseased rice plant (virus source).
8. Insect cages: Cylindrical cages of different dimensions are used to encage leafhoppers on diseased as well as healthy rice seedlings. The body of the cages are provided with air vents on opposite sides being covered with 60 mesh nylon net and on the top too, for ventilation and to prevent accumulation of moisture. After encaging the plants, leafhoppers are released through a circular hole and plugged with cotton. Big cages (8 cm diameter) can be used for mass inoculation while the small ones (2–3 cm diameter) are used for single plant inoculation.
9. Insect sweep net.
10. Insect transfer chamber.
11. Nymphs and newly emerged leafhopper adults (Fig. 12.3).

12.3 Method (Transmission of Rice Tungro Spherical Virus (RTSV) by *Nephotettix* spp.)

12.3.1 Raising Plants and Rearing of Insects

1. Grow the test plants (rice) in pots in glasshouse.
2. Allow plants to grow.
3. Collect the adult and/or eggs of leafhoppers from the field with sweep net.
4. Release the identified species of leafhoppers into the chamber with an aspirator containing healthy grown up virus-free rice

Fig. 12.3 Nymphs (**a**) and adults (**b**) of *Nephotettix virescens* on rice plant

plants and allow them for egg laying. Transfer egg-laden leaves to 40–45-day-old healthy grown up rice plants (eggs hatch in about 6–10 days) to serve as fresh food source for hatching nymphs in the rearing cage.

5. Allow the hatched nymphs to go through all five instars (Fig. 12.1) before moving them to the new test cages for transmission experiments.
6. Allow the first generation of leafhoppers to oviposit on fresh healthy plants for the new colonies.
7. Add more new healthy plants to cages as required for proper maintenance of insect populations.
8. Allow eggs to hatch, nymphs to go through all five instars and eclose (to emerge from the eggshell or pupal case). At this stage, allow the first colony created to die (Whitcomb 1972).

12.3.2 Generating Virus-Infected Leafhopper Colonies and Virus Transmission

9. Test the emerging nymphs for the absence of virus by PCR assays.
10. Transfer the virus-free adult (Fig. 12.3) and/or eggs to the plant under cages in glasshouse for the maintenance of insect population.
11. Collect required number (10–30) of newly emerged adults (Fig. 12.3) with an aspirator from the rearing cage and transfer them to a transmission cage with tungro-diseased plants.
12. Allow an acquisition access period (AAP) of 4 days.
13. Introduce these leafhoppers onto the healthy rice plants. Cover immediately with cylindrical cage and label them.
14. Allow an inoculation access period (IAP) of 1 day (for semi-persistent viruses) to 50 days (for persistent viruses).

15. Kill leafhoppers by insecticide spray (0.05% imidacloprid) and keep inoculated seedlings in the glasshouse for symptom development.
16. Infected plants show typical symptoms of the disease within a few weeks or months depending on the latent period of the virus. Latent period may vary with the age of the plants and temperature.

12.4 Notes (Maramorosch 1999)

1. It is always a good idea to have a dedicated separate glasshouse for rearing each species of leafhoppers. Provide ideal conditions (temperature and humidity) for proper development and multiplication of leafhoppers. Use virus-free plants of susceptible crop to raise healthy culture of leafhoppers. All operations such as planting, watering, etc. should be done first in this glasshouse to avoid contamination with any viruliferous leafhoppers.
2. Inoculate test plants at primary leaf stage using 10–12 leafhoppers per plant.
3. Do not use too much suction through aspirator during collection and release of leafhoppers.
4. The rice seedlings are raised well before collection of insects from the field.
5. Microcages used for handling leafhoppers should be dry.
6. A temperature of 28 °C is suitable for many leafhopper species. In experimental transmission studies, acquisition of the virus by leafhoppers from many hosts is low, so prolonged feeding may be necessary for successful acquisition.
7. In general, nymphs of third and fourth instar would be ideal for transmission. As latent period can vary from 0 to 50 days, vectors should be transferred to healthy test plants immediately after the acquisition and may be held on test plants as long as possible.

References

Carter W (1974) Insect in relation to plant diseases. Wiley Interscience, New York

Ling KC, Tiongco ER (1979) Transmission of rice tungro virus at various temperatures-a transitory virus-vector interaction. In: Maramorosch K, Harris KF (eds) Leaf hopper vectors and plant disease agents. Academic, New York

Liu Y, Du Z, Wang H, Zhang S, Cao M, Wang X (2018) Identification and characterization of wheat yellow striate virus, a novel leafhopper-transmitted nucleorhabdovirus infecting wheat. Front Microbiol 9:468

Maramorosch K (1999) Leafhopper and plant hopper rearing. In: Maramorosch K, Mahmood F (eds) Maintenance of human, animal, and plant pathogen vectors. Science Publishers Inc., Enfield, NH

Nault LR, Ammar ED (1989) Leaf hopper and plant hopper transmission of plant viruses. Annu Rev Entomol 34:503–529

Sastry KS (2013) Plant virus and viroid diseases in the tropics. Springer, Heidelberg

Whitcomb RF (1972) Transmission of viruses and mycoplasma by the auchenorrhynchous homoptera. In: Kado CI, Agrawal HO (eds) Principles and techniques in plant virology. Van Nostrand Reinhold Company, New York

Chapter 13

Transmission of Viruses by Whiteflies

Abstract

The whiteflies (*Bemisia tabaci*, Hemiptera: Aleyrodidae) are efficient vectors of many plant viruses (*Begomovirus*, *Carlavirus*, *Crinivirus*, *Ipomovirus*, *Torradovirus*). Whiteflies are known to transmit more than 80 virus diseases mainly in the tropical and subtropical countries. Many of the whitefly transmitted viruses cause yellow or golden mosaic or leaf curl symptoms. Whitefly transmitted viruses have non-persistent, semi-persistent and persistent relationships.

Key words Non-persistent viruses, Semi-persistent viruses, Circulative transmission, *Begomovirus*, *Carlavirus*, *Crinivirus*, *Ipomovirus*, *Torradovirus*

13.1 Introduction

Whiteflies are an important group of vectors transmitting begomoviruses that cause diseases such as yellow or golden mosaics and leaf curl especially in the tropical and subtropical plants (Costa 1976; Muniyappa 1980; Nayudu 2008). Whiteflies are small piercing and sucking insects belonging to the family Aleyrodidae in the order Homoptera. Since whiteflies are not always white and are not flies, the name is a misnomer. Among all, the species *Bemisia tabaci* is important (Fig. 13.1a, b). *B. tabaci* though originally known in tropical and subtropical regions has now spread also to temperate regions. This spread has led to the economic importance of begomoviruses that are known to infect important agricultural crops including legumes, cotton, melon, pepper, potato, squash, tobacco, tomato and watermelon. Two biotypes (A and B) have been reported in *B. tabaci*. They suck plant juices through their slender stylets. Eggs are laid on the lower surface of leaves. There are four nymphal instars.

Nymphs are firmly attached to the host plant except for a short period of the first instar. Total developmental time (egg to adult) is variable from 14 to 107 days. Faster development takes place at a higher temperature. Differences in developmental rates have also been observed on different hosts. The longevity may range from 6 to 34 days (in males) and 15 to 55 days (in females). Adults are

Alangar Ishwara Bhat and Govind Pratap Rao, *Characterization of Plant Viruses*, Springer Protocols Handbooks, https://doi.org/10.1007/978-1-0716-0334-5_13,

Fig. 13.1 Whitefly population on a leaf (**a**), adults of whitefly laying eggs on a leaf surface (**b**)

winged, and many generations may be produced in a year. They penetrate plant tissues intercellularly and feed in the phloem. Virus particles ingested with phloem sap in the gut are translocated to the haemocoel and then to salivary glands. Viruses that pass salivary gland cells to the salivary duct are inoculated to plants during feeding. Though acquisition and inoculation feeding by *B. tabaci* could occur within 6 min, longer feeding times are more efficient. The latent period is 8–12 h. Viruses do not multiply in the vector. Although whiteflies are reported to transmit closteroviruses and criniviruses semi-persistently, transmission of begomoviruses is paramount (Nayudu 2008).

In this chapter, the transmission of begomovirus through the whitefly (*Bemisia tabaci*) is discussed (Polston and Capobianco 2013; Czosnek et al. 2001).

13.2 Materials

1. Aspirator.
2. Cages.
3. Healthy plants (test plants).
4. Healthy whitefly (*Bemisia tabaci*) culture.
5. Insecticide (to kill whiteflies after transmission).
6. Test tube.
7. Virus-infected plant (source plant).

13.3 Method

13.3.1 Maintenance of Healthy Whitefly Culture

1. Collect adult whiteflies (*Bemisia tabaci* Genn) from natural hosts and release them on healthy tobacco (*Nicotiana tabacum* cv. Xanthi), cotton (*Gossypium hirsutum* cv. Varalakshmi) or bottle gourd (*Lagenaria siceraria*) raised in an insect-proof glasshouse.

Fig. 13.2 Transmission of begomovirus through whitefly. Healthy whitefly (**a**), microcage and aspirator (**b**), whiteflies are allowed to feed infected twigs for virus acquisition (**c**), inoculated seedlings are covered with microcage (**d**), inoculated grown up plants covered with polystyrene cage (**e**), expression of symptoms in inoculated plant (**f**). (Courtesy: Dr. K.K. Biswas, ICAR-IARI, New Delhi)

2. Maintain the culture of whiteflies in tobacco, cotton or bottle gourd by regular transfer after 6 weeks to a fresh batch of plants.

13.3.2 Virus Transmission

3. Collect required number of adult non-viruliferous whiteflies with the help of an aspirator in a suitable test tube.
4. Give them an acquisition access period (AAP) of 12–14 h on the leaves of infected source plant held inside a cylindrical cage (Fig. 13.2).
5. Collect viruliferous whiteflies through aspirator and release them on the test (healthy) plants for an inoculation access period of 24 h. For better transmission, inoculate test plants at primary leaf stage using 10–12 viruliferous whiteflies per plant. Cover the plant for preventing the escape of whiteflies.
6. Kill the whiteflies by spraying dimethoate @ 0.1% or any other suitable systemic insecticide (imidacloprid).
7. Keep the test plants under the insect-proof condition and watch for symptom expression.

8. Susceptible plants will show typical symptoms in about 10–15 days after inoculation (Fig. 13.2). Non-symptomatic plants need to be back-inoculated onto the indicator host to check for any latent infection in these plants.

13.4 Notes (Muniyappa 1980; Nayudu 2008)

1. It is always a good idea to have a dedicated separate glasshouse for rearing whiteflies. Provide ideal conditions (temperature, photoperiod and humidity) for fast multiplication of the whiteflies. Use virus-free plants of susceptible crop to raise healthy culture of whiteflies. A temperature of 28 °C, 30–50% relative humidity and a 14-h photoperiod will yield a colony that develops from egg to adult emergence in 18 days (this time varies with the plant host). All operations such as plating, watering, etc. should be done first in this glasshouse to avoid contamination with any viruliferous whiteflies.
2. Inoculate test plants at primary leaf stage using 10–12 whiteflies per plant.
3. Whiteflies are tiny delicate insects and hence do not use too much suction through aspirator during collection.
4. Test tubes and microcages used for handling whiteflies should be dry.
5. For easier handling, hold test tubes/microcages with whiteflies upside down (i.e. the closed portion facing light as whiteflies are phototropic).
6. Plants free of both whiteflies and virus must be produced to introduce into the whitefly colony each week. Whitefly cultures must be kept free of whitefly pathogens, parasites and parasitoids that can reduce whitefly populations and/or reduce the transmission efficiency of the virus.
7. There are two basic types of whitefly colonies that can be maintained: a non-viruliferous and a viruliferous whitefly colony. The non-viruliferous colony is composed of whiteflies reared on virus-free plants and allows the weekly availability of whiteflies which can be used to transmit viruses from different cultures. The viruliferous whitefly colony, composed of whiteflies reared on virus-infected plants, allows weekly availability of whiteflies which have acquired the virus, thus omitting one step in the virus transmission process.
8. Whiteflies can be introduced by aspiration of known numbers or by gently shaking whiteflies from another source plant depending on the demands on the colony.

9. During week 1, whiteflies are laying eggs on the underside of plant leaf surfaces.
10. An approximate population size can be predicted for each emergence based on the number of adults used to lay eggs and the average number of eggs laid by the female whitefly on the host plant in question.
11. To ensure high transmission rates: (1) Whiteflies must be handled as gently as possible to prevent damage to the insect which will reduce transmission rates; (2) There must be adequate leaf area available for whiteflies to either probe or feed. An increase in acquisition time or an increase in the number of acquisition host plants may increase rates of infectivity in test plants due to crowding in either the acquisition or inoculation access period.
12. Newly emerged adult whiteflies (1–3 days post emergence) are highly active and feed often, so they tend to give the highest transmission rates. Older whiteflies still transmit but at a lower frequency. Multiple whiteflies per test plant (15–40 per plant) should be used for high rates of transmission since a ratio of 1 whitefly per plant often results in unacceptably low transmission rates.
13. Never aspirate whiteflies that are feeding on plants. Their stylets are embedded in the plant while feeding so pulling them off the plants will break their stylets and rendering them unable to either acquire or transmit virus.
14. The inoculation access period is ended by killing the whiteflies with approved chemicals. Apply two insecticides one after the other: a contact insecticide to quickly terminate adult whiteflies and a systemic insecticide to terminate any whiteflies that develop in the following weeks and those missed by the contact insecticide.

References

Costa AS (1976) Whitefly- transmitted plant diseases. Annu Rev Phytopathol 14:429–449

Czosnek H, Morin S, Rubinstein G, Fridman V, Zeodan M, Ghanim M (2001) Tomato yellow leaf curl virus: a disease sexually transmitted by whiteflies. In: Harris KF, Smith OP, Duffus JE (eds) Virus-insect-plant interactions. Academic, San Diego

Muniyappa V (1980) Whiteflies. In: Harris KF, Maramorosch K (eds) Vectors of plant pathogens. Academic, New York

Nayudu MV (2008) Plant viruses. Tata McGraw-hill Publishing Company Limited, New Delhi

Polston JE, Capobianco H (2013) Transmitting plant viruses using whiteflies. J Vis Exp (81): 4332

Chapter 14

Transmission of Viruses by Thrips

Abstract

Thrips are small insects difficult to see by naked eye, cosmopolitan and highly polyphagous. Tospoviruses are transmitted by thrips in a persistent propagative manner. Only first stage larvae of the thrips can acquire the virus while adults can transmit the virus throughout its life. In majority of the cases especially ilarviruses (passively) and tospoviruses transmitted by thrips cause necrotic symptoms in affected plants.

Key words Persistent propagative transmission, Tospoviruses, Ilarviruses, Culturing of thrips

14.1 Introduction

Thrips are important sucking pests which transmit the members of the genus, *Tospovirus* (Fam: *Tospoviridae*) in persistent and propagative manner. About 16 species of thrips (Family: Thripidae; Order Thysanoptera) are known as vectors of 29 Tospoviruses worldwide (Rotenberg et al. 2015; Turina et al. 2016). Thrips are mostly yellow, orange, black or whitish-yellow in colour. Adult thrips are usually very small in size, 1–2 mm in length and slender in shape (Fig. 14.1). They are not good fliers and usually make short flights; however, long distance migration occurs through the wind, atmospheric convection and turbulence. The life cycle of thrips consists of egg, first and second instar larva, pre-pupa, pupa and adult (Ullman et al. 1997; Hull 2002). Thrips are extremely polyphagous and have the ability to reproduce on a broad range of host plants. It reproduces mainly parthenogenetically. The larvae are rather inactive while adults are winged and very active. In general, adults live up to about 4–5 weeks. Several generations can develop in a year. The successful transmission of the plant virus by thrips involves the acquisition of the virus followed by establishment of virus in thrips and inoculation of the virus in a susceptible host. Thrips are known to transmit viruses belonging to at least four virus genera, *Carmovirus, Ilarvirus, Sobemovirus* and *Tospovirus* (Ghosh et al. 2017). Among these, Tospoviruses have a persistent propagative relationship with their thrips vector while ilar, sobemo and carmoviruses are pollen-borne, and thrips are

Alangar Ishwara Bhat and Govind Pratap Rao, *Characterization of Plant Viruses*, Springer Protocols Handbooks,
https://doi.org/10.1007/978-1-0716-0334-5_14,

Fig. 14.1 Morphology of adult thrips, *Frankliniella* fusca (Courtesy: Dr. Rajagopalbabu Srinivasan, University of Georgia, USA)

generally essential in facilitating transmission by carrying the pollen on their body. Some of the important thrips vector species and viruses they transmit include: *Frankliniella occidentalis* (groundnut ringspot tospovirus, impatiens necrotic spot tospovirus, pea early browning virus, prunus necrotic ringspot virus), *F. schultzei* (tomato chlorotic tospovirus), *Thrips palmi* (groundnut bud necrosis orthotospovirus, watermelon silver mottle orthotospovirus) and *T. tabaci* (prunus necrotic ringspot virus, sowbane mosaic virus, tobacco streak ilarvirus, tomato spotted wilt tospovirus).

We describe here the procedure involved in the transmission of an orthotospovirus, namely, *Groundnut bud necrosis virus* (GBNV) by thrips. In the case of the thrips–orthotospovirus relationship, only the first instar larvae can acquire the virus. Adults could transmit the virus if it acquired in the larval stage. Tospoviruses are also known to replicate in the vector body. The virus is not known to pass through the egg of the thrips. The eggs hatch in 3 days, and the first instar larvae appear for only 1 day followed by the second instar that lasts for 2–3 days. The third and fourth instar (pre-pupae and pupae) do not feed and lasts for 3–5 days. This is followed by the emergence of winged adults which starts feeding on plants.

14.2 Materials

1. Aspirator.
2. Centrifuge tube.
3. Diseased plants (source plants).
4. Healthy plant (test plants).
5. Incubator with temperature and light control.

6. Painting brush.
7. Petri dishes.
8. Small glass vials.

14.3 Method

14.3.1 Rearing of Thrips (Thrips palmi)

1. Collect adults of *T. palmi* from healthy plants (preferably melons) with the help of aspirator.
2. Transfer the thrips with a painting brush into glass vials containing groundnut leaves.
3. Close the vials with rubber cork and place them in the incubator at about 25 °C (12 h of light followed by 12 h of dark) for 5–7 days.
4. Open the rubber cork and remove the adults from each vial on fourth or sixth day.
5. By 7 days, the first instar larvae start coming out which can be used in transmission studies (Fig. 14.2).

14.3.2 Virus Transmission Studies

6. Take a systemically infected leaf showing symptoms and float it on water in a Petri plate.
7. Transfer the required number of first instar larvae (0–12 h old) with the help of a brush (Fig. 14.2b).
8. Allow them an acquisition access period (AAP) of 24–48 h at room temperature.
9. After AAP, transfer each larvae individually to vials containing fresh, healthy groundnut leaves (5–10 larvae per vial). Leave them at 25 °C for 7–9 days till they become adults (latent period for the vector to become viruliferous will also be covered during this period).
10. Transfer the viruliferous adults from above tubes to healthy seedlings placed inside centrifuge tubes at the rate of 5–10 adults per seedlings. Allow the adults an inoculation access period (IAP) of 24 h (Fig. 14.2d).
11. Keep the seedlings for symptom development in an insect-proof chamber at 26–28 °C with proper lighting (14 h light and 8 h dark). Symptoms (local lesions either chlorotic/necrotic) may appear in 3–5 days in some assay plants while systemic symptoms may appear within 2–3 weeks after inoculation (Fig. 14.2e).

Fig. 14.2 Thrips rearing (**a**), symptomatic infected groundnut source plant for virus culture (**b**), raising groundnut seedlings in pots (**c**), stage of groundnut for thrips transfer (**d**), symptoms appearance on groundnut test plant after thrips feeding (**e**). (Courtesy: Dr. YB Basavraj, ICAR-IARI, New Delhi)

14.4 Notes (Ullman et al. 1997; Hull 2002)

1. It would be always better if the adults are taken from aseptically maintained homogenous culture of known species.

2. Virus culture under test should be pure and identified before transmission test.
3. Only the first instar larvae should be used for acquisition feeding.

References

Ghosh A, Dey D, Timmanna, Basavaraj, Mandal B, Jain RK (2017) Thrips as the vectors of tospoviruses in Indian agriculture. In: Mandal B, Rao GP, Baranwal VK, Jain RK et al (eds) A century of plant virology in India. Springer Nature Singapore Pvt. Ltd, Singapore

Hull R (2002) Matthews' plant virology, 4th edn. Academic, San Diego

Rotenberg D, Jacobson AL, Schneweis DJ, Whitfield AE (2015) Thrips transmission of tospoviruses. Curr Opin Virol 15:8–88

Turina M, Kormelink R, Resende RO (2016) Resistance to tospoviruses in vegetables crops: epidemiological and molecular aspects. Annu Rev Phytopathol 54:347–371

Ullman DE, Sheerwood JL, German TL (1997) Thrips as vectors of plant pathogens. In: Lewis T (ed) Thrips as crop pests. CAB International, Wallingford

Chapter 15

Transmission of Viruses Through Mealybugs

Abstract

Mealybugs are white, soft-bodied, cottony-looking insects equipped with piercing/sucking mouth parts under the order hemiptera. They are like plant scale insects and aphids in sucking the fluids from leaves and stems from the plants for the requirement of essential nutrients. Mealybugs are sluggish insects that transmit viruses in semi-persistent manner. They are known to transmit ampelo, badna and clostero genera of plant viruses.

Key words Mealybugs, *Ferrisia virgata*, *Planococcus* sp., Ampelovirus, Badnavirus, Closterovirus, Semi-persistent transmission

15.1 Introduction

Mealybugs (Fam: Pseudococcidae), the false scale insects can be found everywhere especially in the subtropics and tropics, and like warmer dry environments. These pests have a life cycle of about 30 days. They are much less mobile as compared to other vectors. Mealybugs tend to live together in clusters in protected parts of plants, such as on the leaf axils, leaf sheaths, between twining stems and under loose bark. Their wax covering and preference to stay tucked away out of sight make them difficult to eradicate. Several species of mealybug occur in greenhouses or on house plants. These include *Pseudococcus calceolariae* (glasshouse mealybug), *P. longispinus* (long-tailed mealybug) and *Planococcus citri* (citrus mealybug). Mealybugs derive their name from the fact that from the third larval stage onwards, the females are covered with a white wax-like substance. The adult females have flattened oval-shaped soft bodies between 0.10 and 0.16 in. long and 0.08 and 0.12 in. wide. They are sometimes whitish pink in colour but usually appear whitish due to the white, waxy powder that covers their bodies. Waxy filaments project from the edges of their bodies. Male mealybugs are small, winged fly-like insects that are rarely seen (Roivainen 1980).

They spread from one plant to another plant which is in contact, and the crawling nymphs move more readily than the adults.

Alangar Ishwara Bhat and Govind Pratap Rao, *Characterization of Plant Viruses*, Springer Protocols Handbooks, https://doi.org/10.1007/978-1-0716-0334-5_15,

Fig. 15.1 Morphology of *Ferrisia virgata* (**a**) and *Planococcus citri* (**b**)

Occasionally long distance dispersal by the wind may occur. Mealybugs generally feed on the phloem of plant tissues. About six species of mealybugs are known to act as vectors of different viruses of banana, black pepper, citrus, cocoa, grapevine, pineapple, sugarcane and cherry. The virus vector relationship is of semi-persistent type. They are known to transmit many of the ampeloviruses, badnaviruses, closteroviruses and trichoviruses. The principal genera of mealybug involved in the transmission of badnaviruses include *Pseudococcus*, *Planococcus* and *Ferrisia* (Fig. 15.1a, b). The important badnaviruses transmitted by mealybugs include *banana streak virus, citrus yellow mosaic virus, cocoa swollen shoot virus, dioscorea bacilliform virus, piper yellow mottle virus, and sugarcane mosaic virus.* Mealybugs acquire badnaviruses within 20 min and virus persists in the vector for 3 h (Herrbach et al. 2013). The *Ampelovirus, pineapple mealybug wilt associated virus* (PMWaV) is transmitted by the pink pineapple mealybug, *Dysmicoccus brevipes*, while the grapevine leaf roll-associated virus 3 is transmitted semi-persistently by its mealybug vector, *Planococcus citri* (Tsai et al. 2010).

In this chapter, we have described the transmission of badnaviruses and closteroviruses through mealybug species, *Ferrisia virgata* and *Planococcus ficus*.

15.2 Materials

1. Black cloth.
2. Insect cages.
3. Diseased plant (source plants).
4. Healthy mealybug culture.
5. Healthy plants (test plants).
6. Incubator.

7. Insecticide.
8. Painting brush.
9. Petri dishes.

15.3 Method

*15.3.1 Rearing of Mealybug (*Ferrisia virgata, Planococcus *sp.) and Transmission of* Piper yellow mottle virus *(Genus:* Badnavirus*) (Bhat et al. 2003)*

1. Collect the adult mealybugs from healthy plants with the help of a brush. Transfer them onto a matured pumpkin fruit @ 5–10 per fruit.
2. Allow them to multiply on the pumpkin. A temperature of around 26 °C is optimum for the multiplication. Within 4 weeks many young ones are born parthenogenetically and feed on the pumpkin (Fig. 15.2a).
3. Take a systemically infected leaf showing symptoms and place it (upside down) on a Petri plate lined with moist blotting paper.
4. Transfer required number of young crawling nymphs with the help of a brush onto the leaf, place the lid and cover the Petri plate with a black cloth (Fig. 15.2b).
5. Allow them an acquisition access period (AAP) of 24 h at room temperature.
6. After AAP, transfer each mealybug from above to healthy seedlings placed inside an insect-proof cage at the rate of 10–15 per seedlings. Allow the adults an inoculation access period (IAP) of 24 h (Fig. 15.2c).
7. Kill the mealybugs by spraying with insecticide (chlorpyriphos @ 0.1%).
8. Keep the seedlings for symptom development under insect-proof conditions. Symptoms (mild chlorosis and chlorotic flecks) may appear in about 30 days.

15.3.2 Transmission of Grapevine Leaf Roll-Associated Virus *(Genus:* Ampelovirus*; Fam: Closteroviridae) by* Planococcus ficus *(Tsai et al. 2010)*

1. Allow the mealybug (*Planococcus ficus*) to move onto virus-infected source vine cuttings (20 cm) laid on mealybug colonies.
2. Remove the cutting from the mealybug colonies after 2 h and maintain in flasks of water.
3. After an AAP of 24 h, insects are gently shaken off the source tissue onto paper discs (0.5 cm in diameter).
4. Transfer the viruliferous mealybugs on paper discs to healthy test plants by caging the insects on leaf blades using clip cages.
5. Inoculate the healthy leaves of grapevines with groups of five mealybugs per vine.
6. Remove the mealybugs from the test plants after a 24-h IAP, with a fine brush.

Fig. 15.2 Culturing of mealybug on pumpkin (**a**), acquisition of the virus (**b**) and inoculation of the test plant by viruliferous mealybug (**c**)

7. Maintain the test plants in a greenhouse and spray with insecticide until tested for the virus.
8. Confirm the presence of virus in inoculated plants by any of the assays such as ELISA, RT-PCR or RT-LAMP.

15.4 Notes (Roivainen 1980; Bhat et al. 2003; Tsai et al. 2010)

1. Regular subculturing of mealybug should be done once in a month.
2. The young crawling nymphs should be used in the transmission studies.
3. Mealybugs are difficult to control with insecticides because of protected wax-like exterior coverings. They can be controlled quite easily at crawler stage when they are most vulnerable.
4. Insecticidal soaps will help to reduce numbers of mealybugs as the soap reacts with the waxy layer, dissolving it and drying out the insect.

References

Bhat AI, Devasahayam S, Sarma YR, Pant RP (2003) Association of a badnavirus in black pepper (*Piper nigrum* L.) transmitted by mealy bug (*Ferrsia virgata*) in India. Curr Sci 84:1547–1550

Herrbach E, Le Maguet J, Hommay G (2013) In: Brown JK (ed) Virus transmission by mealybugs and soft scales (Hemiptera: Coccoidea). APS Press, Minnesota

Roivainen O (1980) Mealybugs. In: Harris KF, Maramorsch K (eds) Vectors of plant pathogens. Academic, New York

Tsai CW, Rowhani A, Golino DA, Daane KM, Almeida RPP (2010) Mealybug transmission of grapevine leafroll viruses: an analysis of virus–vector specificity. Phytopathology 100:830–834

Chapter 16

Transmission of Viruses Through Beetles

Abstract

Although majority of plant viruses are transmitted by insects with sucking mouth parts, some viruses are transmitted by insects with biting mouth parts. Beetles (Fam: Chrysomelidae) have the biting type of mouth parts and lack salivary glands. Plant viruses in the genera, *Tymovirus, Comovirus, Bromovirus* and *Sobemovirus* are transmitted by beetles. More than 74 species of beetle are known to transmit viruses which infect economically important vegetable and grain crops. Beetles transmit approximately 11% of the insect-borne viruses.

Key words *Tymovirus*, *Comovirus*, *Bromovirus*, *Sobemovirus*, *Epilachna varivestis*, Beetle

16.1 Introduction

The beetles (order: Coleoptera) form the largest group of insects worldwide. Important species of beetle vectors belongs to the family Chrysomelidae (leaf beetles). Adults and larval stages of some chrysomelid beetles are known to transmit plant viruses. Unlike other vectors, beetles have the biting type of mouth parts and lack the salivary glands. Beetles lay eggs on the surface of leaves or in soil near roots of their host plants. Emerging larvae are not very active and mobile and often used to feed on the same plants where their parent fed before egg laying. They eat parenchyma tissues between vascular bundles and leave holes in the leaf. They regurgitate during feeding that bathes the mouth parts with plant sap along with virus in infected tissues of the plants (Smith et al. 2017).

Although initially thought that transmission by beetles involved a simple mechanical process of wounding in the presence of the virus, later studies clearly indicated the precise relationship between the beetle and the virus they transmit. Viruses transmitted and not transmitted by beetles were found in the regurgitant and hemolymph of beetles fed on virus-infected plants indicated they have the circulative type of relationship with the viruses. Regurgitant is a complex mixture of enzymes containing ribonuclease and proteases that do not inactivate viruses transmitted by beetles.

Alangar Ishwara Bhat and Govind Pratap Rao, *Characterization of Plant Viruses*, Springer Protocols Handbooks, https://doi.org/10.1007/978-1-0716-0334-5_16,

Viruses transmitted by beetles belong to the genera: *Tymovirus, Comovirus, Bromovirus* and *Sobemovirus* (Walters 1969). All these viruses have stable small isometric particles with ssRNA genome and are readily sap transmissible. Beetles can acquire the virus in much shorter periods of time (may become viruliferous after a single bite) and percentage of beetles that becomes viruliferous increases with increase in feeding time. The retention time of virus varies from 1 to 10 days without any latent period after the acquisition. There is no evidence for virus replication in beetle vectors (Fulton et al. 1987).

Early studies reported that beetles transmit the virus to plants by a simple mechanical process involving the contamination of mouth parts. However, both the foregut-borne and circulative types of virus transmission are reported in some beetles. Beetle transmission of plant viruses is multifaceted and can include larvae–plant host, beetle–beetle (adults) and beetle–plant host interactions. Larvae of *Oulema melanopus* and *Phaedon cochleariae* transmit plant viruses more efficiently than adults. Both the adults and larvae of Mexican bean beetles, *Epilachna varivestis* (Nault et al. 1978) feed on leaves and transmit viruses (Jansen and Staples 1970; Fulton and Scott 1974) and the beetles may indirectly become infected by ingesting virus-contaminated faeces (Fulton and Scott 1980; Fulton et al. 1987). *E. varivestis* is reported to successfully transmit cowpea severe mosaic virus, southern bean mosaic virus and blackgram mottle virus (Fulton and Scott 1974).

Little information is available on the mechanism of virus transmission by beetles and it remains poorly understood. There is no research evidence available for virus propagation in beetles. Some of the circulative viruses move into the insect hemolymph immediately after ingestion (Sanderlin 1973; Scott and Fulton 1978). During feeding, the inoculative beetles deposit the active virus in regurgitant on the surface of the wounded leaf which are readily translocated in the xylem and can infect fresh plant tissues (Gergerich and Scott 1988). However, little is known about the infection process after translocation occurs.

In this chapter, transmission of bean pod mottle virus infecting soybean by Mexican bean beetle (*Epilachna varivestis*) is described (Smith et al. 2017).

16.2 Materials

1. Soybean plants infected with bean pod mottle virus (virus source).
2. Heathy soybean seedlings of the same variety.
3. Potassium phosphate buffer (10 mM, pH 7.4): Dissolve 1.74 g di-potassium hydrogen orthophosphate (K_2HPO_4) in 1 L

distilled water (10 mM solution). To this, slowly add 10 mM solution of potassium dihydrogen orthophosphate (KH_2PO_4) prepared by dissolving 0.34 g of KH_2PO_4 in 250 mL distilled water till the desired pH of 7.4 is obtained.

4. Pestle and mortar.
5. Carborundum (Silicon carbide) (SiC): Use 600 mesh powder.
6. Insect cages: cylindrical cages of different dimensions are used to encage beetle on soybean seedlings.
7. Insect sweep net.
8. Insect transfer chamber.
9. The adult populations of test beetle (*Epilachna varivestis*).
10. Glasshouse.

16.3 Method

16.3.1 Rearing Beetle Colony and Virus Culture

1. Establish a laboratory colony of *Epilachna varivestis* (Fig. 16.1) from field collections.
2. Rear the beetles in growth chambers under controlled conditions of 25 ± 3 °C, 65% RH with a 14 h light and 10 h dark cycle.
3. Maintain the beetle colonies on soybean seedlings placed in 47.5 cm × 47.5 cm × 47.5 cm cages.
4. Maintain the BPMV isolate in soybean through mechanical inoculation with inoculum made from leaves of infected plants.
5. To generate BPM-infected experimental plants, inoculum is made by grinding infected leaf tissue in 10 mM potassium phosphate buffer pH 7 (1:4 w/v) with carborundum (600 mesh) to induce wounding.
6. The inoculum is mechanically inoculated onto 10-day-old soybean plants.

Fig. 16.1 An adult *Epilachna varivestis* beetle

7. After visual appearance of symptom, the presence of the virus infection is confirmed using enzyme-linked immunosorbent or PCR assays (for details, please *see* Chapters 30 and 35).

16.3.2 Beetle-Virus Transmission Assays

1. Newly emerging adult beetles are fed on BPMV-infected soybean for a 5 days acquisition access period (AAP).
2. Transfer the individual beetle to uninfected seedlings for a 2 days inoculation access period (IAP).
3. Remove and kill the beetles from the seedlings.
4. Transfer the experimental plants to a growth chamber for 10 days.
5. Observe the virus symptom development.
6. Presence of virus is checked with RT-PCR assays (as discussed in Chapter 36).

16.4 Notes (Smith et al. 2017)

1. All operations such as planting, watering, etc. should be done first in this glasshouse to avoid contamination with any viruliferous beetles.
2. Inoculate test plants at primary leaf stage using one or two beetles per plant.
3. Raise soybean seedlings well before collection of insects from the field.
4. Microcages used for handling beetles should be dry.
5. Only adult beetle should be used for the transmission studies.
6. Proper identification of beetle species is important before starting the experiment.

References

Fulton JP, Scott HA (1974) Virus vectoring efficiencies of two species of leaf feeding beetles. Proc Am Phytopathol Soc 1:159

Fulton JP, Scott HA (1980) Beetles. In: Harris KF, Maramorosch K (eds) Vectors of plant pathogens. Academic, San Diego

Fulton JP, Gergerich RC, Scott HA (1987) Beetle transmission of plant viruses. Annu Rev Phytopathol 25:111–123

Gergerich RC, Scott HA (1988) Evidence that virus translocation and virus infection of non-wounded cells are associated with transmissibility by leaf-feeding beetles. J Gen Virol 69:2935–2938

Jansen WP, Staples R (1970) Transmission of cowpea mosaic virus by the Mexican bean beetle. J Econ Entomol 65:1719–1720

Nault LR, Styer WE, Coffey ME, Gordon DT, Negi LS, Niblett CL (1978) Transmission of maize chlorotic mottle virus by chrysomelid beetles. Phytopathology 68:1071–1074

Sanderlin RS (1973) Survival of bean pod mottle and cowpea mosaic virus in beetles following intrahemocoelic injections. Phytopathology 63:259–226

Scott HA, Fulton JP (1978) Comparison of the relationships of Southern bean mosaic virus and the cowpea strain of tobacco mosaic virus with the bean leaf beetle. Virology 84:207–209

Smith CM, Gedling CR, Wiebe KF, Cassone BJ (2017) A sweet story: bean pod mottle virus transmission dynamics by Mexican bean beetles (*Epilachna varivestis*). Genome Biol Evol 9:714–725

Walters HJ (1969) Beetle transmission of plant viruses. Adv Virus Res 15:339–363

Chapter 17

Transmission of Viruses Through Mites

Abstract

Members of mite families, Eriophyidae and Tetranychidae, are known to transmit plant viruses in a semi-persistent and persistent manner. They are very small in size (<250 μm) and hence easily dispersed by wind. Eriophyidae mites are important vectors of pigeonpea sterility mosaic virus, wheat streak mosaic virus, fig mosaic virus, peach mosaic virus and ryegrass mosaic virus and cause serious losses to the economy of these crops.

Key words Eriophyid mite, *Aceria tosichella*, *Aceria cajani*, Pigeonpea sterility mosaic virus

17.1 Introduction

Mites belonging to the families, Eriophyidae and Tetranychidae, are efficient vectors of plant viruses (Dhooria 2016). They feed (preferably on fresh tissue or buds) and survive only in living tissues. They feed by piercing the plant cells and sucking the contents. Members of the genus *Rymovirus* are transmitted by Eriophyid mites while a few unsubstantiated claims for vectors in the Tetranychid mites are also known. Eriophyid mites are worm-like, colourless and tiny in size (<250 μm in length). They have two pairs of legs, very few setae with modified piercing and sucking mouth parts. They have no eyes, spiracles or tracheae. They have four developmental stages—egg, first nymph, second nymph and adult. They have a simple life cycle in which males and females occur throughout the year. Some species can complete a generation in 6 days during summer months. They are highly host specific. Their movement is limited due to their inability to fly. They depend on wind, insects and birds for long distance movements.

Aceria tosichella Keifer is one of the best studied mite vectors which transmits the wheat streak mosaic virus (WSMV) in wheat (Slykhuis 1955) (Fig. 17.1) and corn (Sill and del Rosario 1959). The wheat curl mites are cigar shaped and white in colour with very small size (<0.3 mm or ~1/100 in.) invisible to unaided eyes (Fig. 17.1a) (Jeppson et al. 1975). A heavy infestation of the wheat curl mite induces curling of the leaf blades towards the

Alangar Ishwara Bhat and Govind Pratap Rao, *Characterization of Plant Viruses*, Springer Protocols Handbooks, https://doi.org/10.1007/978-1-0716-0334-5_17,

Fig. 17.1 (**a**) *Aceria tosichella* feeding on wheat leaves; (**b**) symptoms of wheat streak mosaic virus

midribs and distortion of the leaf lamina. Their dispersal in nature is primarily wind assisted (Jeppson et al. 1975). The life cycle of *A. tosichella* is of 1 week consisting of an egg, two instars of nymph and an adult. The mites overwinter on the crown of winter wheat, maize or other perennial grass hosts as eggs, nymphs or adults. An adult mite can acquire the wheat streak mosaic virus from an infected host in 15 min of acquisition feeding; however, transmission efficiency is below <2%, when acquisition time is short. Longer acquisition feeding periods result in higher transmission efficiency of the mite (Orlob 1966; Hadi et al. 2011).

The second important mite transmitted plant virus, pigeonpea sterility mosaic virus (PPSMV) causes mosaic and sterility symptoms in pigeonpea (Kulkarni et al. 2002; Kubota 2017) (Fig. 17.2a) and is transmitted by the eriophyid mite vector, *Aceria cajani*, in semi-persistent manner (Seth 1962; Reddy et al. 1998). *A. cajani* measures about 200–250 μm in length bearing two pairs of legs. It has short life cycle of less than 2 weeks comprising an egg, two nymphal and adult stages (Oldfield and Proeseler 1996). The transmission efficiency of single *A. cajani* is up to 53% but can be 100% when >5 mites per plant are used (Seth 1962). *A. cajani* can acquire PPSMV after a minimum acquisition access period (AAP) of 10–15 min and inoculate the virus after a minimum inoculation access period (IAP) of 60–90 min at 20–30 °C (Reddy et al. 1989). No latent period is observed. PPSMV has an incubation period of 8 days for expression of symptoms on pigeonpea (Latha and Doriaswamy 2004). The mite retains infectivity through moults and remains up to 9 days. *A. cajani* possesses a short stylet of ~2 μm that permits penetration only into epidermal tissue and, at most, underlying mesophyll cells (Reddy et al. 1995).

In this chapter, transmission of pigeonpea sterility mosaic by *A. cajani* is described.

Fig. 17.2 (**a**) Symptoms of pigeonpea sterility mosaic disease; (**b**) the mite, *Aceria cajani*

17.2 Materials

1. Acaricide.
2. Diseased plant (source plant).
3. Healthy mite source (*Aceria cajani*).
4. Healthy pigeonpea plant (test plants).
5. Mite-free cages.
6. Petri dishes.
7. Stereo binocular microscope.
8. Sterile distilled water.
9. Wooden stick fixed with a human eyelash.

17.3 Method

17.3.1 Rearing of Healthy Aceria cajani

1. Handling of mites needs to be done under a binocular stereomicroscope using an eyelash fixed to a wooden stick.
2. Take leaflets from healthy pigeonpea plants that support mite multiplication but immune to sterility mosaic virus (for example cv. ICP8136) and float them on sterile distilled water in a Petri dish.
3. Transfer mites from infected pigeonpea leaf onto the floating leaflet.
4. After 2 days, move them to another set of new floating leaflet from healthy pigeonpea plant.
5. Next day, transfer a group of 50 mites onto healthy pigeonpea plant to ensure that the mite colony is free from virus.

17.3.2 Transmission

1. Establish the non-viruliferous *A. cajani* colonies on detached healthy pigeonpea leaflets of a PPSMV-immune pigeonpea cultivar floating on water.
2. Collect required number of mites and allow them for AAP of 24 h on the virus-infected pigeonpea leaflet floating in a Petri dish.
3. Transfer group of 10 mites onto healthy pigeonpea plants and allow an IAP of 24 h.
4. Spray the inoculated plants with dicofol (kelthane) @ 0.2% to kill the mites.
5. Label the plants and keep them under mite-free cages for symptom development.
6. Confirm the virus presence with ELISA or PCR assays.

17.4 Notes (Slykhuis 1955; Reddy et al. 1989; Kulkarni et al. 2002)

1. Cages of desired design and dimensions for potted plants can be made with plastic or glass to allow light and muslin cloth for ventilation.
2. Stock mites should be placed as near as possible to growing points of the host plants.
3. Caging is necessary to prevent the spread of mites to or from other plants and to avoid infestation by predators of mites of the host plants.
4. A human eyelash affixed to a 6 cm long, 2 mm wide wooden stick may be used for transferring mites from source plant/ leaflet onto test plants.
5. Eggs may be transferred to healthy seedlings of host and covered with a cage to initiate culturing of mites. Eggs would hatch, and mites develop on seedlings that can be directly used in transmission studies.
6. The acquisition of the virus may be done by placing mites (desired stage) on leaves from infected mite-free plants. The leaves are placed on moist filter paper in a Petri dish, so that mites can be observed under a stereomicroscope. After acquisition access period, mites may be transferred singly or in groups to test plants.

References

Dhooria MS (2016) Fundamentals of applied acarology. Springer, Heidelberg

Hadi BAR, Langham MAC, Osborne L, Tilmon KJ (2011) *Wheat streak mosaic virus* on wheat: biology and management. J Integr Pest Manag 2:J1–J5

Jeppson LR, Keifer HH, Baker EW (1975) Mites injurious to economic plants. University of California Press, Berkeley

Kubota K (2017) Emergence of emaraviruses, the eriophyoid mite-transmitted viruses in plants. Uirusu 67:37–48

Kulkarni NK, Kumar PL, Muniyappa V, Jones AT, Reddy DVR (2002) Transmission of *Pigeon pea sterility mosaic virus* by eriophyid mite, *Aceria cajani* (Acari: Arthropoda). Plant Dis 86:1297–1302

Latha TKS, Doriaswamy S (2004) Interaction studies between pigeonpea sterility mosaic virus and mite vector. Indian J Plant Protect 32:105–110

Oldfield GN, Proeseler G (1996) Eriophyid mites as vectors of plant pathogens. In: Lindquist E, Sabelis MW, Bruin J (eds) Eriophyid mites, their biology, natural enemies and control. Elsevier, Amsterdam, pp 259–275

Orlob GB (1966) Feeding and transmission characteristics of *Aceria tulipae* Keifer as vector of *Wheat streak mosaic virus*. Phytopathol Z 55:218–238

Reddy MV, Beniwal SPS, Sheila VK, Sithanathan S, Nene YL (1989) Role of eriophyid mite (*Aceria cajani*) in transmission and spread of sterility mosaic of pigeonpea. In: Channabasavanna GP, Virakthmath CA (eds) Progress in acerology. Oxford and IBH Publishing Co., New Delhi, pp 121–127

Reddy MV, Sheila VK, Murthy AK, Padma P (1995) Mechanism of resistance to *Aceria cajani* in pigeonpea. Int J Trop Plant Dis 13:51–57

Reddy MV, Raju TN, Lenne JM (1998) Diseases of pigeonpea. In: Allen TDJ, Lenne JM (eds) Pathology of food and pasture legumes. CAB International, Wallingford, pp 540–558

Seth ML (1962) Transmission of pigeonpea sterility by an eriophyid mite. Indian Phytopathol 15:225–227

Sill WH, del Rosario MS (1959) Transmission of *Wheat streak mosaic virus* to corn by the eriophyid mite, *Aceria tulipae*. Phytopathology 49:396

Slykhuis JT (1955) *Aceria tulipae* Keifer (Acarina: Eryophydae) in relation to the spread of wheat streak mosaic. Phytopathology 45:116–128

Chapter 18

Transmission of Viruses Through Fungi

Abstract

A number of plant viruses in different genera are transmitted through the soil. Thirty soil-borne viruses are known to be transmitted by five species of fungal vectors. Soil-borne viruses are difficult to study, and their biology is relatively poorly understood at present. Fungi belonging to *Olpidium* spp., *Polymyxa graminis*, *Spongospora subterranea* and *Synchytrium endobioticum* have shown to be involved in the transmission of viruses belonging to the genera *Bymovirus, Furovirus, Tombusvirus* and *Varicosavirus* through either non-persistent or persistent means. When these fungi parasitize roots of virus-infected plants, the zoospores of these fungi acquire the virus either on the surface or internally. These zoospores upon infection of roots of healthy plants introduce the virus and cause infection.

Key words *Olpidium* spp., *Polymyxa graminis*, *Spongospora subterranean*, *Synchytrium endobioticum*, Virus transmission

18.1 Introduction

Five species of obligate soil inhabiting fungi belonging to two classes have been shown to transmit plant viruses. They are *Olpidium brassicae* and *O. radicale* belonging to Chytridiales and *Polymyxa graminis*, *Spongospora subterranean* and *Synchytrium endobioticum* belonging to the order Plasmodiophorales. These vectors are distributed worldwide and have wide host range. The two fungal classes are distinguished by the morphology of their zoospores (chytrid zoospores have a single long flagellum while those of plasmodiophorids contain two flagella of unequal lengths). Fungi infect epidermal cells or root hairs near the root tips and produce zoosporangia containing many motile zoospores. Zoosporangia produce exit tubes through which zoospores are released into soil water around the infected root. These zoospores swim to the surface of another root, withdraw their cilia and may produce thin walled zoospore cysts that later produce an infection tube through root cell wall by which the zoospore protoplasm enters the cell. After a few days, the zoospore contents develop into a thin walled zoosporangium producing many zoospores. Sometimes instead of zoosporangium, zoospores may produce a thick-walled resting zoosporangium that are resistant to drying and may remain

Alangar Ishwara Bhat and Govind Pratap Rao, *Characterization of Plant Viruses*, Springer Protocols Handbooks, https://doi.org/10.1007/978-1-0716-0334-5_18, © Springer Science+Business Media, LLC, part of Springer Nature 2020

viable for many years. Under suitable conditions, these spores germinate to release zoospores (Teakle 1972; Hull 2014).

Virus transmission occurs by zoospores in two ways: (1) Non-persistent transmission: viruses transmitted by *O. radicale* (cucumber leaf spot virus, cucumber necrosis virus) and tobacco necrosis virus and its satellites are acquired from soil water on the outer membrane of zoospores. The virus particles appear to enter the zoospore through infection canal, but they do not enter the resting spore during its formation. (2) Persistent transmission: Viruses transmitted by other fungi (*Olpidium brassicae*, *Polymyxa graminis*, *Spongospora subterranean* and *Synchytrium endobioticum*) are acquired from plant cells and enter the resting spores of vectors and are released with zoospores on germination (Teakle 1962, 1972).

Thirty soil-borne viruses or virus-like agents are transmitted by five species of fungal vectors. Ten polyhedral viruses, of which nine are in the family *Tombusviridae*, are acquired through in vitro manner and do not occur within the resting spores of their vectors, *Olpidium brassicae* and *O. bornovanus*. Eighteen rod-shaped viruses belonging to the furo- and bymovirus groups survive within the resting spores of their vector, *O. brassicae*, *P. graminis*, *P. betae* and *Spongospora subterranean* (Campbell 1996).

Some of the examples include lettuce big vein virus (*O. brassicae*), beet necrotic yellow vein virus (*P. betae*), barley mild mosaic virus (*P. graminis*), Indian peanut clump virus (*P. graminis*), wheat soil-borne mosaic virus (*P. graminis*), rice necrosis mosaic virus (*P. graminis*) and potato mop top virus (*S. subterranea*). The isometric viruses (cucumber necrosis virus and cucumber leaf spot virus) in the family *Tombusviridae* are transmitted by *Olpidium radicale* in non-persistent manner (Hull 2014).

Most of the viruses are soil-borne and enter plants though roots. Hence, indexing roots is an often a more reliable test of virus infection than indexing the leaves. Most of these viruses depend on their vector for the spread. They are not readily transmitted by seeds except in the case of Indian peanut clump virus and cucumber leaf spot virus.

18.1.1 Isolation and Culturing of Fungal Vectors

The important steps in the isolation and culturing of fugal vectors include:

- Collect the soil samples near around the roots of virus-infected plants.
- Bait plants are used to recover both the virus and the fungal vector.

- Mix the soil samples with sterile sand and place it in the sterile pot to minimize abiotic transmission.
- Maintain adequate growth of the host plant to ensure adequate multiplication of the vectors.
- After 3–4 weeks, the roots of bait plants are washed free from soil, and they can be used for establishing further cultures and for the identification of the fungi present. A 100× magnification microscope is needed for the identification of fungal zoospores. Subcultures may be derived from small root portions or zoospore suspensions obtained from root washings. This may be further subjected to purification to obtain a pure culture of the fungal isolate by using any of the known methods. Once unifungal cultures have been established, they can be maintained by transferring zoospore suspensions to fresh bait seedlings.
- Preferred growing media are sterile sand, Molochite or small glass beads as they allow active root growth and provide ample scope for zoospore movement.
- This procedure provides enough zoospores for experimental work without disturbing the plant roots; plants are watered, and leachate from the pots containing zoospores is collected.
- A virus-free culture of the fungus is necessary to demonstrate transmission. Such a culture may be obtained from soil samples taken from outside diseased area. Alternatively, virus-infected cultures may be freed from infection by serially culturing the fungus on a non-host of the virus. All fungal vectors can be preserved in the form of resting spores. Alternatively, root pieces infected with the fungus may be air dried and stored in airtight vials.

18.1.2 Transmission Studies

Zoospores of the fungi are used for transmission experiments.

The important steps in the transmission of virus through fungi include

- Pour the suspension of zoospores around the root of virus-infected plants. Incubate the plants for 3–4 weeks to allow time for fungal colonization of the roots and acquisition of the virus.
- Collect the zoospore suspension from the roots of the plants and use to inoculate healthy virus-free seedlings of bait plants growing in sterile medium in pots.
- In the case of non-persistent viruses, virus particles are adsorbed to the exterior of zoospores.
- Keep the inoculated bait plants for 3–8 weeks at room temperature and observe for symptom development and virus multiplication in roots.

- The methods of detecting the virus in bait plants are based on either symptoms or other methods such as electron microscope, ISEM or by mechanical transmission by sap from bait plant parts.

18.2 Materials

1. Healthy lettuce plants grown in sterile soil.
2. Healthy test plants of bean or cowpea.
3. *Olpidium* culture on lettuce plants.
4. Pestle and mortar.
5. Phosphate buffer 0.01 M, pH 7.0: Dissolve 1.74 g dipotassium hydrogen orthophosphate (K_2HPO_4) in 1 L distilled water (0.01 M solution), and to this, add slowly 0.01 M solution of potassium dihydrogen orthophosphate (KH_2PO_4) prepared by dissolving 0.34 g of KH_2PO_4 in 250 mL distilled water till the desired pH of 7.0 is obtained.
6. Sterile soil.
7. Tobacco necrosis virus (TNV)-infected plants.

18.3 Method (Transmission of Tobacco Necrosis Virus by *Olpidium* sp.)

1. Prepare virus suspension using pestle and mortar by grinding TNV-infected bean or cowpea leaves in 0.01 M phosphate buffer pH 7.0. Alternatively purified virus particle may also be used.
2. Prepare zoospore suspension by placing *Olpidium*-infected lettuce roots in phosphate buffer for 10–20 min.
3. Gently mix virus and zoospore suspension and leave for 15 min.
4. Pour the suspension over healthy lettuce seedlings growing in sterile soil and observe for symptom development.
5. Confirm the transmission by grinding roots of lettuce in phosphate buffer and inoculating the sap extract into bean or cowpea leaves.

18.4 Notes (Teakle 1972; Hull 2014)

1. Transmission tests should always include control treatments such as inoculation with zoospore suspension alone, virus suspension (without zoospores) and untreated bait plants to show

that they are virus-free. The bait plant to be used are selected based on the host range of both virus and the fungus vector.

2. The host plant from which fungal vector was originally obtained should be suitable for the multiplication of the fungus.
3. Before transmission studies, the test fungus species to be used should be identified properly on the basis of morphology and molecular means.
4. Appropriate cultural conditions of nutrition, moisture, aeration, temperature, pH and absence of other pathogens must be ascertained for the host plant. Abundant moisture is needed for the release and movement of zoospores. All the vectors are highly active at low temperatures (10–20 °C).

References

Campbell RN (1996) Fungal transmission of plant viruses. Annu Rev Phytopathol 34:87–108

Hull R (2014) Matthews' plant virology, 5th edn. Academic, San Diego

Teakle DS (1962) Transmission of tobacco necrosis virus by a fungus, *Olpidium brassicae*. Virology 18:224–231

Teakle DS (1972) Transmission of plant viruses by fungi. In: Kado CI, Agrawal HO (eds) Principles and techniques in plant virology. Van Nostrand Reinhold Company, New York

Chapter 19

Transmission of Viruses Through Nematodes

Abstract

The appearance of a patch or patches of diseased plants in a crop is often the first indication of the existence of nematode–borne virus infection in the field. Ectoparasitic nematodes in the genera *Longidorus*, *Paratrichodorus*, *Trichodorus* and *Xiphinema* are involved in the transmission of nepo- and tobra-group of viruses. These viruses are also mechanically transmissible through infected sap. Infection of a bait plant is the standard method used for detecting nematode transmission in a test soil. Plants are grown in the soil to allow any viruliferous nematodes to feed and transmit the virus. Sap extracted from roots and leaves of the bait plants are then inoculated mechanically with a range of indicator hosts to confirm and identify the associated virus.

Key words *Longidorus*, *Paratrichodorus*, *Trichodorus*, *Xiphinema*, Virus transmission

19.1 Introduction

Nematode transmitted viruses infect a broad range of hosts, and a few of them are seed- and pollen-borne. They are readily transmitted by mechanical inoculation and are relatively stable. Fan leaf of grapes was the first viral disease shown to be transmitted by nematodes. Two genera each of Longidoridae (*Longidorus* and *Xiphinema*) and Trichodoridae (*Trichodorus* and *Paratrichodorus*) have proved to be the vectors of the viruses belonging to the genera *Nepovirus* and *Tobravirus*, respectively (Fig. 19.1). More than 20 nematode species are known to transmit plant viruses, and these species are relatively cosmopolitan and widely distributed throughout the world and have broad host range. Soil type and water holding capacity of the soil may limit their distribution. They are obligate migratory ectoparasites feeding on the roots. Both adult and larval stages of most of the nematode vectors can transmit virus with equal efficiency. The largest numbers of nematodes are likely to be found in the region of young plant roots of the host plant to a depth of about 60 cm from the top. These nematodes are ectoparasitic and feed on epidermal cells of the root. They acquire the virus within minutes of feeding and remain viruliferous for several weeks or months. The nematode-transmitted viruses are carried as extracellular contaminants, do not persist after moulting

Alangar Ishwara Bhat and Govind Pratap Rao, *Characterization of Plant Viruses*, Springer Protocols Handbooks,
https://doi.org/10.1007/978-1-0716-0334-5_19,

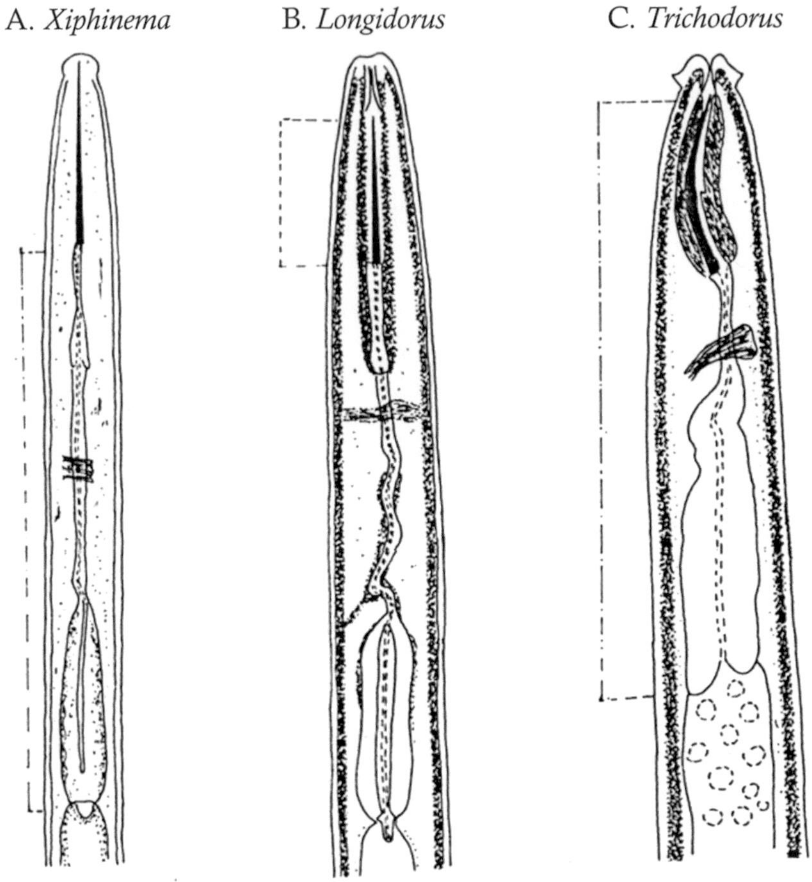

Fig. 19.1 Anterior portion of nematode vectors of plant viruses. Broken lines show part of alimentary tract where virus particles are retained. (Figure modified from Singh and Sitaramaiah (1994), plant pathogens: the parasitic nematodes)

and are not known to pass to progeny via eggs. No replication of virus takes place in nematode body. The distribution and use of infected planting material are the important way in which nematode-borne viruses are disseminated. The appearance of a patch or patches of diseased plants in a crop is often the first indication of the existence of nematode-borne virus infection. The spread of the disease is very slow (Raski and Hewitt 1963; Taylor 1972; Nayudu 2008).

Infection of a bait plant is the standard method used for detecting nematode transmission in a test soil. Plants are grown in the soil to allow any viruliferous nematodes to feed and transmit the virus. Sap extracted from roots and leaves of the bait plants are then inoculated mechanically with a range of indicator hosts.

19.1.1 Collection, Isolation, Handling and Culturing of Nematodes

19.1.1.1 Sampling

The soil samples are collected from the infested area using spade or screw auger with minimum disturbance to avoid injury to nematodes. It should be noted that a large number of nematodes are usually found around young roots of host plants, but nematode species differ in their vertical distribution and abundance in different crops, soil and moisture conditions. The *Longidorus* spp. is found in top 20 cm of soil while *Xiphinema* spp. is found at 20–60 cm depths. *Trichodorus* and *Paratrichodorus* spp. occur in sandy and light soils. Samples (about 200–300 g) may be collected in polyethylene bags or containers with the help of spade or shovel. Samples should be kept cool until use as nematodes are prone to injury and may be killed by drying, overheating or suffocation.

19.1.1.2 Isolation

Many methods are available for separating nematodes from the soil. For *Longidorus* and *Xiphinema* spp., modified Cobb's (1918) decanting and sieving technique is the most ideal method (Barker 1985). The method consists of a period of elutriation or flotation with nematodes, and particles of similar size and density are separated from soil mass in water and removed from the bulk of the fluid by sieving (Fig. 19.2). The sieving may then be examined directly or after separation of nematodes from debris by Baermann funnel technique (Fig. 19.3).

19.2 Materials (Isolation of Nematodes)

1. 150 μ aperture sieves (three numbers).
2. 2 mm aperture sieve.
3. 90 μ aperture sieve.
4. Baermann funnel: Funnel made of glass with a piece of soft polyethylene tube attached to the stem and closed with a spring or screw clip (Fig. 19.3).
5. Buckets (1 and 5 L): Diameter of the bucket should fit into the rim of sieves.
6. Compound microscope.
7. Microscope slide.
8. Stirring rod.
9. Wash bottle.

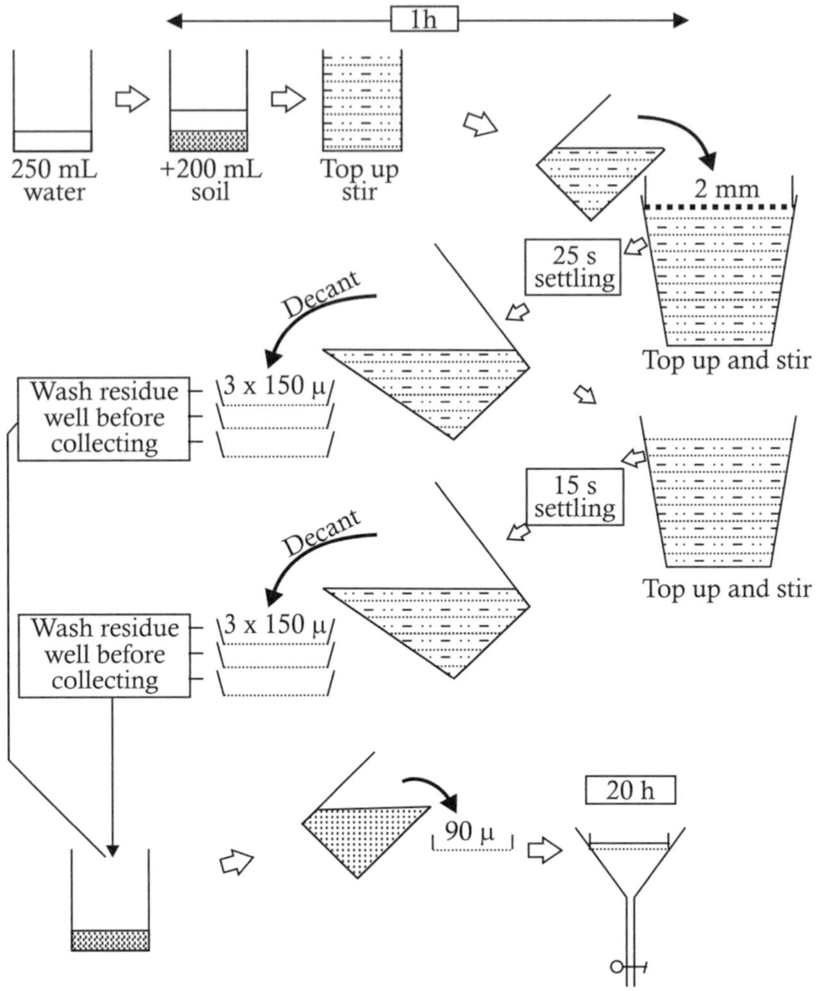

Fig. 19.2 Schematic diagram of Flegg's modified Cobb's decanting and sieving method for the isolation of Longidorid nematodes

19.3 Method (Isolation of Nematodes by Flegg Modified Cobb's Decanting and Sieving Technique)

1. Each soil sample (approximately 200 mL) is mixed with about 250 mL of water in a 1 L bucket and allowed to stand for 30–60 min, stir 2–3 times during soaking period.
2. The suspension is then passed through a 2 mm aperture sieve (to remove larger stones and debris).
3. Filtrate after addition of water agitated by hand and after 20 s supernatant is passed through a set of three 150 μm sieves (Fig. 19.2).
4. The material on the sieve is collected in a beaker.

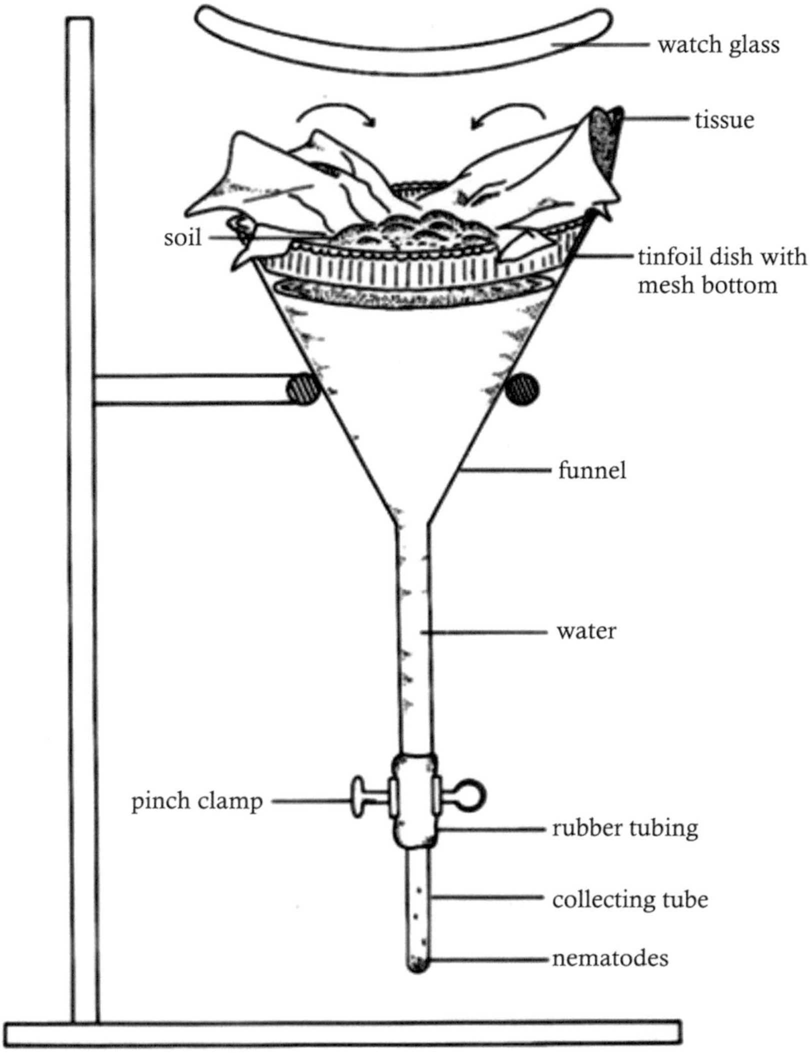

Fig. 19.3 Baermann funnel apparatus for nematode extraction from soil

5. The filtrate is again added with water and mix thoroughly. After 20 s, decant the supernatant through the same set of three 150 μm aperture sieves.
6. Residue on the sieves is added to the beaker collected previously.
7. Pour the complete content of the beaker onto a 90 μm aperture sieve, and the materials on the sieve are washed and collected in a beaker.
8. The nematodes are then separated from debris in the Baermann funnel (Fig. 19.3).

9. The contents of the beaker are poured onto a 90 μm aperture nylon screen supported by a rigid polyethylene ring, which is placed in a 15 cm diameter Baermann funnel with sufficient water to submerge debris in the screen.
10. The nematodes are then collected by running off about 25 mL of water from the Baermann funnel after 20 h of standing (Fig. 19.3).
11. Concentrate the nematodes into a minimum volume of water and examine in a glass dish under a binocular stereo microscope with a magnification ranging 20–40×. Eyelash or metal tool is used to handle and pick nematodes of interest. The taxonomic identification of the species may be done with the help of established keys. Specimens can be preserved by gently heating in water to 65 °C for 2 min to kill them and then fixing them using suitable fixatives.

19.3.1 Culturing Nematodes

When field soil is used as a source for culturing nematodes, it is important to determine the nematode species present, whether they are viruliferous and also whether other soil-borne viruses (tobacco necrosis virus) are present. The use of bait plants can determine the presence of an infective virus. *Xiphinema* spp. can be cultured easily in pots in glasshouse to reach large numbers within 4–6 months as compared to other species where multiplication rate is very slow requiring several years to double in size. Temperature, soil moisture and host plant species significantly influence nematode survival and reproduction. Fig and grapevine are used to culture *X. index*, perennial ryegrass for *Longidorus* and *Trichodorus* spp., raspberry, rose and strawberry for *X. diversicaudatum* and *L. macrosoma*. A temperature of about 20 °C is used for culturing most nematodes. Most nematodes prefer relatively low moisture conditions and die following long periods of very wet conditions. Hence, sand may be added to the field or sterilized soil used for nematode culturing.

19.3.2 Transmission Tests

Transmission test involves the addition of virus-free nematodes to pot containing infected source plant for acquisition. After 1 month, the plant is removed, washed and observed for a number of root galls. The potential viruliferous nematodes are then extracted, counted and transferred to a pot containing healthy bait plants for inoculation. After 1 month, carefully remove the plant, wash roots and observe for the number of root injuries/lesions. The roots of bait plants are also subjected to virus assay by electron microscopy, ISEM or reverse transcription polymerase chain reaction (RT-PCR) tests to confirm the presence of virus particles. To determine the presence of infection in nematodes, nematodes are ground in a small volume of water or buffer between two glass slides. The extract can be either manually inoculated to indicator test plants or used in ISEM or RT-PCR.

19.4 Notes (Barker 1985; Nayudu 2008)

1. Avoid collecting soil samples from dry areas.
2. There should be no lump or clod, and the soil should be thoroughly mixed with water.
3. Sieve should always be held at 45° angle with respect to the ground and not horizontally. The water should be drained by tapping the sides of the sieve and not by putting fingers into the middle.
4. The ambient temperature should not go above 28 °C after the nematodes are obtained in a water suspension.
5. Nematodes should be picked from suspension by a soft pick to avoid damaging them. Expert advice should be taken for proper identification if required.
6. Juvenile stage of nematodes should be used for virus transmission.
7. Proper identification of general species and identification of the nematodes study in test should be confirmed before starting and planning the transmission experiments.

References

Barker KR (1985) Nematode extraction and bioassays. In: Barker KR, Carter CC, Sasser JN (eds) An advanced treatise on *Meloidogyne*, vol 2. Methodology. North Carolina State University, Raleigh

Cobb NA (1918) Estimating the nema populations of soil. US Dep Agr Tech Circular No.1

Nayudu MV (2008) Plan viruses. Tata McGraw-Hill Publishing Company Limited, New Delhi

Raski DJ, Hewitt WB (1963) Plant parasitic nematodes as vectors of plant viruses. Phytopathology 53:39–47

Singh RS, Sitaramaiah K (1994) Plant pathogens: the plant parasitic nematodes. Inter Sci Publ, The University of Michigan, p 320

Taylor CE (1972) Transmission of viruses by nematodes. In: Kado CI, Agrawal HO (eds) Principles and techniques in plant virology. Van Nostrand Reinhold Company, New York

Chapter 20

Storage and Preservation of Plant Virus Cultures

Abstract

Viruses are obligate pathogens that cannot be cultured without living in host. Plant viruses require special preservation methods to ensure long-term viability/infectivity of the sample and to maintain the original characteristics. Usually symptomatic leaf tissue of virus-infected plants is taken and kept as original sample for future study purpose. Many different methods are employed for the storage of virus cultures like desiccation, freeze drying (lyophilization), freezing and low temperature storage in liquid nitrogen, which are discussed in this chapter.

Key words Desiccator, Freeze-drying, Low-temperature storage, Virus culture, Liquid nitrogen, Lyophilization

20.1 Introduction

Viruses exhibit a wide variety of structural and chemical differences, but, in general, their infectivity may be destroyed by degradative enzymes that expose the nucleic acids at higher temperatures or breakdown capsid proteins by chemicals. Preserving the viruses at low or ultra-low temperatures, and/or in the absence of water, slows down these destructive processes sufficiently to increase the length of time that the virus can be stored in infectious state. Virus-infected material is preserved by freeze-drying of virus-infected leaves or as homogenates of virus-infected plants. Storage and preservation of virus-infected plant materials (fresh/dried, leaves/stem cuttings/root stocks) are an essential prerequisite for studying the biological, serological and molecular properties of a virus. It also allows for inoculating the virus in original host to maintain a live culture of the virus. Infectivity of viruses varies with different species and genera with different methods of storage. It is, therefore, necessary to store viruses by more than one method to have better chances of virus recovery. The storage procedures involve removal of moisture from the virus-infected plant tissues or extracts by the addition of protectant materials, desiccation, storage at low temperature or a combination of these methods. For quick transport between regions or countries, fresh leaf samples can be placed

Alangar Ishwara Bhat and Govind Pratap Rao, *Characterization of Plant Viruses*, Springer Protocols Handbooks,
https://doi.org/10.1007/978-1-0716-0334-5_20,

between pieces of dry paper, sealed in polyethylene bags and transported in a cool box. For perishable tissues, leaves may be soaked in 50% glycerol for a few hours before being sealed in bags. The samples stored in liquid nitrogen serve as nuclear virus seed stocks in cases where infectivity is lost after prolonged storage, or when stock to be replenished fails authentication due to contamination with other viruses (Holling and Stare 1970; Bos 1973; Dijkstra and deJager 1998; Hull 2014). Some of the simplest and most reliable storage protocols for the different types of preservation methods are discussed in the present chapter.

20.1.1 Desiccation of Samples

McKinney et al. (1965) developed a simple method to store sap-transmissible viruses that remain infective after desiccation by placing finely chopped pieces of infected leaves over calcium chloride ($CaCl_2$) or silica gel and stored in refrigerator (Fig. 20.1).

20.2 Materials

20.2.1 Desiccation of Samples

1. Absorbent paper.
2. Airtight vessel with anhydrous $CaCl_2$ or silica gel.
3. Cotton.
4. Desiccator.
5. Infected tissue.
6. Parafilm.
7. Refrigerator.

20.3 Method

20.3.1 Desiccation of Samples

1. Wrap the infected plant tissues (leafs, stalks, roots) with absorbent (blotting) paper to remove excess water.
2. Place the tissue in a desiccator vessel containing, at least, double the leaf weight of anhydrous $CaCl_2$ or silica gel (Fig. 20.1).
3. Seal the container with cap and keep the jar at room temperature for 2–3 weeks. If vacuum can be applied, this will speed up the drying process.
4. Remove the samples after 2–3 weeks, break into small pieces and immediately place the tissue in labelled glass bottles with a layer of anhydrous $CaCl_2$ and cotton. Cover the tissue with another layer of cotton and $CaCl_2$. Close the bottles and seal with parafilm and store at 4 °C.
5. Before using the samples, allow the bottle to return to room temperature to minimize condensation.

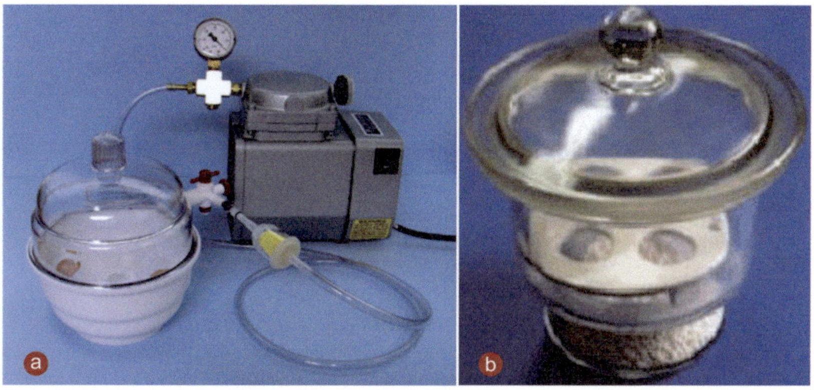

Fig. 20.1 Desiccator showing drying of virus-infected tissues

20.3.2 Preservation of Virus-Infected Rice Leaves and Recovery of the Virus Through Leafhopper Vector

20.3.2.1 Materials

1. Absorbent cotton.
2. Anhydrous $CaCl_2$.
3. Healthy culture of leafhopper, *Nephotettix virescens.*
4. Infected paddy leaves.
5. Phosphate buffer (0.1 M pH 7.0): Dissolve 17.4 g dipotassium hydrogen orthophosphate (K_2HPO_4) in 1 L distilled water (0.1 M solution). To this, add slowly 0.1 M solution of potassium dihydrogen orthophosphate (KH_2PO_4) prepared by dissolving 3.4 g of KH_2PO_4 in 250 mL distilled water till the desired pH of 7.0 is obtained.
6. Refrigerator.
7. Sodium sulphate (0.1 M): Prepared by dissolving 14.2 g of sodium sulphate salt in 1 L of water. Autoclave at 121 °C for 20 min before use.
8. Test tube with rubber stoppers.
9. Wax.

20.3.2.2 Method

1. Harvest virus-infected young leaves of rice and keep each leaf in a test tube half filled with $CaCl_2$ powder covered with absorbent cotton.
2. Close the tube with a rubber stopper and seal with wax.
3. Store the tubes in the refrigerator. The virus remains viable for more than 1 year at preservation.
4. For inoculation through its vector leafhopper *Nephotettix virescens*, the individual leaf is soaked overnight in a solution of sodium sulphate.
5. Next morning wash the leaves with several changes in distilled water (leaves turn green and succulent as a freshly plucked leaf).

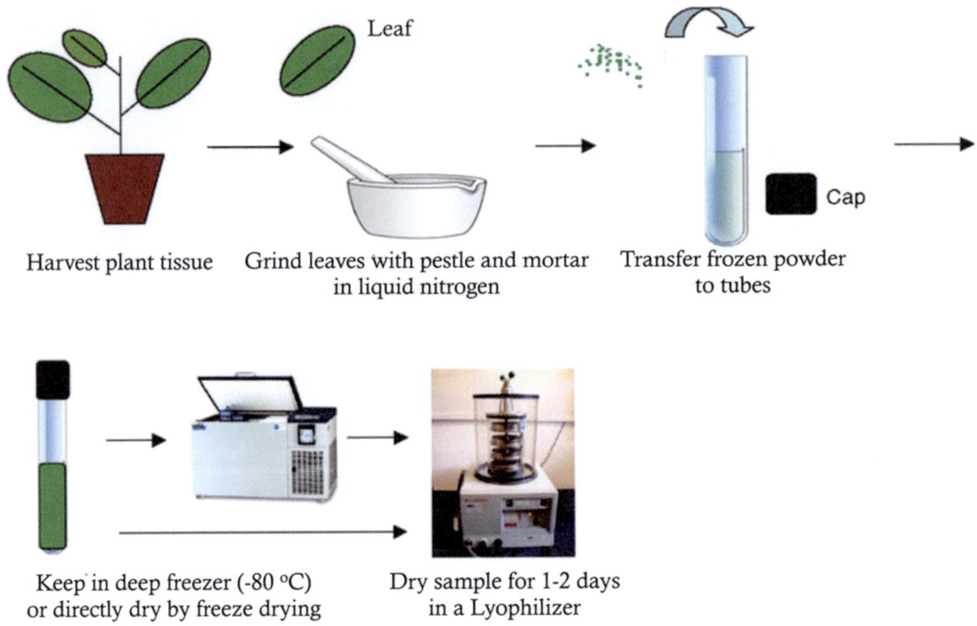

Fig. 20.2 Preservation of samples by freeze-drying

6. Keep a portion of leaf covered with cotton soaked pad in phosphate buffer, pH 7.0, and leafhopper can directly feed on this leaf for the acquisition of the virus (details of transmission by leafhoppers is provided in Chapter 12.

20.3.3 Freeze Drying

Freeze drying under vacuum has been used for storing virus isolates. The procedure can be used to preserve most viruses except those that cannot withstand freeze drying and lose their infectivity rapidly (e.g. bean yellow mosaic virus, potato virus X). For the preparation of freeze-dried virus-infected plant homogenates, symptomatic leaves are excised and a virus suspension is prepared by homogenizing the material 1:10 (w/v) in 0.1 M sodium phosphate buffer, pH 7.0, with a sterile pestle and mortar (Fig. 20.2). After filtering the homogenate through three layers of muslin to remove plant debris, 0.5 mL aliquots are pipetted into 5 mL vials and processed as stated below by freeze-drying. The labelled virus reference vials are stored at −20 °C.

20.3.3.1 Materials

1. Ampoule.
2. Freeze dryer (lyophilizer).
3. Freezer (−80 °C) or liquid nitrogen.
4. Glucose and peptone 7%: Dissolve 7 g each of glucose and peptone in 100 mL water and filter sterilize.
5. Infected tissue.
6. Pasteur pipette.

7. Pestle and mortar.

20.3.3.2 Method

1. Cut the leaves and macerate in 1:2 (w/v) of a freshly prepared solution containing 7% glucose and 7% bacteriological peptone to a fine solution (Fig. 20.2).
2. Place about 0.25–0.5 mL of this solution in a freeze-drying ampoule using a Pasteur pipette. Avoid contact of the solution with the neck of the ampoule. Label them.
3. Freeze samples in liquid nitrogen or in a −80 °C freezer.
4. When the samples are thoroughly frozen, transfer them to the lyophilizer (freeze-drying machine) with the vacuum pump on (Fig. 20.2).
5. Continue until samples are dry. Seal ampoules under vacuum and store them in −20 or −80 °C.

20.3.4 Storage at −80 °C

In this method, virus-infected tissues are cut into small bits and stored in a cryovial at −80 °C.

20.3.4.1 Materials

1. Cryovials.
2. Freezer (−80 °C).
3. Infected tissue.

20.3.4.2 Method

1. Cut infected leaves into small bits and place them into cryovials.
2. Label vials and place them in −80 °C freezer.

20.3.5 Preservation of Virus-Infected Plant Material in Liquid Nitrogen

Original virus-infected samples can be stored in liquid nitrogen for longer duration. Material preserved in liquid nitrogen serves as nuclear seed stock of virus inoculum to infect propagation in host plants which are subsequently used to replenish stocks preserved by freeze-drying. Virus culture preservation in liquid nitrogen is also useful for viruses which lose their infectivity upon prolonged storage or which are sensitive to other preservation and storage conditions.

20.3.5.1 Materials

1. Cryovials.
2. Liquid nitrogen.
3. Infected tissue.

20.3.5.2 Method

1. A small sample of symptomatic plant tissue infected with virus is excised and placed directly in a 2 mL cryovial and subjected to liquid nitrogen without any prior manipulation to the material.

2. The infected tissue is stored in liquid nitrogen and can be used again for inoculation when required. Infectivity tests are performed after a maximum of 3 years storage in liquid nitrogen.

20.3.6 Colour Preservation of Virus-Infected Tissues

Colour preservation is mainly used to store the infected tissue in their natural colour mainly for showing the original nature of viral symptoms. This will be useful for exhibition, teaching, farm extension workers and farmers to get acquaint with symptoms of various viral diseases in different hosts.

20.3.6.1 Materials

1. Burner.
2. Cupric acetate (saturated solution prepared in 50% glacial acetic acid): Dissolve cupric acetate salt in 50% glycerol while stirring continuously until no more cupric acetate can be dissolved in the solvent.
3. Formaldehyde: Appropriate personal protection equipment (PPE) such as a lab coat, disposable gloves, and safety glasses or goggles must be worn when handling formaldehyde.
4. Glacial acetic acid (50%): Add the water to the bottle first so that it can serve as a heat sink. To prepare the solution, pour 50 mL of distilled water into the appropriate glass bottle followed by addition of 50 mL of glacial acetic acid. Stir or shake for proper mixing.
5. Glass rod.
6. Paraffin wax.
7. Specimen jar with lid.
8. Virus-infected plant tissues.

20.3.6.2 Method

1. Dilute saturated solution of cupric acetate prepared in 50% glacial acetic acid four times (1:4).
2. Boil the infected leaf tissue in diluted saturated solution of cupric acetate for 10 min in a beaker. First the leaf gets bleached then turns brown and gradually regains its colour.
3. Drain the acid solution and wash leaf tissue thoroughly with tap water several times to remove all traces of acid.
4. Spread the tissue on the glass plate of the specimen jar and tie with a thread.
5. Put the glass plate inside the jar.
6. Prepare 4% formaldehyde solution and fill in the jar.
7. Label the jar by putting the name of the host, disease, causal virus involved, place of collection, date of collection, collector's name, etc.

20.4 Notes (Dijkstra and deJager 1998; Hull 2014)

1. Leaf samples infected with some of the highly stable viruses like TMV, CMV or PVX, etc. could be dried between the folds of blotting sheets with frequent change of sheets till the leaves dry. Then they are crushed into powder and preserved in small vials with tight caps and stored in the refrigerator in small packets of blotters.
2. Retention of infectivity vary with virus, host and method of storage used for preservation of virus culture.
3. It is always a good idea to store virus cultures through more than one method so that it could be retained for longer period.
4. Virus materials for distribution may be stored in aliquots.
5. Only stable viruses can retain infectivity when stored in desiccators.
6. Storage under liquid nitrogen is the best method as it retains infectivity.
7. Be sure that one should handle acetic acid under appropriate ventilation and remember that combining acid and water is an exothermic process.

References

Bos L (1973) Handout demonstration. Second International Congress of Plant Pathology, Minneapolis

Dijkstra J, deJager CP (1998) Practical plant virology-protocols and exercises. Springer, Heidelberg

Holling M, Stare OW (1970) The long-term survival of some plant viruses preserved by lyophilization. Ann Appl Biol 65:411–418

Hull R (2014) Matthews' plant virology, 5th edn. Academic Press, San Diego

McKinney HH, Silber G, Greeley LW (1965) Properties of virus in crude sap. Phytopathology 55:1043–1044

Chapter 21

Purification of Plant Viruses

Abstract

Purification refers to the separation of virus particles from host components in a biologically active state. Purified virus is required for the production of antibodies, physical, biochemical and molecular characterization of virus isolates. Purification of virus involves several steps such as propagation of the virus in the host, extraction of sap, clarification, concentration and further purification. Purity of purified preparation can be checked through UV absorption spectra and its infectivity by inoculating to a susceptible host under optimal environmental conditions in an insect-proof glasshouse. Purification methods vary with different viruses, and there are no universal methods of virus purification. Procedures that are effective for one virus may not work with the other. Stable viruses that reach high concentration in their propagation hosts are easy to purify compared to viruses that are less stable and occur in low concentration in their hosts. In general herbaceous hosts (*Chenopodium*, *Cucumis*, *Nicotiana*, *Petunia* and *Vigna*) that have less of virus inhibitors are more suitable for purification than perennial hosts that contain inhibitors. In this chapter, detailed description of purification procedure for some of the important viruses is discussed.

Key words Purified virus, Propagation host, Extraction, Clarification, Centrifugation, Density gradient centrifugation, Equilibrium density centrifugation, High-speed centrifugation, Low-speed centrifugation, Isolation of virus

21.1 Introduction

Isolation and purification of plant viruses are performed to separate virus particles from plant constituents. Purified virus preparations are required to study the chemical, physical, biological, serological and molecular properties of a virus. The purpose of purification is to produce a preparation containing only infective virus particles. Plants infected with some viruses contain a variable virus products (particles of different length and composition), and some virus particles lose their infectivity during purification.

The first step in the virus purification is to crush the plant cells and extract the virus containing sap (Fig. 21.1). Most of the larger constituents of sap such as chloroplasts, mitochondria, starch grains and fragments of cell walls sediment quickly and can be removed by short low *g* centrifugation. Sap constituents such as plant proteins, ribosomes and microsomes resemble virus particles in size, composition and stability, and they are the most troublesome host

Alangar Ishwara Bhat and Govind Pratap Rao, *Characterization of Plant Viruses*, Springer Protocols Handbooks, https://doi.org/10.1007/978-1-0716-0334-5_21,

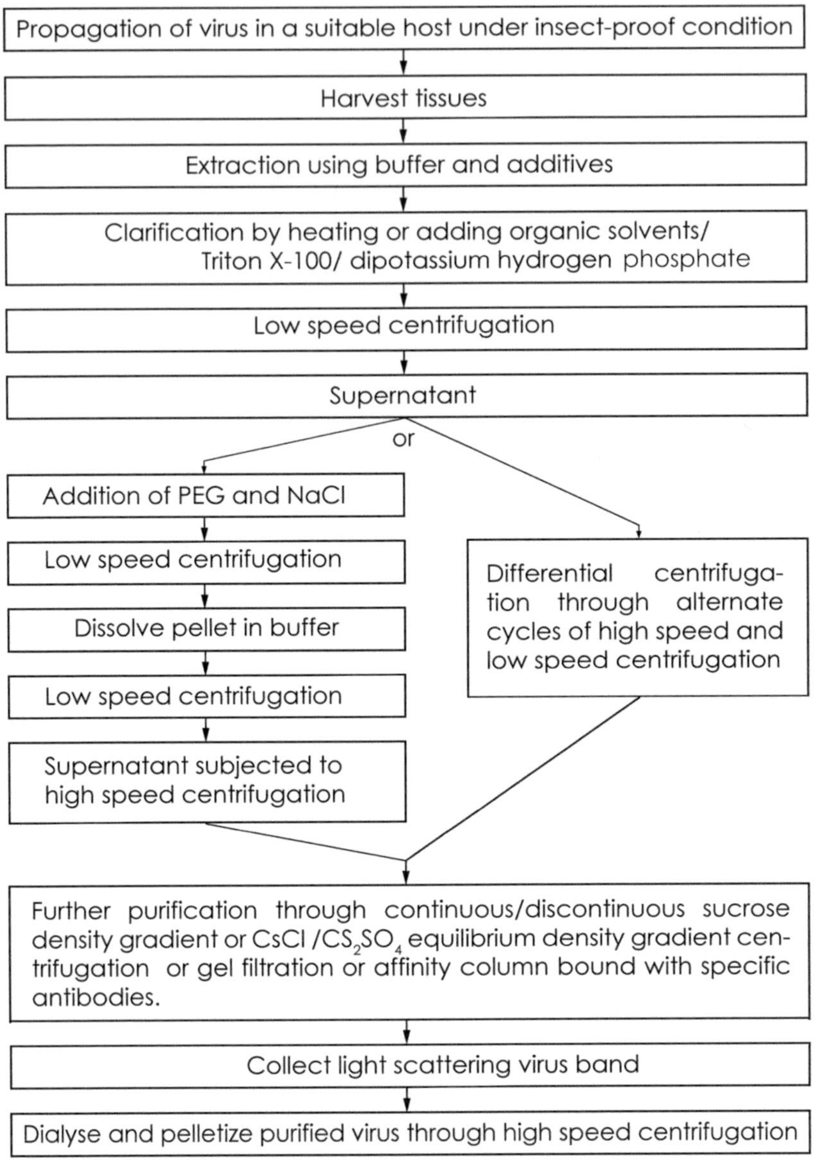

Fig 21.1 Various stages in the purification of a virus

contaminants to remove during purification of plant virus (Matthews 1991; Hull 2014).

21.2 Stages in Purification

21.2.1 Propagation of Virus

Selection of the right kind of a host plant for the propagation of virus to be purified is of paramount importance (Fig. 21.1). A good propagation host should be susceptible and contain high concentration of virus in the plant and should not contain a high amount of phenolic compounds, mucilage, latex or other inhibitors. Some of the plant species in the genera *Chenopodium*, *Cucumis*, *Nicotiana*, *Petunia* and *Vigna* have been found to be suitable propagation hosts for purification of a large number of plant viruses. The conditions under which plants to be grown and the time at which it is harvested should be standardized to maximize the starting concentration of the virus. Similarly, distribution of the virus within the host plant also varies. Hence, it is important to use plant tissues containing a high titre of the virus. Freshly harvested tissues always give better virus yields compared to stored and frozen tissues (Francki 1972; Hull 2002).

21.2.2 Extraction of Virus

Many viruses require a particular pH and ion strength for their stability. Buffer solutions with pH values in the range of 7.0–8.0 are used during extraction to keep the virus particles in suspension. Some of the commonly used buffers include borate, citrate, phosphate and tris. Oxidation affects the stability of a virus during extraction, often leading to loss of infectivity. Hence, reducing agents such as sodium sulphite, sodium thioglycolate, 2-mercaptoethanol and cystine hydrochloride are added to the extraction buffers. Additives such as EDTA (sodium salt) is added to break down ribosomes. However, this can be used only for viruses that do not require divalent metal ions for stability. The particles of many viruses aggregate, and many compounds have been used to reduce this tendency. Detergents such as Triton X-100 and Tween-80 are used. For potyviruses, use of urea was found useful in reducing aggregation. Use of enzymes like pectinase and cellulase aids in release of virus particles that would otherwise remain in the fibre fraction.

A variety of procedures are used to crush or homogenize the virus-infected tissues like pestle and mortar, blenders and juice extractors, roller mills or meat mincer. The crushed tissue is usually expressed through a muslin cloth. For many viruses, extraction is done at 4 °C. The standard extraction procedure involves grinding the tissues in 2–10 volumes (w/v) of extraction buffer at 4 °C and passing homogenate through a muslin cloth (Matthews 1991). The filtrate obtained is then subjected to clarification as explained below.

21.2.3 Clarification of the Extract

The first step in virus purification is to remove host macromolecular materials, leaving the virus in solution. This can be achieved by the addition of organic solvents, heating (50–60 °C) or addition of dipotassium hydrogen phosphate (Fig. 21.1). The method has to be chosen on a case-by-case basis as it may damage the virus particles. The treated extract is subjected to centrifugation at low speed (10,000 × *g* for 10–20 min). Centrifugation sediments cell debris and coagulated host material. Solvents and detergents cannot be used for viruses with membrane while for some viruses, organic solvents such as chloroform, carbon tetrachloride or *n*-butanol give adequate precipitation of host components.

21.2.4 Concentration of the Virus

From the clarified extract, the virus can be concentrated by following any of the methods like high-speed centrifugation, precipitation with polyethylene glycol, precipitation with salt or precipitation at isoelectric point. Centrifugations at high speed for a sufficient duration (2–4 h) will sediment the virus. Virus particles may be brought back into solution by dissolving the pellet (sediment) using an appropriate dilution of buffer. Some viruses remain as an aggregate when pellets are redissolved. The procedure involves subjecting the clarified extract to centrifugation (40,000–100,000 × g) for 2–4 h at 4 °C in a fixed angle rotor. During the process, virus particles will sediment at the bottom of the centrifuge tube in the form of pellet leaving low MW materials in the supernatant. After the centrifugation, the supernatant is discarded, and the pellet is redissolved after the addition of a small volume of suitable buffer.

Concentration by the addition of polyethylene glycol (PEG) is suitable for the extraction of many viruses. The conditions for precipitation depend on pH, ionic strength and concentration of macromolecules. The procedure usually involves the addition of PEG (4–10%) and sodium chloride to the clarified extract followed by stirring to dissolve the contents (for about 30 min) and further incubation up to 4 h. The contents are then subjected to low-speed centrifugation for 10–20 min. The supernatant is discarded, and the pellet is redissolved in a suitable buffer.

Salt precipitation may be used only for viruses that are not inactivated by salt solutions. Ammonium sulphate (up to 1/3rd saturation) is normally used. After incubating the clarified extract with salt for a few hours or days, the contents are centrifuged at low speed and redissolved in a small volume of suitable buffer.

Isoelectric precipitation can be used for viruses that are stable at the isoelectric point. The pH of clarified extract is brought to the isoelectric point to allow the virus particles to precipitate. The precipitate is then collected by centrifugation and dissolved in a suitable buffer.

Of the different methods available, precipitation by centrifugation and PEG are the most commonly used method while purifying many plant viruses.

21.2.5 Further Purification of the Virus

Even after suitable concentration, virus preparations still contain some low and high molecular weight host contaminants for which further purification is required which would remove most of them. Further purification can be achieved by differential centrifugation (alternate cycles of high and low speed centrifugation), density gradient centrifugation, gel filtration or affinity column bound with specific antibodies (Fig. 21.1).

Differential centrifugation involves subjecting the partially purified preparation into several alternate cycles of high- and low-speed centrifugation. Loss of virus often occurs with this procedure as a part of the virus under purification may remain insoluble or because of insufficient time allowed for re-suspension.

Of all the methods, density gradient centrifugation (Brakke 1960) is the most common accepted method used for further purification. Sucrose, cesium chloride or cesium sulphate is used to form a gradient. The centrifuge tube is filled with the solution having a decreasing density from bottom to the top of the tube. An ultracentrifuge with gradients may be used in different ways—(a) rate-zonal centrifugation in which the partially purified virus preparation is loaded at the top of the pre-formed gradient before centrifugation. During centrifugation, virus particles move as a zone/band through the gradient depending on their shape, size and density. The centrifugation is stopped while particles are still moving, and zone/band containing the virus is identified and collected with the help of a syringe. (b) Isopycnic gradient centrifugation—in this procedure, centrifugation is continued until all the particles in the gradient have reached a position where their density is equal to that of the medium. Here the separation of the particle is based on its density. In both the methods, after centrifugation process, the zone/band containing the virus is identified either by passing vertical light through the tube (to locate the light scattering virus band) or by moving contents of the tube through gradient fractionator. Dialysis may remove sucrose or salt in the preparation with 2–3 changes in between (Hull 2002).

Filtration through agar gel or Sephadex can also be used for further purification of viruses that are unstable to the pelleting by centrifugation. Affinity column bound with virus-specific antibodies can form a column that will specifically bind to the virus when the partially purified virus preparation is passed through the column. After thoroughly washing the column to remove all unbound materials, the virus bound to the column can be eluted by using a low pH buffer.

Electron microscopy, ultraviolet absorption spectra, SDS-PAGE and electro-blot immunoassay (EBIA) are some of the methods used to assess the purity of purified virus preparations. Virus yield is usually expressed as weight of purified virus obtained from a given weight of starting material. It can be determined by UV absorption spectrum of the preparation or by immuno-dot blots or hybridization dot-blots. The detailed procedures of all these methods are given in later chapters.

Purified preparations may be stored after the addition of low concentration of sodium azide or thymol or by the addition of an equal volume of glycerol or by freezing. However, it is advisable to avoid long-term storage as it may lead to loss of infectivity.

In this chapter, we have discussed protocols for the purification of some of the important representative plant viruses.

21.3 Materials

1. Aluminium foil.
2. Balance.
3. Beakers: Different capacities such as 10, 20, 50, 100, 250, 500 and 1000 mL capacity.
4. Centrifuge tubes: Different capacities such as 500 mL buckets, 100, 50, 30, 20, 2, 1, 0.5 and 0.2 mL.
5. Cold room: Required for incubation at 4 °C.
6. Dark room with light source: This is required for locating virus band in the centrifuge tube after sucrose density gradient centrifugation or equilibrium density gradient centrifugation.
7. Dialysis tube preparation: Cut dialysis tube of appropriate length and boil in 10 mM EDTA (prepared by adding 2.92 g disodium EDTA in 1 L of water) for 10–20 min followed by washing several times with 96% ethanol. Store at 4 °C in closed container with 50% ethanol and 10 mM EDTA. Rinse well with H_2O before use.
8. Filter paper.
9. Flasks.
10. Glass rod.
11. Gradient fractionator: Equipment used to locate virus zone/band after subjecting the partially purified preparation for further purification through sucrose density gradient centrifugation or equilibrium density centrifugation using cesium chloride or cesium sulphate. After the centrifugation run, the centrifuge tube is subjected to fractionation to locate virus containing fraction using this equipment. The solution of the tube is slowly passed through using a chase solution, and the peak corresponding to nucleoprotein (virus) is detected through the detector fitted with a UV filter at 256 nm. The fraction showing peaks are collected separately, dialysed and concentrated to get purified virus preparation.
12. High-speed refrigerated centrifuge: This instrument is available with maximum rotor speeds in the region of 25,000 rev per min (rpm) generating a relative centrifugal field of about 60,000 × *g*. It generally has a large capacity and a range of inter changeable fixed angle and swinging bucket rotors. These instruments are most often used to collect cellular debris, larger cellular organelles, ammonium sulphate and polyethylene glycol protein precipitates.
13. Ice flaking machine.
14. Magnetic stirrer.
15. Measuring cylinders.

16. Micropipettes.
17. Muslin cloth.
18. Organic solvents such as chloroform, carbon tetrachloride, *n*-butinol, *n*-amylalcohol, acetone.
19. Pasteur pipette.
20. Pestle and mortar.
21. pH meter.
22. Polyethylene glycol.
23. Polyvinyl pyrrolidone (PVP).
24. Preparation of continuous sucrose density gradient (10–40%) (Fig. 21.2): Prepare 10, 20, 30 and 40% sucrose of required volume (w/v) separately in beakers using appropriate buffer as diluent. Keep the prepared solutions at 4 °C for at least 1 h. Continuous sucrose gradient is then prepared in the centrifuge tube (e.g. 12 mL tube) by pipetting 3 mL of 40%, into the bottom of the tube. Following this, add slowly 3 mL of 30% sucrose solution through the side of the tube with the help of pipette. Care should be taken not to mix with 40% sucrose solution already added to the tube. After this, slowly add 2 mL of 20% sucrose solution and then 2 mL of 10% sucrose solution. Make sure that each concentration of sucrose is not mixed while loading. Place the tube in a cold room or in a refrigerator overnight to form a continuous gradient. The gradient tube should be used within 14–16 h after layering. Volume and per cent sucrose may vary depending on protocol used for

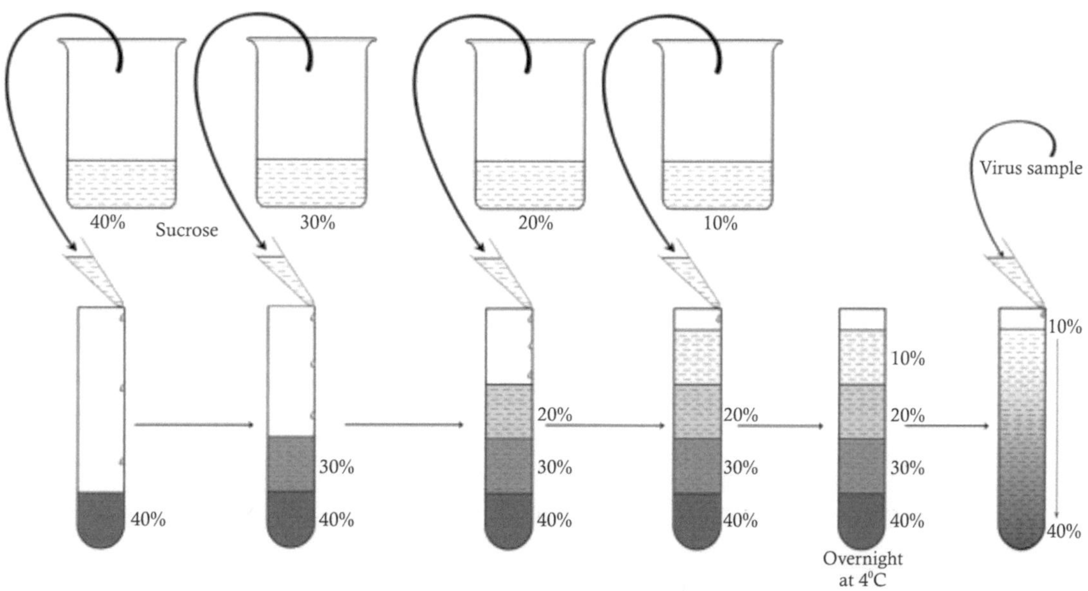

Fig. 21.2 Preparation of continuous sucrose density gradient

purification. Gradient tubes can also be prepared using gradient former equipment.

25. Reducing agents: Diethyl dithio carbamic acid (DIECA), ethylene diamine tetra acetic acid (EDTA), β-mercaptoethanol, sodium sulphite, thioglycolic acid.
26. Sodium chloride.
27. Sucrose.
28. Triton X-100.
29. Ultracentrifuge: Ultracentrifuges are capable of spinning rotors to a maximum speed of 80,000 rev per min (rpm) and can produce a relative centrifugal field of up to 600,000 × *g*. Ultracentrifugation is used for concentrating as well as final purification of the viruses. The aqueous phase containing the virus is centrifuged in a fixed angle rotor (about 75,000–100,000 × *g* for 90–120 min). During centrifugation, virus particles sediment against the sloping outer wall of the tubes in the angled rotor, and slide down to form a pellet in the bottom of the tube. The low molecular plant material is left in solution. After centrifugation run is over, the aqueous solution is discarded and the virus pellet is resuspended in a small volume of dilute buffer. This concentrated partially purified preparation may be further clarified by low-speed centrifugation, after which the aqueous phase containing the virus is carefully removed. The preparation may be subjected to further cycle of high- and low-speed centrifugation (differential centrifugation) to increase the purity of the virus preparation.
30. Urea.
31. Virus-infected tissue.
32. Waring blender: For grinding tissues used for purification of viruses.

21.3.1 Purification of Potato Virus Y *(PVY) (Genus: Potyvirus) (Moghal and Francki 1976; Bhat et al. 1997)*

21.3.1.1 Materials

1. Borate buffer (0.5 M, pH 8.0): First, prepare 0.5 M borate solution by dissolving 47.68 g of sodium tetra borate ($Na_2{\cdot}B_4O_7$) (synonym: Borax) in about 800 mL of water and make up the volume to 1 L. Also make a solution of 0.5 M boric acid (H_3BO_3) by dissolving 15.45 g of boric acid in about 400 mL water and make up the volume to 500 mL. Starting with the sodium borate solution, place a pH electrode in it while stirring slowly add the boric acid solution until the pH is 8.0.
2. Extraction buffer (0.5 M borate buffer pH 8.0 containing 0.2 M urea and 0.15% thioglycolic acid): Prepare 0.5 M borate buffer pH 8.0 as mentioned above. To this add required quantity of urea salt and thioglycolic acid to get a final concentration of 0.2 M and 0.15%, respectively (e.g. add 12 g of urea and 1.5 mL of thioglycolic acid for 1 L of borate buffer).

3. Resuspension buffer (0.05 M borate buffer containing 5 mM EDTA): Dissolve 146 mg of sodium EDTA in 10 mL of 0.5 M borate buffer, pH 8.0 and make up the volume to 100 mL.

21.3.1.2 Method

1. Harvest tissues 2–3 weeks after inoculation with the virus (*Nicotiana tabacum* or *N. glutinosa*, which serve as good propagation host for the virus) and homogenize them in a waring blender with two volumes (w/v) of 0.5 M borate buffer pH 8.0 containing 0.2 M urea and 0.15% thioglycolic acid.
2. Pass the homogenate through a double layer of muslin cloth and emulsify with one-half volume (w/v) each of chloroform and carbon tetrachloride by stirring for 15 min.
3. Centrifuge the mixture at 5000 × *g* for 10 min.
4. Pass the supernatant through Whatman No. 1 filter paper, polyethylene glycol (PEG 6000) and sodium chloride (NaCl) are added to a final concentration of 4 and 1.75% (w/v), respectively.
5. Stir the mixture for 30 min and allow standing at 4 °C for 1 h.
6. Centrifuge the contents at 8000 × *g* for 12 min.
7. Resuspend the precipitate in one fifth of the original 0.5 M borate buffer pH 8.0 with slow overnight agitation at 4 °C.
8. Next day, centrifuge the material at 5000 × *g* for 10 min and discard the pellet.
9. Layer the supernatant over 5 mL of 30% sucrose (w/v) in 0.5 M borate buffer per centrifuge tube and pelletize at 80,000 × *g* for 2 h in an ultracentrifuge (Beckman 55.2 Ti rotor).
10. Resuspend the pellets in 0.05 M borate buffer containing 5 mM EDTA (re-suspension buffer).
11. Give one more cycle of differential centrifugation if required.
12. Further purification is carried out by sucrose density gradient centrifugation (10–40% prepared in resuspension buffer) for 2 h at 80,000 × *g*.
13. Collect virus peak fractions by passing the contents of the tube through density gradient fractionator. Alternatively, virus containing zone can be collected with the help of a syringe after locating the same in a dark room with a light source.
14. Dialyze virus containing fractions overnight at 4 °C (with three changes) against 0.05 M borate buffer pH 8.0 containing 5 mM EDTA.
15. Pelletize the virus at 80,000 × *g* for 2 h in an ultracentrifuge.
16. Resuspend the pellets in a small volume of re-suspension buffer.
17. Store at 4 °C.

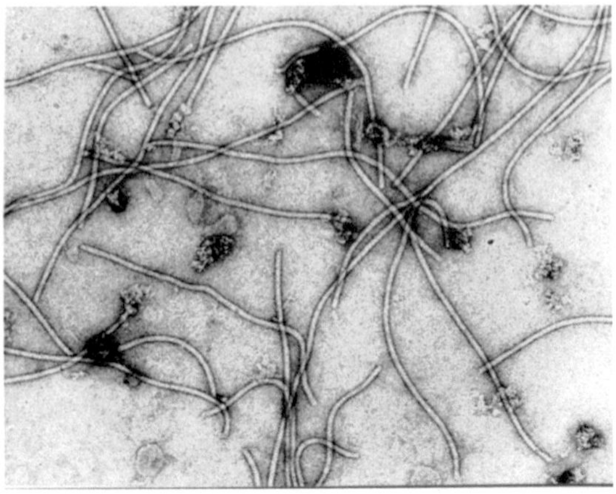

Fig. 21.3 Purified preparation of potato virus Y (magnification: ×80,000)

18. Observe the particle morphology under electron microscope (EM) (details available in the Chapter 23) (Fig. 21.3).

21.3.2 Purification of Sugarcane Mosaic Virus (SCMV) (Genus: Potyvirus) (Rao et al. 1998)

21.3.2.1 Materials

1. Extraction buffer (0.5 M borate buffer, pH 8.0 + 0.01 M diethyl dithio carbamic acid sodium salt, DIECA-Na + 0.15% thioglycolic acid): to 1 L of 0.5 M borate buffer pH 8.0 (prepared as mentioned above), add 2.25 g of DIECA-Na and 1.5 mL of thioglycolic acid.
2. Resuspension buffer (0.05 M borate buffer containing 5 mM EDTA): Dissolve 146 mg of sodium EDTA in 10 mL of 0.5 M borate buffer, pH 8.0 and make up the volume to 100 mL.

21.3.2.2 Method

1. Place 100 g virus-infected (sorghum) leaves at −20 °C (12–14 h), thaw, grind in 250 mL pre-cooled 0.5 M borate buffer, pH 8.0 + 0.01 M DIECA-Na + 0.15% TGA (extraction buffer) or powder frozen virus-infected leaves in liquid nitrogen, blend with extraction buffer (as above) for 10 min.
2. Stir continuously 2 h at 4 °C.
3. Strain sap through cheese cloth (3–4 layers).
4. Centrifuge at 1500 × *g* for 10 min and discard the pellet.
5. Emulsify the supernatant with 5% chloroform + 5% butanol.
6. Centrifuge at 4000 × *g* for 10 min and discard the pellet.
7. To the supernatant, add Triton X-100 (0.5%) + Tween-20 (0.5%), stir 10–15 min 4 °C.
8. Centrifuge (if required) at 1500 × *g* for 10 min.
9. Load supernatant onto 30% sucrose pad prepared in 0.5 M borate buffer, centrifuge at 81,000 × *g* for 2 h at 4 °C and discard the supernatant.
10. Dissolve the pellet in a small volume of 0.1 M borate buffer and stir for 15 min.

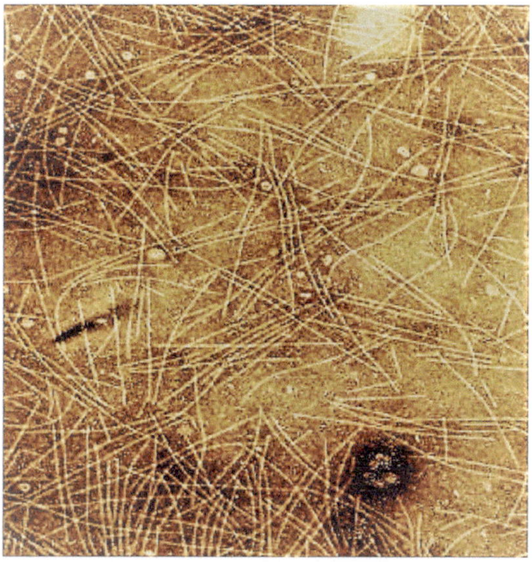

Fig. 21.4 Purified preparation of sugarcane mosaic virus showing flexuous filamentous particles (magnification: ×30,000)

11. Centrifuge at 1500 × *g* for 10 min and discard the pellet.
12. Layer the supernatant onto 10–40% linear sucrose gradient prepared in resuspension buffer.
13. Centrifuge at 83,500 × *g* for 90 min in swing out (SW) rotor.
14. After centrifugation, observe the tube for light scattering zone by passing single source light from top of the tube in a dark room and collect band with the help of a syringe.
15. Dilute the virus containing band with resuspension buffer and centrifuge at 90,000 × *g* for 2 h.
16. Dissolve the pellet in about 2 mL of resuspension buffer.
17. Centrifuge at 300 × *g* for 2 min.
18. Collect the supernatant and observe under electron microscope (Fig. 21.4).

21.3.3 Purification of Cucumber Mosaic Virus *(CMV) (Genus: Cucumovirus)* (Lot et al. 1972; Bhat et al. 2004)

21.3.3.1 Materials

1. Borate buffer (5 mM pH 9.0): First, prepare 5 mM borate solution by dissolving 0.4768 g of sodium tetra borate ($Na_2B_4O_7$) (Synonym: Borax) in about 800 mL of water and make up the volume to 1 L. Also make a solution of 5 mM boric acid (H_3BO_3) by dissolving 0.1545 g of boric acid in water and make up the volume to 500 mL. Starting with the sodium borate solution, place a pH electrode in it while stirring add slowly the boric acid solution until the pH is 9.0.
2. Extraction buffer (0.5 M sodium acetate buffer, pH 6.4 containing 5 mM ethylene diamine tetra acetic acid (EDTA)

disodium salt and 0.5% thioglycolic acid): To 1 L of 0.5 M sodium acetate buffer, pH 6.4, add 1.86 g of disodium EDTA and 5 mL of thioglycolic acid.

3. Resuspension buffer (borate buffer 5 mM pH 9.0 containing 0.5 mM EDTA): To 100 mL of 5 mM borate buffer, add 186 mg of sodium EDTA.
4. Sodium acetate buffer (0.5 M pH 6.4): Prepare 0.5 M sodium acetate solution by dissolving 40 g of sodium acetate (CH_3COONa) in water and make up the volume to 1 L. Prepare 0.5 M of NaOH solution by dissolving 1 g of NaOH in water and make up the volume to 50 mL. Add slowly NaOH solution on to 0.5 M sodium acetate solution, with constant stirring till desired pH of 6.4 is obtained.

21.3.3.2 Method

1. Homogenize the leaves in a blender in two volumes (w/v) of extraction buffer (0.5 M sodium acetate buffer, pH 6.4 containing 5 mM ethylene diamine tetraacetic acid (EDTA) disodium salt and 0.5% thioglycolic acid).
2. Filter the extract through a muslin cloth and emulsify with an equal volume of chloroform by stirring for 30 min at 4 °C.
3. Centrifuge the extract at 12,000 × *g* for 10 min.
4. Discard the pellet and add to the supernatant 10% of polyethylene glycol (PEG 8000) in the presence of sodium chloride (1.75% w/v) with continuous stirring for 1 h.
5. Centrifuge the mixture at 12,000 × *g* for 15 min.
6. Discard the supernatant and resuspend the pellet overnight in the one-fifth volume of 5 mM sodium borate buffer containing 0.5 mM EDTA, pH 9.0.
7. Add Triton X-100 to a final concentration of 2% (v/v) and stir for 30 min.
8. Centrifuge at 18,000 × *g* for 25 min.
9. Discard the pellet and the supernatant subjected to centrifugation at 45,000 × *g* for 3 h.
10. Discard the supernatant and suspend the pellet in 5 mM borate buffer containing 0.5 mM EDTA, pH 9.0 (resuspension buffer).
11. Further purification is carried out by layering partially purified preparation on a linear 10–40% (w/v) pre-formed sucrose density gradient (prepared by layering 5 mL of 10% and 10 mL each of 20, 30 and 40% sucrose prepared using resuspension buffer at 4 °C).
12. Centrifuge the gradients at 55,000 × *g* for 2 h using swing out rotor (SW 28, Beckman).

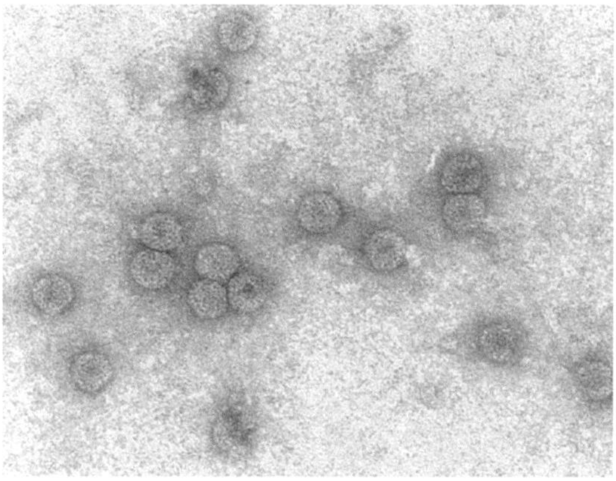

Fig. 21.5 Purified preparation of cucumber mosaic virus (magnification: ×100,000)

13. After the run, remove the gradient tube and locate the virus containing light scattering band with the help of light vertically passing through the tube in a dark room.
14. Collect the virus band using either a syringe or a Pasteur pipette, or a fraction collector.
15. Dilute the virus preparation with resuspension buffer and pellet the virus by centrifugation at 45,000 × *g* for 3 h.
16. Discard the supernatant and resuspend the pellet in a small volume of resuspension buffer.
17. Observe the virus particles under EM (Fig. 21.5).

21.3.4 Purification of Piper Yellow Mottle Virus (PYMoV) (Genus: Badnavirus) (deSilva et al. 2002)

21.3.4.1 Materials

1. $CsCl_2$/sucrose step gradient (0–40%): Step gradients are prepared by successively layering solutions of different density in the centrifuge tube and then layering the sample to be fractionated on top of the last "step." Prepare 0, 10, 20, 30 and 40% sucrose or $CsCl_2$ of required volume (w/v) separately in beakers using resuspension buffer. The step gradient is then prepared in the centrifuge tube (e.g. 30 mL tube) by pipetting 7 mL of each of different solution carefully layering one step on another, beginning with the densest step in the bottom of the tube. Make sure that there is no mixing of steps while loading. Particles are trapped in abrupt density steps during centrifugation that results in discrete particle layers at each step.
2. Extraction buffer [0.25 M Tris–HCl, pH 7.4 containing 0.5% (w/v) sodium sulphite, 4% (w/v) polyvinyl pyrrolidone (PVP 40), 0.5% (v/v) 2-β-mercaptoethanol, 0.25% diethyldithio carbamic acid (DIECA), 2% (w/v) celluclast]: To 250 mL of 1 M Tris pH 7.4, add 5 g of sodium sulphite, 40 g of polyvinyl pyrrolidone (PVP 40), 5 mL of 2-β-mercaptoethanol, 2.5 g

diethyldithio carbamic acid (DIECA), 22 g celluclast, dissolve and make up the volume to 1 L.

3. Resuspension buffer (50 mM Tris–HCl, 150 mM NaCl pH 7.4): To 5 mL of 1 M Tris pH 7.4, add 0.876 g of NaCl, dissolve and make up the volume to 100 mL.
4. Tris–HCl (1 M pH 7.4): Dissolve 121.14 g Tris in 800 mL distilled H_2O. Adjust the pH to 7.4 with the appropriate volume of concentrated HCl. Bring final volume to 1 L with distilled water.

21.3.4.2 Method

1. Cut leaves into pieces and homogenize with 10 volumes (w/v) of extraction buffer [0.25 M Tris–HCl pH 7.4 containing 0.5% (w/v) sodium sulphite, 4% (w/v) polyvinyl pyrrolidone (PVP 40), 0.5% (v/v) 2-β-mercaptoethanol, 0.25% diethyldithio carbamic acid (DIECA), 2% (w/v) celluclast] in waring blender at 4 °C.
2. Pass the homogenate through a double-layer muslin cloth.
3. Add Triton X-100 (to a final volume 2%) to the filtrate and stir at room temperature for 35 min.
4. Centrifuge the extract at 12,000 × *g* for 20 min at 10 °C.
5. Discard the pellet and add 4% polyethylene glycol (PEG, MW 6000) and 0.2 M NaCl to the supernatant.
6. Stir for 4.5 h at 4 °C.
7. Centrifuge at 12,000 × *g* for 20 min at 10 °C.
8. Resuspend the pellets overnight in about 60 mL of resuspension buffer (50 mM Tris–HCl, pH 7.4, 150 mM NaCl) at 4 °C.
9. Next day overlay the suspension in a 5 mL of 30% sucrose solution prepared in resuspension buffer and centrifuge for 3 h at 51,000 × *g* at 10 °C.
10. Resuspend the pellets in about 25 mL of resuspension buffer at 4 °C.
11. Final purification is done by using $CsCl_2$/sucrose step gradient (0–40%). Partially purified preparation (2 mL) is overlaid on preformed $CsCl_2$/sucrose step gradient (0–40%) and centrifuged at 35,000 × *g* for 5 h.
12. Pass the contents of the tube through gradient fractionator and collect fractions containing virus peaks.
13. Pool the fractions containing virus particles and dialyse against three changes of resuspension buffer.
14. Pelletize by centrifuging at 51,000 × *g* for 3 h.
15. Observe under EM (Fig. 21.6).

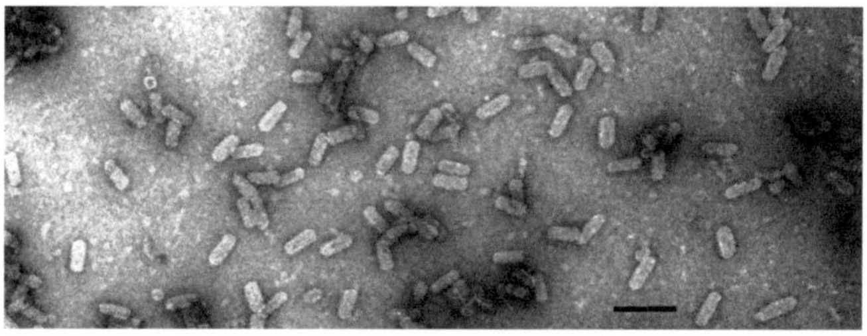

Fig. 21.6 Purified preparation of piper yellow mottle virus (magnification: ×80,000)

21.3.5 Purification of Cucumber Green Mottle Mosaic Virus (CGMMV) (Genus: Tobamovirus) (Mandal et al. 2008)

21.3.5.1 Materials

1. Extraction buffer (0.5 M phosphate buffer pH 7.0 containing 0.1% β-mercaptoethanol): To 1 L of 0.5 M phosphate buffer, pH 7.0, add 1 mL of β-mercaptoethanol
2. Phosphate buffer (0.5 M pH 7.0): Prepare 0.5 M of dipotassium hydrogen phosphate (K_2HPO_4) solution by dissolving 87.1 g of K_2HPO_4 in water and make up the volume to 1 L. Also, prepare 0.5 M of potassium dihydrogen phosphate (KH_2PO_4) by dissolving 34 g of KH_2PO_4 in water and make up the volume to 500 mL. To the prepared 0.5 M of K_2HPO_4 solution, add KH_2PO_4 solution slowly while stirring till desired pH of 7.0 is obtained.
3. Resuspension buffer (0.01 M phosphate buffer pH 7): Take 10 mL of 0.5 M phosphate buffer pH 7.0 and make up the volume to 500 mL with distilled water.

21.3.5.2 Method

1. Sap inoculate CGMMV to new emerging leaves of bottle gourd, cucumber or *N. benthamiana*. Collect leaves after 2–3 weeks of inoculation that show good symptoms.
2. Grind 100 g symptomatic leaf with liquid nitrogen in a pestle and mortar to a fine powder.
3. Homogenize the powdered tissues in a blender using 300 mL of extraction buffer (0.5 M phosphate buffer (PB) pH 7.0 containing 0.1% β-mercaptoethanol).
4. Strain the extract through two layers of muslin cloth, add *n*-butanol (10%) and stir for 15 min in a cold room.
5. Centrifuge the mixture of 10,000 × *g* for 30 min at 4 °C and collect the supernatant.
6. Add polyethylene glycol (PEG MW 6000, 4 g/100 mL) to the supernatant and stir for 2 h at 4 °C.
7. Centrifuge at 10,000 × *g* for 15 min, collect the pellet and dissolve in 80 mL of resuspension buffer (0.01 M phosphate buffer pH 7.0).

8. Centrifuge for 5 min at 10,000 × *g* and collect the supernatant.
9. Add NaCl (0.2 M final concentration) and PEG 6000 (4 g/ 100 mL) and stir overnight at 4 °C.
10. Collect the pellet by centrifugation at 10,000 × *g* for 15 min and resuspend the pellet in 2 mL of resuspension buffer, vortex and centrifuge at 10,000 × *g* for 2 min.
11. Collect the diffused white supernatant containing virus particles.
12. If the PEG precipitation is not suitable, the clarified supernatant can be subjected for precipitation of virus through sucrose cushion or padding in ultracentrifugation.
13. Prepare sucrose density gradient (10, 20, 30, and 40%) by dissolving sucrose in resuspension buffer. Carefully load sucrose solution one above the other each: 3 mL of 40%, 3 mL of 30%, 2 mL of 20% and 2 mL of 10% (from higher to lower concentration) in a 12 mL centrifuge tube. Make sure that each concentration of sucrose is not mixed while loading. Place the tube in a cold room or in refrigerator overnight to form a continuous gradient.
14. Just prior to centrifugation, load 2 mL of partially purified virus preparation on the gradient tube and place the tube gently in a swinging bucket rotor tube and centrifuge at 142,500 × *g* for 1 h at 4 °C.
15. Collect light scattering fraction zone with the help of a pasture pipette or syringe from the tubes.
16. Dialyse to remove the sucrose and examine in an electron microscope (Fig. 21.7).

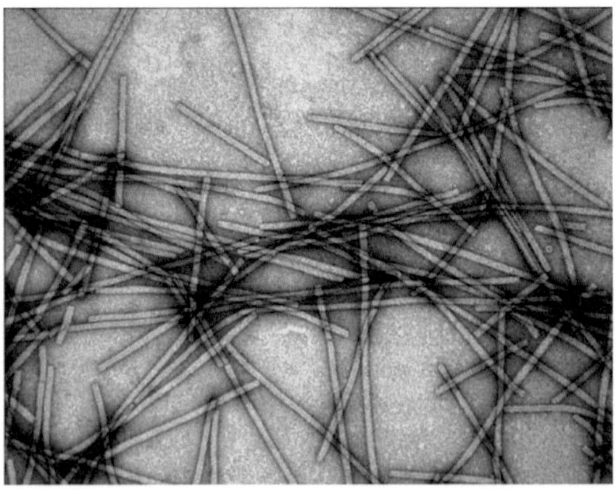

Fig. 21.7 Purified preparation of a tobamovirus (magnification: ×80,000)

21.3.6 Purification of Groundnut Bud Necrosis Virus (GBNV) (Genus: Tospovirus) (Reddy et al. 1992)

21.3.6.1 Materials

1. Extraction buffer (0.1 M potassium phosphate buffer, pH 7.5 containing 0.01 M sodium sulphite): To 1 L of 0.1 M potassium phosphate buffer, pH 7.5, add 1.26 g of sodium sulphite and get it dissolved.
2. Potassium phosphate buffer (0.1 M pH 7.5): Prepare 0.1 M of dipotassium hydrogen phosphate (K_2HPO_4) solution by dissolving 17.4 g of K_2HPO_4 in water and make up the volume to 1 L. Also prepare 0.1 M of potassium dihydrogen phosphate (KH_2PO_4) by dissolving 6.8 g of KH_2PO_4 in water and make up the volume to 500 mL. To the prepared 0.1 M K_2HPO_4 solution, add KH_2PO_4 solution slowly while stirring till desired pH of 7.5 is obtained.
3. Resuspension buffer (0.01 M potassium phosphate buffer, pH 7.5 containing 0.001 M sodium sulphite): Dissolve 0.126 g of sodium sulphite in 100 mL of 0.1 M potassium phosphate buffer, pH 7.5, and make up the volume to 1 L.

21.3.6.2 Method

1. Homogenize the tissue (4 mL/g of tissue) in chilled extraction buffer (0.1 M potassium phosphate buffer, pH 7.5 containing 0.01 M sodium sulphite).
2. Pass through the muslin cloth and centrifuge the filtrate at 400 × *g* for 10 min.
3. To the supernatant, add sodium chloride and polyethylene glycol at 11.7 and 40 g/L, respectively.
4. Dissolve and hold the mixture at 4 °C for 90–120 min.
5. Centrifuge at 8000 × *g* for 15 min and suspend the pellet in resuspension buffer (0.3 mL/g of tissue).
6. Centrifuge at 4000 × *g* for 15 min.
7. Layer the supernatant onto 20–60% preformed sucrose gradients (prepared by layering 8 mL each of 20 and 30% and 12 mL of 60% sucrose in resuspension buffer and held overnight at 4 °C).
8. Centrifuge at 90,000 × *g* for 45 min.
9. Collect the diffuse light scattering zone, at 2.6–3.1 cm from the bottom of the tube. Also, suspend the pellets in resuspension at the rate of 2 mL per pellet.
10. Mix sucrose zones and resuspend pellets, and stir at 4 °C for 1 h.
11. Layer onto 30–60% preformed linear (continuous) sucrose gradients (prepared by layering 7 mL each of 30, 40, 50 and 60% sucrose in resuspension buffer and leaving overnight at 4 °C).
12. Centrifuge at 90,000 × *g* for 2.5 h.
13. Collect the light scattering zone at 2.5–3.1 cm from the bottom of the tube and dilute in resuspension buffer.

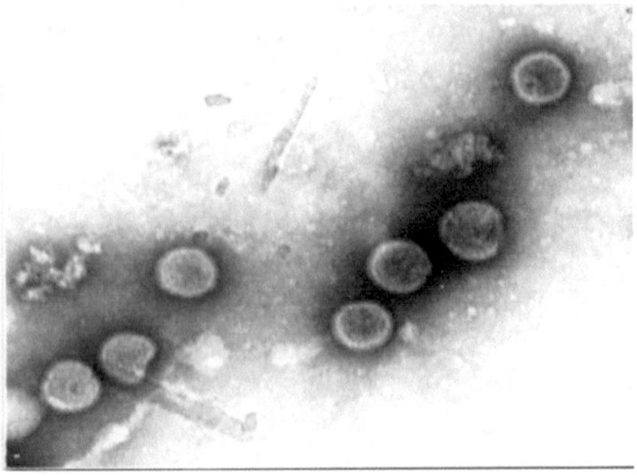

Fig. 21.8 Purified preparation of groundnut bud necrosis virus (magnification: ×80,000)

14. Layer onto 30–60% preformed linear sucrose gradients (prepared as described in **step 11**) and centrifuge at 90,000 × *g* for 2.5 h.
15. Collect light scattering zone found approximately between 2.7 and 3.0 cm from the bottom of the tube.
16. Dilute in resuspension buffer and pellet the virus at 80,000 × *g* for 2 h.
17. Resuspend the final pellet in a small quantity of buffer.
18. Observe the virus particles under EM (Fig. 21.8).

21.3.7 Purification of Tobacco Streak Virus (Genus: *Ilarvirus*) (Ramiah et al. 2001)

21.3.7.1 Materials

1. Extraction buffer (0.06 M sodium phosphate buffer pH 7.2 containing 0.1% 2-mercaptoethanol and 0.001 M EDTA): To 1 L of 0.06 M sodium phosphate buffer pH 7.2, add 1 mL of 2-mercaptoethanol and 0.372 g of sodium EDTA.
2. Sodium phosphate buffer 0.06 M pH 7.2: Prepare 0.06 M of disodium phosphate (Na_2HPO_4) by dissolving 8.58 g of Na_2HPO_4 in water and make up the volume to 1 L. Similarly prepare 0.06 M of sodium dihydrogen phosphate (NaH_2PO_4) by dissolving 3.6 g of NaH_2PO_4 in water and make up the volume to 500 mL. To Na_2HPO_4 solution, add NaH_2PO_4 solution slowly while stirring till desired pH of 7.2 is obtained.

21.3.7.2 Method

1. Harvest tissue showing typical symptoms and triturate at 1:2 ratio (w/v) in extraction buffer (0.06 M sodium phosphate buffer pH 7.2 containing 0.1% 2-mercaptoethanol and 0.001 M EDTA).
2. Express the sap through muslin cloth and emulsify with one-fifth volume of cold chloroform by stirring continuously for 15 min.
3. Centrifuge the mixture at 10,000 × *g* for 10 min.

4. Remove the aqueous layer and precipitate the virus by adding polyethylene glycol (PEG 6000, 6% w/v) and sodium chloride (1.75%, w/v) with continuous stirring for 20 min.
5. Incubate the mixture for 1 h at 4 °C and sediment the virus by centrifugation at 10,000 × *g* for 30 min.
6. Resuspend the sediment (pellet) in one-fifth volume of original sodium phosphate buffer (0.06 M, pH 7.2) and stir slowly for 2 h at 4 °C.
7. Centrifuge at 10,000 × *g* for 10 min.
8. Pellet the supernatant at 80,000 × *g* for 90 min in an ultracentrifuge.
9. Dissolve the pellet in 0.06 M sodium phosphate buffer (pH 7.2) and repeat the differential centrifugation at 10,000 × *g* for 10 min and 80,000 × *g* for 90 min.
10. Dissolve the pellet in 0.06 M sodium phosphate buffer and layer them on a 10–40% preformed continuous sucrose density gradient (prepared by layering 5 mL of 10% and 10 mL each of 20, 30 and 40% sucrose in 0.06 M sodium phosphate buffer and incubating at 4 °C overnight) and centrifuge at 72,000 × *g* for 3 h.
11. Collect virus peak fraction by passing the contents of the centrifuge tube through density gradient fractionator. Alternatively, virus containing zone is located with the help of monochromatic light passing vertically from top of the centrifuge tube and collect with the help of a syringe.
12. Dialyse the fractions overnight at 4 °C (with three changes) against 0.06 M phosphate buffer.
13. Concentrate the virus by subjecting to pelletizing at 80,000 × *g* for 90 min and dissolve the final pellet in a small volume of buffer and observe under EM (Fig. 21.9).

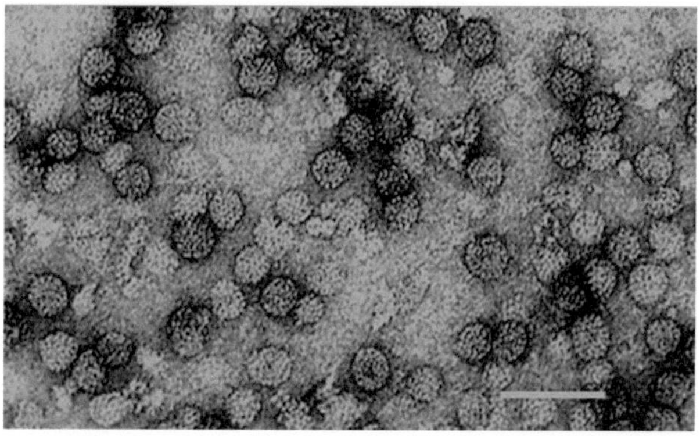

Fig. 21.9 Purified preparations of tobacco streak virus (magnification: ×80,000)

21.3.8 Purification of Potato Leaf Roll Virus (PLRV) (Genus: *Luteovirus*) (Rowhani and Stace-Smith 1979)

21.3.8.1 Materials

1. Extraction buffer (0.5 M potassium phosphate buffer pH 7.4 containing 0.1% 2-mercaptoethanol and sodium diethyl dithiocarbamate): To 1 L of 0.5 M potassium phosphate buffer, pH 7.4, add 1 mL of 2-mercaptoethanol and 1 g of sodium diethyldithiocarbamate, dissolve well and use.
2. Phosphate buffer (0.5 M pH 7.4): Prepare 0.5 M of dipotassium hydrogen phosphate (K_2HPO_4) solution by dissolving 87.1 g of K_2HPO_4 in water and make up the volume to 1 L. Also prepare 0.5 M of potassium dihydrogen phosphate (KH_2PO_4) by dissolving 34 g of KH_2PO_4 in water and make up the volume to 500 mL. To the prepared 0.5 M of K_2HPO_4 solution, add KH_2PO_4 solution slowly while stirring till desired pH of 7.4 is obtained.
3. Resuspension buffer (0.07 M phosphate buffer containing 0.75 M urea, pH 7.4): Dissolve 31.5 g urea in 100 mL of 0.5 M phosphate buffer, adjust the pH to 7.4 and make up the volume to 700 mL.

21.3.8.2 Method

1. Use fresh potato foliage and stems infected with PLRV. Grind the tissue in extraction buffer, 0.5 M potassium phosphate buffer pH 7.4 containing 0.1% 2-mercaptoethanol and 0.1% sodium diethyldithiocarbamate (2 mL of buffer per gram of tissue).
2. Add *n*-butanol-chloroform (1:1) gradually at the rate of 1 mL per 10 mL of extract and emulsify for 1 h with a magnetic stirrer.
3. Centrifuge the sample at 5000 × *g* for 10 min and discard the pellet.
4. Add 7% polyethylene glycol (PEG) to the supernatant and stir the mixture for 1 h.
5. Recover the precipitate by centrifuging at 8000 × *g* for 30 min.
6. Resuspend the pellet in 1/10th the original volume with resuspension buffer (0.07 M phosphate buffer containing 0.75 M urea, pH 7.4).
7. Clarify the suspension by centrifugation at 6000 × *g* for 10 min.
8. Supernatant subjected to ultracentrifugation at 80,000 × *g* for 2.5 h.
9. Resuspend the pellet in resuspension buffer.
10. Further, purify by subjecting the above preparation to sucrose density gradient centrifugation at 180,000 × *g* for 90 min.
11. Collect the UV-absorbing band and dilute twofold with resuspension buffer.
12. Centrifuge at 115,000 × *g* for 4 h.

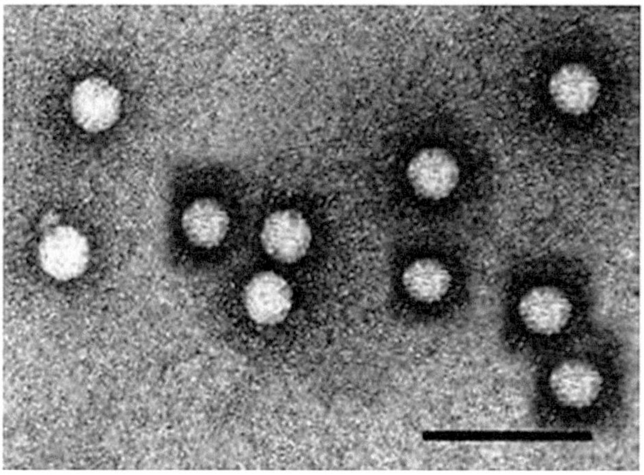

Fig. 21.10 Purified preparation of potato leaf roll virus (magnification: ×80,000)

13. Resuspend the purified virus pellet in 0.07 M phosphate buffer, pH 7.2 and stored at 4 °C.
14. Observe under EM for virus concentration (Fig. 21.10).

21.3.9 Purification of Banana Bunchy Top Virus *(BBTV) (Genus: Babuvirus) (Thomas and Dietzgen 1991)*

21.3.9.1 Materials

1. Extraction buffer (0.2 M potassium phosphate pH 7.4, containing 0.5% w/v sodium sulphite): To 1 L of 0.2 M potassium phosphate, pH 7.4, add 5 g of sodium sulphite and use.
2. Potassium phosphate buffer (0.2 M pH 7.4): Prepare 0.2 M of dipotassium hydrogen phosphate (K_2HPO_4) solution by dissolving 34.8 g of K_2HPO_4 in water and make up the volume to 1 L. Also prepare 0.2 M of potassium dihydrogen phosphate (KH_2PO_4) by dissolving 13.6 g of KH_2PO_4 in water and make up the volume to 500 mL. To the prepared 0.5 M K_2HPO_4 solution, add KH_2PO_4 solution slowly while stirring till desired pH of 7.4 is obtained.
3. PB (sodium phosphate buffer 0.07 M pH 7.2): Prepare 0.07 M of disodium phosphate (Na_2HPO_4) by dissolving 11.41 g of Na_2HPO_4 in water and make up the volume to 1 L. Similarly prepare 0.07 M of sodium dihydrogen phosphate (NaH_2PO_4) by dissolving 4.2 g of NaH_2PO_4 in water and make up the volume to 500 mL. To Na_2HPO_4 solution, add NaH_2PO_4 solution slowly while stirring till desired pH of 7.2 is obtained.

21.3.9.2 Method

1. Cut banana leaf midrib and petiole tissue into pieces of less than 1 cm in diameter, place them immediately into liquid nitrogen and pulverize into a fine powder.
2. Thaw the powder in two volumes of extraction buffer (0.2 M potassium phosphate pH 7.4, containing 0.5% w/v sodium sulphite) and stir for 30 min at 5 °C.

3. Filter the extract through cheese cloth and re-extract the fibres with one volume of extraction buffer using mortar and pestle and acid washed sand.
4. Mix the filtrate and add 0.1 vol of chloroform : butinol (1:1) mixture, stir for 1 h at 5 °C.
5. Centrifuge at 8000 × *g* for 10 min.
6. Collect aqueous phase and subject to centrifugation at 170,000 × *g* for 1 h.
7. Resuspend the pellet in PB (0.07 M sodium phosphate, pH 7.2) at 1 mL per gram of original tissue. Stir for 2 days and allowed to stand for a further 2 days at 5 °C.
8. Clarify the extract by centrifuging at 8000 × *g* for 10 min.
9. Supernatant subjected to centrifugation at 270,000 × *g* for 1 h and pellets resuspended in about 1 mL PB/100 g of the original tissue.
10. Layer them on a 10–40% continuous sucrose gradient prepared in PB and centrifuge at 70,000 × *g* for 4 h.
11. Locate and collect the zone containing virus particles with the help of density gradient fractionator.
12. Dilute the collected fraction in PB and concentrate by centrifugation at 270,000 × *g* for 1 h.
13. Further, purify the above preparation by subjecting them to equilibrium density centrifugation in cesium sulphate (3 mL gradient of 1.325 g/mL, overlaid with paraffin oil) at 180,000 × *g* for about 20 h.
14. Locate the virus zone with the help of vertical beam of light and remove by puncturing the side of the tube with a needle.
15. Dilute in PB and concentrate by centrifugation at 270,000 × *g* for 1 h.
16. Observe under EM (Fig. 21.11).

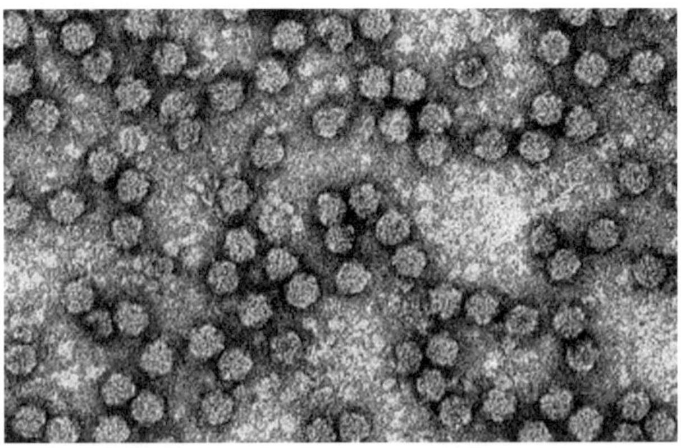

Fig. 21.11 Purified preparation of banana bunchy top virus (magnification: ×120,000)

21.3.10 Purification of Cymbidium Mosaic Virus (CymMV) (Genus: Potexvirus) (Frowd and Tremaine 1977)

21.3.10.1 Materials

1. Extraction buffer (0.5 M sodium citrate buffer, pH 6.5 containing 0.002 M sodium EDTA, 0.1% mercaptoethanol and chloroform): To 500 mL of 0.5 M sodium citrate buffer, pH 6.5, add 0.168 g of sodium EDTA and 0.5 mL of mercaptoethanol, mix and dissolve properly. To this add 500 mL of chloroform and use for extraction.
2. Sodium citrate buffer 0.5 M pH 6.5: Prepare 0.5 M of sodium citrate solution by dissolving 147 g of citric acid trisodium salt (trisodium citrate) in about 800 mL of water, adjust the pH to 6.5 and make up the volume to 1 L.
3. Sodium borate buffer 0.1 M, pH 8.5: First, prepare 0.1 M borate solution by dissolving 9.53 g of sodium tetra borate ($Na_2 B_4O_7$) (synonym: Borax) in about 800 mL of water and make up the volume to 1 L. Also make a solution of 0.1 M boric acid (H_3BO_3) by dissolving 3.09 g of boric acid in about 400 mL water and make up the volume to 500 mL. Starting with the sodium borate solution, place a pH electrode in it while stirring and slowly add boric acid solution until the pH is 8.5.
4. Tris–HCl 0.01 M buffer pH 7.0: Dissolve 1.21 g of Tris in about 900 mL of water, adjust pH to 7.0 with the help of concentrated HCl and make up the volume to 1 L.

21.3.10.2 Method

1. Homogenate infected leaves in a mixture of 0.5 M sodium citrate buffer pH 6.5 containing 0.002 M sodium EDTA, 0.1% mercaptoethanol and chloroform in the ratio 1:1:1 (w/v/v).
2. Centrifuge at 540 × *g* for 15 min.
3. Clarify the aqueous phase at 5200 × *g* for 25 min.
4. Collect aqueous phase and add polyethylene glycol 6000 (4%) and sodium chloride (0.25%), stir for 1 h.
5. Centrifuge at 5200 × *g* for 25 min.
6. Resuspend the pellet in 0.1 M sodium borate buffer pH 8.5 (1/10th of the original volume used for extraction).
7. After overnight incubation, centrifuge at 68,000 × *g* for 1.5 h and dissolve the pellet in 2 mL of borate buffer.
8. On the next day, centrifuge the contents at 3140 × *g* for 15 min.
9. Layer the supernatant onto 10–40% continuous sucrose density gradient prepared in borate buffer and centrifuge at 46,000 × *g* for 1.5 h.
10. Locate and collect light scattering virus band with a syringe and dialyze overnight against 0.01 M Tris–HCl buffer pH 7.0.

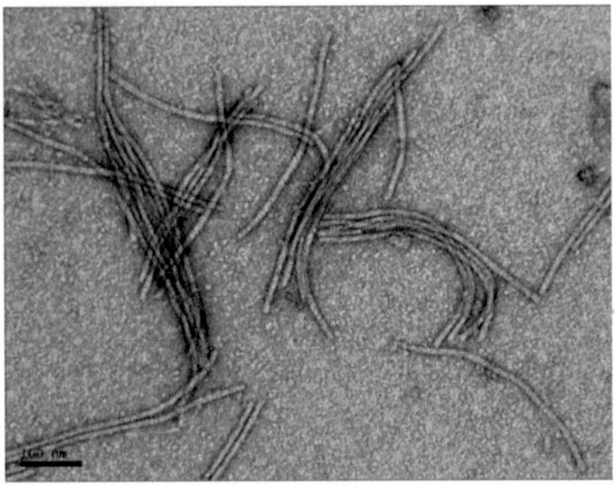

Fig. 21.12 Purified preparation of a *potexvirus* (magnification: ×60,000)

11. Pellet the virus by centrifugation at 68,000 × *g* for 1.5 h and dissolve the final purified virus pellet in a small volume of 0.01 M Tris–HCl, pH 7.0.
12. Observe under EM (Fig. 21.12).

21.3.11 Purification of Barley Yellow Dwarf Virus *(BYDV) (Genus: Luteovirus) (Geske et al. 1996)*

21.3.11.1 Materials

1. 2-[*N*-morpholino] ethanesulfonic acid (MES) (0.1 M): Make a 0.1 M solution of 2-(*N*-morpholino) ethanesulfonic acid by dissolving 19.5 g of this chemical in water and make up the volume to 1 L (solution A). Make separately 0.1 M solution of 2-(*N*-morpholino) ethanesulfonic sodium salt by dissolving 21.72 g of this salt chemical in water and make up the volume to 1 L (solution B). Add solution B slowly to solution A while stirring till desired pH of 6.2 is obtained.
2. Extraction buffer: To 1 L of 0.1 M of MES solution, add 3.36 g of Na-EDTA and 5 g of sodium sulphite (pH 6.2), dissolve nicely and use it.
3. Extractase P20X enzyme.

21.3.11.2 Method

1. Grind leaves in liquid nitrogen with a mortar and pestle and then in a blender to a fine powder.
2. Add cold buffer (0.1 M 2-[*N*-morpholino] ethanesulfonic acid (MES), 0.02 M EDTA, 0.5% sodium sulphite (pH 6.2) at rate of 1:2 (w/v) and homogenize on ice.
3. Add extractase P20X enzyme at the rate of 20 mg/mL of mixture. Mix and incubate on ice for 30 min followed by 2.5 h at room temperature. Periodically homogenize for 15 s.
4. Add Triton X-100 to a final volume of 0.5% and stir slowly at room temperature for 30 min.
5. Add 1:1 mixture of cold chloroform and *n*-amyl alcohol at the rate of 3.3 mL per 5 mL of homogenate. Blend rapidly for 30 s and then stir slowly at room temperature for 30 min.

6. Centrifuge at 1210 × *g* for 10 min at 4 °C and pass the supernatant through filter paper.
7. Add one-third volume of 30% polyethylene glycol containing 0.6 M NaCl with vigorous mixing and held at 4 °C overnight.
8. Centrifuge at 27,000 × *g* for 10 min at 4 °C and resuspend the pellets in 4–5 mL MES buffer and layered onto a 30% sucrose pad before centrifuging at 100,000 × *g* for 3 h at 4 °C.
9. Resuspend the pellets in 1 mL of MES buffer for 1 h.
10. Layer on 10–40% continuous sucrose gradients in MES buffer and centrifuge at 85,000 × *g* for 3 h at 4 °C.
11. Scan the gradient through gradient fractionator and fractions containing virus are centrifuged at 150,000 × *g* for 3 h.
12. Dissolve the final pellet in a small volume of buffer and observe under EM.

21.3.12 Purification of Bean Common Mosaic Virus (BCMV) (Genus: Potyvirus) (Alberio et al. 1979)

21.3.12.1 Materials

1. Extraction buffer (0.5 M potassium phosphate buffer, pH 7.5 containing 0.5% sodium sulphite): To 1 L of 0.5 M potassium phosphate buffer, pH 7.5, add 5 g of sodium sulphite and use.
2. Phosphate buffer (0.02 M pH 7.2) containing 0.1% mercaptoethanol: Prepare 0.02 M of dipotassium hydrogen phosphate (K_2HPO_4) solution by dissolving 3.48 g of K_2HPO_4 in water and make up the volume to 1 L. Also prepare 0.02 M of potassium dihydrogen phosphate (KH_2PO_4) by dissolving 1.36 g of KH_2PO_4 in water and make up the volume to 500 mL. To the prepared 0.02 M of K_2HPO_4 solution, add KH_2PO_4 solution slowly while stirring till desired pH of 7.2 is obtained. To this solution, add mercaptoethanol at the rate of 1 mL per 100 mL of buffer and use.
3. Potassium phosphate buffer 0.5 M, pH 7.5: Prepare 0.5 M of dipotassium hydrogen phosphate (K_2HPO_4) solution by dissolving 87.1 g of K_2HPO_4 in water and make up the volume to 1 L. Also prepare 0.5 M of potassium dihydrogen phosphate (KH_2PO_4) by dissolving 34.0 g of KH_2PO_4 in water and make up the volume to 500 mL. To the prepared 0.5 M K_2HPO_4 solution, add KH_2PO_4 solution slowly while stirring till desired pH of 7.5 is obtained.
4. Tris buffer 0.02 M pH 8.2: Dissolve 2.42 g of Tris base in about 900 mL of water, adjust the pH to 8.2 with the help of concentrated HCl and make up the volume to 1 L.

21.3.12.2 Method

1. Grind infected leaf tissues in two volumes of 0.5 M potassium phosphate buffer pH 7.5 containing 0.5 or 1.0% sodium sulphite.
2. Pass the resulting extract through a double layer of cheese cloth and add *n*-butinol to make a final concentration of 8% (v/v).

3. Stir the mixture overnight at 4 °C and centrifuge at 11,700 × *g* for 10 min.
4. To the supernatant, add polyethylene glycol (6–8%) followed by stirring for 1 h at 4 °C.
5. Centrifuge at 13,200 × *g* for 10 min and the resulting pellet dissolved in 0.02 M phosphate buffer, pH 7.2 containing 0.1% mercaptoethanol.
6. Subject the mixture to equilibrium density gradient centrifugation at 150,000 × *g* for 16–18 h in 30% cesium chloride prepared in the same buffer.
7. Locate and collect the virus zone present at 11–15 mm from the bottom of the tube by puncturing the tube at the bottom.
8. Dilute with 0.02 M phosphate buffer pH 7.2 containing 0.1% mercaptoethanol and subject to centrifugation at 11,700 × *g* for 10 min.
9. The supernatant concentrated by centrifugation at 120,000 × *g* for 90 min.
10. The final pellet is resuspended in 0.02 M Tris buffer pH 8.2 and virus yield is determined by assuming an extinction coefficient of 2.4.

21.3.13 Purification of Lily Symptomless Virus (LSV) (Genus: Carlavirus) (Wang et al. 2007)

21.3.13.1 Materials

Extraction buffer (0.02 M phosphate buffer pH 7.2 containing 0.1% β-mercaptoethanol and 1 mM PMSF): To 1 L of 0.02 M phosphate buffer, pH 7.2, add 1 mL of β-mercaptoethanol and 0.174 g of PMSF (phenyl methyl sulphonyl fluoride).

1. Potassium phosphate buffer (0.02 M, pH 7.2): Prepare 0.02 M of dipotassium hydrogen phosphate (K_2HPO_4) solution by dissolving 3.48 g of K_2HPO_4 in water and make up the volume to 1 L. Also prepare 0.02 M of potassium dihydrogen phosphate (KH_2PO_4) by dissolving 1.36 g of KH_2PO_4 in water and make up the volume to 500 mL. To the prepared 0.02 M of K_2HPO_4 solution, add KH_2PO_4 solution slowly while stirring till desired pH of 7.2 is obtained.
2. Ammonium sulphate.
3. Superdex 2000 HR GFC.
4. Sephacryl S-100 SF GFC column.
5. Elution buffer (0.02 M PBS pH 7.2 containing 0.1% β mercaptoethanol and 1 mM PMSF).

21.3.13.2 Method

1. Infected leaf tissues of Lily frozen and crushed into powder in liquid nitrogen with a mortar.
2. Homogenize quickly at 4 °C by adding two volumes of extraction buffer (0.02 M phosphate buffer pH 7.2 containing 0.1% beta-mercaptoethanol and 1 mM PMSF) and two volumes of chloroform.

3. Centrifuge at 5000 × *g* for 20 min and filter supernatant through filter paper.
4. Carry out further purification through gel filtration chromatography (GFC) using either Sephacryl or Superdex as indicated below.

By Sephacryl S-1000 SF GFC

Sample preparation (carried out at 4 °C)

(a) Add solid ammonium sulphate slowly with a 25% saturation to the supernatant obtained from **step 3** above, and store overnight.
(b) Centrifuge at 5000 × *g* for 20 min.
(c) Add solid ammonium sulphate to the collected supernatant at 50% saturation and store overnight.
(d) Centrifuge at 5000 × *g* for 30 min.
(e) Resuspend the pellet in 5 mL of PB (phosphate buffer) pH 7.2 and dialyze against the same buffer with three changes.
(f) Centrifuge at 8000 × *g* for 5 min and collect the supernatant.

Gel filtration chromatography (GFC)

(g) Equilibrate Sephacryl S-1000 SF GFC column (100 cm × 1.6 cm) by passing about 800 mL of elution buffer (0.02 M PB pH 7.2 containing 0.1% β mercaptoethanol, 1 mM PMSF) using a flow rate of 2 mL/min.
(h) Load about 5 mL of the supernatant obtained from the above step and elute with 160 mL of elution buffer.
(i) Collect fractions of 5 mL each and determine absorbance at 254 nm and also calculate $A_{260/280}$ ratio for each of the fractions.
(j) Fractions showing $A_{260/280}$ ratio of about 1.17 containing virus particles may be pooled and used as purified virus preparation.

By Superdex-2000 HR GFC

Sample preparation (carried out at 4 °C)

(a) To the supernatant obtained from **step 3**, add polyethylene glycol (PEG 6000) to 8% (w/v) and store overnight.
(b) Centrifuge at 5000 × *g* for 20 min.
(c) Resuspend the pellet in 5 mL of elution buffer (0.1 M phosphate buffer pH 7.2 containing 0.1% β-mercaptoethanol and 1 mM PMSF).

Gel filtration chromatography (GFC)

(d) Equilibrate Superdex 200 HR GFC column (35 cm × 1.0 cm) by passing about three times the volume of elution buffer

(0.1 M phosphate buffer pH 7.2 containing 0.1% β mercaptoethanol and 1 mM PMSF) using a flow rate of 0.5 mL/min.

(e) Load about 10 mL of the supernatant obtained from the above step and elute with about 15 mL of elution buffer.

(f) Collect fractions of 2 mL each and determine absorbance at 254 nm and also calculate $A_{260/280}$ ratio for each of the fractions.

(g) Fractions showing the $A_{260/280}$ ratio of about 1.19 containing virus particles may be pooled and used as purified virus preparation.

21.3.14 Purification of Indian Tomato Leaf Curl Virus (Genus: Begomovirus) (Muniyappa et al. 1991)

21.3.14.1 Materials

1. Sodium citrate buffer (0.1 M, pH 6.0): Prepare 0.1 M of sodium citrate solution by dissolving 29.4 g of citric acid trisodium salt (trisodium citrate) in about 800 mL of water, adjust the pH to 6.0 and make up the volume to 1 L.

21.3.14.2 Method

1. Harvest systemically infected symptom-bearing leaves of tomato cv. Pusa Ruby or Moneymaker 4–6 weeks after inoculation.
2. Disrupt the tissues in two volumes of extraction buffer containing 0.1 M sodium citrate buffer, pH 6.0 and 0.05% thioglycol using Waring blender.
3. Emulsify the extract with chloroform (1 mL/3 g leaf) and centrifuge for 10 min at 15,000 × *g*.
4. Add solid polyethylene glycol 6000 (7%) and sodium chloride (0.2 M) to the aqueous phase and stir the mixture for 3 h at 5 °C.
5. Collect the precipitate by centrifuging at 15,000 × *g* for 25 min.
6. Resuspend the pellet in resuspension buffer (0.1 M sodium citrate pH 6.0).
7. Centrifuge at 15,000 × *g* for 10 min.
8. Supernatant subjected to two cycles of differential centrifugation (2 h at 230,000 × *g* and 15,000 × *g* for 10 min).
9. Further purification carried out by layering partially purified preparation obtained after differential centrifugation on a 10–40% linear sucrose density gradient and centrifuging at 230,000 × *g* for 1 h.
10. Subject the virus containing fraction to equilibrium density centrifugation in a cesium sulphate gradient (of initial refractive index about 1.36) at 160,000 × *g* for 24 h.

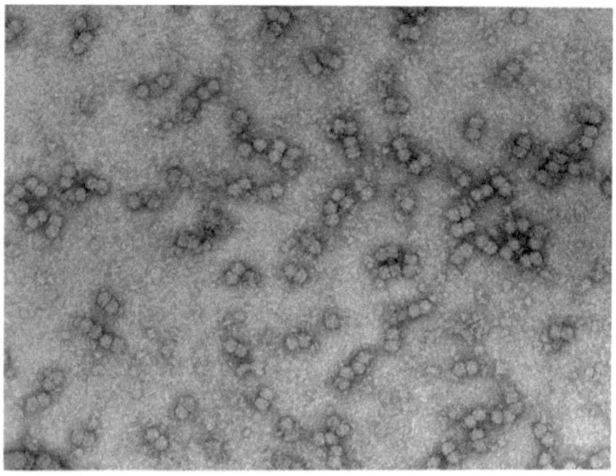

Fig. 21.13 Purified preparation of Indian tomato leaf curl virus (magnification: ×80,000)

11. Identify virus containing fractions from cesium sulphate gradient by passing the contents through a gradient fractionator and dilute in resuspension buffer.
12. Concentrate the virus particles by centrifugation at 300,000 × *g* for 1.5 h and dissolve the final pellet in a small volume of citrate buffer.
13. Observe the virus particles under EM (Fig. 21.13).

21.3.15 Purification of Citrus Tristeza Virus *(Genus: Closterovirus) (Bar-Joseph et al. 1985; Ozturk and Cirakoglu 2003)*

21.3.15.1 Materials

1. Separation buffer (Tris–HCl 0.05 M, pH 7.8, 1 mM $MgCl_2$ and 0.2% Na azide): To 1 L of Tris–HCl 0.05 M, pH 7.8, add 95 mg of $MgCl_2$, 2 g of sodium azide, dissolve and use.
2. Sodium phosphate buffer 0.04 M pH 8.2: Prepare 0.04 M of disodium phosphate (Na_2HPO_4) by dissolving 5.72 g of Na_2HPO_4 in water and make up the volume to 1 L. Similarly prepare 0.04 M of sodium dihydrogen phosphate (NaH_2PO_4) by dissolving 2.4 g of NaH_2PO_4 in water and make up the volume to 500 mL. To the prepared 0.04 M of Na_2HPO_4 solution, add NaH_2PO_4 solution slowly while stirring till desired pH of 8.2 is obtained.
3. Tris–HCl buffer 0.1 M pH 7.8: Dissolve 12.1 g of Tris base in about 900 mL of water, adjust the pH to 7.8 with the help of concentrated HCl and make up the volume to 1 L.
4. Tris–HCl 0.05 M pH 7.8: To 100 mL of 0.1 M Tris, pH 7.8, add 100 mL of water.

21.3.15.2 Method

1. Grind infected bark samples in a mortar with a pestle in the presence of liquid nitrogen.
2. Thaw the frozen powder by placing it in five volumes (w/v) of 0.1 M Tris–HCl buffer (pH 7.8) and filter by passing through four layers of wet cheese cloth.

3. Re-extract the pulp with the same buffer and combine filtrates and centrifuge at 4000 × *g* for 10 min.
4. Collect the supernatant and again subject to centrifugation at 8000 × *g* for 5 min to remove debris.
5. Pass the supernatant fluid through paper wipes and add 20 mL of a solution of 30% polyethylene glycol 6000 (PEG 6000), 0.6 M NaCl plus additional 2 mL of 20% (w/v) NaCl to each 100 mL of the filtrate.
6. Allow the mixture stand for 30–60 min at 4 °C and collect the precipitate by centrifugation at 19,500 × *g* for 15 min.
7. Resuspend the pellet by gentle agitation in 25 mL of 0.04 M sodium phosphate buffer (pH 8.2) for at least 1 h.
8. Centrifuge at 5900 × *g* for 10 min.
9. The supernatant that contains partially purified viral extract is then subjected to further purification by gel filter chromatography.
10. Equilibrate the Sephacryl 300 (S 300) column for gel filtration (1.5 × 70 cm) with separation buffer (0.05 M Tris–HCl pH 7.8, 1 mM $MgCl_2$ and 0.2% Na azide).
11. Load the partially purified virus extract obtained from the above step and elute with separation buffer.
12. Collect fractions of 1 mL each and test them for the presence of virus through indirect ELISA system (coat ELISA plate with 50 μL of 1/10 diluted fractions. Use fractions of healthy citrus extract after gel filtration on a Sephacryl S 300 column as negative control) (Note: Details of indirect ELISA is provided in Chapter 30).
13. Fractions containing virus particles (as determined by ELISA) may be pooled and used as purified virus preparation.
14. Observe under EM (Fig. 21.14).

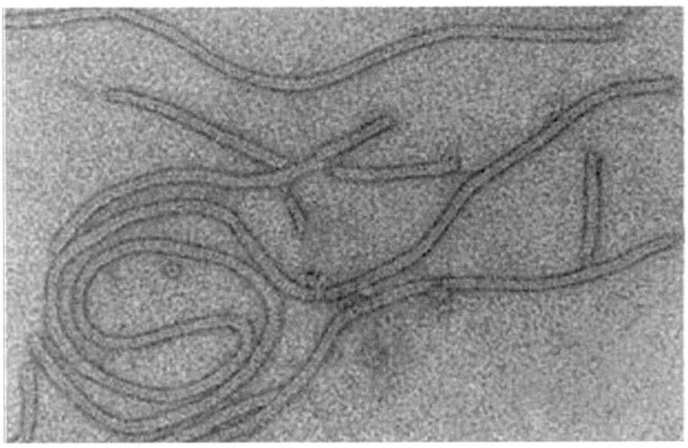

Fig. 21.14 Purified preparation of *Citrus tristeza virus* (magnification: ×60,000)

21.3.16 Purification of Squash Leaf Curl Virus (Genus: Curtovirus) (Cohen et al. 1983)

21.3.16.1 Materials

1. Extraction buffer (0.1 M phosphate buffer, pH 7.2 containing 0.002 M disodium EDTA and 0.01 M sodium sulphite): To 1 L of 0.1 M phosphate buffer, pH 7.2, add 0.672 g of disodium EDTA salt and 1.26 g of sodium sulphite, dissolve and use.
2. Resuspension buffer (0.1 M phosphate buffer containing 0.002 M EDTA): To 1 L of 0.1 M phosphate buffer, pH 7.2, add 0.672 g of disodium EDTA salt, dissolve and use.
3. Phosphate buffer (0.1 M pH 7.2): Prepare 0.1 M of disodium phosphate (Na_2HPO_4) by dissolving 14.32 g of Na_2HPO_4 in water and make up the volume to 1 L. Similarly prepare 0.1 M of sodium dihydrogen phosphate (NaH_2PO_4) by dissolving 6.0 g of NaH_2PO_4 in water and make up the volume to 500 mL. To Na_2HPO_4 solution, add NaH_2PO_4 solution slowly while stirring till desired pH of 7.2 is obtained.

21.3.16.2 Method

1. Harvest infected squash leaves 4–7 days after the appearance of symptoms and grind in a blender using 0.1 M phosphate buffer (pH 7.2) containing 0.002 M disodium EDTA and 0.01 M sodium sulphite at the rate of 100 mL per 100 g of plant material.
2. Filter the homogenized extract through a muslin cloth and stir with 1% (v/v) Triton X-100 for 2 h at room temperature or overnight at 4 °C.
3. Clarify the sap by adding 10% (v/v) cold chloroform, stir for 10 min and centrifuge at 10,000 × *g* for 10 min.
4. Add 12% polyethylene glycol (MW 7000–9000) and 0.2 M sodium chloride to the aqueous phase and stir for 1–2 h at 4 °C.
5. Centrifuge the mixture at 10,000 × *g* for 15 min and resuspend the pellets in one-tenth the original volume of 0.1 M phosphate buffer containing 0.002 M EDTA (resuspension buffer) and 1% Triton X-100 (v/v).
6. Stir the virus suspension for a minimum of 1 h at 4 °C and centrifuge for 10 min at 10,000 × *g*.
7. Ultracentrifuge the supernatant for either 3 h at 75,000 × *g* or 1 h at 260,000 × *g*.
8. Resuspend the pellet in 10 mL of resuspension buffer and centrifuge at 10,000 × *g* for 10 min.
9. Subject the supernatant to centrifugation at 90,000 × *g* for 2 h.
10. Dissolve the pellet in 4 mL of resuspension buffer and incubate overnight at 4 °C.
11. Next day centrifuge at 4000 × *g* for 5 min and load the supernatant onto a 10–35% linear sucrose gradient prepared in 0.1 M phosphate buffer with 0.001 M EDTA (pH 7.2).

12. Centrifuge at 95,000 × *g* for 3.5 h in an ultracentrifuge.
13. Examine the gradient for light scattering bands followed by analysis with a gradient fractionator.
14. Pool virus containing fractions, dialyse and palletize by centrifugation at 90,000 × *g* for 2 h.
15. Dissolve the final purified virus pellet in a small volume of buffer.
16. Observe under EM.

21.3.17 Purification of Rice Tungro Bacilliform Virus (RTBV) (Genus: Tungroviurs) and Rice Turngo Spherical Virus (RTSV) (Genus: Waikavirus) (Cabauatan and Hibino 1988)

21.3.17.1 Materials

1. Extraction buffer (0.1 M phosphate buffer, pH 7.2 containing 0.002 M disodium EDTA and 0.01 M sodium sulphite): To 1 L of 0.1 M phosphate buffer, pH 7.2, add 0.672 g of disodium EDTA salt and 1.26 g of sodium sulphite, dissolve and use.
2. Resuspension buffer (0.1 M phosphate buffer containing 0.002 M EDTA): To 1 L of 0.1 M phosphate buffer, pH 7.2, add 0.672 g of disodium EDTA salt, dissolve and use.
3. Phosphate buffer (0.1 M pH 7.2): Prepare 0.1 M of disodium phosphate (Na_2HPO_4) by dissolving 14.32 g of Na_2HPO_4 in water and make up the volume to 1 L. Similarly prepare 0.1 M of sodium dihydrogen phosphate (NaH_2PO_4) by dissolving 6.0 g of NaH_2PO_4 in water and make up the volume to 500 mL. To the prepared Na_2HPO_4 solution, add NaH_2PO_4 solution slowly while stirring till desired pH of 7.2 is obtained.
4. Driselase enzyme.

21.3.17.2 Method

RTSV and RTBV isolated and propagated separately on rice plant are to be used for the purification of RTSV and RTBV.

Isolation and Propagation of RTSV

One month after inoculation with tungro, test all 'healthy' looking seedlings for RTSV by latex test using antiserum to RTSV. Propagate RTSV from latex test-positive plants in TN1 rice cultivar by serial transfer using green leaf hopper. At 50–60 days after inoculation, plants may be tested by the latex test and used for virus purification.

Isolation and Propagation of RTBV

Index seedlings showing tungro symptoms for RTSV by latex test to identify RTSV-free plants. RTSV-free plants are then used as RTBV source for further propagation. Identify green leaf hoppers that would transmit only or mostly RTBV by infectivity neutralization techniques as follows.

Feed green leaf hoppers (that fed on plants infected with both viruses) for 16 h through parafilm membranes on anti-RTSV IgG diluted 25 times with 2% sucrose solution. Then transfer green leaf hoppers to RTBV-infected plants for an acquisition access feeding of 0–24 h and transfer to healthy seedlings. One month after inoculation, all plants with tungro symptoms were indexed once more in ELISA to confirm the absence of RTSV 2 days before purification.

Purification of RTSV and RTBV

The same purification procedure is adopted for both RTBV and RTSV.

1. Harvest infected plants (either with RTBV or RTSV) and homogenize in two volumes of extraction buffer (0.1 M phosphate buffer, pH 7.2 containing 0.002 M disodium EDTA and 0.01 M sodium sulphite)
2. Add the homogenate with enzyme driselase (3 g for 500 g of tissues) and incubate for 1 h at room temperature followed by incubation at 40 °C for 1 h in a water bath.
3. Centrifuge at 15,000 × *g* for 10 min and supernatant added with 7% polyethylene glycol (MW 8000), 0.2 M NaCl and 1% Triton X-100 and stir for 1 h at room temperature.
4. Centrifuge at 30,000 × *g* for 30 min and suspend the pellet in 20 mL of cold 0.01 M EDTA (pH 8.0).
5. Centrifuge the suspension at 11,000 × *g* for 10 min.
6. Ultracentrifuge the supernatant at 100,000 × *g* for 1 h and the pellet is dissolved in 0.01 M phosphate buffer (pH 7.4).
7. Centrifuge at 10,000 × *g* for 10 min and layer the supernatant on a 10–50% linear sucrose density gradient prepared in 0.01 M phosphate buffer and centrifuge at 95,000 × *g* for 2.5 h.
8. Scan the tube contents through gradient fractionator.
9. Recover the virus containing fractions, dilute and centrifuge at 130,000 × *g* for 1 h.
10. The final pellet is resuspended in a small volume of phosphate buffer and observe under EM (Fig. 21.15).

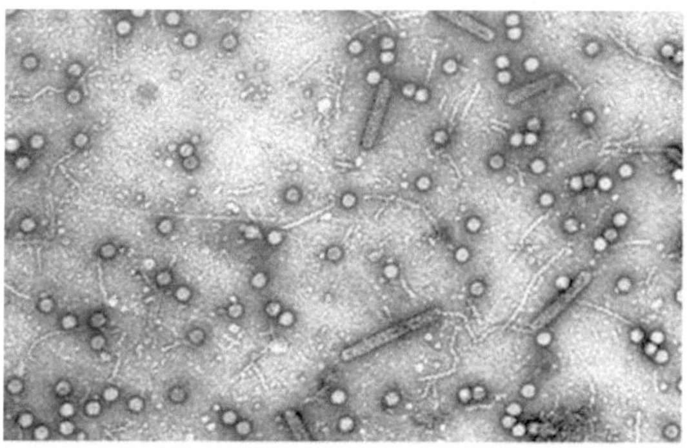

Fig. 21.15 Purified preparation of RTSV and RTBV (magnification: ×40,000)

21.4 Notes (Francki 1972; Brakke 1960; Matthews 1991; Hull 2002)

1. Selection of right kind of propagation host is the first and foremost step in the purification of a virus.
2. Host should be susceptible to the virus, preferably a fast growing herbaceous annual species without any virus inhibitors.
3. Propagation host should not have any seed-borne virus infection.
4. Determining the right time to harvest tissues for purification is important to maximize virus yield. This may vary from virus to virus and host to host. In general most viruses reach high concentration within 2–3 weeks after inoculation in herbaceous hosts.
5. As far as possible avoid using naturally infected plants for purification as they may contain a mixture of virus isolates, and the concentration of virus may also vary depending on several factors.
6. In general young actively growing tissues of plants harbour more virus particles than old or senescent leaves/tissues.
7. It is advisable to use freshly harvested tissues for purification than stored tissues as virus degradation may take place in old tissues.
8. Selection of right buffer, its molarity and pH is important. In addition, use of additives such as reducing agents and enzymes may be required for the purification of certain viruses.
9. Carry out all steps of virus purification at 4 °C unless otherwise mentioned. This will prevent degradation of virus particles.
10. Use correct speed for centrifugation. Always, it is a good idea to mention the speed of centrifugation in '*g*' (also called as RCF, relative centrifugal force) rather than in 'rpm' (revolutions per minute) as same 'rpm' may give different '*g*' values in different rotors. If only 'rpm' is mentioned, then protocol should also mention the rotor number and make of the centrifuge used, so that the value can be converted into '*g*'. Once '*g*' value is obtained, it can be converted into 'rpm' depending on the rotor used. Some of the newer versions of centrifuge have the option to select the speed in either 'rpm' or '*g*'. The relation between '*g*' value and 'rpm' is based on the rotor radius (in mm). Each rotor vendor provides ways to calculate the 'rpm' for a certain '*g*' and *vice versa*, so that one can adapt protocols to different rotors.
11. Use pro-pipette or micropipette to collect supernatant after centrifugation steps. Never attempt mouth pipetting.

12. Whenever supernatant has to be discarded after centrifugation, do it so slowly without disturbing the pellet. Pour off the supernatant by holding the tube with pellet side facing upwards.
13. Whenever pellet obtained from low-speed centrifugation has to be dissolved, do it slowly using a glass rod with appropriate buffer.
14. The pellet obtained after high-speed centrifugation using ultracentrifuge will be very compact and hard to dissolve immediately. Hence, it is advisable to decant the supernatant slowly without disturbing the pellet. Add a very small quantity of resuspension buffer over the pellet and leave the tube in tilted position to submerge the pellet overnight at 4 °C. This will help in the complete release of virus particles into the solution (do not forcibly dissolve the pellet immediately after centrifugation).
15. Aggregation (both side to side and end to end) may occur during purification of certain viruses especially rod-shaped elongated viruses. In such cases use of high molarity buffer and urea in the extraction was found to reduce aggregation.
16. After purification of the virus, it is necessary to store it for future studies. The purified virus can be stored in small aliquots in 10% glycerol at −20 °C. For long-time storage, keep the purified virus preparations at −80 °C.
17. A pinch of sodium azide should be used for long-term preservation of purified virus preparations to avoid any contamination.

References

Alberio J, Lima A, Purcifull DE, Hiebert E (1979) Purification, partial characterization and serology of *Blackeye cowpea mosaic virus*. Phytopathology 69:1252–1258

Bar-Joseph M, Gumpf DJ, Dodds JA, Rosner A, Ginzberg I (1985) A simple purification method for *Citrus tristeza virus* and estimation of its genome size. Phytopathology 75:195–198

Bhat AI, Varma A, Jain RK, Khurana SMP (1997) Differentiation of *Potato virus Y* strains by N-terminal serology and HPLC peptide profiling. Indian Phytopathol 50:89–96

Bhat AI, Faisal TH, Madhubala R, Hareesh PS, Pant RP (2004) Purification, production of antiserum and development of enzyme-linked immunosorbent assay-based diagnosis for *Cucumber mosaic virus* infecting black pepper. J Spices Aromat Crops 13:16–21

Brakke MK (1960) Density gradient centrifugation and its application to plant viruses. Adv Virus Res 7:193–224

Cabauatan PQ, Hibino H (1988) Isolation, purification and serology of rice tungro bacilliform and rice tungro spherical viruses. Plant Dis 72:526–528

Cohen S, Duffus JE, Larsen RC, Liu HY, Flock RA (1983) Purification, serology, and vector relationships of squash leaf curl virus, a whitefly-transmitted geminivirus. Phytopathology 73:1669–1673

deSilva DPP, Jones P, Shaw MW (2002) Identification and transmission of *Piper yellow mottle virus* and *Cucumber mosaic virus* infecting black pepper (*Piper nigrum* L) in Sri Lanka. Plant Pathol 51:537–545

Francki RIB (1972) Purification of viruses. In: Kadoo CI, Agrawal HO (eds) Principles and techniques in plant virology. Van Nostrand Reinhold Company, New York

Frowd JA, Tremaine JH (1977) Physical, chemical and serological properties of *Cymbidium mosaic virus*. Phytopathology 67:43–49

Geske SM, French R, Robertson NL, Carroll TW (1996) Purification and coat protein gene sequence of Montana RMV-like isolate of *Barley yellow dwarf virus*. Arch Virol 141:541–556

Hull R (2002) Matthews' plant virology, 4th edn. Academic Press, San Diego

Hull R (2014) Mathews plant virology, 5th edn. Academic Press, San Diego

Lot H, Marrou J, Quiot J-B, Esvan C (1972) Contribution à l'étude du virus de la mosaïque du Concombre (CMV). II. Méthode de purification rapide du virus. Annales de Phytopathologie 4:25–38

Mandal S, Mandal B, Haq QMR, Varma A (2008) Properties, diagnosis and management of Cucumber green mottle mosaic virus. Plant Viruses 2:25–34

Matthews REF (1991) Plant virology. Academic Press, New York

Moghal SM, Francki RIB (1976) Towards a system for the identification and classification of potyviruses. I. Serology and amino acid composition of six distinct viruses. Virology 73:350–362

Muniyappa V, Swanson MM, Duncan GH, Harrison BD (1991) Particle purification, properties and epitope variability of Indian tomato leaf curl geminivirus. Ann Appl Biol 118:595–604

Ozturk S, Cirakoglu B (2003) Production of a monoclonal antibody specific for *Citrus tristeza virus*. Food Agric Immunol 15:65–73

Ramiah M, Bhat AI, Jain RK, Pant RP, Ahlawat YS, Prabhaka K, Varma A (2001) Partial characterization of an isometric virus causing sunflower necrosis disease. Indian Phytopathol 54:246–250

Rao GP, Tosic M, Ford RE (1998) Tween-20 shortens a protocol for purification of sugarcane mosaic and maize dwarf mosaic potyviruses. Sugar Cane 6:19–22

Reddy DVR, Ratna AS, Sudarshana MR, Paul F, Kirankumar I (1992) Serological relationships and purification of bud necrosis virus, a tospovirus occurring in peanut (*Arachis hypogaea* L.) in India. Ann Appl Biol 120:279–286

Rowhani A, Stace-Smith R (1979) Purification and characterization of potato leafroll virus. Virology 98:45–54

Thomas JE, Dietzgen RG (1991) Purification, characterization and serological detection of virus-like particles associated with banana bunchy top virus disease in Australia. J Gen Virol 72:217–224

Wang R, Wang J, Li J, Wang Y, Xie Z, An L (2007) Comparison of two gel filtration chromatographic methods for the purification of Lily symptomless virus. J Virol Methods 139:125–131

Chapter 22

Ultraviolet Absorption Spectra of Purified Virus Preparation

Abstract

Spectrometry is one of the methods used to check the purity of purified virus preparation. In this method, optical density value of purified virus preparation at 260 and 280 nm and absorption spectra from 245 to 300 nm is determined using a spectrometer. The UV spectra and $A_{260/280}$ ratio provide information on the purity of the virus preparation. A_{260} nm value can be used to estimate the yield of purified virus.

Key words Purity of virus, Spectrophotometer, Extinction coefficient, Virus yield

22.1 Introduction

Viruses, being nucleoproteins, have a characteristic absorption spectra. The absorbance of a virus suspension can be measured in a spectrophotometer. In this apparatus, the UV rays transmitted through a cuvette containing the virus suspension in a buffer is compared with that transmitted through the same cuvette with buffer alone. The log of the ratio between two transmissions is given as the absorbance of the virus sample. The absorbance curve (absorption spectra) of plant viruses has a typical peak around a wavelength of 260 nm and a minimum around 245 nm. The shape of the absorption spectra varies depending on protein/RNA ratio of viruses. The maximum absorption for RNA is at 260 nm and for the proteins at 280 nm. Although virus is encapsidated by protein, it has a maximum absorption at 260 nm. This is due to 20 times higher absorption capacity of nucleic acids than the protein. As virus is a nucleoprotein particle and its nucleic acids contribute more in absorption of UV than the protein, purified virus shows a peak at 260 nm. Therefore, spectrophotometric measurement of virus concentration is calculated at 260 nm. As the absorption for protein is maximum at 280 nm, the ratio of absorbance of a virus at 260 and 280 nm ($A_{260/280}$) reflects its protein content. This ratio is lower for viruses with higher protein content. Both absorption spectra and the $A_{260/280}$ ratio can be used to judge the purity of purified virus preparations (Noordam 1973; Wilson and Goulding 1986).

Alangar Ishwara Bhat and Govind Pratap Rao, *Characterization of Plant Viruses*, Springer Protocols Handbooks, https://doi.org/10.1007/978-1-0716-0334-5_22,

The absorbance values are also used for estimating the virus yield in a preparation. The absorbance (extinction, E) is linearly proportional to the length of the light path, the concentration of virus and its extinction coefficient (Ec). The extinction coefficient of viruses has been established as the extinction of 1 mg/mL (0.1%) suspension at 260 nm wavelength and an optical path length of 1 cm ($E_{0.1\%}$ 1 cm, 260 nm). Whenever a new virus is discovered, its extinction coefficient should be determined. Once extinction coefficient is known, yield of the purified virus in a preparation can be determined by the formula:

Virus yield (mg/mL) = absorbance at 260 nm/extinction coefficient

22.2 Materials

1. Buffer in which purified virus is present.
2. Purified virus preparation.
3. Spectrophotometer.

22.3 Method

1. Take the required quantity of buffer in the cuvette. Using one cuvette as the blank, measure absorbance value of the second cuvette filled with diluent to be used for the virus sample. Adjust the spectrophotometer to give zero value for the second cuvette.
2. Replace the buffer in the second cuvette by appropriately diluted purified virus preparation.
3. Measure the absorbance from 200 to 300 nm using wavelength scan option (if wavelength scan option is not available in the spectrophotometer, the readings are taken at every 5 nm intervals starting from 200 to 300 nm).
4. Take a print out of the absorption spectra (if this option is not available, plot the values in a graph indicating wavelengths on the abscissa and absorbance values on the ordinate).
5. Determine $A_{260/280}$ ratio value for the virus.
6. Determine the virus yield using the formula:

Virus yield (in mg/mL) = absorbance at 260 nm/extinction coefficient

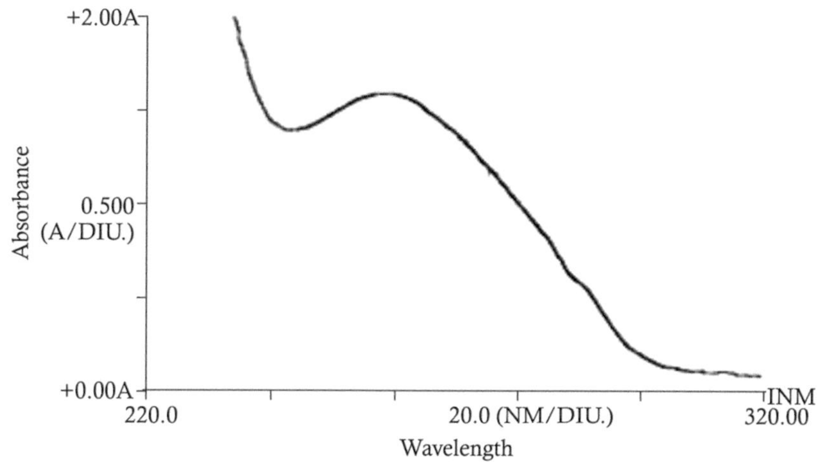

Fig. 22.1 Absorption spectra of purified virus preparation of potato virus Y (PVY)

Judge the purity of preparation based on the absorption spectra and $A_{260/280}$ ratio. A good purified virus preparation provides the absorbance curve (absorption spectra) with peak around a wavelength of 260 nm and a minimum around 245 nm (*see* Fig. 22.1). If satisfied, virus yield in the preparation can be estimated. If purity of the preparation is not up to the mark, subject them again to sucrose density gradient or $CsCl_2$ equilibrium density centrifugation.

22.4 Notes (Noordam 1973; Wilson and Goulding 1986)

1. Cuvette should be handled very carefully. The optical surface should be kept clean and not to be touched by hand. Scratching or dirt on this surface should be avoided.
2. Dilute the purified virus preparation to give an OD value between 0.5 and 1.0.
3. Always remember to blank the OD values with the same buffer in which purified preparations are present.
4. Virus does not absorb radiation of wavelength >315 nm, so any reading in this range is due to light scattering.

References

Noordam D (1973) Identification of plant viruses. Methods and experiments. Centre for Agricultural Publishing and Documentation, Wageningen

Wilson K, Goulding KH (1986) Biologists guide to principles and techniques of practical biochemistry. Edward Arnold, London

Chapter 23

Electron Microscopy and Utramicrotomy

Abstract

Virus particles can be seen only in transmission electron microscope. Transmission electron microscope not only helps to assess the purity of virus preparation but also throws light on the shape, size and surface details of virus particles. It is used to measure the dimension of the virus particles. Observation of virus particle morphology through electron microscopy (EM) is carried out either with purified virus preparations or with crude extracts of infected plant sap after negative staining. Besides, electron microscopy is also used in serological diagnosis of viruses through immunosorbent electron microscopy (ISEM). In the present chapter, process of sample preparation for the observation of shape and size of the virus particles in electron microscope is discussed. The process of ultrathin sectioning of the infected tissues for the observation of the virus particles in situ through EM is also discussed.

Key words Virus particle morphology, Electron microscope, Leaf dip method, Negative stains, Ultramicrotomy, Fixation

23.1 Introduction

In the transmission electron microscope (TEM) (Fig. 23.1), a beam of electrons is used to form an enlarged image of a tiny object through a series of electromagnetic lenses. Tiny negatively charged electrons originated by a heated tungsten wire filament can be accelerated by a high voltage to produce a coherent electron beam that can be focused by an electric field. An electric field of a doughnut-shaped electromagnet surrounding the electron beam acts just like the glass that focuses the light beam on the specimen. A similar electromagnet is used as the objective lens to form a highly magnified image of the specimen, and one/two more electromagnetic lenses further magnify and project the image onto a fluorescent viewing screen or CCD camera attached to a computer. Magnification can be increased by increasing the current through the electromagnets. TEM can provide an image with ×1000 greater spatial resolution than a light microscope (resolution 0.2 nm). Vacuum is maintained in the stack of electromagnetic doughnuts to allow a free path for the electrons from the filament to the viewing screen. The specimen to be viewed must be ultrathin to allow some electrons to pass through to form an image;

Alangar Ishwara Bhat and Govind Pratap Rao, *Characterization of Plant Viruses*, Springer Protocols Handbooks,
https://doi.org/10.1007/978-1-0716-0334-5_23,

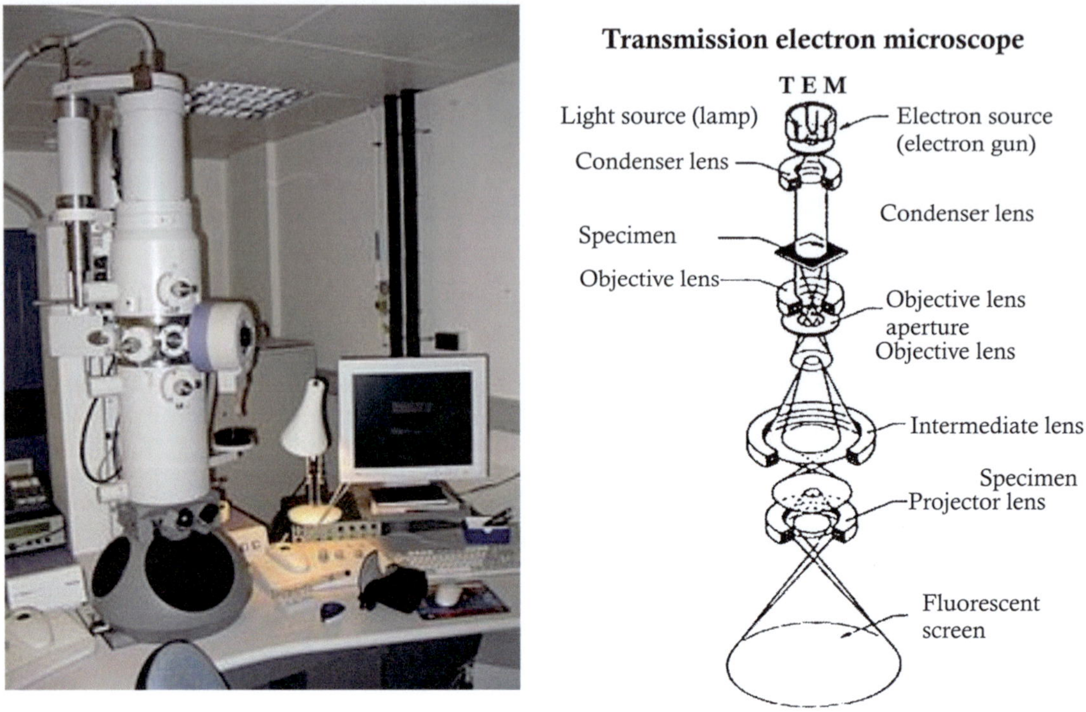

Fig. 23.1 Transmission electron microscope

otherwise, the only outline of the objective is visible. Methods are available to observe tiny particles and solid surfaces. Both particles and ultrathin sections are held in the electron beam on a plastic film thin enough to be transparent to the electrons. This film is held by a small wire mesh screen (grid) about 1/8″ in diameter. The grid is held in a movable holder for observation in the electron microscope (Griffin 1990; Milne 1984).

Knowledge of the size, shape, and any surface features of the virus particles is a basic requirement for virus identification. Electron microscopy can provide this information quickly and reliably with approximate particle dimensions together with morphology, and on this basis, a virus tentatively can be placed to a particular taxonomic group. A representative of different major kinds of particle morphology associated with plant viruses is presented in Fig. 23.2. It can be carried out on purified preparations or crude extracts of infected sap. Negative staining technique allows simple and rapid examination of virus-infected leaf material or purified preparations (Brenner and Horne 1959). It involves the mixing of virus preparation with a solution of electron dense stains such as sodium phosphotungstate (PTA) or uranyl acetate (UA) or ammonium molybdate at a concentration of about 2%. This mixture is placed on electron microscope grid that has been pre-coated with carbon or another support film. The grid is made up of copper, gold

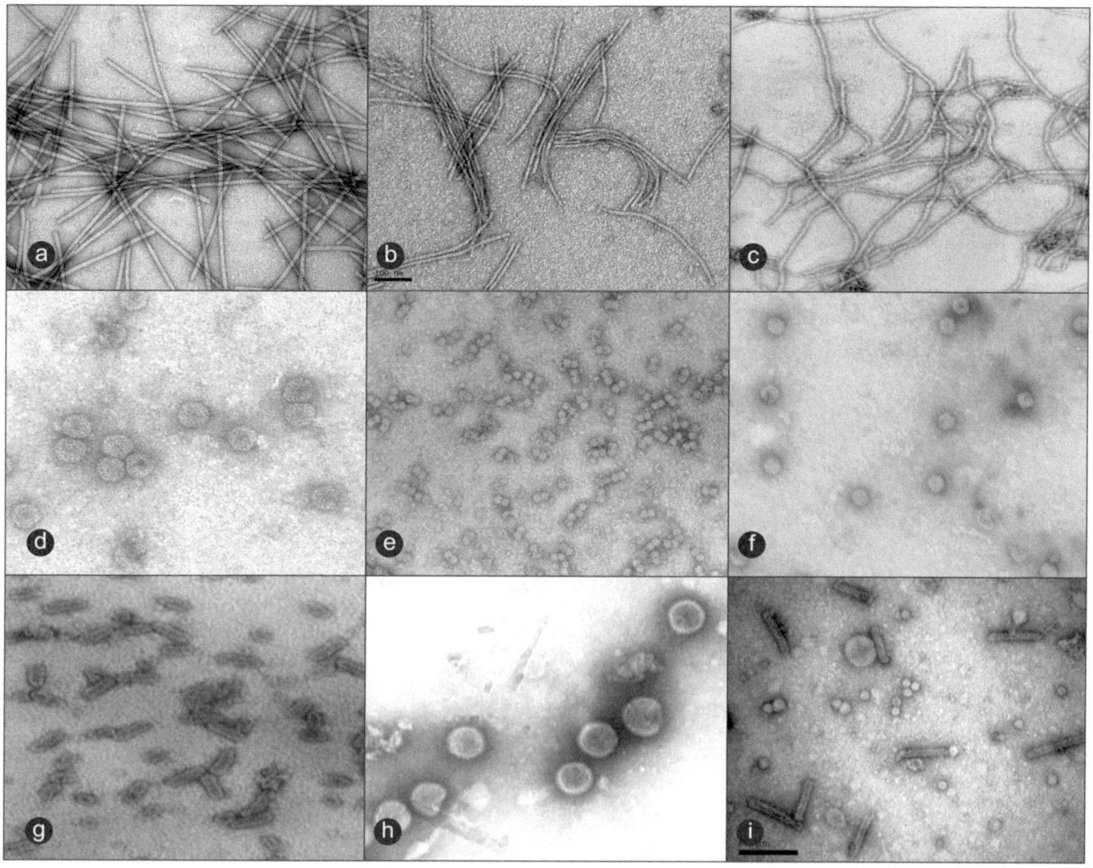

Fig. 23.2 Particle morphology of viruses. (**a**) Rigid rods, 300 × 18 nm, tobamovirus; (**b**) Flexuous rods, 450–550 nm, potexvirus; (**c**) Flexuous rods, potyvirus, 700–750 nm, (**d**) spherical (isometric), cucumovirus, 28–30 nm; (**e**) Geminate virus particles (20 × 30 nm); (**f**) isodiametric, carmovirus, 34 nm; (**g**) Spherical and bacilliform alfamovirus, 30, 35, 43, 56 × 18 nm; (**h**) Enveloped tospovirus, 80–120 nm; (**i**) Spherical and bacilliform rice tungro virus, 110–400 × 30–35 nm

or platinum. Excess liquid is removed with filter paper, and the mixture is allowed to dry. The grid is then examined under an electron microscope, where electrons pass through the object and hit a fluorescent screen, which is lighted up. The metal stains surround the object and penetrate only its interstices, thus making the object visible against a dark background. The technique for examining a virus in crude sap extracts by negative staining is referred to as ‘quick dip’ or ‘quick leaf dip’ (Kado and Agrawal 1972; Hill 1984; Roberts 1986; Matthews 1991).

A resolution of 1 nm or below can be obtained with an electron microscope. Most viruses can be seen at a magnification of ×10,000 to ×30,000, although magnification up to 200,000 may be required to study further finer details. Electron microscope is also used to determine virus particle dimensions. Particles obtained by leaf dip method could be used for determining the particle dimension such as size and shape as the morphology of purified preparations may be affected by the procedure used for virus purification. A minimum of 100 particles should be measured (either directly from an electron microscope and micrometer eyepiece or from a photomicrograph). Measurements of rod-shaped particles are given in the form of the histogram, and particle length is quoted as either modal length (a value that occurs most frequently) or the arithmetical mean (Milne 1984; Roberts 1986).

23.2 Leaf Dip Method

Brandes and Wetter (1959) developed leaf dip method for the quick detection of plant viruses in crude preparations and/or in infected plant materials. This method is used to have a quick look of the virus in the infected plant materials. In this technique, a drop of the infected sap obtained by squeezing the freshly cut surface of a virus-infected leaf or other portion of the plant is stained and examined through electron microscope.

23.2.1 Materials

1. Clean and dry tweezers with very fine tips.
2. Copper grids (400 mesh) with support film of carbon.
3. Electron microscope.
4. Filter paper.
5. Glass Petri dish.
6. Micropipette with disposable tips.
7. Fine forceps.
8. Camlin brush.
9. Parafilm.
10. Pasteur pipette.
11. Infected tissues or purified virus preparations.
12. Phosphate buffer 0.07 M phosphate buffer, pH 6.5: Prepare 0.07 M of disodium phosphate (Na_2HPO_4) by dissolving 11.41 g of Na_2HPO_4 in water and make up the volume to 1 L. Similarly prepare 0.07 M of sodium dihydrogen phosphate (NaH_2PO_4) by dissolving 4.2 g of NaH_2PO_4 in water and make up the volume to 500 mL. To Na_2HPO_4 solution, add NaH_2PO_4 solution slowly while stirring till desired pH of 6.5 is obtained.
13. Phosphotungstic acid (PTA): Potassium or sodium phosphotungstate is used at 1% or 2% aqueous solution, it is stable in light at room temperature, sometimes it is buffered with 0.1 M ammonium acetate or carbonate buffer.

14. Uranyl acetate (UA): A 1% or 2% aqueous solution is prepared and kept in a dark bottle as it precipitates in strong light, the pH could be near 4.2, it gives better resolution than the phosphotungstate.
15. Ammonium molybdate (AM): A 2% solution prepared in water and pH is adjusted to 6.0–8.0 with ammonia and stored at room temperature.

23.2.2 Method

1. With a razor blade chop up a small piece of virus-infected leaf on the microscope slide in a few drops of distilled water (purified virus preparation is used directly).
2. Transfer one drop of the suspension to a clean piece of parafilm.
3. With forceps, place a grid with its coated side upside down on the drop.
4. After 2 min, remove the grid from the drop and blot off the excess fluid with the tip of a filter paper.
5. Transfer a drop of stain, PTA, AM or UA to the parafilm and place the grid, upside down, on this drop.
6. After 15 s, pick up the grid with forceps, blot off the excess stain and allow the grid to dry.
7. Insert the grid into the electron microscope and search for viral particles. If typical virus particles are seen, take photograph and record the magnification factor.
8. For determining the size of virus particles, measure the length or diameter of about 100 virus particles.

23.3 Electron Microscopy of Ultrathin Sections

To perform electron microscopy of ultrathin sections, the tissues have to be fixed. The purpose of fixation is to kill the cells, preserving the structure of the whole cells as nearly as possible to the living state. An excellent fixative will rapidly coagulate or precipitate the cell contents while maintaining the morphology and structure of cell organelles preventing any further damage by staining, washing, dehydration and embedding of the tissues. A well-fixed material should resemble its fresh unfixed counterpart, and organelles should not be broken, displaced or distorted. Some of the commonly used fixatives include glutaraldehyde, acrolein, osmium tetroxide, formaldehyde, chrome-osmium and potassium permanganate. Different kinds of buffers are used for the preparation of fixatives and additives such as sucrose, and salts are added to improve preservation of certain organelles. Fixation time should be as short as possible and vary depending on the tissue and fixative used. Similarly, the temperature of fixation would also change. After fixing, the tissue is dehydrated (using acetone or ethanol) if a water-insoluble resin such as Epon or Araldite is used for embedding. Glass or diamond knives are used for sectioning. Sections should be as thin as possible. Stains such as uranyl acetate and lead citrate are normally used to stain the sections (Edwardson 1966; Milne 1984; Roberts 1986; Griffin 1990).

23.3.1 Materials

1. Same as in Subheading 23.2.1.
2. Copper grids (3 mm diameter with a mesh size of 100 μm) with support film of carbon.
3. Razor blade.
4. Scalpel handle.
5. Uranyl acetate (UA): 5% UA solution is made in double-distilled water and stored in an amber coloured bottle.
6. Lead citrate: Add half a pellet of NaOH to 10 mL of distilled water in a tightly screwed bottle, shake well and add 0.5 g of lead citrate. Shake well and seal it with parafilm and store at 4 °C.
7. Phosphate buffer (0.1 M phosphate buffer, pH 7.0): Prepare 0.1 M of dipotassium hydrogen phosphate (K_2HPO_4) solution by dissolving 17.4 g of K_2HPO_4 in water and make up the volume to 1 L. Also prepare 0.1 M of potassium dihydrogen phosphate (KH_2PO_4) by dissolving 6.8 g of KH_2PO_4 in water and make up the volume to 500 mL. To the prepared 0.1 M K_2HPO_4 solution, add KH_2PO_4 solution slowly while stirring till desired pH of 7.5 is obtained.
8. Glutaraldehyde (GA): It is commercially available at 25% or 70% stock solution, take 1 mL of 25% GA and add 9 mL of phosphate buffer, pH 7.0, make the solution in an amber-coloured bottle and store at 4 °C. As its fumes are toxic, the solution should be handled in fume hood.
9. Osmium tetroxide: It is available in ampoules containing 1 g or 0.1 g, the ampoule is scored using a diamond cutter, and ampoule is transferred to the big amber-coloured bottle. It is pounded using glass rod, and enough phosphate buffer (0.1 M, pH 7.0) is added to make 0.2% osmium tetraoxide solution. The bottle is sealed with parafilm and stored at 4 °C, it is toxic, and the solution should be done under the fume hood.
10. Acetone: Use analytical grade acetone.
11. Propylene oxide: Use analytical grade.
12. Resin: Many types of resins are available. Commonly used ones are Araldite CY 212 and Epoxy epon 812, which are highly viscous resins. Epoxy epon 812 is made by adding four components in a plastic beaker and mixing them thoroughly for

10 min, as it is highly expensive and can be used only once, only required amount should be made fresh:

(a) Methyl nadic anhydrite (MNA): 25 g (it is a softener).

(b) TAAB epon 812 resin: 50 g.

(c) Dodecenyl succinic anhydride (DDSA): 5 g (it is a hardener).

(d) 2,4,6-Tri (dimethyl aminomethyl) phenol DMP: 1.5 g (it is an accelerator).

13. The block can be made hard or soft by varying amount of softener or hardener. The resin is proven carcinogen so disposable gloves should be used while handling.
14. Ultramicrotome.

23.3.2 Method

Fixation

1. Wash the material (leaf, stem or root) to be fixed in distilled water and cut them into small size (about 1.5 × 3 mm) by using a sharp razor blade.
2. Transfer them to a vial containing 2.5% glutaraldehyde (GA) using fine forceps. Vacuum infiltrate the tissues till tissues sink. Keep the vials at 4 °C overnight (give more time if a harder tissue is used).
3. Wash the tissues with 0.1 M phosphate buffer, pH 7.0.
4. Immerse the tissues in 0.2% osmium tetroxide (OSO_4) in vials and keep them at 4 °C overnight.
5. Dehydrate the tissues at room temperature by washing with 25% acetone (two changes for 15 min each), 50% acetone (two changes for 15 min each), 70% acetone (keep tissues in a mixture of half vol of 70% acetone and half vol of 70% acetone saturated with uranyl acetate) at 4 °C overnight followed by 100% acetone (two changes for 15 min each).
6. Place tissues in propylene oxide to remove acetone for 5 min.

Infiltration and embedding

1. Immerse tissue in a mixture of one part resin and three parts propylene oxide at 40 °C for 1 h in a shaker.
2. Transfer to a mixture of one part of resin and one part of propylene oxide at 40 °C for 1 h on a shaker.
3. Transfer the tissues to a mixture of three parts of resin and 1 part of propylene oxide at 40 °C for 1 h on a shaker.
4. Transfer the tissues to the pure resin and give three changes at 1 h interval at 40 °C on a shaker.

5. Embedding of tissue is done in pure resin using flat silicon moulds. Moulds are just filled with resin and tissue is placed at one end either in a transverse or longitudinal orientation. The resin is then polymerized at 60 °C overnight. Moulds are taken out and allowed to cool on the bench. After cooling blocks are taken out of moulds. They will be ready for sectioning after 24 h.

Sectioning

1. Trim the block with a sharp razor blade to a pyramid shape. Cut sections (0.5–2 μm thick) using a razor blade.
2. Check the sections under an optical microscope to know the state of embedded specimen, fixation and to select the area for ultrathin sectioning.
3. Select the area to be examined under the electron microscope and trim the block further if needed.
4. After final trimming, make a fresh face of the block. It should be smooth and shiny.
5. Fix glass or diamond knife at the specified place and adjust its height and clearance angle as per the specifications.
6. Lock at the orientation head containing block in the well in the ultramicrotome. Observe the knife and block through the microscope. The reflection of the knife edge in the block face should be completely parallel and horizontal with the real knife edge.
7. Based on the instruction of the ultramicrotome make and model, the cutting speed is set, and sections are cut. The

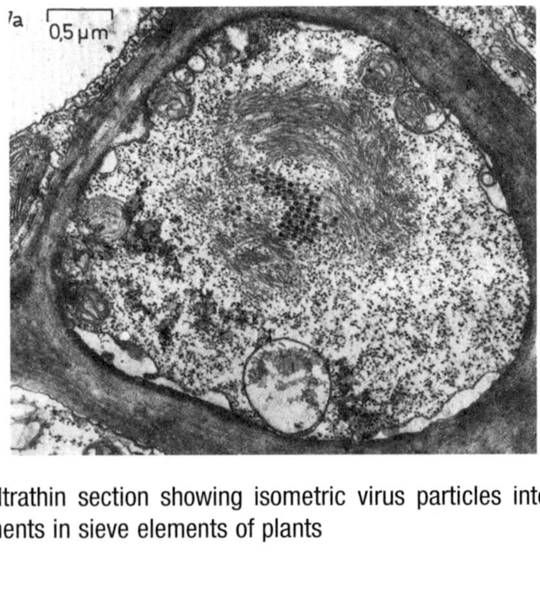

Fig. 23.3 Ultrathin section showing isometric virus particles intermixed with protein filaments in sieve elements of plants

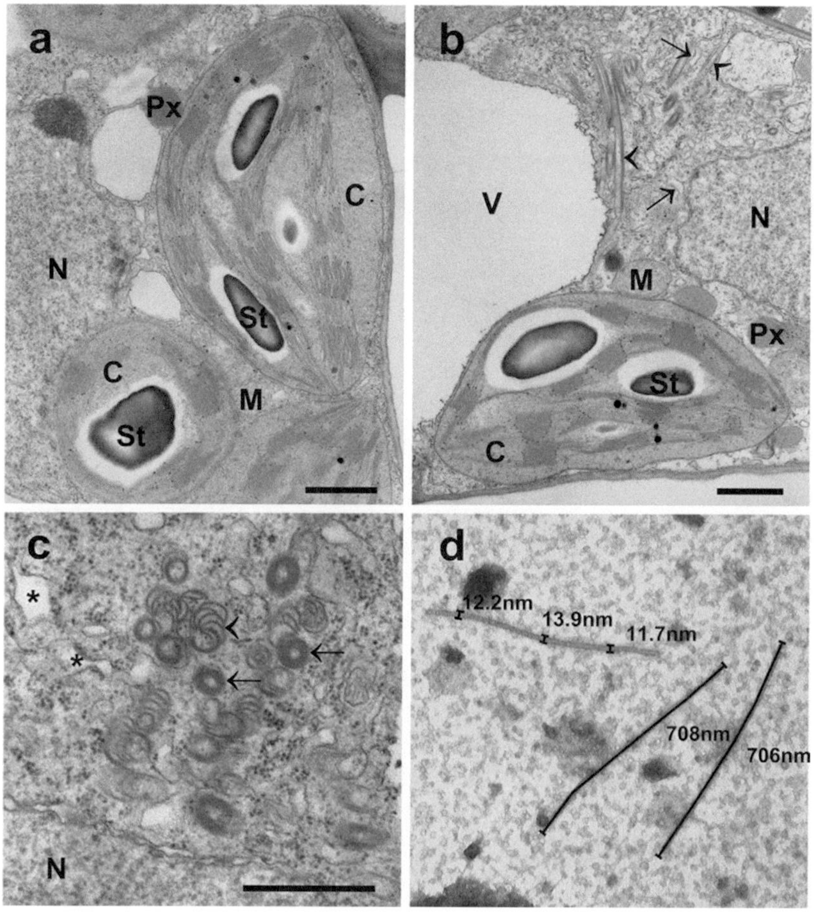

Fig. 23.4 Transmission electron micrographs showing the ultrastructure of control (**a**) and ZYMV-infected (**b**, **c**) mesophyll leaf cells of *Cucurbita pepo* after sample preparation using microwave irradiation. Virions (**d**) were detected by negative staining in the sap of infected leaves. Images a–c show parts of mesophyll cells with chloroplasts (C) containing starch (St), mitochondria (M), nuclei (N), peroxisomes (Px) and vacuoles (V). Note cylindrical inclusions induced by ZYMV in the cytosol of infected leaf cells (**b** and **c**) which appear as long tubular structures (arrowheads in **b**) or scroll-like structures (arrows in **c**) and pinwheels (arrowheads in **c**). Dilated ER as observed commonly in ZYMV-infected samples (arrows in **b** and stars in **c**). Virions (**d**) had an average length of 707 nm and an average width of 12 nm. Bars = 1 m in (**a**) and (**b**) and 0.5 m in (**c**). (Reproduced from Zechmann and Zellnig 2009)

following colours for the sections in the knife trough will be seen:

a. Grey: 60 nm thick (sections are useful for high-resolution work).

b. Silver: 60–90 nm thick (sections are useful for seeing ultrastructures under electron microscope.

c. Gold: 90–150 nm thick (sections are useful for viewing under low magnification for pathogens like phytoplasmas).

8. The sections are manipulated into position using mounted eyelash and then transferred to the grid by either carefully scooping them from below or touching from above. If grids are of large mesh size (100 or 200), then a coating is done using 0.6% formavar (polyvinyl formaldehyde) in ethylene dichloride.

Staining

1. Stain the grids using uranyl acetate and lead citrate.
2. To stain with uranyl acetate, immerse the grids on a drop of this stain on parafilm for 10 min followed by washing with 15–20 drops of distilled water. Grid is drained off excess water.
3. To stain with lead citrate, take 10 μL drop of lead citrate solution and place it on the parafilm-covered with a Petri dish. Four to five pellets of NaOH is kept inside the Petri dish. NaOH keeps the air free of CO_2. CO_2 reacts with the lead citrate to form electron-dense globules of lead carbonate. Grid is floated on this drop for 5 min and rinsed with ten drops of distilled water. Grid is dried by using Whatman paper strip before viewing through the electron microscope (Figs. 23.3 and 23.4).

23.4 Notes (Hill 1984; Milne 1984; Roberts 1986)

1. Always use fresh tissues for electron microscopy.
2. As far as possible, select soft tender tissues as they are easy to fix and sectioning.
3. The chemicals used for fixing, embedding and staining are bio-hazardous, and hence, proper care should be taken while handling them.
4. Experiments should be carried out in dust-free environment.
5. Temperature maintenance is necessary throughout the experiments.

References

Brandes J, Wetter C (1959) Classification of elongated plant viruses on the basis of particle morphology. Virology 8(1):99–115

Brenner S, Horne RW (1959) A negative staining method for high resolution electron microscopy of viruses. Biochim Biophys Acta 34:103–110

Edwardson JR (1966) Electron microscopy of cytoplasmic inclusions in cell infected with rod-shaped viruses. Am J Bot 53:359–364

Griffin RL (1990) Using transmission electron microscope in the biological sciences. Ellis Harwood, New York

Hill SA (1984) Methods in plant virology. Blackwell Scientific, Oxford

Kado CI, Agrawal HO (1972) Principles and technique in plant virology. Van Nostrand Reinhold, New York

Matthews REF (1991) Plant virology, 3rd edn. Academic Press, New York

Milne RG (1984) Electron microscopy for the identification of plant viruses in in-vitro preparation. Methods Virol 7:87–120

Roberts IM (1986) Practical aspects of handling, preparing and staining samples containing plant virus particles for electron microscopy. In: Jones RAC, Torrance L (eds) Developments and applications in virus testing, development and applied biology-I. Association of Applied Biologists, Wells Bourne

Zechmann B, Zellnig G (2009) Rapid diagnosis of plant virus diseases by transmission electron microscopy. J Virol Methods 162:163–169

Chapter 24

Determination of Coat Protein Molecular Weight of Viruses

Abstract

Molecular weight (mw) and number of polypeptides in the coat protein is one of the criteria used for the identification of virus species. The determination of molecular weight of the viral coat protein involves denaturing of protein subunits in the presence of sodium dodecyl sulphate (SDS) by heating in boiling water. The denatured protein is further run on an SDS-poly acrylamide gel electrophoresis (SDS-PAGE) along with protein markers of known molecular weight. The molecular weight of the unknown virus protein is estimated by comparing its mobility in the gel, with that of other proteins of known molecular weight.

Key words SDS-PAGE, Resolving gel, Staking gel, Vertical gel electrophoresis, Purified virus

24.1 Introduction

Molecular weight and number of polypeptides that a virus particle contains provide useful information in identifying the genus to which a new virus may belong. Of the several methods available, SDS-PAGE is the most routinely used technique for determining the molecular weight of coat protein. In this process, the virus is usually dissociated by SDS treatment followed by heating in boiling water for 3–5 min. The denatured proteins become linear molecules and form SDS-polypeptide complexes with net negative charge. Most polypeptides bind a fixed amount of SDS per amino acid. As a result, denatured proteins acquire a fixed charge to mass ratio and differ only in their size. Thus in SDS-PAGE, migration of denatured proteins is determined by their molecular weight. Since they carry net negative charges at neutral pH, the SDS-polypeptide complexes migrate towards the anode. After electrophoresis is complete, the gel is removed from the apparatus and stained to locate polypeptide bands. Molecular weight of the virus coat protein is then determined from the standard curve prepared based on the mobility of different proteins of known molecular weight (Laemmli 1970; Sambrook and Russel 2001).

Alangar Ishwara Bhat and Govind Pratap Rao, *Characterization of Plant Viruses*, Springer Protocols Handbooks, https://doi.org/10.1007/978-1-0716-0334-5_24,

24.2 Materials

1. SDS-PAGE resolving and stacking gel composition:

Solution Component	Resolving gel (mL) 8%	10%	12%	15%	Stacking gel (mL) 5%
Water	4.6	4.0	3.3	2.3	3.4
30% acrylamide mix	2.7	3.3	4.0	5.0	0.83
1.5 M Tris, pH 8.8	2.5	2.5	2.5	2.5	0.63[a]
10% SDS	0.1	0.1	0.1	0.1	0.05
10% ammonium persulphate (APS)	0.1	0.1	0.1	0.1	0.05
TEMED	0.006	0.004	0.004	0.004	0.005
Total volume (mL)	10.000	10.000	10.000	10.000	5.000

[a]Use 0.5 M Tris, pH 6.8

2. 30% acrylamide mix: It is prepared by dissolving 30 g acrylamide and 0.8 g *N*,*N*-methylene bisacrylamide in distilled water, mixture filtered through Whatman No. 1 paper, volume made up to 100 mL and stored in amber coloured bottle.
3. Sodium dodecyl sulphate (SDS): A 10% stock solution prepared in distilled water and stored at room temperature.
4. Ammonium persulphate (APS): A freshly prepared 10% stock in distilled water is used.
5. TEMED: Use electrophoresis grade TEMED.
6. Resolving gel buffer (1.5 M Tris–HCl, pH 8.8): Dissolve 18.15 g of Tris base in about 80 mL water, adjust the pH 8.8 with HCl and make up volume to 100 mL.
7. Stacking gel buffer (0.5 M Tris–HCl, pH 6.8): Dissolve 6.09 g of Tris base in water, adjust the pH to 6.8 with HCl and make up volume to 100 mL.
8. Tris-glycine electrophoresis running buffer (pH 8.3) (5×): Dissolve 25 mM Tris base, 250 mM glycine, 0.1% SDS in water, adjust the pH to 8.3 with HCl and make the volume to 1000 mL.
9. Sample buffer: Distilled water (4.0 mL), 0.5 M Tris–HCl, pH 6.8 (1.0 mL), glycerol (0.8 mL), 10% SDS (1.6 mL), 2 mercaptoethanol (0.4 mL), bromophenol blue (0.2 mL), mix all components.
10. Protein sample preparation: Dissolve the sample (virus) in sample buffer at 1:1 (v/v) and heat in boiling water for 3 min.

11. Protein markers: Standard protein molecular weight marker commercially available from different manufacturers may be used.
12. Staining solution: Dissolve 100 mg of Coomassie brilliant blue in about 50 mL of methanol and then add 10 mL of acetic acid and make up the volume using distilled water to 100 mL.
13. Destaining solution: Mix 40 mL of methanol and 10 mL of acetic acid, and make up the volume to 100 mL with distilled water.
14. Glass tray for staining.
15. Purified virus preparations.
16. Vertical gel apparatus and its accessories.

24.3 Method

1. Assemble the glass plates as per manufacturer's instruction.
2. Prepare an appropriate volume of resolving gel containing desired concentration of acrylamide (12%).
3. Pour acrylamide solution into the gap between the glass plates and leave sufficient space for stacking gel. Place the gel in a vertical position at room temperature for polymerization.
4. After polymerization is complete, pour off the overlay and wash the top of the gel with water to remove any unpolymerized acrylamide.
5. Prepare an appropriate volume of stacking gel (5%) and pour directly onto the surface of the polymerized resolving gel.
6. Immediately insert the comb into the stacking gel solution (avoid air bubbles). Add more stacking gel solution if required to fill the spaces of the comb completely. Allow the gel to polymerize at room temperature.
7. After polymerization is complete, remove the comb and wash the wells with water to remove any unpolymerized acrylamide.
8. Mount the gel in the electrophoresis apparatus and add Tris-glycine electrophoresis running buffer to the top and bottom tanks.
9. Prepare the samples (purified virus preparations) by heating them to 100 °C for 3 min in sample buffer (1:1) to denature the proteins. Load (up to 20–30 μL) each of the samples in a pre-decided order into the bottom of the wells. Also, load protein marker standard in one of the wells.
10. Attach the electrophoresis apparatus to an electric power supply and run the gel at a constant voltage (100 V) till

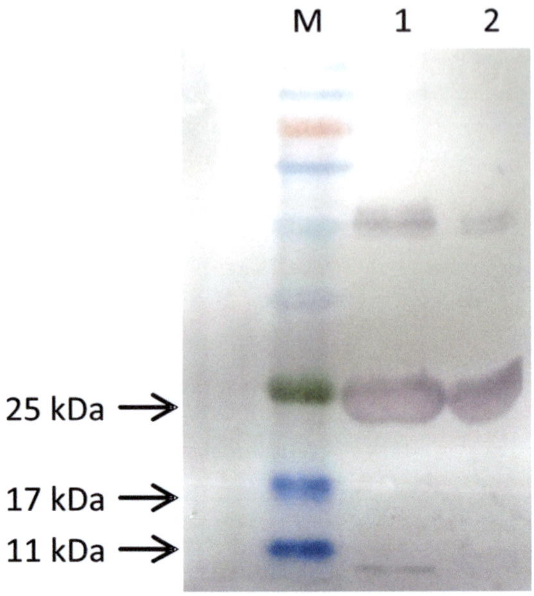

Fig. 24.1 Determination of molecular weight of virus through SDS-PAGE. Relative mobility of marker proteins is plotted against their molecular weight on a logarithm scale graph. Molecular weight of virus is then estimated by comparing its relative mobility with the marker proteins. Lane M; Molecular weight markers; Lane 1–2: purified virus preparations

bromophenol blue (tracking dye) reaches the bottom of the resolving gel.

11. Remove gel from the apparatus and soak overnight in staining solution.
12. Destain the gel by soaking in destaining solution till the background becomes transparent at room temperature. Photograph the gel if necessary by keeping on a light box (Fig. 24.1).
13. Record the distance migrated by the dye front, individual marker proteins and viral protein.
14. Calculate the relative mobility of each protein using the formula:

    Relative mobility = Distance migrated by protein/Distance migrated by the dye front.
15. Plot the mobilities of marker proteins against their molecular weight, expressed on a semi-logarithmic scale.
16. Estimate the molecular weight of the viral coat protein using its mobility (Fig. 24.1).

24.4 Notes (Laemmli 1970; Sambrook and Russel 2001)

1. Acrylamide and bisacrylamide are potent neurotoxins, it is always suggested to wear gloves and mask during their handling.
2. The prepared acrylamide solution can be stored up to 1 month in an amber coloured bottle at 4 °C.
3. Vertical gel apparatus of different dimensions from different firms are available. Depending on the requirement such as resolution and number of samples, one can choose mini, midi or large apparatus.
4. The acrylamide gel volume mix required would vary with the apparatus used.
5. Before pouring the gel, it is always important to check for any leak in the set up by pouring water. If leak is noticed, rectify the same before pouring gel mix.
6. While preparing acrylamide gel mix, ammonium persulphate should be added at the end. The polymerization starts immediately after the addition of ammonium persulphate.
7. Once the acrylamide gel mix is prepared, pour it immediately.
8. After pouring the resolving gel mix, overlay with a small amount of water to get a uniform surface of the gel. Pour off the water after polymerization is complete before adding the stacking gel mix.
9. Polymerization process is quicker at higher temperature.
10. Stacking gel helps to concentrate the proteins while actual separation of the proteins takes place in the resolving gel. Hence, it is always better to have more area for resolving gel (at least two thirds of the gel).
11. The comb should be inserted immediately after pouring the stacking gel mix to avoid air bubbles.
12. Remove the comb from the stacking gel after polymerization carefully without disturbing wells.
13. Wash each well with buffer to remove any unpolymerized acrylamide solution, fill upper and lower tank with running buffer and give a pre-run for about 10–20 min before loading protein samples.
14. Choose appropriate protein markers—low, medium or high range depending on the expected molecular weight of test protein.
15. After the run, cut and remove stacking gel with the help of a spacer, give a slanting cut at the left top of the resolving gel (to know the orientation) and put for staining.

16. Staining solution can be re-used for at least staining another 2–3 gels. Destaining of gels is done till the background is clear, so that the protein bands can be seen clearly.

References

Laemmli UK (1970) Cleavage of structural proteins during the assembly of the head of bacteriophage T4. Nature 227:680–685

Sambrook J, Russel DW (2001) Molecular cloning, vol I-III, 3rd edn. Cold Spring Harbor Laboratory Press, New York

Chapter 25

Isolation of Nucleic Acid from Purified Virus and Determination of Its Nature

Abstract

Unlike higher organisms, genome of viruses is made up of DNA or RNA, which further may be either single or double stranded. The first step in the isolation of nucleic acid from purified virus preparation involves removal of coat protein of the virus particle followed by precipitation of the nucleic acid by ethanol in the presence of a salt. Yield of nucleic acid is determined using spectrophotometer, and quality can be judged by gel electrophoresis. The nature of virus genome is determined by treating the isolated nucleic acid with various enzymes followed by electrophoresis.

Key words Viral DNA, Viral RNA, Yield, Type of nucleic acid

25.1 Introduction

Most plant viruses have single-stranded (ss) RNA genomes enclosed in a protective protein made up of many copies of one or few protein molecules. Some viruses have double-stranded (ds) RNA genomes while others have ss DNA or ds DNA. The separation of the viral genome is needed to study viruses at the genomic level that would help in the identification, molecular characterization, and development of sensitive diagnostics. The genome can be cloned, sequenced, expressed and used to transform plants to get virus-resistant transgenic plants. Many methods are available to separate protein coat from the virus particles so as to isolate the nucleic acid in infectious state. In general, the method involves incubation of purified virus preparation with mixture of proteases (called as pronase) to digest the protein. The digested protein is then separated by phenol-chloroform extraction, and the nucleic acid is precipitated by ethanol. Many methods are available for the isolation of nucleic acid from different viruses. Two procedures used for the isolation of nucleic acid are described herein. Users may refer to the articles listed in the references for the other protocols (Brakke and van Pelt 1970; Brisco et al. 1985; Yeh and Gonsalves 1985; Hewings et al. 1986; Wu and Su 1990).

Alangar Ishwara Bhat and Govind Pratap Rao, *Characterization of Plant Viruses*, Springer Protocols Handbooks,
https://doi.org/10.1007/978-1-0716-0334-5_25, © Springer Science+Business Media, LLC, part of Springer Nature 2020

Correct measurement of quality and quantity of the nucleic acid is important for several applications such as cDNA synthesis, restriction digestion and amplification by polymerase chain reaction. Quantity and quality of isolated nucleic acid is determined by spectrophotometric measurements or by agarose gel electrophoresis. Absorption spectroscopy is used to measure the amount of DNA and RNA in the solution. It is a rapid, simple and non-destructive method. The nitrogenous bases the have an absorption maximum at about 260 nm. The yield of the nucleic acid is determined by spectrophotometer by taking optical density (OD) readings at 260 nm. An OD of 1 is considered as 50 μg/mL of ds DNA or 40 μg/mL of ss-DNA and RNA. The purity of nucleic acid is determined based on the ratio between the OD readings at 260 and 280 nm. $A_{260/280}$ values of 1.8 and 2.0 indicate contamination-free preparations of DNA and RNA respectively while, values less than these indicate contamination with protein or phenol (Brown 1980; Burrin 1986; Manchester 1995).

About 90% of known plant viruses have single-stranded (ss) RNA as their genome, and 4% are known to have double-stranded (ds) RNA as their genome while another 4% and 2% of viruses contain ssDNA and dsDNA as their genome. Thus, it is important to know the nature of viral genome whenever new viruses are discovered. To determine the nature of viral genome, its nucleic acid is isolated from purified viral preparations. The nucleic acids are then treated with various modifying enzymes such as RNase A, DNase I, S1 nuclease and run on the gel. S1 Nuclease, which is more active on DNA than on RNA cuts only single-stranded nucleic acids releasing 5′-phosphoryl mono- or oligonucleotides. DNase I non-specifically cuts DNA to release 5′-phosphorylated di-, tri-, and oligonucleotide products. RNase A specifically degrades ssRNA 3′ pyrimidine residues and degrades the phosphate linkage to the adjacent nucleotide.

25.2 Isolation of Nucleic Acid from Purified Virus Preparation

25.2.1 Materials

25.2.1.1 Method I

1. Chloroform: Use DNase, RNase and proteinase-free chloroform.
2. Ethanol.
3. Phenol: chloroform: isoamyl alcohol (25:24:1): Mix equal volume of equilibrated phenol and chloroform: isoamylalcohol (24:1) and store under 0.1 M Tris–HCl (pH 8.0) in amber coloured bottle at 4 °C.
4. Proteinase K (commercially available product).
5. Purified viral preparation.
6. RNase-free water.

7. Sodium acetate (3 M, pH 5.2): Prepare by dissolving sodium acetate (408.1 g) in about 800 mL of water, add glacial acetic acid to adjust the pH to 5.2 before making up the volume to 1 L with water and sterilize by autoclaving.
8. Sodium dodecyl sulphate (SDS) (10%): Dissolve 10 g of SDS in 90 mL of water, heated to dissolve and adjust the pH to 7.2 with HCl, volume made up to 100 mL and dispense into aliquots.
9. Tris-saturated phenol: Melt liquefied phenol stored at −20 °C at 65 °C and add hydroxyquinone to a final concentration of 0.1%. To this add equal volume of buffer (0.5 M Tris–HCl, pH 8.0) at room temperature, stir for 15 min and allow standing for separation of phases. Aspirate upper aqueous phase using a glass pipette (avoid mouth sucking) and discard. To the phenol, add an equal volume of 0.1 M Tris–HCl pH 8.0, stir for 15 min, turn off the stirrer and collect the upper aqueous phase. Repeat extraction until the pH of the phenol phase is >7.8. After final extractions, add 0.1 volume of 0.1 M Tris–HCl (pH 8.0) that contain 0.2% β mercaptoethanol. Store the solution in an amber-coloured bottle at 4 °C.
10. Tris–HCl (pH 9.0) (1 M): Prepare by dissolving 121 g of Tris base in 800 mL of water, add concentrated HCl to bring down the pH to 9.0; make up the volume to 1 L and sterilize by autoclaving.
11. Waterbath.

25.2.1.2 Method II

1. Same as in subheading 25.2.1.1.
2. EDTA (0.5 M, pH 8.0): Weigh and dissolve 186.1 g of ethylene diamine tetra acetic acid-disodium salt in about 800 mL of water, stir vigorously and adjust the pH to 8.0 with NaOH and sterilize by autoclaving.
3. Phenol: chloroform (1:1): Prepared by mixing equal volume of equilibrated phenol and chloroform; store in 0.1 M Tris–HCl (pH 8.0) in amber-coloured bottle at 4 °C.
4. Sodium phosphate buffer (0.07 M, pH 7.2): Add 12.18 g di-potassium hydrogen orthophosphate (K_2HPO_4) in 1 L distilled water (0.1 M solution). To this, add slowly 0.07 M solution of potassium dihydrogen orthophosphate (KH_2PO_4) prepared by dissolving 2.38 g of KH_2PO_4 in 250 mL distilled water till the desired pH of 7.0 is obtained.

25.2.2 Methods

25.2.2.1 Method I

1. Incubate purified viral preparation (in 400 μL) with proteinase K (50 μg/mL) and 0.1% SDS at 37 °C for 30 min.
2. Add an equal volume of Tris buffer-saturated phenol and mix well by inverting the tube gently on ice for 15 min.
3. Centrifuge at 8000 × *g* for 10 min at 15 °C.
4. Aspirate aqueous phase, add an equal volume of phenol : chloroform:isoamyl alcohol (25:24:1) and mix by inverting the tube for 5 min at room temperature.
5. Centrifuge at 8000 × *g* for 5 min at 15 °C.
6. Collect aqueous phase, add three volumes of chloroform and mix by inverting the tube for 2 min.
7. Repeat centrifugation as in **step 3**.
8. Transfer the supernatant to a fresh tube and add 0.1 volume of 3 M sodium acetate, pH 5.2 and two volumes of 100% ethanol.
9. Incubate at −70 °C for 1 h or overnight at −20 °C.
10. Centrifuge at 15,000 × *g* for 20 min at 4 °C.
11. Decant the supernatant, wash the pellet by adding 70% ethanol and centrifuging for 2 min at maximum speed.
12. Decant the supernatant and air dry the pellet.
12. Resuspend the pellet in RNase-free sterile water and freeze at −70 °C immediately.

25.2.2.2 Method II

1. Mix purified viral preparation (in 0.07 M sodium phosphate buffer pH 7.2) with same volume of 0.02 M Tris–HCl, pH 9.0 containing 0.001 M EDTA and 4% SDS.
2. Keep at 60 °C for 15 min.
3. Add an equal volume of phenol : chloroform (1:1) and mix well.
4. Centrifuge at 8000 × *g* for 10 min.
5. Collect supernatant and add with an equal volume of chloroform.
6. Repeat centrifugation as in **step 4**.
7. Collect supernatant; add 0.1 volume of 3 M sodium acetate, pH 5.2 and three volumes of ethanol.
8. Keep the tube overnight at −20 °C.
9. Centrifuge at 16,000 × *g* for 20 min at 4 °C.
10. Throw the supernatant and add 1 mL of 70% ethanol to the pellet, centrifuge at maximum speed for 2 min.
11. Discard supernatant, air dry the pellet, resuspend in RNase-free sterile water and store at −70 °C.

25.3 Quantification of Nucleic Acid

25.3.1 Materials

1. Spectrophotometer.
2. Diluent buffer.
3. Purified nucleic acid.

25.3.2 Method

1. Calibrate the spectrophotometer at 260 nm and at 280 nm wavelength using 995 μL diluent buffer in a cuvette.
2. Add 2–5 μL of appropriately diluted nucleic acid (DNA or RNA) isolated from the above method to the cuvette, mix thoroughly and take the optical density (OD) readings at both 260 and 280 nm.
3. Calculate the DNA or RNA concentration using the following formula.

$$\text{DNA concentration}(\mu g/\mu L) = (OD_{260}) \times \text{dilution factor} \times (50\,\mu g\ \text{DNA}/\mu L)/(1\,OD_{260}\text{unit}).$$

$$\text{RNA concentration}(\mu g/\mu L) = (OD_{260}) \times \text{dilution factor} \times (40\,\mu g\ \text{RNA}/\mu L)/(1\,OD_{260}\text{unit}).$$

4. Assess the quality of the nucleic acid from the ratio of the OD values recorded at 260 and 280 nm.
   (a) An A_{260}/A_{280} value of 2.0 is considered as pure RNA.
   (b) An A_{260}/A_{280} value of 1.85–1.95 is considered as pure DNA.
   (c) An A_{260}/A_{280} value of 0.6 is considered as pure protein.

25.4 Determination of the Nature of Viral Genome

25.4.1 Materials

1. Agarose.
2. Horizontal agarose gel apparatus with power pack.
3. Transilluminator and gel documentation unit.
4. Water bath.
5. 10× DNase I buffer: 50 mM Tris–HCl, 10 mM $MgCl_2$, pH 8.0
6. 10× S1 nuclease buffer: 0.2 M NaCl, 50 mM sodium acetate, pH 8.0, 1 mM $ZnSO_4$
7. Ethidium bromide: Weigh and dissolve 1 g of ethidium bromide to 100 mL of water, stir vigorously and transfer to dark bottle and store at room temperature.
8. Loading dye.

9. Modifying enzymes (RNase A, DNase I and S1 nuclease): Commercially available.
10. Purified nucleic acid samples.
11. Sterile distilled water.

25.4.2 Method

1. Set up the digestion reaction in an Eppendorf tube as follows
2. **For DNase I**

(a) Viral nucleic acid (1 μg)	10.0 μL
(b) 10× DNase I buffer	2.0 μL
(c) DNase (7 units/μL)	1.0 μL
(d) Sterile water	7.0 μL
Total	20.0 μL

3. **For RNase I**

(a) Viral nucleic acid (1 μg)	10.0 μL
(b) RNase A (10 mg/mL)	1.0 μL
(c) Sterile water	9.0 μL
Total	20.0 μL

4. **For S1 nuclease treatment**

(a) Viral nucleic acid (1 μg)	10.0 μL
(b) 10× S1 nuclease buffer	2.0 μL
(c) S1 nuclease (10 units/μL)	1.0 μL
(d) Sterile water	7.0 μL
Total	20.0 μL

5. Incubate all reactions at 37 °C for 1 h.
6. Keep appropriate controls for each of the modifying enzymes used.
7. Stop the reactions by adding 2 μL of gel loading dye to each of the tubes.
8. Electrophorase the reactions in 1% agarose gel in 1× TAE and visualize the bands by ethidium bromide staining (details of electrophoresis is provided in the Chapter 26).
   (a) Viruses containing ssRNA genome do not show any bands in RNase I and S1 nuclease-treated nucleic acids. Bands of expected size are seen only in DNase I treatment.

(b) Viruses containing dsRNA genome show bands at expected positions in DNase I, RNase I and S1 nuclease treatments.

(c) Viruses containing ssDNA genome do not show bands in S1 nuclease and DNase I treatments while bands at expected size would be seen in RNase I treatments.

(d) Viruses containing dsDNA genome do not show bands in DNase I treatment while bands of expected size would be observed in RNase I and S1 nuclease treatments.

25.5 Notes (Burrin 1986; Manchester 1995)

1. Pronase consists of mixture of enzymes that are isolated from *Streptomyces griseus*. It can act on both denatured and native proteins leading to complete digestion of proteins into individual amino acids.
2. Spectrophotometer method is less sensitive as most spectrophotometers need a concentration of at least 1 μg/mL of nucleic acid to get reliable estimates at A_{260}. However, nanophotometer is highly sensitive and can be used to quantify nucleic acid even in a very minute quantity.
3. Most currently available instruments can directly determine the concentration of both DNA and RNA.
4. Always calibrate the spectrophotometer using the same solution used for diluting the nucleic acid.
5. Nucleic acids isolated from some plant samples may contain contaminants such as polysaccharides and other metabolites that can interfere with OD readings. If such contaminations are suspected in the isolated nucleic acids, recording absorbance from 220 to 320 nm is recommended as this will indicate whether contaminants are interfering with absorbance at 260 and 280 nm.
6. Ensure that the cuvette used for taking spectral readings is RNase-free. In order to achieve this, wash cuvettes with 0.1 M NaOH, 1 mM EDTA followed by washing with RNase-free water.
7. If estimating concentration through gel electrophoresis, load different known concentrations of nucleic acid (standards) and estimate the concentration of test nucleic acid by comparing intensity of the band with standards.

References

Brakke MK, van Pelt N (1970) Properties of infectious ribonucleic acid from wheat streak mosaic virus. Virology 42:699–706

Brisco MJ, Hull R, Wilson TMA (1985) The effect of extraction protocol on yield, purity, and translation products of RNA from an isometric plant virus. J Virol Methods 10:195–202

Brown SB (1980) An introduction to spectroscopy for biochemists. Academic Press, London

Burrin DH (1986) Spectroscopic techniques. In: Wilson K, Goulding KH (eds) A biologist's guide to principles and techniques of practical biochemistry. Edward Arnold, London

Hewings AD, Damssteegt DD, Tolin SA (1986) Purification and some properties of two strains of soybean dwarf virus. Phytopathology 76:759–763

Manchester KL (1995) Value of A260/A280 ratios for measurement of purity of nucleic acids. BioTechniques 19:208–210

Wu RY, Su HJ (1990) Purification and characterization of banana bunchy top virus. J Phytopathol 128:153–160

Yeh SD, Gonsalves D (1985) Translation of papaya ringspot virus RNA in vitro: detection of a possible polyprotein that is processed for capsid protein, cylindrical inclusion protein, and amorphous-inclusion protein. Virology 143:260–270

Chapter 26

Agarose Gel Electrophoresis for Nucleic Acids

Abstract

Agarose gel electrophoresis is the easiest, most popular and effective way of separating and analysing nucleic acid fragments to assess the quality and quantity of DNA and RNA on the basis of charge by applying an electric field to the electrophoretic apparatus. The DNA and RNA can be visualized in the gel by the addition of ethidium bromide or other dyes under UV light. The use of agarose gel electrophoresis revolutionized the separation of nucleic acids of different sizes. The rate of migration of a nucleic acid molecule through a gel is determined by the size, agarose concentration, nucleic acid conformation, voltage applied and electrophoresis buffer. The different steps involved in the process of agarose gel electrophoresis are discussed in detail in this chapter.

Key words Agarose, Horizontal gel electrophoresis, Integrity of nucleic acid, Quality of nucleic acid

26.1 Introduction

Agarose [(1,4)-3,6-anhydro-α-L-galactopyranosyl-(1,3)-β-D-galactopyranan], is a highly purified form of the polysaccharide extracted from the seaweed genera, *Gelidium* and *Gracilaria*, and consists of repeated agarobiose (L- and D-galactose) subunits. During gelation, agarose polymers associate non-covalently and form a network of bundles whose pore sizes determine a gel's molecular sieving properties. For separation of nucleic acids using agarose gel electrophoresis, the nucleic acid is loaded into pre-cast wells in the gel, and an electric current is applied. The phosphate backbone of the DNA (and RNA) molecule is negatively charged; therefore, when placed in an electric field, nucleic acid fragments migrate towards the positively charged anode. Because the nucleic acid (DNA and RNA) has a uniform mass/charge ratio, these molecules are separated by size within an agarose gel in a pattern such that the distance travelled is inversely proportional to the log of its molecular weight (Simpson and Whittaker 1983; Sambrook and Russel 2001).

Alangar Ishwara Bhat and Govind Pratap Rao, *Characterization of Plant Viruses*, Springer Protocols Handbooks,
https://doi.org/10.1007/978-1-0716-0334-5_26,

26.2 Agarose Gel Electrophoresis for DNA

The agarose gel is prepared and immersed in the buffer. The DNA fragments are loaded onto a well at one end of the gel and made to move through the gel by the application of electric current. DNA is negatively charged and so will move towards the anode. The rate of movement of DNA in the gel depends on (1) molecular weight of the DNA (higher molecular weight DNA move slowly compared to low molecular weight DNA), (2) agarose concentration (migration rate is inversely proportional to the agarose concentration), (3) conformation of the DNA (linear form travels slowest, and the supercoiled form travels fastest) and (4) voltage applied. The DNA is visualized by adding ethidium bromide (EtBr), a fluorescent molecule that intercalate with the DNA bases, extending the length of linear and nicked circular DNA molecules and making them more rigid. When EtBr is added, UV radiation at 254 nm is absorbed by the DNA and transmitted to the bound dye. The energy is re-emitted at 590 nm in the red–orange region of the spectrum. The DNA bands can be visualized under UV light, and the result image data is recorded by gel documentation apparatus.

Agarose is usually used at 0.5–2% concentration. Agarose gels have a large range of separation, but relatively low resolving power. By varying the concentration of agarose, fragments of DNA from about 200–50,000 bp can be separated. For separation of molecules of <500 bp, polyacrylamide gel electrophoresis (PAGE) is used. A 12% PAGE can be used to resolve DNA fragments between 20 and 100 bp and 3.5% gels for the range 80–1000 bp. Agarose gels have the advantage of being simple to prepare and are easy to handle and use in electrophoresis. Purified agarose is insoluble at room temperature but dissolves in boiling water when it starts to cool, and it undergoes polymerization causing the solution to gel into the semisolid matrix. Thus, preparation of agarose gel involves melting a required quantity of agarose in the buffer by heating and pouring it into the casting tray (Fig. 26.1a), when it is still hot. A desired comb is placed in the gel to form wells when the gel solution get solidifies (Fig. 26.1b). After the gel has solidified, the comb is removed carefully, and the casting tray along with gel is placed in the electrophoresis chamber and filled with buffer which contains a positive electrode at one end and a negative electrode at the other end. The DNA samples are prepared for electrophoresis by mixing them with dye, which contains glycerol or sucrose and some colouring dyes such as bromophenol blue and xylene cyanol. Samples containing DNA mixed with loading buffer are then loaded into wells with micropipette, the lid and power leads are placed in the apparatus, and current is applied. DNA will migrate towards the positive electrode (usually coloured red). When adequate migration has occurred, DNA fragments are visualized by

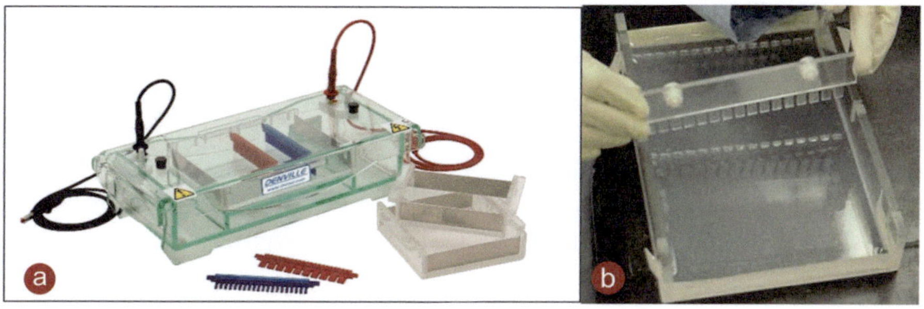

Fig. 26.1 Agarose gel casting apparatus, combs (**a**) and a casted gel showing wells used for loading nucleic acid samples (**b**)

staining with ethidium bromide. To visualize DNA, the gel is placed on a UV transilluminator and photographed using a gel documentation system (Simpson and Whittaker 1983; Sambrook and Russel 2001).

26.2.1 Materials

1. Agarose (good quality purified form of agarose should be used).
2. Electrophoresis unit consisting gel casting tray, comb, gel running tank with lid and electrodes.
3. Power pack.
4. Gel documentation unit.
5. Microwave.
6. 1× TAE buffer: Prepare 10× TAE by dissolving Tris base (48.4 g), glacial acetic acid (11.42 mL), 0.5 M EDTA (pH 8.0) (20 mL) and water to make 1000 mL. Dilute 1:10 with water to get 1× TAE.
7. DNA samples.
8. Ethidium bromide: Add 1 g of ethidium bromide to 100 mL of water, stir vigorously and transfer to dark or amber coloured bottle and store at room temperature.
9. Gel-loading dye: Prepare gel-loading dye by mixing 1% bromophenol blue (200 μL), glycerol (200 μL), 10% SDS (60 μL), 0.5 M EDTA (50 μL), 10× TAE (60 μL), H_2O (30 μL).
10. Marker DNA size standards (commercially available).

26.2.2 Method

1. Prepare 1% agarose gel in Tris-acetate-EDTA buffer (1× TAE) containing ethidium bromide. To 1 g of agarose, add 100 mL of 1× TAE, heat until dissolved, cool to about 50 °C and add EtBr (0.5 μg/mL) before pouring into the gel apparatus.
2. Clean and place the gel casting tray on a level surface, check with the spirit level and adjust the level.
3. Choose appropriate comb and fix into position.

4. Pour the gel onto the casting tray and allow it to cool and set.
5. After the gel has set firmly, pour a little amount of 1× TAE buffer and remove the comb gently. Take care not to drag the comb and break the gel.
6. Immerse the gel along with casting tray slowly into the gel tank. Add sufficient amount of 1× TAE buffer. Connect to the electrode and check the current.
7. Load the samples into wells as per the pre-decided order carefully. Always load an aliquot of standard molecular weight marker along with the samples to assess the size of DNA fragment in the samples.
8. Place the lid on the unit and attach the leads to a power supply. Turn on the power supply to start the run at the required volts (5 V per cm of the gel) for a specified period (1–2 h).
9. Confirm the current flow by observing bubbles coming from the electrodes. Monitor the distance migrated by monitoring the migration of tracking dye.
10. Disconnect the power, remove the electrodes. Carefully lift the gel along with casting tray and place the gel on a transilluminator (use used X-ray sheet or scoop to lift the gel).
11. Put on the transilluminator and view the gel through UV-ray-protected goggles (do not see the gel directly through the naked eye) and locate the presence of band(s) and assess their size by comparing the mobility of the standards used. Depending upon its conformation and size, the DNA fragments will move as discrete bands (Fig. 26.2). The EtBr intercalated with the DNA fragments will make it emit pink fluorescence under UV lamps, which is recorded in any gel documentation system.

26.3 Denaturing Formaldehyde Gel for the Analysis of RNA

The integrity and size of RNA can be checked by denaturing agarose gel electrophoresis and ethidium bromide staining. Unlike DNA, RNA molecules quickly form secondary and tertiary structures that can impede its separation by electrophoresis. Hence regular agarose gel cannot be used to separate them. To keep them in linear form, they should be denatured and run on a denaturing gel. Denaturing of the RNA sample by heat before electrophoresis is insufficient, as secondary structures will just reform unless a denaturing system such as formaldehyde is used. The formaldehyde agarose gel is a simple denaturing electrophoresis system that allows good size separation and resolution of single-stranded RNA. Hence, electrophoresis of RNA is done in two steps: (1) RNA is heat denatured before electrophoresis, and (2) denatured RNA is run on electrophoresis where conditions are

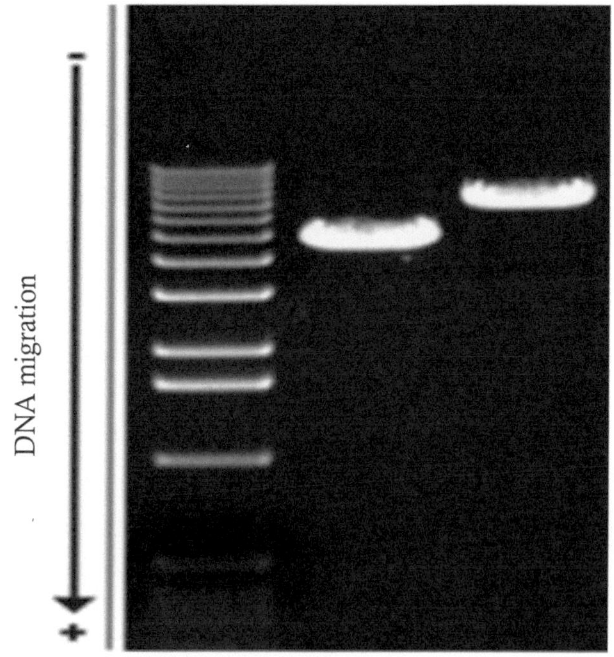

Fig. 26.2 DNA migration in an agarose gel

established that maintain the RNA in a denaturing state. The RNA thus resolved can be subsequently transferred to nitrocellulose membrane from the gel and hybridized with a labelled probe of interest (Sambrook and Russel 2001).

26.3.1 Materials

1. Same as in Subheading 26.2.1.
2. Fume hood.
3. Hydrogen peroxide (3% solution).
4. Methanol.
5. 10× MOPS buffer: 200 mM 3-(*N*-morpholino) propane sulfonic acid (MOPS), 50 mM sodium acetate and 10 mM EDTA.
6. RNA sample.
7. Formaldehyde solution (37%).
8. Formaldehyde gel loading buffer: Mix 720 μL of formamide, 160 μL of 10× MOPS, 260 μL of 37% formaldehyde, 100 μL of sterile water, 100 μL of EtBr (10 mg/mL), 80 μL of sterile glycerol and 80 μL of saturated bromophenol blue in the sterile water. The buffer is not stable and should be made fresh on the day of use.
9. RNA size standards: Can be made or purchased (total eukaryotic RNA contains 28S and 18S rRNA corresponding to 5.1 and 2.0 kb, respectively).

26.3.2 Method

1. Clean beakers, gel casting tray and combs by rinsing them with 3% (v/v) hydrogen peroxide. Rinse the hydrogen peroxide off with methanol and allow the equipment to air dry.
2. Melt 1 g agarose in 74 mL sterile water in a microwave and cool to about 60 °C.
3. Add 10 mL of 10× MOPS buffer and 16 mL of 37% formaldehyde.
4. Mix the agarose solution by swirling and pour the solution into the gel mould.
5. After solidifying, remove the comb and submerge the gel in 1× MOPS buffer. Do not add ethidium bromide to gel or running buffer.
6. Prepare RNA samples by adding 10 μL of formaldehyde gel loading buffer to the RNA pellet or a solution of RNA (about 2 μL). Boil the sample for 2 min.
7. Load the samples into the wells on the gel in a predetermined order along with suitable RNA size markers.
8. Run the gel at 100–200 V. The ethidium bromide will migrate to the negative electrode, and the bromophenol blue will travel to the positive electrode. To prevent unnecessary background, remove the free ethidium bromide before it exits the top of the gel by turning off the power supply and by slicing the top most section off, which contains the ethidium bromide, with a razor.
9. When bromophenol blue has run half the distance of the gel, the RNA has migrated sufficiently to allow the examination of the size distribution relative to the standards.
10. Examine the gel under UV illumination in gel documentation system. RNA molecules can be seen when viewed on UV transilluminator. Approximate length of bands in base pair can be estimated by comparing with the RNA size markers. The gel can be photographed using gel documentation system.

26.4 Notes (Simpson and Whittaker 1983; Sambrook and Russel 2001)

1. The loading dye for DNA contains glycerol, which gives density to help the sample sink to the bottom of the well and marker dyes.
2. Avoid overloading of samples which may cause contamination of neighbouring wells.
3. Bromophenol blue moves on par with 300–400 bp DNA and xylene cyanol with 2–3 kb DNA.

4. A 0.8% agarose gel is good for DNA size ranging from 1 to 20 kb. For smaller DNA (0.2 bp to 1.0 kb), use 1.5% agarose and for DNA size >20 kb, use 0.5% agarose.
5. The DNA is visualized by adding EtBr, a fluorescent molecule that intercalates with the DNA bases. UV radiation at 254 nm is absorbed by the DNA and transmitted to the bound dye. The energy is emitted at 590 nm in the red–orange region of the spectrum.
6. The sensitivity of detection of ethidium bromide is 20 ng per single band of cloned DNA.
7. Ethidium bromide is a powerful mutagen and hence the gel should be handled carefully with the gloves.
8. All ethidium bromide-containing gels and solutions should be decontaminated before their disposal with special care.
9. Always wear gloves when working with RNA.

References

Sambrook J, Russel DW (2001) Molecular cloning, vol I–III, 3rd edn. Cold Spring Harbor Laboratory Press, New York

Simpson CF, Whittaker M (eds) (1983) Electrophoretic techniques. Academic Press, London

Chapter 27

In Vitro Expression of Viral Coat Protein in Prokaryotic System and Its Purification

Abstract

Generally, the quality of polyclonal antibodies (PAbs) depends on the purity of the virus preparation used for immunization. But in most cases, PAbs raised against the purified virus cross reacts with host proteins which limit their use in serological assays. To overcome these limitations, in vitro expression of viral coat protein in a bacterial system is a good alternative to produce large quantity of the quality coat protein. The purified in vitro expressed protein has been used for the production of polyclonal antiserum for large number of viruses. Bacterial expression of viral proteins is simple, fast and economical for generating high amount of purified proteins with minimal post-translational modifications. The method involves amplification, cloning and expression of the coat protein of the virus in a prokaryotic system followed by purification of expressed coat protein that are then used as antigen for antiserum production for immunodiagnosis of plant viruses.

Key words Expression vector, Cloning, Coat protein, Antigen, Purification, Antiserum

27.1 Introduction

Usually purified virus is used as antigen for immunization for the production of polyclonal antibodies. However, it is very difficult to obtain sufficient quantity of purified virus preparation in certain viruses especially in phloem-borne viruses and from certain hosts due to the presence of inhibitors. To overcome these difficulties, viral coat protein is cloned with some tags such as histidine- or maltose-binding protein in a prokaryotic expression vector and used to transform *Esherichia coli*. Some of the commonly used expression vectors include pMAL, pQE, pET, pRSETA, pET28a, pMAL-c2x vector (Agarwal et al. 2009; Sharma et al. 2014; Kapoor et al. 2014) (Fig. 27.1). The transformed *E. coli* is then induced to overexpress the fusion protein containing viral coat protein. The expressed viral coat protein is harvested, purified and used as antigen for the production of polyclonal antiserum. Bacterial expression of protein is simple and fast for generating high yield of purified proteins. The bacterial expressed plant viral proteins have been used as a recombinant antigen for raising antibodies for

Alangar Ishwara Bhat and Govind Pratap Rao, *Characterization of Plant Viruses*, Springer Protocols Handbooks, https://doi.org/10.1007/978-1-0716-0334-5_27, © Springer Science+Business Media, LLC, part of Springer Nature 2020

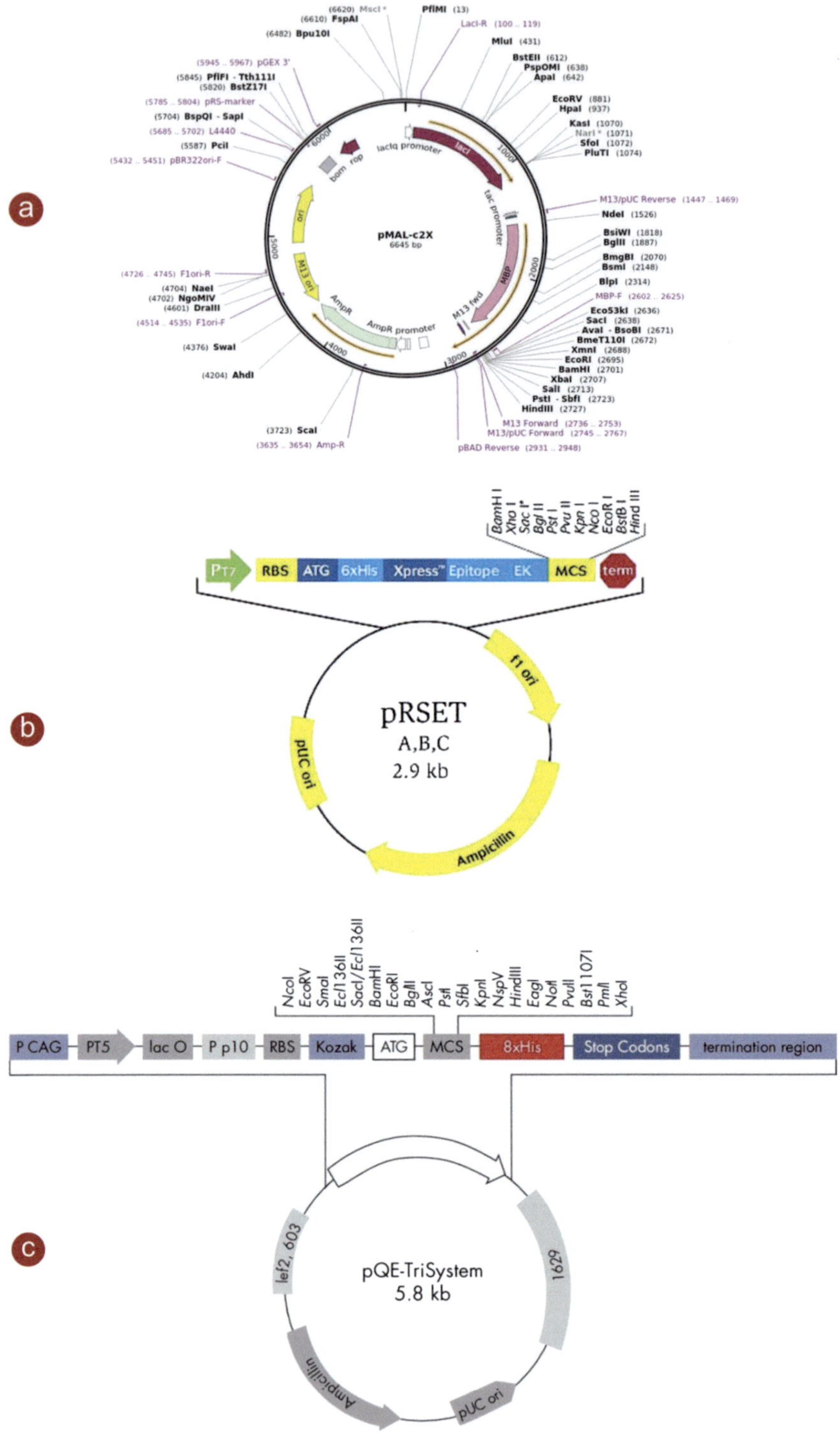

Fig. 27.1 Map of the in vitro fusion expression plasmids (**a**) pMAL-c2X (www.neb.com), (**b**) pRSET (www.invitrogen.com), (**c**) pQE (www.qiagen.com), showing restriction sites

serological diagnosis of many plant viruses (Folwarczna et al. 2008; De Sa et al. 2000; Sambrook and Russel 2001; Kapoor et al. 2014; Sharma et al. 2014).

In the following, overexpression of coat protein gene of a test virus in the pMAL-c2x vector is provided (Agarwal et al. 2009).

27.2 Materials

1. SDS-PAGE vertical gel apparatus.
2. Western blot transfer apparatus.
3. Sonicator.
4. Shaker incubator.
5. In vitro expression vector.
6. *E. coli* DH 5 alpha and TBI strains.
7. Horizontal gel electrophoresis unit with power pack.
8. Thermal cycler.
9. Luria broth: To 950 mL of water add 10 g of bacto-tryptone, 5 g bacto-yeast extract and 10 g NaCl; adjust the pH to 7.4 with 1N NaOH, make up the volume to 1 L and sterilize by autoclaving.
10. Luria agar: To 950 mL of water added 10 g of bacto-tryptone, 5 g bacto-yeast extract and 10 g NaCl; adjust the pH to 7.4 with 1N NaOH; add 15 g agar, make up the volume to 1 L and sterilize by autoclaving.
11. Antibiotics.
12. IPTG.
13. Spectrophotometer.
14. Protein markers.
15. Antiserum against the test virus or maltose binding protein or histidine.
16. Coat protein gene of the test virus cloned in a plasmid vector.
17. RT-PCR and PCR: The details of chemicals/reagents and their preparation for RT-PCR and PCR are provided in Chapter 35.
18. Restriction enzymes.
19. SDS-PAGE analysis: The details of chemicals/reagents and their preparation for SDS-PAGE is provided in the Chapter 24.
20. Western blotting: Details of chemicals/reagents and their preparation for western blotting (electro-blot immunoassay) is provided in Chapter 30.

27.3 Method

27.3.1 Subcloning of Virus Coat Protein (CP) in Expression Vector

1. Amplify coat protein gene of the test virus from the clone already available in the laboratory through PCR using the appropriate primers—both forward and reverse having appropriate restriction enzyme sites. The PCR contains 1× PCR buffer, 2.5 mM $MgCl_2$, 200 μM dNTPs, 50 ng each of forward and reverse primers, 1.5 units of *Taq* polymerase, 1 μL of template DNA and sterile water to a final volume of 25 μL.
2. Programme the thermal cycler initially for a denaturation step at 94 °C for 4 min, followed by denaturation at 94 °C for 30 s, primer annealing at Tm of the primer minus 2 °C for 30 s, primer extension at 72 °C for 1 min (40 cycles) and finally an extension at 72 °C for 20 min.
3. Analyse the PCR products on 1% agarose gel electrophoresis, and examine by the gel documentation system.
4. Excise the band of interest (coat protein) using a sterile razor blade and purify DNA using appropriate commercially available agarose gel purification kit (*see* Chapter 42 for details of gel elution and purification of DNA).
5. Quantify the DNA yield using a spectrophotometer (*see* Chapter 22 for details).
6. Clone this product into pGEM-T Easy (Promega) vector or any other suitable TA vector using TA cloning strategy as described in Chapter 42.
7. Isolate plasmid from recombinant clones and confirm the presence and orientation of the insert using PCR with specific and universal primers present in the vector (details provided in Chapter 42).
8. Digest recombinant plasmid (pGEM-T) containing CP region and expression vector pMAL-C2X (1 μg) simultaneously with *Xba*I and *Bam*HI (restriction enzyme can vary as per need) (Fig. 27.1). Run the restricted products of CP clone and pMAL-C2X vector on 1% agarose gel. Cut the fragment corresponding to coat protein (CP) of the test virus, elute and purify the DNA as described in Chapter 42. Also, purify the linearized band corresponding to pMAL-C2X.
9. Ligate the eluted CP fragment into pMAL-C2X in frame with the fragment of MBP using T_4 DNA ligase as follows: Vector (pMAL-C2X) (50 ng/μL)—1.0 μL; 2× ligase buffer—5.0 μL; PCR product (CP insert)—3.0 μL; T4 DNA ligase (5 Weiss unit/mL)—1.0 μL. Incubate the reaction at 4 °C overnight.

10. On the next day, transform *E. coli* competent cells (strain TBI) with ligated plasmid (detailed procedure of transformation is provided in Chapter 42).
11. Plate the transformed *E. Coli* on Luria agar containing ampicillin 100 μg/mL and keep at 37 °C overnight.
12. Confirm the presence of the CP gene in transformants by colony PCR using appropriate primers and also by restriction with *Bam*HI and *Xba*I.
13. Confirm orientation of the CP gene in the recombinant pMAL-C2X vector using PCR and sequencing.
14. Maintain positive clones on Luria agar plate containing ampicillin100 μg/mL at 4 °C and also as liquid culture at −80 °C or lyophilized culture.

27.3.2 Induction of CP Clones in Expression Cell Line

1. Inoculate single colony of CP-positive recombinant clone of pMAL-2× harbouring fused CP gene (identified in step 14 above) in 5 mL of LB with 100 μg/mL ampicillin.
2. Grow culture at 37 °C overnight at 200 rpm.
3. Next day inoculate 5 mL LB (with 100 μg/mL ampicillin) with 100 μL of the overnight culture and grow at 37 °C with vigorous shaking at 200 rpm until the OD_{600} value of the culture reach 0.5.
4. Collect 1 mL sample as non-induced control.
5. Induce cells for expression of protein by adding 1 mM IPTG and incubating for 3 h with vigorous shaking at 200 rpm.
6. Pelletize cells at 10,000 × *g* and re-suspend in 20 μL of 1× SDS-PAGE sample buffer.

27.3.3 SDS-PAGE Analysis

1. Analyse protein expression by sodium dodecyl sulphate polyacrylamide gel electrophoresis (SDS-PAGE) in 12% resolving gel and 5% stacking gel using discontinuous Tris-glycine SDS system (details of SDS-PAGE is provided in Chapter 24).
2. Denature the sample from **step 6** above after adding equal volume of SDS-PAGE sample buffer by placing in boiling water for 5 min.
3. Load the denatured protein sample in the SDS-PAGE gel along with known protein molecular weight markers (pre-stained protein markers can also be used).
4. Electrophorase the gel at constant voltage of 50 V in stacking gel and at 90 V in resolving gel.
5. Stain the gel with 0.1% Coomassie blue overnight.
6. Destain the gel in destaining solution till the background become clear at room temperature. A protein band at expected size of the CP of the test virus should be visible in the lane loaded with induced protein (Fig. 27.2).

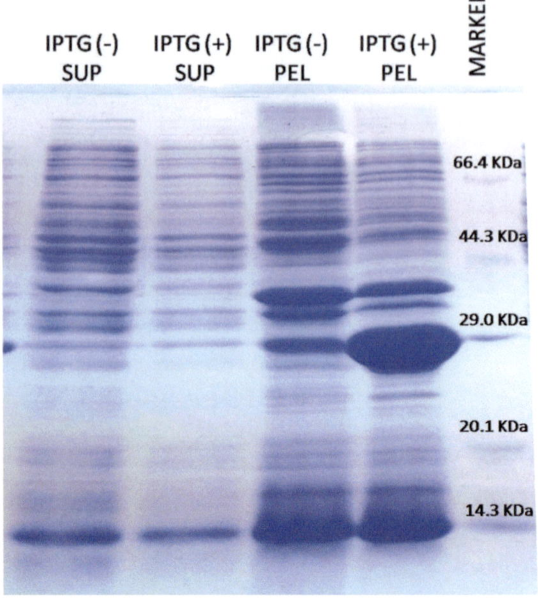

Fig 27.2 Analysis of in vitro expressed coat protein of a virus in SDS-PAGE. Supernatant (SUP) and pellet (PEL) obtained after centrifugation of the total protein of recombinant bacteria with (+) and without (−) addition of IPTG were subjected to SDS-PAGE. Note over-expression of the recombinant protein in the pellet fraction of IPTG added tube. Numbers on the right indicate molecular weight of marker proteins

27.3.4 Protein Purification

1. Once the expression of desired protein is confirmed, purify the virus protein fused to maltose binding protein (MBP) using manufacturer's instructions.
2. Analyse the purified fusion protein by SDS-PAGE (12% polyacrylamide).
3. Confirm the specificity of fusion protein by western blotting using anti-MBP antibodies or virus-specific antiserum following standard protocol (please refer Chapter 30 for detailed procedure on western blotting or electro-blot immunoassay).
4. The purified and confirmed CP of the virus is then used as antigen for polyclonal antiserum production as indicated in Chapter 28.

27.4 Notes (Sambrook and Russel 2001; Agarwal et al. 2009)

1. Bacterial expression of coat protein of plant viruses is simple and fast for generating high yield of purified proteins for high titre production of polyclonal antibodies for their efficient use in virus screening of large number of plant samples.

2. As choice of fusion tags, one can select expression vector, which can produce desired protein. Normally expression vectors with either histidine or maltose binding proteins are available.
3. It would be better to confirm the presence of virus coat protein gene and its orientation in the transformants by PCR, restriction analysis and finally by sequencing.
4. The expressed protein with *His* tag is generally purified through Ni-NTA affinity column. The procedure involves loading lysates of the bacteria onto the column followed by washing the column to remove unbound proteins. The bound protein (expressed protein of interest) is then eluted from the column using elution buffer with low pH.
5. The expressed protein can be purified under either native or denaturing conditions depending on the protein location and solubility. If purification under native conditions is preferred, the expressed tagged protein must be soluble. If the *His* tag of expressed protein is hidden by structure of the native protein, purification should be done under denaturing conditions.
6. If the tagged proteins are purified under denaturing conditions, it can be used directly or may have be re-natured and refolded before using as antigen for antiserum production.

References

Agarwal S, Reddy MK, Jain RK (2009) Production of polyclonal antibodies using recombinant coat protein of *Papaya ring spot virus* and their use in immunodiagnosis. J Plant Biochem Biotech 8:109–111

De Sa PB, Hiebert E, Purcifull DE (2000) Molecular characterization and coat protein serology of *Watermelon leaf mottle virus* (Potyvirus). Arch Virol 145:641–650

Folwarczna J, Plchova H, Moravec T, Hoffmeisterova H, Dědič P, Čeřovská N (2008) Production of polyclonal antibodies to a recombinant coat protein of *Potato virus Y*. Folia Microbiol 53:438–442

Kapoor R, Mandal B, Paul PK, Chigurupati P, Jain RK (2014) Production of cocktail of polyclonal antibodies using bacterial expressed recombinant protein for multiple virus detection. J Virol Methods 196:7–14

Sambrook J, Russel DW (2001) Molecular cloning, vol I–III, 3rd edn. Cold Spring Harbor Laboratory Press, New York

Sharma SK, Kumar PV, Baranwal VK (2014) Immunodiagnosis of episomal *Banana streak MY virus* using polyclonal antibodies to an expressed putative coat protein. J Virol Methods 207:86–94

Chapter 28

Production of Polyclonal Antiserum

Abstract

Polyclonal antibodies (pAbs) are a heterogeneous mixture of antibodies which are usually produced by different B-cell lymphocytes of an animal and directed towards different antigenic determinants or epitopes of the virus protein with varying affinities. Antigenically, the active part of plant virus is its coat protein, and serological tests depend on the epitopes (antigenic determinants) in the coat protein. These antibodies can recognize and bind to many different epitopes of a single virus antigen and hence can form lattices with the antigens. Rabbit, rat, mouse, guinea pig, goat, chicken and sheep are the most commonly used animals in the laboratory for pAb production. Rabbits are more preferred due to their size and relatively long-life span. The polyclonal antiserum production involves injecting purified virus preparation or coat protein of the virus either intramuscularly or intravenously to animals at weekly intervals. Antiserum containing antibodies are collected by bleeding the animal and used in serological tests. pAbs are widely used for the detection of viruses in several serological procedures. In this chapter, different steps involved in the production of pAb are discussed.

Key words Antigen, Purified virus, Coat protein, Intramuscular injection, Intravenous injection, Antibodies, Serum

28.1 Introduction

Antiserum is produced by the immunizing animals with the purified virus or coat protein of the virus expressed under in vitro condition. Antiserum against plant viruses is produced in suitable laboratory animals such as rabbits, rats, mice, guinea pigs and chickens (van Regenmortel 1982; Dijkstra and deJagar 1998). Rabbits are most commonly used in antibody production because of the amount of serum obtained from a single animal is high and relative ease of handling and keeping. The antigen is usually introduced intravenously or intramuscularly into the animal. Intravenous injections are given into the marginal vein of one ear, and the corresponding vein of the other ear is used for bleedings. Intramuscular injections are generally given into the thigh muscles. For this, antigen is emulsified with an equal volume of Freund's complete or incomplete adjuvant. Adjuvant prevents rapid translocation of the antigen in the animal body. The general immunization scheme consists of 4–5 injections at weekly intervals intramuscularly/intravenously.

Alangar Ishwara Bhat and Govind Pratap Rao, *Characterization of Plant Viruses*, Springer Protocols Handbooks, https://doi.org/10.1007/978-1-0716-0334-5_28,

The production of high-titred antiserum depends on the amount of virus injected, viral antigenicity and disposition of the animal (Diano et al. 1987; Hampton et al. 1990).

Bleedings are collected when the antibody level has reached its maximum that can be tested by taking small test bleedings at different time intervals. When an intravenous immunization schedule is followed, the maximum antiserum will generally be achieved 1 week after the last injection while, with intramuscular immunization, 2 or 3 weeks is required after the last injection. The animals should be fasted 12 h before bleeding. The serum can be collected by severing the lateral ear vein with a razor blade. Blood samples can also be obtained by direct heart puncture. Blood samples are allowed to coagulate for about 2 h at room temperature and placed overnight in a refrigerator. On the next day, the serum is clarified by centrifuging at low speed. Preservatives such as sodium azide or glycerol may be added. Sera can be stored under refrigeration or in a deep freezer or as lyophilized powders (Rao et al. 2006).

28.2 Materials

1. Rabbit: Healthy New Zealand white rabbit (about 1 year of age) (Fig. 28.1).
2. Sterile needle (20–22 gauge) with syringe.
3. Adjuvant: Freund's complete or incomplete adjuvant are used. Freund's adjuvant is a solution of antigen emulsified in mineral oil and used as a booster (immunopotentiator). The Freund's Complete Adjuvant (FCA or CFA) is composed of inactivated and dried mycobacteria (usually *Mycobacterium tuberculosis*), whereas the incomplete form (FIA or IFA) lacks the mycobacterial components (hence just the water in oil emulsion).

Fig. 28.1 New Zealand white rabbit (*Oryctolagus cuniculus*) used for polyclonal antibody production

4. Purified virus preparation or purified coat protein of the virus.
5. Restraint box.

28.3 Method

28.3.1 Intramuscular Injection

1. Take the required volume of purified virus suspension or purified coat protein of the virus (usually about 0.5 mg virus in 500 μL) in a vial.
2. Add an equal volume of adjuvant (e.g. Freund's incomplete adjuvant).
3. Make the immunogen–adjuvant mix into a thin creamy emulsion by repeatedly withdrawing and expelling.
4. For intramuscular injection, immunogen preparation is administered into the thigh muscle of hind leg (Fig. 28.2).
5. After keeping the rabbit immobile, swab the area to be injected with ethanol.
6. Aspirate the syringe to ensure proper placement. Hold the leg in the palm of one hand and slowly inject the immunogen preparation at one or more sites of the thigh muscle (Fig. 28.2). Administer the contents in a steady fluid motion. Do not administer rapidly, as this may cause tissue trauma.

28.3.2 Subcutaneous Injection

1. Dilute the immunogen to 0.5 μL with sterile saline and combine with 0.5 μL of the appropriate adjuvant.
2. The immunogen and adjuvant are mixed thoroughly to form a stable emulsion.
3. Fill the syringe with the appropriate amount of immunogen to be administered.
4. Restrain and place the animal on a firm surface.

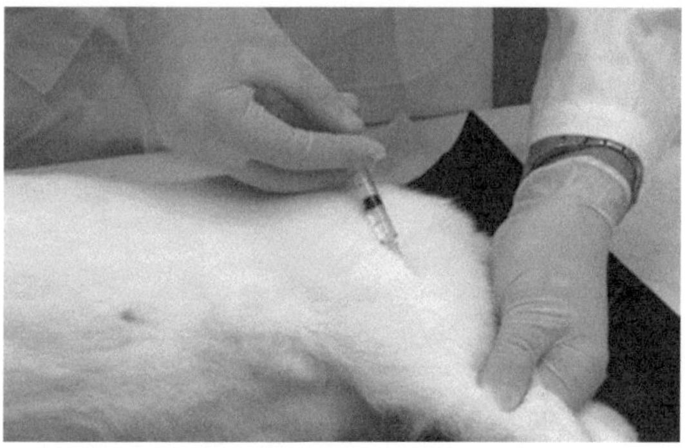

Fig. 28.2 Intramuscular injection in thigh muscle of rabbit

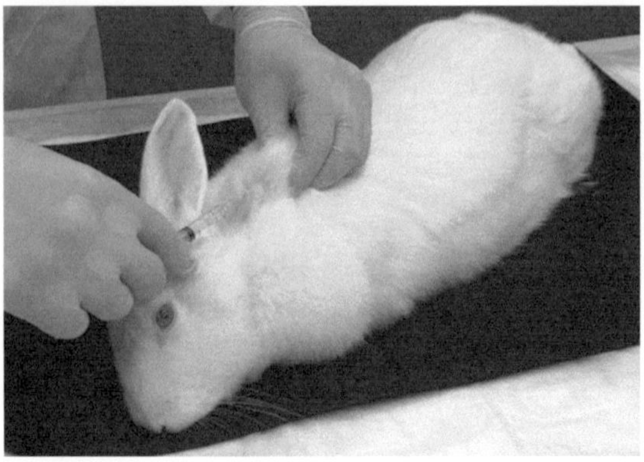

Fig. 28.3 Subcutaneous injection of a rabbit

5. Shave the fur with razor blade, and then sterilize the area with an alcohol swab to disinfect the skin.
6. Insert the needle at the base of the skin fold between your thumb and forefinger (Fig. 28.3), keeping the needle straight.
7. Aspirate the syringe to ensure proper placement and administer the immunogen in a steady fluid motion.

28.3.3 Intravenous Injection

1. To make the marginal vein stand out strongly, hold the ear straight and press on the base of the ear (use xylene if needed to dilate blood vessels).
2. Clean the injection area with 70% ethanol and insert 22–25 gauge needle into the vein and pull back the plunger.
3. Fill the syringe with appropriate amount of sample to be administered.
4. Restrain the rabbit, remove the hair with scalpel blade.
5. Wash the ear with an alcohol swab to disinfect the skin.
6. Insert the needle into the marginal ear vein at a 20° angle (Fig. 28.4).
7. Aspirate the syringe to ensure proper placement.
8. If blood appears in the needle hub, slowly inject the immunogen (without adjuvant). Remove the needle, place cotton over the injection site and hold the site tightly for a few seconds till bleeding stops.
9. Repeat injections at weekly intervals for a total of at least four doses.

28.3.4 Blood (Serum) Collection

1. Restrain the rabbit.
2. Prepare the area with an alcohol swab. Raise the artery (located in the centre of the external ear) by rubbing across the external ear with gentle, quick motions.

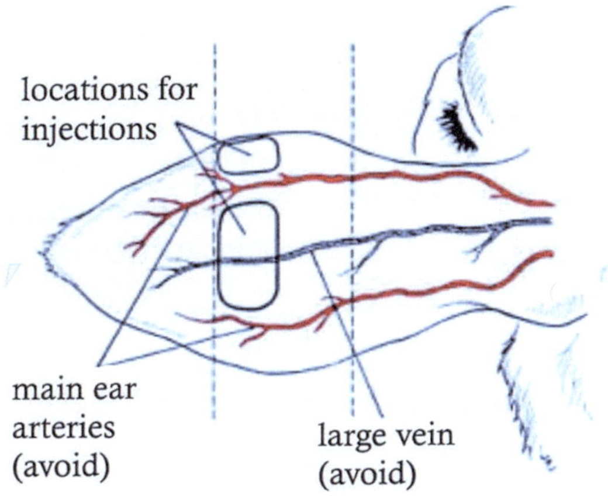

Fig. 28.4 Intravenous injection utilizing marginal ear vein

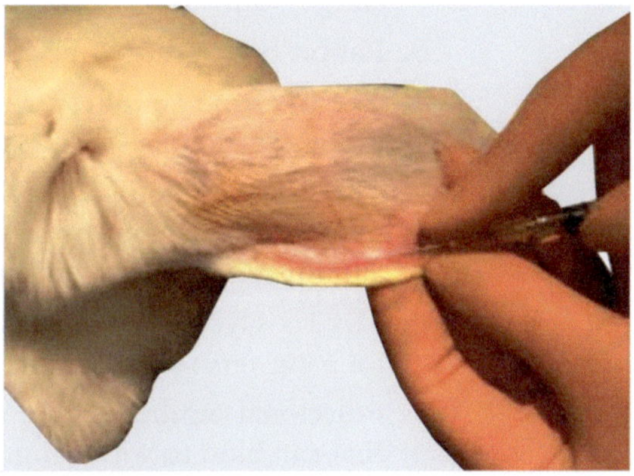

Fig. 28.5 Blood withdrawal from marginal ear vein

3. Before bleeding from the marginal ear vein, remove the hair from the area around the bleeding site with a blade.
4. Swab the ear with xylene to dilate the marginal veins.
5. Swab the bleeding area with 70% ethanol.
6. With a new sterile razor blade, make a small cut in the marginal ear vein.
7. Let the blood flow freely into the collection tube (Fig. 28.5).
8. Bend the ear downward and collect the flowing blood in a dry and sterile beaker.
9. About 20–40 mL blood may be collected depending on the size of the rabbit.
10. After getting the required volume of blood, hold the ear upward and stop the bleeding by pressing the cut area with a cotton swab, to ensure that the blood flow has stopped before returning the animal to its cage.

28.3.5 Processing of Antiserum

1. Leave the blood for 2 h at room temperature to clot.
2. Place the beaker at 4 °C overnight.
3. Centrifuge the contents at 5000 × *g* for 30 min.
4. Decant the serum and make aliquots of 1 mL or less.
5. The serum can be stored at 4 °C for a few weeks. For long-term storage add an equal volume of glycerol and store at −20 °C or lyophilize the antiserum.

28.4 Notes (Hampton et al. 1990; Dijkstra and deJager 1998)

1. For good antiserum production, the rabbit should be 1 year old.
2. One rabbit should be used only for one particular virus/antigen only.
3. Rabbit house should be temperature controlled.
4. The hygienic condition should be maintained to avoid infection to the animals.
5. The virus antigen should be injected slowly in a steady, fluid motion at one or more sites of the thigh muscle. Do not administer rapidly, as this may cause tissue trauma.
6. Mix 1:1 glycerol with the antiserum or lyophilize the antiserum for long-term storage.
7. Polyclonal antibodies are a heterogeneous mix of antibodies that can bind to wide range of antigenic epitopes. Hence, small change in the epitopes of an antigen is less likely to affect pAbs.
8. Polyclonal antibodies are very stable across a wide range of salt concentrations and pH values.
9. Polyclonal antibodies have a higher specificity compared to monoclonal antibodies (mAbs) as they have specific antibodies for all epitopes of an antigen.
10. pAbs are relatively easy to prepare in less time and is cost-efficient.
11. The amount of antiserum produced is limited by the size and lifespan of the animal.
12. The purity and concentration levels of a specific antibody are lower in pAbs than that in mAbs.

References

Diano M, LeBivic A, Hirn M (1987) A method for production of highly specific polyclonal antibodies. Anal Biochem 166:224–229

Dijkstra J, deJager CP (1998) Practical plant virology—protocols and exercises. Springer, Berlin

Hampton R, Ball E, De Boer S (1990) Serological methods for detection and identification of viral and bacterial plant pathogens—a laboratory manual. American Phytopathological Society, St Paul

Rao GP, Valverde RA, Dovas CI (2006) Techniques in diagnosis of plant viruses. Studium Press LLC, Houston, p 245

Van Regenmortel MHV (1982) Serology and immunochemistry of plant viruses. Academic Press, New York

Chapter 29

Production of Monoclonal Antibody

Abstract

Monoclonal antibodies (MAb) are identical immunoglobulins targeted to a single epitope, generated from a single B-cell clone. These antibodies recognize and bind to a single epitope in an antigen. The steps involved in the generation of MAb include production of polyclonal antibodies against the antigen in a mouse. Instead of collecting antiserum (as in the case of polyclonal antibodies), in monoclonal antibody generation, the animal is sacrificed and cells that make antibodies (lymphocytes) are harvested. The lymphocytes are fused with myelomas, cancerous B cells, to create an immortalized hybrid that can undergo many passages in vitro. Hybridomas are then screened against the antigen to identify MAbs specific for the antigen. A protocol with different steps involved in the development of hybridoma and production of monoclonal antibody is discussed in this chapter.

Key words Epitope, Virus detection, Serology, ELISA, Hybridoma, Monoclonal antibody

29.1 Introduction

Antibodies or immunoglobulins are protein molecules produced by a specialized group of cells called B-lymphocytes (plasma cells) in mammals. Antibodies are a part of the defense system to protect the body against the invading foreign substances namely antigens. Each antigen has specific antigen determinants (epitopes) located on it. The antibodies have complementary determining regions (CDRs) which are mainly responsible for the antibody specificity. In response to an antigen (with several different epitopes), B-lymphocytes produce many different antibodies. The types of antibodies that can react with the same antigen are designated as polyclonal antibodies. Monoclonal antibody (MAb) is a single type of antibody that is directed against a specific antigenic determinant/epitope. In the early years, animals were immunized against a specific antigen, B-lymphocytes were isolated and cultured in vitro for producing MAbs. This approach was not successful since culturing normal B-lymphocytes is difficult, and the synthesis of MAb was short-lived and very limited. George Köhler and Cesar Milstein (1975) who got the Nobel Prize, in the year 1984 for hybridoma technology achieved large-scale production of MAbs by

Alangar Ishwara Bhat and Govind Pratap Rao, *Characterization of Plant Viruses*, Springer Protocols Handbooks,
https://doi.org/10.1007/978-1-0716-0334-5_29,

using this technology. They could successfully hybridize antibody-producing B-lymphocytes with myeloma cells in vitro and create a hybridoma. The result is that the artificially immortalized B-lymphocytes can multiply indefinitely in vitro and produce MAbs. The hybridoma cells possess the growth and multiplying properties of myeloma cells but secrete antibody of B-lymphocytes. The production of monoclonal antibodies by the hybrid cells is referred to as hybridoma technology.

Hybridoma technology involves couple of in vivo and in vitro steps, namely antigen selection, immunization, fusion, screening, single cell cloning and production of monoclonal antibodies. The term hybridoma denotes the fusion of immortal myeloma cell with antigen-specific B lymphocyte. Wherein, the inherent immortal character of myeloma cell and antibody secreting ability of B cell are exploited to obtain hybridoma. Moreover, myeloma cell deficient in hypoxanthine-guanine phosphoribosyl transferase (HGPRT) enzyme which is required for salvage pathway is used for the selection of hybridoma. In the fusion step, B cell to B cell, B cell to myeloma cell and myeloma cell to myeloma cell fusion may occur. But, hypoxanthine-aminopterin-thymidine (HAT) blocks the de novo synthesis, so the cell must use salvage pathway for survival. Hybridoma cell uses salvage pathway for survival but due to the deficiency of HGPRT enzyme myeloma cell dies spontaneously. In such a way, monoclonal antibodies produced by a single clone of hybridoma cell are specific to single epitope of an antigen based on theory of clonal selection. Whereas, earlier known polyclonal antibodies are against multiple epitopes of antigen.

MAbs are often considered superior to polyclonal antibodies in virus disease diagnosis (van Regenmortel 1986; Torrance 1995; Van Regenmortel and Dubs 1993; Singh et al. 2004). Since the MAb-secreting hybridoma cells are immortal, they can be stored for long periods at low temperature and regenerated as and when needed, thereby achieving a continued supply of antibodies with constant specificity and titre. Many of the problems associated with polyclonal antibodies can thus be overcome by using MAbs, allowing detection and discrimination of an increasing number of plant viruses in infected plants and vectors at the strain, species and genus level (Torrance et al. 1986; D'Arcy et al. 1989; Jordan and Hammond 1991; Smith et al. 1991; Macintosh et al. 1992; Konaté et al. 1995; Franz et al. 1996; Naidu et al. 1997).

29.1.1 Steps Involved in Monoclonal Antibody Production

The protocol of hybridoma technology for MAb production can be divided as immunization, fusion, screening, MAb production and characterization (Singh et al. 2004) (Fig. 29.1).

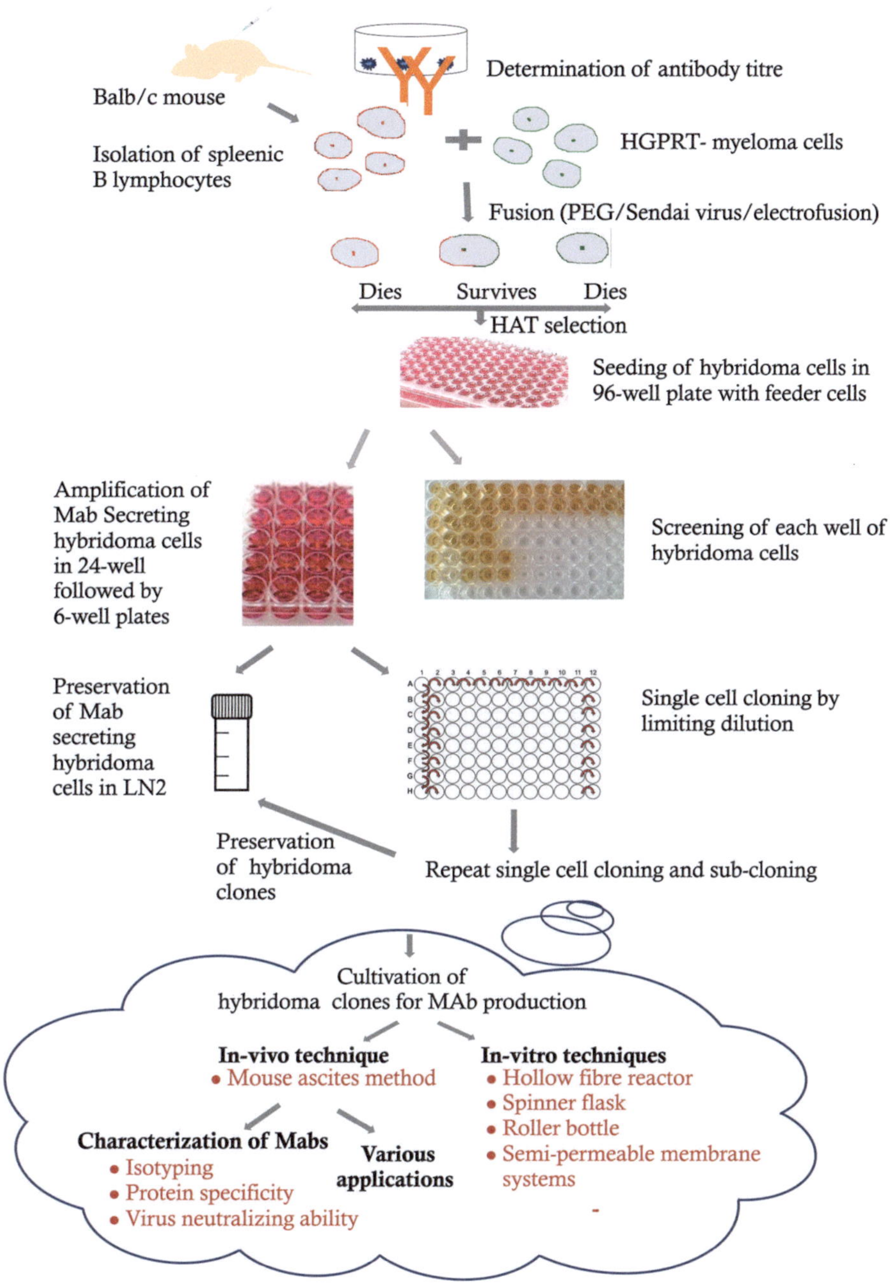

Fig. 29.1 Diagrammatic representation of hybridoma technology for MAb production. (Courtesy: Dr. R.P. Singh, ICAR-IVRI, Bareilly, India)

29.2 Materials

29.2.1 Special Reagents for Hybridoma Technology

1. Hybridoma tested foetal bovine serum.
2. Hybridoma tested polyethylene glycol (PEG-1300-1600).
3. Iscove's modified Dulbecco's medium (IMDM).

4. Hypoxanthine-aminopterine-thymidine (HAT).
5. Myeloma cells (sp2/0) (characteristics: does not secrete immunoglobulins, resistant to 6-thioguanine or 8-azaguanine, HGPRT deficient and dies in HAT medium).
6. Balb/c female mice (4–6 weeks old).
7. Freund's complete (FCA) and Freund's incomplete adjuvant (FIA).
8. Refrigerated micro centrifuge.
9. Bench top centrifuge.
10. Ultracentrifuge.
11. Vortex mixer/Cyclo mixer.
12. Laboratory incubator.
13. CO_2 incubator.
14. ELISA plate and microplate reader.
15. Inverted microscope.
16. Cell culture plates (96 wells, 24 wells and 6 wells).
17. Cell culture flasks.
18. Micropippetts (0–200 μL) single and multichannel.
19. Hybridization oven/shaker.
20. Orbital shaker.
21. Freeze drier.
22. Immune wash—12 channels.
23. Cell culture facility with class II type A2 biological safety cabinet.
24. Liquid nitrogen (LN_2) storage facility with uninterrupted supply of LN_2.

29.3 Immunization

The very first step in hybridoma technology is to immunize an animal (usually a mouse), with appropriate virus antigen. The antigen, along with Freund's complete adjuvant is injected intraperitoneally. The injections are repeated with antigen along with Freund's incomplete adjuvant several times. This enables increased stimulation of B-lymphocytes which are responding to the antigen. Three days prior to the collection of spleen from the animal, a final dose of antigen is administered intravenously. The concentration of the desired antibodies is assayed in the serum of the animal at frequent intervals during the course of immunization. The spleen is collected from the animal, when the serum concentration of the antibodies is optimal. The spleen is aseptically removed and disrupted by mechanical methods to release the spleenocytes. The lymphocytes of the spleen are separated from the rest of the cells by centrifugation.

A wide variety of antigen preparations have been used successfully to produce monoclonal antibodies. In this protocol, emulsified antigen is injected intraperitoneally into the species of choice. A booster injection is administered 10–14 days after the primary immunization. Three days after the booster injection, the spleenocytes are ready for fusion with myeloma cells.

29.3.1 Materials

1. As in Subheading 29.2.
2. Antigen.

29.3.2 Method

1. Prepare antigen using 1–50 μg protein per animal to be immunized in normal saline.
2. Draw up antigen into a sterile 1–2 mL glass syringe. Connect syringe to a three-way stop cock.
3. Completely resuspend FCA to disperse the *Mycobacterium tuberculosis* bacilli which settle to the bottom of the container with time. Draw up a volume of FCA equal to the antigen volume in a syringe and connect to the antigen-containing syringe.
4. Emulsify antigen and FCA by discharging antigen into FCA, then discharging back and forth until a thickened mixture results. Test whether the emulsion is stable (a stable emulsion will not disperse when a drop of it is placed in water).
5. Transfer all of the FCA/antigen emulsion to one syringe and remove the other syringe and stop cock. Attach a sterile 20-G needle to the syringe containing the emulsion.
6. Inject 0.2 mL emulsion intraperitoneally into the mouse.
7. Boost animal after 10–14 days with approximately the same dose of antigen as in **step 6**. If cell fusion is planned for 3 days after boosting, immunize with antigen alone in aqueous solution. If a fusion is not immediately planned, boost the animal with antigen emulsified in FIA (which does not contain *Mycobacterium tuberculosis* bacilli).

29.4 Cell Fusion

Every step in fusion experiment must be performed following sterile precautions in biosafety cabinet Class II type A2.

29.4.1 Materials

1. SP2/0-Ag14 myeloma cell line, complete IMDM with 10 mM HEPES.
2. 1 mM sodium pyruvate.
3. Complete IMDM.

4. Serum-free 50% polyethylene glycol (PEG).
5. Ammonium chloride solution.
6. Complete IMDM, 1 mM sodium pyruvate and 1× HAT or 1× HT supplement.
7. Fine-mesh metal screen.
8. 50 mL conical polypropylene centrifuge tubes.
9. Centrifuge.
10. 96-Well flat-bottom microtiter plates.
11. Reagents and equipment for counting cells and assessing cell viability by trypan blue exclusion.
12. All culture incubations should be performed in a humidified 37 °C, 5% CO_2 tension unless otherwise specified.

29.4.2 Method

29.4.2.1 Preparation of Feeder Cells

Feeder cells are believed to supply growth factors required for the growth of hybridoma cells after fusion. Growth factors secreting cells such as peritoneal macrophages, splenocytes, thymocytes and myeloma cells of mice could be used as feeder cells (Fuller et al. 2001). Peritoneal macrophages as feeder cells serve dual purpose for providing growth factors and also avoid low-grade contaminations due to their property as scavenger. Recently, these feeder cells were replaced by commercially available preparation of growth factors. Feeder cells from peritoneal washings has to be prepared 3 days ahead of fusion as per the following procedure:

1. Following aseptic precautions, collect the peritoneal macrophages by repeatedly washing the peritoneal cavity of two healthy Balb/c mice using sterile PBS.
2. Keep the collection tube on ice to prevent the adherence of macrophages to the inner surface of the collection tube.
3. Centrifuge the peritoneal washing at 1000 × *g* for 10 min and discard the supernatant.
4. Treat the pellet containing peritoneal macrophages with RBC lysis buffer (0.83% ammonium chloride solution) to remove contaminated RBCs.
5. Suspend the cells in 100 mL of Iscove's modification of Dulbecco's medium (IMDM) containing antibiotics (penicillin and streptomycin) and 20% foetal bovine sera.
6. Subsequently seed the cells in 96-well cell culture plate as 100 μL cell suspension per well.
7. Place the plate in incubator under 5% CO_2 tension.

29.4.2.2 Preparation of Myeloma Cells

In the area of biomedical research and diagnostics, Balb/c mouse-derived myeloma cells are commonly used as fusion partner for mouse spleen cells. However, the need of hetero-hybridoma raised for therapeutics has been realized due to the stimulation of anti-mouse antibodies in other species. But, these hetero-hybridoma cells are genetically unstable and obtaining viable hybridoma cells are difficult due to poor fusion efficiency. Hence, rat and human myeloma cell lines have also been developed for hybridoma production. The preferred characteristics of myeloma cells are that these should not express/secrete immunoglobulins and be sensitive to HAT. Owing to the lymphocyte origin of myeloma cell, it would secrete antibodies. Hence, a myeloma cell has to be selected for lack of antibody secretion. They are made sensitive to HAT by cultivating with 8-azaguanine containing media. Although several characterized non-producing mouse myeloma cell lines (X63-Ag8.653, FO, NSO/1) are available for the development of hybridoma, cell line Sp2/0-Ag14 is commonly used. Myeloma cell partner can be prepared for fusion as described below:

1. Revive the myeloma cells from LN_2 before a week of fusion to get at least 10^7 cells at the time of fusion.
2. Subculture the cells regularly to get highly multiplying viable cells.
3. In order to obtain log phase cells for fusion, subculture the cells by 48 and 24 h prior to fusion.
4. Subculture once again in the morning of the day of fusion.

29.4.2.3 Preparation of Spleenic B Lymphocytes for Fusion

The site of B-cell collection depends on the kind of antigen, i.e. B cells from lymphnodes, payers, patches or any lymphoid organs could be of choice. Usually spleen lymphocytes are obtained from immunized mice for fusion as follows:

1. Anesthetize the mouse using volatile ether
2. Aseptically perform the laparotomy and collect spleen in pre-warmed IMDM (37 °C).
3. Wash the spleen once with IMDM and transfer it to sterile Petri dish.
4. Perforate the spleenic capsule at several sites with 18 G needle.
5. Collect the spleen cells by flushing the IMDM through the pre-formed holes using 2 mL syringes.
6. Wash the cells twice with IMDM and subsequently lyse the RBCs using 0.83 % ammonium chloride.

29.4.2.4 Fusion Experiment

Fusion of spleenocyte with myeloma cells and screening strategy of hybridomas are critical procedure for the production of MAbs and their application in diagnosis. Among the agents used for fusion, i.e. Sendai virus, PEG and electrofusion, PEG is now the most commonly used fusogen.

1. Wash the myeloma cells twice with IMDM (without serum) by centrifugation at 1000 × *g* for 5 min.
2. Count the myeloma cells and spleenic B lymphocytes using Neubauer WBC counting chamber.
3. Subsequently, mix the myeloma cell with B cell at the ratio of 1:10 (myeloma: spleenic B cell).
4. Wash the above mixture of cell suspension with calcium- and magnesium-free pre-warmed (37 °C) PBS (0.01 M, pH 7.4).
5. Remove the PBS completely from cell pellet and maintain the cells at 37 °C.
6. Perform fusion using 50% PEG MW. 1300–1600 in PBS at 37 °C as described below:
   (a) Add 1 mL of 50% PEG solution slowly to the mixture of B cells and myeloma cells over a period of 2 min with constant agitation at 37 °C.
   (b) Subject the cells to slow agitation for further 2 min and immediately add 1 mL of IMDM at 37 °C for a period of 1 min.
   (c) Subsequently add 9 mL of IMDM at 37 °C for 1 min and incubate the cell suspension at 37 °C for 5 min.
   (d) Centrifuge the above cell mixture at 500 × *g* for 5 min and discard the supernatant.
   (e) Resuspend the cells slowly in 100 mL of IMDM containing 20% FBS and 2× HAT at 37 °C.
   (f) Seed the cell suspension in feeder cell pre-seeded 96-well cell culture plate as 100 μL per well.
   (g) Incubate the cells at 37 °C with 5% CO_2 tension and observe regularly under inverted microscope for the growth of hybridoma cells.

29.5 Screening the Hybridoma

Screening is the procedure performed to determine the wells that are containing antibody-secreting hybridoma cells specific for immunized/desired antigen. The supernatant media in the cell culture wells has to be taken for screening the clones; hence, it must be performed at appropriate stage of hybridoma growth. Usually, bunch of 8–10 cells would appear on day 4 of seeding, it is preferable to start screening by 7th day and complete by 12th day because beyond that cells may die off in the well, so the positive wells has to be expanded without delay. At least by 14th day of fusion, positive wells need to be transferred to 24-well plates. The antibody secreted by the hybrid cell is referred to as monoclonal antibody; however, monoclonality has to be ensured by at least two single-cell cloning steps.

29.5.1 Materials

1. Growing hybridomas.
2. Candidate hybridoma line.
3. Cloning/expansion medium.
4. Reagents and equipment for ELISA.

29.5.2 Method

29.5.2.1 Screening Strategy

Screening strategy for hybridoma should be similar to the intended use of MAbs in future. ELISA is the most popular and quick screening technique. Kind of ELISA should depend on the intended use of monoclonal antibody in future. Various kinds of ELISAs used for screening are indirect-ELISA, competitive-ELISA, sandwich-ELISA and cell-ELISA. If the intention of the MAb production is to apply in fluorescent antibody technique, clones have to be screened by FAT or the FACS technique as the case may be. If virus-neutralizing clones are desired, virus neutralization tests need to be performed for screening. Similarly, while MAbs produced for therapeutic purpose, use similar recommended technique for screening. The suitable technique should be easy and quick (within 12 h) to perform for large sample size as many wells need to be screened at times. The screening technique has to be optimized prior to fusion experiments because once fusion is performed, within few days screening would be needed. Optimize the screening technique using immunizing antigen and test bleed sera of mice in place of MAbs. Indirect ELISA for screening of hybridoma for virus as a model is described below:

1. Coat the ELISA plate with antigen and incubate at 37 °C for 1 h under constant shaking.
2. Wash the wells thrice using diluted PBS (1:4) in distilled water to remove unbound antigen.
3. Add 25 μL of blocking buffer (PBS with 0.1% tween-20) to all the wells.
4. Add 25 μL of hybridoma culture supernatant to each well by following the similar format of 96-well cell culture plate.
5. Incubate the plate at 37 °C for 1 h and wash the wells as described above.
6. Add 50 μL of anti-mouse HRP (horseradish peroxidase) conjugate diluted at 1:1000 or recommended dilution in blocking buffer and repeat **step 5.**
7. Add 50 μL of *o*-phenylene diamine (OPD) containing H_2O_2 (4 μL of 3% H_2O_2 per mL of OPD) to each well and observe for the development of colour for 10 min.

8. Add an equal volume of stopping solution of 1 M H_2SO_4 to each well and measure the optical densities at 492 nm in a microplate reader.

According to the optimized indirect ELISA, wells with twice or more the A_{492} value in antigen coated wells compared to mock-infected wells could be considered as positive. The cells in the positive wells are amplified further using 24-well plate to 6-well plate with appropriate volume of medium. From the 6-well plate, preserve the cells to avoid the risk of losing the positive clones. Preserve the cells at gaseous phase of LN_2 using preservation media containing following components or use isopropanol bath for cell preservation.

Preservation media	Quantity (mL)
IMDM	7
Foetal bovine serum	2
Dimethyl sulphoxide	1

29.5.2.2 Single Cell Cloning and Sub-cloning

The single hybrid cells producing the desired antibody are isolated and cloned. Two techniques are commonly employed for cloning hybrid cell-limiting dilution method and soft agar method. In **limiting dilution method,** the suspension of hybridoma cells is serially diluted and the aliquots of each dilution are put into micro culture wells. The dilutions are so made that each aliquot in a well contains only a single hybrid cell. This ensures that the antibody produced is monoclonal. However, in **soft agar method,** the hybridoma cells are cultured in soft agar. It is possible to simultaneously grow many cells in semisolid medium to form colonies. These colonies will be monoclonal in nature. In actual practice, both the above techniques are combined and used for maximal production of MAbs. In this chapter we describe the limiting dilution method, wherein the antibody-secreting hybridoma cells are diluted in such a way to obtain single cell in each well. Perform twofold dilutions of hybridoma cells in 96-well plate as follows:

1. Add 100 μL of growth media to all the wells and add 100 μL of hybridoma cells to A1 well of 96-well plate.
2. Make the dilutions of hybridoma cells from A1 to H1 and subsequently from A1 → A12, B1 → B12 and so on using multichannel micropipette.
3. Observe the wells regularly under inverted microscope for the growth of cells.
4. Screen the wells when 25% of well surface is covered with growing cells as described.

5. Repeat the single cell cloning procedure at least twice so as to prove the monoclonality of the hybridoma culture supernatants.
6. Select the hybridoma clones derived from single cells and expand using cell culture flasks and preserve in LN_2.

29.6 Production of MAbs from Hybridoma Clones

Both in vivo and in vitro methods are used for the production of MAbs from hybridoma clones for further characterization and various applications. In vitro methods are batch-culture (cell culture flask, spinner flask, roller bottle and gas permeable bag) and semi-permeable membrane-based systems (hollow fibre and low molecular weight cutoff membrane tubing). Owing to the high cost and requirement of technical expertise, many of the in vitro methods are mainly applied in commercial production of MAbs by industries.

29.6.1 Laboratory Methods of Hybridoma Cultivation (Batch Culture Method: Cell Culture Flask)

Hybridoma clone can be cultivated in cell culture flask using 20% foetal bovine serum containing IMDM. However, to prevent the risk of contamination of immunoglobulins of serum origin, serum-free media are available commercially. Also, hybridoma cells can be revived and grown using 10–20% serum containing media, and once the 70–80% of cell culture flask is covered by growing healthy cells, spent media can be replaced with regular IMDM without foetal bovine serum. Wherein the healthy cells might not divide significantly but would secrete antibody in the media. Depending on the condition of cells, MAbs can be collected at least twice from these flasks. Even though this method requires more media, the MAb concentration obtained would be few micrograms per mL. If purified MAb is required, ammonium sulphate precipitation of antibodies followed by protein A- or protein G-based chromatographic techniques can be performed. This is the commonly used method of cultivation of hybridoma in laboratory for research and development purpose.

29.6.2 Mouse Ascites Method

In vivo method is the mouse ascites method, in which hybridoma cells at log phase of growth are inoculated intraperitoneally into the Balb/c mice. The hybridoma cells would implant in the peritoneal cavity and multiply as tumour cells and the MAbs secreted by the hybridoma cells in the peritoneal cavity would lead to ascites. Due to the ethical issues on use of laboratory animals, mouse ascites method is used only with adequate justifications such as inadaptability of hybridoma clone in vitro, requirement of high concentration of MAb and lack of cell culture facilities in poor resource countries. In vivo production of MAbs can be done as per the following protocol:

1. Prime the animal with 0.1 mL of pristine or 0.25–0.5 mL of FIA by intraperitoneal route.
2. A day after FIA priming or 14 days after pristane priming, inoculate the mice with 10^5–10^7 cells in 0.1–0.5 mL volume.
3. Observe the animal twice a day and when the animal undergoes discomfort due to abdominal pain, perform paracentesis using 18-G needle (4–5 mL).
4. Usually paracentesis performed two to three times during survival, and terminal tap should be performed after euthanasia.

The MAb concentration of 3–5 mg/mL could be obtained by this method. Hence, it obviates the MAb concentration step prior to use.

29.7 Characterization of Monoclonal Antibodies

Characterization of MAbs is carried out using parameters which are most important as per intended use of MAbs. The important parameters for the characterization of MAbs against viral proteins include protein specificity, isotyping, ELISA reactivity with closest possible antigens/viruses, virus neutralization ability and utility as diagnostic kit for antigen and antibody detection. However, the ELISA is the most commonly used method for routine characterization of monoclonal antibodies, which is described herein.

29.7.1 Reactivity in ELISA

It is a most important criterion which determines the usefulness of MAb in diagnostic assays. The process of ELISA is described below:

1. Coat the ELISA plate with 50 μL of virus antigen per well and incubate at 37 °C for 1 h.
2. Wash the wells of ELISA plate thrice using 1:4 diluted PBS.
3. Prepare twofold serial dilutions of hybridoma culture supernatants (up to 1:4096) in blocking buffer (PBS with 0.1% tween-20) and dispense as 50 μL/well.
4. Incubate the ELISA plate at 37 °C for 1 h and wash thrice as described above.
5. Add 50 μL/well of anti-mouse HRPO conjugate of appropriate dilution in blocking buffer and repeat **step 4.**
6. Finally, dispense 50 μL/well of OPD containing H_2O_2 in all the wells and incubate at 37 °C for 10 min or till colour development.
7. Stop the reaction using equal volume of 1 M H_2SO_4 and measure the optical densities of the wells at 492 nm.

The dilutions of the MAbs which give 75% A_{492} of the plateaus (maximum A_{492}) are considered as titre of the MAbs for their application in plant virus diagnostics (Saliki et al. 1993).

29.8 Notes (van Regenmortel 1986; Torrance 1995)

1. The advantages of hybridoma technology over polyclonal sera are monoclonality, reproducibility and indefinite quantity of MAb could be obtained without repeated use of laboratory animals.
2. Define the objectives of hybridoma production, whether to develop diagnostic kit or for the research and development. If the objective is to develop diagnostic kit, aim for specific clone while screening and single cell cloning. However, to develop clones for research and development, a variety of clones against different epitopes/antigen of interest might be preferred for wider application.
3. Do not compromise with the quality of reagents at any cost. Success in hybridoma production mainly depends on the quality of the reagents/chemicals and their standard source. Prefer the reagents from standard sources only.
4. Ensure that every step of hybridoma production is performed properly as per the procedure and do not consider any of the step unimportant or more important.
5. Immunizing antigen is the critical starting material for hybridoma production. Antigen could be of virion preparation (native antigen), recombinant protein, synthetic peptide, etc.
6. Test the immunogen/antigen of interest with at least two assays before immunization.
7. Choose purified or at least partially purified antigen because contaminated proteins of impure antigen could sensitize the B lymphocytes, and it might become fusion partner for myeloma cell. It might further lead to less chance for obtaining desired clone and tedious screening procedure.
8. Do not initiate immunization until the confirmation of quality of the antigen.
9. Testing/verification protocol of antigen should coincide with the intended use of monoclonal antibodies.
10. Finalize screening strategy before fusion to obtain suitable clones.
11. Have defined quality and quantity of antigen (immunizing antigen) for screening of the hybridoma clones.
12. Points to consider while planning for immunization schedule:

(a) Priming has to be done with double the dose of antigen than booster.

(b) Avoid too frequent boosters as the peak of antibody which circulates in blood would bind the antigen and may make booster ineffective.

(c) Lower immunization dose and longer schedule would result in high-affinity antibody than higher dose with short schedule.

(d) Bleed the mouse before each immunization for serum collection for the determination of antibody titre.

(e) When 100- to 1000-fold rise in antibody titre is obtained, perform booster without adjuvant.

(f) Usually, 100- to 1000-fold rise in antibody titre is obtained in three immunizations, otherwise do more immunization with adjuvant.

(g) Once desired antibody titre is established in mouse, final booster has to be performed through intra-peritoneal or intra-venous route before 3 days of fusion to get high probability of sensitized B lymphocytes in spleen.

(h) Hybridoma technology is laborious and time consuming. MAbs are produced against a single antigenic determinant; therefore, they cannot differentiate the molecule as a whole. Sometimes, they may be incapable of distinguishing groups of different molecules also.

13. Points to consider during hybridoma production:

(a) Successful hybridoma production requires extreme precautions of cell culture techniques.

(b) Always use feeder cells/peritoneal macrophages.

(c) Always use freshly prepared feeder cells from animal.

(d) Prefer winter season over summer for hybridoma work.

(e) It is always better to share the clones with others rather than losing the clone due to lack of LN_2 or contamination.

(f) Always ensure the availability of preserved back-up for every clone.

(g) Always use myeloma cells in log phase which is sub-cultured in the morning on the day of fusion.

References

D'Arcy CJ, Torrance L, Martin RR (1989) Discrimination among luteoviruses and their strains by monoclonal antibodies and identification of common epitopes. Phytopathology 79:869–873

Franz A, Makkouk KM, Katul L, Vetten HJ (1996) Monoclonal antibodies for the detection and differentiation of Faba bean necrotic yellows virus isolates. Ann Appl Biol 128:255–268

Fuller SA, Takahashi M, Hurrel JGR (2001) Fusion of myeloma cells with immune spleen cells. In: Ausubel FM (ed) Current protocols in molecular biology. John Wiley and Sons, Danvers

Jordan R, Hammond J (1991) Comparison and differentiation of potyvirus isolates and identification of strain-, virus-, subgroup-specific and potyvirus group-common epitopes using monoclonal antibodies. J Gen Virol 72:25–36

Köhler G, Milstein C (1975) Continuous cultures of fused cells secreting antibody of predefined specificity. Nature 256:495–497

Konaté G, Barro N, Fargette D, Swanson MM, Harrison BD (1995) Occurrence of whitefly transmitted geminiviruses in crops in Burkina Faso, and their serological detection and differentiation. Ann Appl Biol 126:121–129

Macintosh S, Robinson DJ, Harrison BD (1992) Detection of three whitefly-transmitted geminiviruses occurring in Europe by tests with heterologous monoclonal antibodies. Ann Appl Biol 121:297–303

Naidu RA, Mayo MA, Reddy SV, Jolly CA, Torrance L (1997) Diversity among the coat proteins of luteoviruses associated with chickpea stunt disease in India. Ann Appl Biol 130:37–47

Saliki JT, Libeau G, House JA, Mebus CA, Dubovi EJ (1993) Monoclonal antibody-based blocking enzyme-linked immunosorbent assay for specific detection and titration of peste-des-petits-ruminants virus antibody in caprine and ovine sera. J Clin Microbiol 31:1075–1082

Singh RP, Sreenivasa BP, Bandyopadhyay P, Dhar PK (2004) Production and characterization of monoclonal antibodies to Peste des Petits Ruminants (PPR) virus. Vet Res Commun 28:623–639

Smith HG, Stevens M, Hallsworth PB (1991) The use of monoclonal antibodies to detect beet mild yellowing virus and beet western yellows virus in aphids. Ann Appl Biol 119:295–302

Torrance L (1995) Use of monoclonal antibodies in plant pathology. Eur J Plant Pathol 101:351–363

Torrance L, Pead MT, Larkins AP, Butcher GW (1986) Characterization of monoclonal antibodies to a UK isolate of barley yellow dwarf virus. J Gen Virol 67:549–556

van Regenmortel MHV (1986) The potential for using monoclonal antibodies in the detection of plant viruses. In: Jones RAC, Torrance L (eds) Developments and applications in virus testing. Association of Applied Biologists, Wellesbourne

van Regenmortel MHV, Dubs MC (1993) Serological procedures. In: Matthews REF (ed) Diagnosis of plant virus diseases. CRC Press, Boca Raton

Chapter 30

Serological Tests

Abstract

Serological tests are based on the reaction of antibody with antigen. Antibody specifically reacts with the antigen against which it is produced. A large number of serological tests are available for the detection and diagnosis of plant viruses. Traditional methods such as tube precipitin, microprecipitin, chloroplast agglutination and gel diffusion involved direct observation of specific precipitates of virus and antibody, either in liquid media or agar gels. These methods require large quantities of reactants—antiserum and antigens for successful detection. With the advent of enzyme-linked immunoassay (ELISA) during 1980s, traditional methods are replaced by this method. Various formats of ELSA such as double-antibody sandwich (DAS)-ELISA, $F(ab)_2$-ELISA, plate-trapped ELISA, and direct antigen-coated (DAC)-ELISA are used widely for the detection and assay of plant viruses. It is very economical in the use of antiserum, adaptable for large-scale testing and quantitative measurement. ELISA performed on a nitrocellulose or nylon membrane is referred to as dot immunobinding assay (DIBA). Tissue blotting immunoassay (TBIA) is another simple method where crude sap from plants are directly blotted and used for detection through immunology. Immunosorbent electron microscopy (ISEM) combines electron microscopy and serology for the detection and diagnosis of viruses. Electro-blot immunoassay (EBIA) though cumbersome is a confirmatory assay that involves separation of viral antigens on a gel followed by its detection through serology. Lateral flow assay (LFA) or dip stick assay is the latest and the only serological method used for the detection of viruses without the aid of a laboratory and skilled technical personnel. It is an onsite and quick assay that can be performed by anyone and suitable for the detection of viruses that occur in high concentration in plants.

Key words Tube precipitin, Chloroplast agglutination, Micro precipitin, Enzyme-linked immunosorbent assay, Dot immunobinding assay, Tissue blot immunoassay, Electro-blot immunoassay, Immunosorbent electron microscopy, Immunofluorescence, Lateral flow immune assay, Antisera, Antigen

30.1 Introduction

Antibodies produced against a virus preparation can be used for the detection of viruses in plants. It is one of the most widely used methods for assay, detection and diagnosis of plant viruses. It is also used for studying the relationships between related virus isolates and strains. The entire serological tests are based on the specificity of the antibodies to its homologous antigen. A variety of methods have been developed to demonstrate the antigen–antibody reaction. The main advantages of serological tests include: (1) specificity of the reaction allows virus to be measured in presence of host material, (2) results are obtained in few hours or overnight, (3) quantification

Alangar Ishwara Bhat and Govind Pratap Rao, *Characterization of Plant Viruses*, Springer Protocols Handbooks, https://doi.org/10.1007/978-1-0716-0334-5_30, © Springer Science+Business Media, LLC, part of Springer Nature 2020

possible, (4) sensitive method than infectivity assays and (5) test can be performed any time of the year as long-term storage of antisera possible. However, the major disadvantage of serological tests is that it measures only the virus protein not the amount of infective virus. The specificity of all serological reactions is checked using suitable controls. Antisera are tested against the sap from healthy plants and concentrated preparation of normal plant proteins. Antigens are tested against sera of non-immunized rabbits (normal sera) to check if any cross-reactions. In some cases, antisera against unrelated viruses have to be used to determine the specificity of an antiserum (Noordam 1975; Hill 1984; Walkey 1985; Hampton et al. 1990).

In the following chapters, protocols for the commonly used techniques like tube precipitin, microprecipitin, agar double-gel diffusion, enzyme-linked immunosorbent assay (ELISA), dot immunobinding assay (DIBA), electro-blot immunoassay (EBIA), immunosorbent electron microscopy (ISEM), immunofluorescence and lateral flow immunoassay (LFIA) are given.

30.2 Precipitin Tests

Plant viruses are polyvalent, i.e. each particle can combine with many antibody protein molecules to form a lattice-structured precipitate. When virus particles are allowed to react with their antibodies, precipitates are formed as a result of the reaction between antibodies and multivalent virus antigens. A large number of virus antigens become linked together to form a network (lattice), when this complex reaches a certain size, they become insoluble. The ratio of the concentration of the reactants is necessary for the formation of precipitates. No precipitation will occur when there is an extreme excess of one of the reactants. With elongated viruses, precipitate zones are broad, whereas, with isometric viruses, they are narrow. Factors such as electrolyte content, pH, temperature and concentration of foreign substances influence the test. In these tests, antisera are usually diluted in saline and antigens in a weak buffer or buffered saline. Tests are performed with clarified plant extracts or purified virus preparations. Different methods of precipitin tests are available for the detection of plant viruses (Noordam 1975; Hill 1984). Two commonly used methods namely tube precipitin and micro precipitin tests are described here.

30.2.1 Tube Precipitin

Tube precipitin test is usually used for determining the titre of an antiserum and rarely for comparing relationships. Small thin-walled glass tubes with a twofold dilution of antiserum and antigen are used. Tubes are kept half immersed in a water bath to allow convection current to speed up the rate of precipitation. Precipitation is observed at regular intervals (Fig. 30.1). Isometric viruses produce a fine granular precipitate and elongated virus, a more flocculent precipitate (Hill 1984; Walkey 1985). The highest dilution at

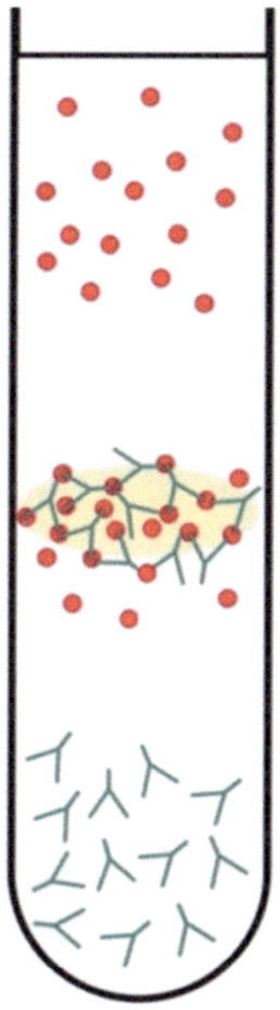

Fig. 30.1 Principle of tube precipitin test showing precipitate formation

which precipitation has occurred gives the titre. It should be noted that titre value varies depending on the method used for its determination. Hence the method used for its determination should always be given. Also, titres are not absolute measures of antigen or antibody concentrations. More antibodies are necessary to precipitate small isometric viruses than long rod-shaped viruses.

30.2.1.1 Materials

1. Beaker or tubes for preparing dilutions.
2. Lightbox.
3. Pipette (1 or 5 mL).
4. Thin-walled tubes (precipitin tubes) 7 mm diameter.
5. Tube racks.
6. Water bath.
7. Antiserum and normal serum (healthy serum).
8. Clarified sap from virus-infected and healthy plant or purified virus preparation: Clarification can be carried out by treating extracted plant sap with an equal volume of chloroform followed by low-speed centrifugation. The upper aqueous layer is the clarified sap. Saline (0.8%) or phosphate-buffered saline (PBS) can be used as diluents.
9. Phosphate-buffered saline (PBS) (10×): Dissolve 8.00 g NaCl, 0.20 g KH_2PO_4, 2.90 g Na_2HPO_4 (or 1.15 g anhydrous), 0.20 g KCl, 0.20 g NaN_3 in distilled water and make up the volume to 1000 mL. For working solution of 1× PBS, mix 100 mL of 10× PBS with 900 mL of distilled water.

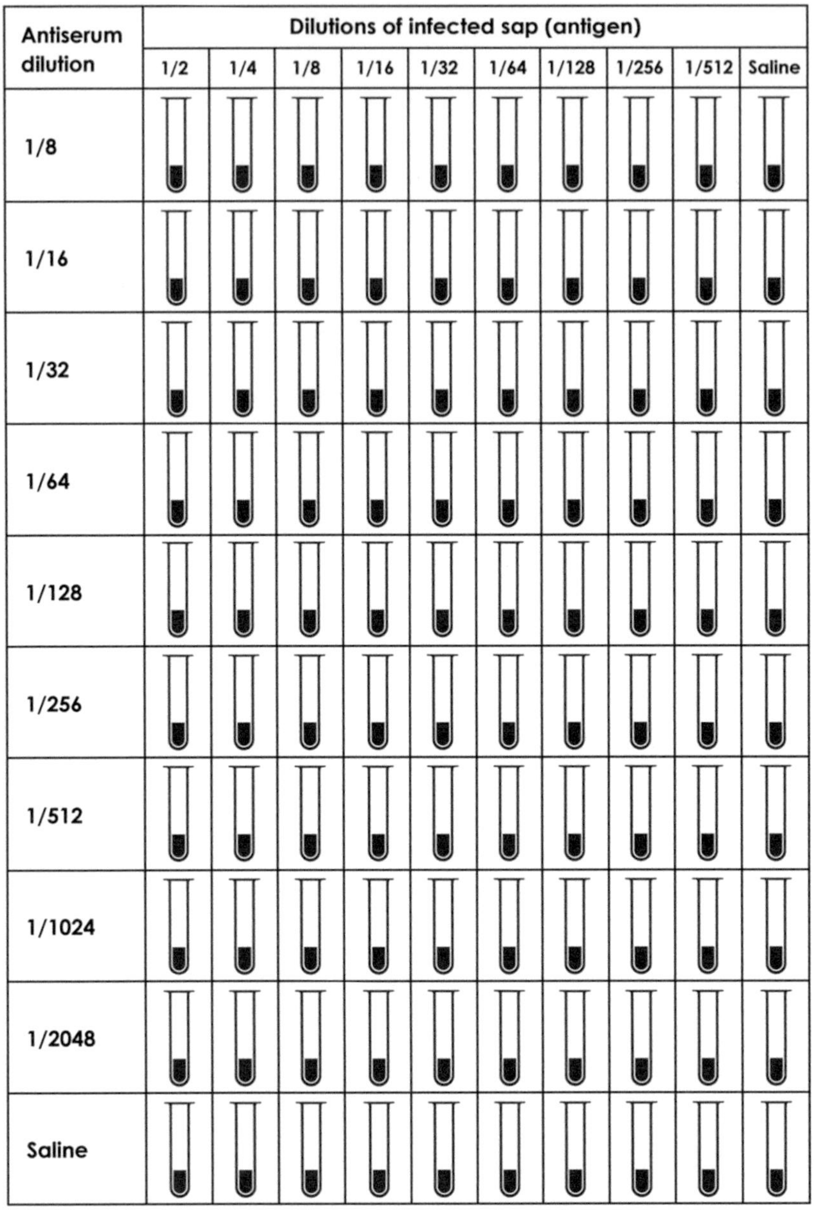

Fig. 30.2 Layout for conduct of the tube precipitin test to determine optimum concentration of antibody and antigen

30.2.1.2 Method

1. Arrange the thin-walled (precipitin) tubes in ten rows with ten tubes in each row (Fig. 30.2).
2. Make dilutions of antigen: add 5 mL undiluted clarified sap (or purified preparation) to the tube 1; add 5 mL saline or PBS to give ½ dilution. Mix thoroughly; take 5 mL from this, and add 5 mL saline or PBS to give ¼ dilution. Continue dilutions like this till 1/512 (Fig. 30.2).

3. For antiserum dilution, add 1.25 mL of antiserum, to a tube and make up the volume to 10 mL to give 1/8 dilution. Transfer 5 mL to the next tube containing 5 mL of saline or PBS to give 1/16 dilution. Continue twofold dilutions till 1/2048 (Fig. 30.2).
4. Load precipitin tubes with antiserum dilutions beginning with the most dilute (1/2048), 0.5 mL to each of the tubes in row 9 as in Fig. 30.2. Repeat with dilution 1/1024 to the tubes in row 8 and so on until all the tubes in row 1 are filled with 1/8 dilution.
5. Starting with most dilute sap (1/512), transfer 0.5 mL to each tube in column 9, repeating with decreasing antigen dilutions in each column until all is added. Add 0.5 mL saline in tubes in row 10 and column 10 with 0.5 mL normal serum. Mix thoroughly.
6. Incubate the tubes, half immersed in the water bath at 37 °C.
7. Read the test by holding tubes over the light box, after 1, 2, 4, 8 min up to 2 h, rotating tubes rapidly to determine the amount of precipitate.
8. Precipitate may be ranked as: −: no reaction, ±: barely visible precipitation, +: slight, ++: moderate, +++: heavy, ++++: very heavy. The tubes that precipitate most rapidly contain an antigen and antibody in optimal proportions. The greatest antigen dilution to produce a precipitate is the antigen or virus dilution end-point and the greatest antiserum dilution to give a precipitate is the antiserum dilution end-point or antiserum titre.

30.2.2 Micro-Precipitin Test

The micro-precipitin test is a modification of tube precipitin, which is economical in the use of both antiserum and antigen while retaining its sensitivity. Varying dilutions of antigen and antiserum are mixed, and the quality and rate at which precipitate is formed are recorded. As only few drops of the reactants are mixed, the quantities used are minimal. A grid titration of the same kind as described for tube precipitation may be set up in a Petri dish. Drops of mixed reactants, placed within the grid, are then covered with a layer of mineral oil to avoid drying out during incubation. The results are read with a dark field microscope or a stereo microscope against a dark background (Van Slogteren 1955; Hill 1984; Walkey 1985).

30.2.2.1 Materials

1. Same as in Subheading 30.2.1.1.
2. Formvar-coated Petri dish.
3. Incubator.
4. Refrigerator.
5. Stereo microscope.

6. Wax pencil or graph paper.
7. Mineral oil.

30.2.2.2 Method

1. Prepare a grid consisting of 8 mm squares with a wax pencil inside the Petri dish, or on a separate graph paper to be attached to the base of the dish.
2. Prepare a separate record grid in which location of dilutions of each component can be registered as shown in the Fig. 30.3.
3. Make an appropriate series of dilutions of antigen and antiserum using saline or PBS as diluent.

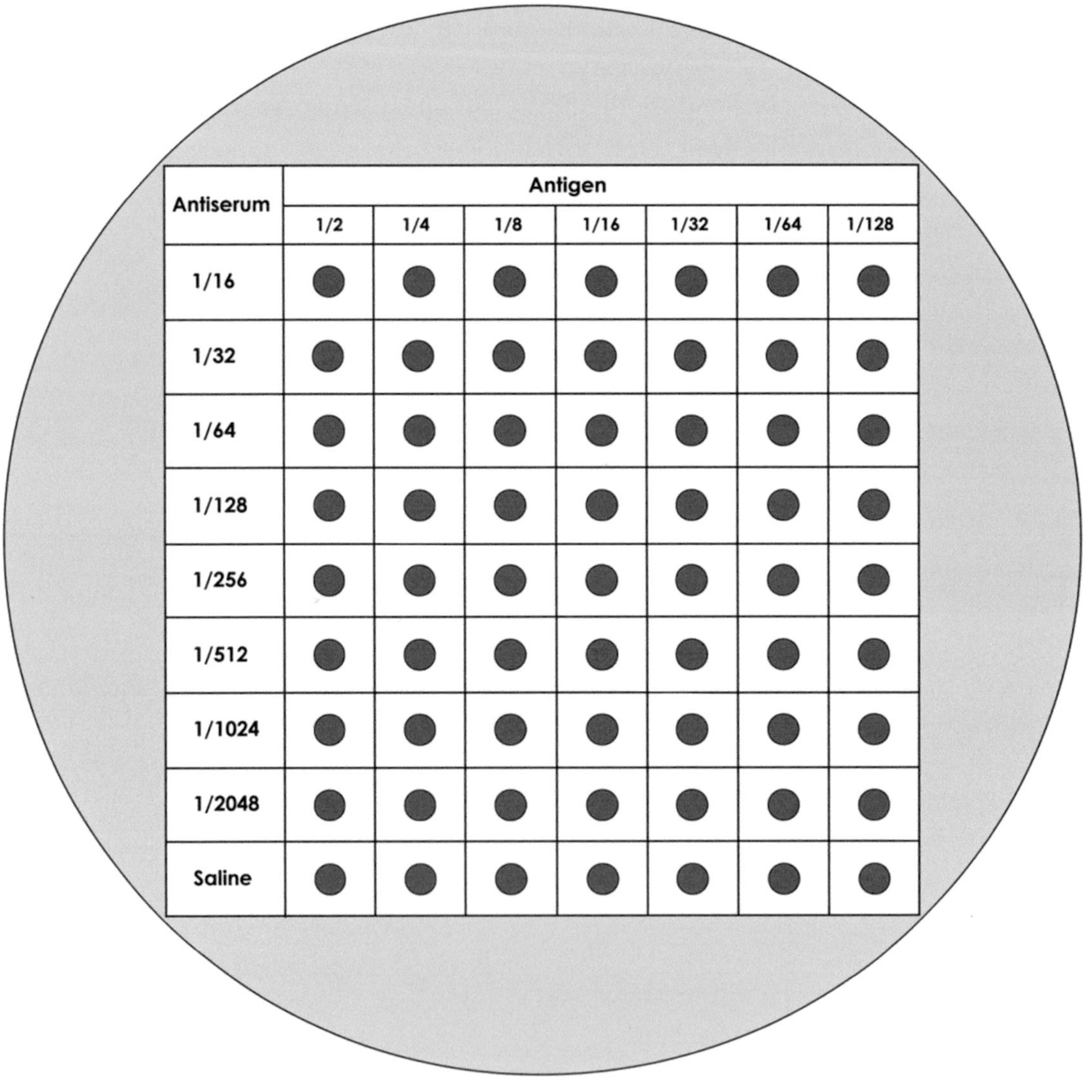

Fig. 30.3 Layout of the grid for conduct of micro-precipitin test

4. Add a 10 μL drop of the right dilution of antigen and antiserum to each space in the grid as designated on the separate record grid (Fig. 30.3).
5. Carefully flood the dish with mineral oil from one edge until all is covered, taking care to avoid disturbing the drops. Viscous parafilm oil should be used to avoid spreading of the drops at incubation.
6. Carefully transfer the Petri dish to the incubator and incubate at 37 °C for 2 h.
7. Examine the drops after 2 h using stereomicroscope with top light and black background.
8. Transfer the dish to a refrigerator and read reactions again after overnight storage.
9. Rank the intensity of the reactions as done for the tube precipitin test.

30.2.3 Chloroplast Agglutination Test

The chloroplast agglutination test, used for diagnosis and detection of plant viruses, is simple but not very sensitive. It is most effective for viruses with elongate particles, such as tobamo-, carla- and potexviruses, which usually occur in high concentrations in the plant. In this test, drops of antiserum against the virus in question are added to drops of crude plant extract. If the plant is infected, the antibodies will bind to virus particles adsorbed to the chloroplasts. If enough antibodies are present for extensive linking of chloroplasts, these organelles will be agglutinated. This is visible, even to the naked eye, as a clumping of the chloroplasts in the drop. It should be noted that this type of precipitation is called agglutination, as large particles such as cell organelles are reacting. The term precipitation is restricted to reactions by molecular antigens, including viruses. Another modification of the microprecipitin test that is more sensitive than chloroplast agglutination is called 'latex agglutination test'. Here the sensitivity is increased by using latex particles sensitized with specific antiserum. When a sample containing specific antigen (virus particles) is mixed with latex particles sensitized with specific antiserum, it causes visible agglutination. Latex particles are used to magnify the antigen–antibody complex. The method is cheap and simple and does not require any special equipment. Kit based on latex serological agglutination test is available especially for the detection of viruses infecting potato and other horticultural crops (Talley et al. 1980; Gella et al. 1991).

30.2.3.1 Materials

1. Same as in Subheadings 30.2.1.1 and 30.2.1.2.
2. Mortar and pestle.
3. Microscope slides.
4. Pasteur pipettes.

30.2.3.2 Method

1. Grind a leaf of the virus-free plant in a mortar, squeeze the homogenate through cheese cloth, and collect the green extract in a beaker.
2. Repeat **step 1** with an infected leaf, using another mortar and pestle.
3. With a Pasteur pipette, place a drop of virus-free leaf extract on each half of a microscope slide.
4. With another Pasteur pipette, place a drop of virus-containing leaf extract on each half of a second microscope slide.
5. Add three drops of normal serum to the extract drops on the right half of both the microscope slides.
6. Add three drops of antiserum to the extract drops on the left half of both the microscope slides.
7. Examine the slides in a stereomicroscope and look for clumping or agglutination in the infected samples.

30.2.4 Agar Double Diffusion (Ouchterlony) Test

In this test, antibody–antigen reaction is carried out in a gel instead of liquid. As antigen and antibody diffuse into the medium, a precipitation line is formed at a point where they meet in optimum proportions. The test is performed in agar gels in Petri plates or on glass slides. Favourable pH and salt conditions should be selected for each system along with a preservative like sodium azide. The antigen can be a crude sap, clarified extract or purified virus. In this case, antigen-containing wells surround a central well of antiserum. In certain genera of viruses (rod-shaped particles), SDS is used to dissociate virions into antigenically active diffusible fragments. It may also be included as a component of the agar gel medium.

The location of the initial precipitin line depends mainly on the diffusibility of the reactants and their concentration. If the reactants diffuse to the reaction site at equivalent concentrations, the system is called balanced. Balanced conditions are required for the formation of sharp stationary precipitin lines. Under balanced conditions, the precipitin lines form characteristic patterns, and conclusions may be drawn from various antigenic relationships. Under unbalanced conditions, the reactant in excess moves across the initial precipitin line and re-precipitation occurs on the side of low concentration reactant. Sometimes, band splitting and the formation of multiple bands will be observed under unbalanced conditions (Ouchterlony 1968; Noordam 1975; Hill 1984; Walkey 1985).

30.2.4.1 Materials

1. Cork borer: To cut wells in the gel.
2. Moist chamber.
3. Pestle and mortar.
4. Petri plates.
5. Agarose.

6. Antiserum.
7. Phosphate buffered saline (PBS): For preparation refer Sub-heading 30.2.1.1.
8. Sodium azide.
9. Virus infected samples or purified virus preparations.

30.2.4.2 Method

1. Prepare 0.7% agarose in phosphate-buffered saline (PBS).
2. Add sodium azide (0.5%).
3. Place Petri plates on levelling table. When agarose is cooled to about 60 °C, dispense the appropriate amount into plates and allow setting undisturbed.
4. Using the cork borer, punch out wells and carefully remove the plugs by slight suction. All the wells should be cut at an equal distance from each other as shown in Fig. 30.4.
5. Make a twofold dilution of the antigen in PBS. Add aliquots of each dilution to the peripheral wells, if it is to be tested against one antiserum or vice versa. Add appropriate dilution of antiserum depending on the titre to the central well (if the test is performed to determine the serological relationship of one antiserum with different antigens, the peripheral wells are loaded with different antigens).
6. Incubate the plates at 37 °C in a moist chamber for 24–48 h.
7. Observe the development of precipitation zone in the space between central and peripheral wells from 24 to 48 h. A white precipitation zone will be visible in between wells containing antigen and antiserum (Fig. 30.4). Depending on the type of precipitation zone, serological relationship between antigen and antibody can be determined (for details *see* Hill 1984).

30.3 Enzyme-Linked Immunosorbent Assay (ELISA)

ELISA is one of the most widely used serological tests for the detection of plant viruses because of its simplicity, adaptability, and sensitivity. In this, the specificity of the antigen-antibody reaction is recognized through the action of the associated enzyme label

Fig. 30.4 Ouchterlony serological reaction in Petri plates using antigen and antibody for the detection and diagnosis of virus

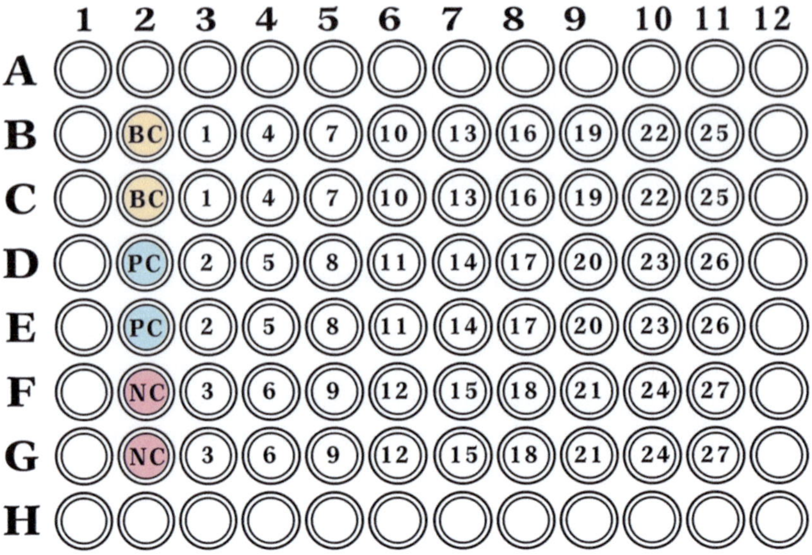

Fig. 30.5 Schematic diagram of an ELISA plate showing test samples, buffer control, negative control and positive controls. *BC* buffer control, *PC* positive control, *NC* negative control; 1, 2, 3. . . . . .27, test samples. Buffer wells are used to confirm the procedure. Positive controls are used to confirm the test. The OD of positive controls must be obviously superior to the O.D values of healthy controls. Healthy controls indicate the level of 'background'. The OD values of healthy controls must be close to the buffer ones. It is advised not to use wells on the edge of the microtitre plate. 'Edge effect' is due to the possible difference in temperature and humidity

on a suitable substrate unlike observing the formation of an insoluble antigen–antibody complex in precipitin reactions. ELISA is performed in a plate referred to as microtitre or ELISA plate. The plate consists of 96 wells (1 to 12 and A to H length wise and width wise, respectively) (Fig. 30.5). Reactants are added to each of the wells in a stepwise manner as per pre-decided plan, and finally positive reaction is identified by the development of colour, the intensity of which is directly proportional to the amount of viral antigen present in the sample. Whenever ELISA is performed, it is essential to keep known positive, known negative and buffer controls in addition to the samples to be tested. An ELISA plate showing reaction with known positive control and test samples (wells with yellow colour), buffer and healthy control (no colour) is given in Fig. 30.6.

ELISA method is very economical in the use of reactants and can be applied to viruses of different morphology in both purified and crude preparations. The method is very useful when a large number of samples are to be tested and is very sensitive in detecting up to 1–10 ng per mL of the virus. The sensitivity of detection of antibody–antigen reaction is increased by attaching the antibody to a minute quantity of enzyme. As enzyme substrate is then added, and the resulting colour response may be quantitatively measured

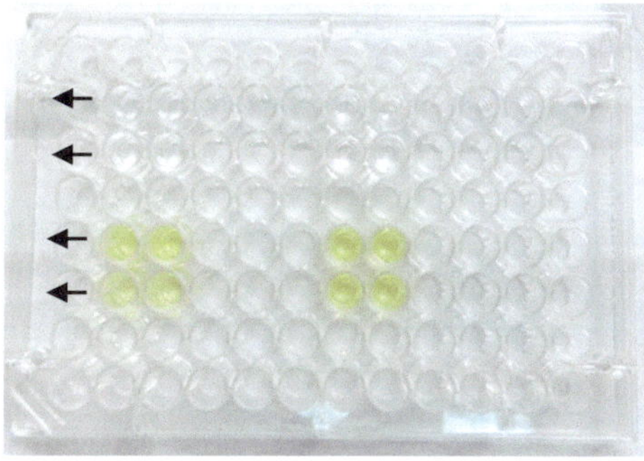

Fig. 30.6 A microtitre plate showing results of an ELISA assay for virus detection after substrate addition. Yellow colour indicates positive result. Note that wells with buffer control and negative controls did not show any colour. Healthy leaf was used as negative control

(Clark and Adams 1977; Barbara and Clark 1982; Clark et al. 1986).

Two forms of ELISA are used: (1) direct ELISA wherein the virus (antigen) is trapped by antiviral antibody and detected using enzyme-labelled virus-specific antibody, (2) in the indirect ELISA, the enzyme used in the final detection and assay step is conjugated to an antiglobulin antibody. (For example, if the virus antibodies were raised in rabbit, a chicken/horse/mouse/goat anti-rabbit globulin might be used.) One of the commonly used techniques of direct ELISA is double-antibody sandwich ELISA (DAS-ELISA) while direct antigen-coated ELISA (DAC-ELISA), $F(ab')_2$ ELISA and triple-antibody sandwich ELISA (TAS-ELISA) are some of the techniques under indirect ELISA (Koenig 1981; Koenig and Paul 1982; Clark et al. 1986). In the following, protocols for DAS-ELISA, DAC-ELISA, $F(ab')_2$-ELISA and TAS-ELISA are described.

30.3.1 Double-Antibody Sandwich (DAS) ELISA

In DAS-ELISA, immunoglobulin G (IgG) extracted from virus-specific antiserum is used for coating the solid surface to trap the antigen, and the same IgG is labelled with an enzyme and is employed for detection. The antigen gets sandwiched between IgGs and thus is referred to as the double-antibody sandwich (DAS) form of ELISA. The reaction is visualized by the addition of substrate of the enzyme used for labelling leading to the development of colour (Fig. 30.7) (Clark and Adams 1977; Koenig 1981). The intensity of the colour is proportional to the concentration of antigen present. The procedure requires isolation of immunoglobulins (IgG) from crude antiserum and preparation of immunoglobulin–enzyme conjugate. The protocol for the same is given below.

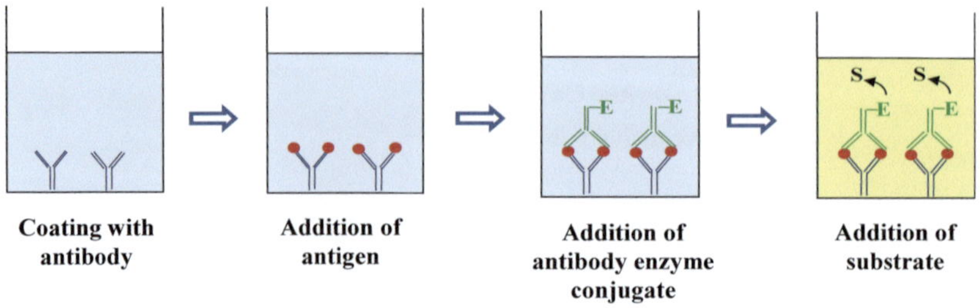

Fig. 30.7 Basic steps involved in DAS-ELISA

30.3.1.1 Preparation of Immunoglobulins (IgG) (Avrameas 1969)

Materials

1. Centrifuge.
2. Spectrophotometer.
3. Dialysis tubing: Cut the tubing into required length. Boil for 10 min in a large volume of 2% sodium bicarbonate and 1 mM EDTA (pH 8.0). Rinse thoroughly in distilled water. Then boil for 10 min in 1 mM EDTA (pH 8.0). Allow the tubing to cool and store at 4 °C. Care should be taken to see that tubing is always submerged and handle the tubing with gloves.
4. Crude antiserum.
5. Phosphate-buffered saline (PBS): For preparation refer Subheading 30.2.1.1.
6. Saturated solution of ammonium sulphate: Add ammonium sulphate salt in a required volume of water with continuous stirring, heat gently to dissolve and let cool to working temperature on a magnetic stirrer. Formation of crystals of ammonium sulphate indicates that the solution is fully saturated.
7. DE 52 cellulose.

Method

1. Take 1 mL of antiserum and add 9 mL distilled water.
2. Add 10 mL saturated neutralized ammonium sulphate and mix thoroughly.
3. Incubate for 1 h at room temperature.
4. Centrifuge at 2000 × *g* for 10 min.
5. Discard the supernatant and dissolve the pellet in 2 mL of 0.5× PBS.
6. Dialyse against three changes of 500 mL of 0.5× PBS over a period of 24 h.
7. Layer the dialysed IgG (2 mL) on top of DE 52 cellulose column and elute with 0.5× PBS.
8. Collect fractions of 1 mL each and measure the OD values at 280 nm.
9. Fractions showing maximum OD are mixed and adjusted with 0.5× PBS to get 1.4 OD (i.e. 1 mg/mL).

30.3.1.2 Preparation of IgG–Enzyme Conjugate

Alkaline phosphatase (ALP) and horseradish peroxidase (HRP) are the two most widely used enzyme labels. Subsequently, urease and penicillinase have been introduced. Reaction kinetics of HRP is not linear, and some of its substrates are hazardous to health. Urease and isozymes of peroxidase are known to be present in seeds and plant extracts, thus limiting their application in plant virus detection. In the following protocols for the preparation of conjugates using ALP, HRP and penicillinase are given.

Conjugation with Alkaline Phosphatase (ALP)

One-step, glutaraldehyde method of Avrameas (1969) is described below.

Materials

1. Cold room.
2. Alkaline phosphatase: Commercial preparation is available from many firms.
3. Bovine serum albumin (BSA).
4. Glutaraldehyde (electron microscope grade).
5. Immunoglobulin (IgG).
6. Phosphate-buffered saline (PBS): Preparation of this is explained earlier, *see* Subheading 30.2.1.1.

Method

1. Add 0.2 mL (equivalent to 2 mg or 5000 units) of alkaline phosphatase solution to 2 mg (2 mL) of IgG.
2. To this, add freshly diluted glutaraldehyde to reach a final concentration of 0.05% and incubate for 4 h at room temperature.
3. After incubation, dialyse the conjugate to remove glutaraldehyde against three changes of 500 mL each of 1× PBS at 4 °C for 24 h.
4. Add bovine serum albumin (BSA) to the dialysed conjugate to a final concentration of 4 mg/mL and store the conjugate at 4 °C.

Conjugation with Horseradish Peroxidase (HRP)

Materials

1. Same as in "Materials" in Subheading "Conjugation with Alkaline Phosphatase".
2. Ammonium sulphate.
3. Carbonate ($NaHCO_3$) buffer (0.01 M) (pH 9.6): Dissolve 0.33 g of sodium carbonate and 0.58 g of sodium bicarbonate in 1 L of water.
4. Horseradish peroxidase (HRP): Use good commercial preparation.
5. Sodium borohydride.

6. Sodium hydroxide (0.1 mM): Dissolve 40 mg of NaOH in 1 L of water.
7. Sodium metaperiodate ($NaIO_4$) (0.1 M): Dissolve 21.39 g of sodium metaperiodate in 1 L of water.

Method

1. Dissolve 4 mg of HRP in 1.0 mL of distilled water.
2. Add 0.2 mL of 0.1 M sodium metaperiodate, which must be freshly prepared. If the HRP does not turn a greenish colour, start over with fresh periodate.
3. Shake for 20–30 min at room temperature.
4. Dialyse at 4 °C overnight against distilled water adjusted to pH 4.4 with acetic acid.
5. Transfer to a glass tube and add 50 μL of 0.01 M carbonate buffer, pH 9.6.
6. Add 8 mg of IgG in 1.0 mL of 0.01 M carbonate buffer, pH 9.6.
7. Incubate at room temperature with occasional shaking.
8. Add 0.8 mL of freshly prepared sodium borohydride solution (5 mg/mL) made in 0.1 mM NaOH and incubate for 2 h at 4 °C.
9. Add an equal volume of saturated ammonium sulphate and leave at room temperature for 30 min.
10. Collect the precipitate by centrifugation, dissolve in 4 mL of PBS and dialyse against three changes of 500 mL PBS.
11. Store at 4 °C or freeze dried under vacuum.

Conjugation with Penicillinase

Materials

1. Same as above.
2. Bovine serum albumin (BSA).
3. Glutaraldehyde (electron microscope grade).
4. IgG.
5. Penicillinase.

Method

1. Place 1 mL IgG (500 μg) in a dialysis bag and add 1 mL (250 μg) of penicillinase.
2. Dialyse against PBS for 1 h at room temperature.
3. Place the dialysis bag (containing IgG and penicillinase) in a beaker containing PBS with 0.06% glutaraldehyde and dialyse for 3–4 h at room temperature.
4. Replace the buffer containing glutaraldehyde with 500 mL of PBS containing 0.02% sodium azide and dialyse for 18 h at 4 °C with at least three changes of buffer.

5. Transfer the conjugate into a new vial and add bovine serum albumin (BSA) at 5 mg/mL. Store the conjugate in small aliquots (100 μL) at 4 °C. Do not freeze the conjugate. If stored properly, the shelf life should exceed 1 year.

30.3.1.3 Direct Antibody Sandwich ELISA (DAS-ELISA) Utilizing Alkaline Phosphatase

Materials

1. Microtitre (ELISA) plate: Polystyrene make of any good brand may be used.
2. Micropipette and tips: Use adjustable volume micropipette.
3. Multichannel pipette (adjustable 100 to 300 μL) (Optional): Multichannel pipette that can hold 4, 8 or 12 micro tips, thus permitting dispensing of volumes simultaneously into several wells can be used to add antigen, antiserum and conjugates.
4. Incubator.
5. ELISA plate reader.
6. IgG (against virus of interest).
7. Coating buffer (carbonate buffer): Dissolve 1.59 g Na_2CO_3, 2.93 g $NaHCO_3$ and 0.20 g NaN_3 in 1 L of distilled water.
8. Phosphate-buffered saline (PBS): Details provided previously in Subheading 30.2.1.1.
9. PBS-T: Add 0.05% Tween 20 in 1× PBS.
10. Antigen extraction/antibody buffer: Add 2% polyvinyl pyrrolidone (PVP, MW 40,000) and 0.2% albumin in PBS-T.
11. IgG-alkaline phosphatase conjugate: Preparation explained in "Method" in Subheading "Conjugation with Alkaline Phosphatase".
12. Substrate buffer: Dissolve 97 mL diethanolamine in 800 mL distilled water, adjust pH to 9.8 with concentrated HCl, make up the volume using distilled water to 1000 mL.
13. Para-nitrophenyl phosphate (PNPP): Commercial preparation available either in powder or tablet forms may be used. Protect from light while storing.
14. Virus-infected and healthy tissue.

Method

1. Add 200 μL of IgG diluted in coating buffer to each well of the microtitre plate as per the pre-decided plan and incubate at 37 °C for 2 h.
2. Well contents are discarded and washed with PBS-T (flooding with three changes of PBS-T for 3 min each time). After final washing, plates are shaken dry over a paper towel.
3. Test samples are extracted in antigen extraction buffer in 1:10 (w/v) using pestle and mortar. Extracts of healthy plants and the extraction buffer are to be used as negative control, whereas extracts from known infected plant should be used as positive

control. About 200 μL aliquots of the test sample are added, and the plates are incubated overnight at 4 °C.

4. The plates are washed as mentioned in **step 2** and diluted enzyme conjugate (200 μL per well) in antibody buffer is added and incubated at 37 °C for 2 h.
5. After washing as described in **step 2**, 200 μL of substrate (p-nitro phenyl phosphate 0.6 mg/mL of substrate buffer) is added to each well and incubated at room temperature to develop colour (Fig. 15.5).
6. Results are recorded by measuring the absorbance at 405 nm in an ELISA reader after 30, 60 and 120 min after substrate addition.
7. Stop the reaction if required by adding 100 μL of 3 M NaOH solution to each well.

30.3.1.4 DAS-ELISA Using Penicillinase

Materials

1. As listed in DAS-ELISA using alkaline phosphatase.
2. Substrate buffer: Dissolve 15 mg bromothymol blue (BTB) in 50 mL 0.01 M NaOH, neutralize by adding HCl and make up the volume to 100 mL. Add sodium penicillin G at 0.5 mg/mL and adjust the pH to 7.2 using HCl or NaOH.
3. IgG-Penicillinase conjugate: Preparation is explained in "Method" in Subheading "Conjugation with Penicillinase".

Method

1. Follow the **steps 1–4** as indicated in DAS-ELISA using alkaline phosphatase system except that in **step 4**, IgG-penicillinase conjugate is used.
2. Wash the plate in distilled water containing 0.05% Tween-20.
3. Dispense the substrate mixtures (penicillin + BTB) at 200 μL per well and incubate at room temperature. Observe the reaction for 30 min to 2 h and record the results. The blue colour of BTB first turns green, and then to an orange yellow colour. A green colour indicates a weak positive and orange yellow, a strong positive. Results can be quantified by measuring loss in absorbance of BTB at 620 nm. Generally 0.2 mg/mL BTB gives an OD of over 2 units, and a positive reaction (orange yellow) gives less than 0.1 O.D units.

30.3.2 F(ab')$_2$-ELISA

Immunoglobulin molecule has a 'Y'-shaped structure. It has three protein domains. Two of the domains that form the arms of the 'Y' are identical and are termed as 'Fab' domain. They each contain an antigen-binding site at the end, making the IgG molecule bivalent. The third domain, 'Fc' domain, forms the stem of the 'Y'. The three domains may be separated from one another by cleavage with the protease, pepsin or papain. The 'Fc' region has the high affinity for binding to protein A, a protein obtained from the cell wall of bacteria, *Staphylococcus aureus.* This property of 'Fc' region is

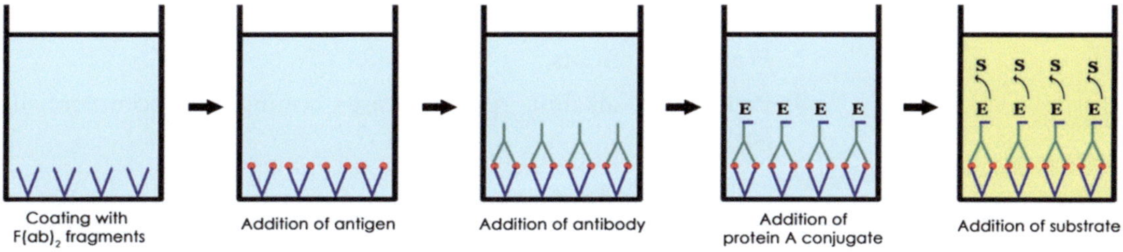

Fig. 30.8 Basic steps involved in F(ab)$_2$-ELISA

exploited in F(ab')$_2$-ELISA. To perform F(ab')$_2$-ELISA, 'Fab' domains are separated from 'Fc' region by pepsin treatment. The F(ab') fragments thus obtained are used to coat the wells of ELISA plate to trap the antigens. The same IgG is used as primary detecting antibody followed by the addition of protein A–alkaline phosphatase conjugate and after that the substrate (Fig. 30.8). The advantage of this technique lies in its ability to enhance the sensitivity of detection levels as more virus particles can be trapped compared to DAC-ELISA (Clark et al. 1986). Here the procedure for isolation of F(ab')$_2$ fragments from antiserum by digestion with pepsin and F(ab')$_2$-ELISA are given.

30.3.2.1 Preparation of F (ab')$_2$ Fragments

Materials

1. Dialysis bag: For preparation refer Subheading 30.3.1.1.
2. IgG.
3. Sodium acetate buffer (0.07 M, pH 4.0): Dissolve 5.74 g of sodium acetate (anhydrous) and 4.2 g of acetic acid in about 800 mL of water, adjust pH to 4.0 using HCl or NaOH and make up the volume to 1 L.
4. Sodium chloride (0.05 M): Dissolve 2.92 g of NaCl in water and make up the volume to 1 L.
5. Pepsin: Use good commercially available source.
6. Phosphate-buffered saline (PBS): For preparation refer Subheading 30.2.1.1.
7. Sodium azide (NaN_3).

Method [Preparation of F (ab')$_2$ Fragments]

1. Dialyse 2.5 mg/mL IgG overnight at 4 °C in three changes of 0.07 M sodium acetate buffer (pH 4.0) containing 0.05 M sodium chloride.
2. After dialysis, transfer the IgG to a clean glass tube and digest by adding pepsin in acetate buffer at a concentration of 50 μg of pepsin per mg of IgG and incubate at 37 °C for 18 h.
3. Dialyse overnight at 4 °C in three changes of PBS to remove low molecular weight products of enzyme hydrolysis.
4. Add sodium azide (0.02%) and store at –20 °C.

30.3.2.2 $F(ab')_2$-ELISA test

Materials

1. As listed in DAS-ELISA.
2. $F(ab')_2$ fragments.
3. Protein A–alkaline phosphatase conjugate (commercially available).

Method

1. Add 200 μL of $F(ab')_2$ to each well of the ELISA plate that has been appropriately diluted in coating buffer and incubate for 2 h at 37 °C or overnight at 4 °C.
2. Wash the plate with PBS-T as in **step 2** of DAS-ELISA procedure.
3. Add test antigen as in **step 3** of DAS-ELISA.
4. Wash the plate with PBS-T as described earlier.
5. Add 200 μL of blocking solution (1% ovalbumin in PBS-T) per well and incubate at 37 °C for 1 h.
6. Wash the plate as described earlier.
7. Add 200 μL/well of antiserum diluted in antibody buffer and incubate for 2 h at 37 °C.
8. Wash the plate as described earlier.
9. Add 200 μL per well of commercial protein A conjugated with alkaline phosphatase appropriately diluted in antibody buffer. Incubate at 37 °C for 2 h or overnight at 4 °C.
10. Wash the plate as described earlier.
11. Add 200 μL of substrate (p-nitro phenyl phosphate, 0.6 mg/ mL of substrate buffer) to each well and incubate at room temperature.
12. Stop reaction if desired by adding 50 μL per well of 3 N NaOH.
13. Measure the absorbance at 405 nm in an ELISA reader.

30.3.3 Plate-Trapped or Triple-Sandwich ELISA

Plate-trapped or triple-sandwich ELISA is also a form of indirect ELISA wherein antibody produced in two different animal systems against the same virus is used. The method is routinely used in monoclonal antibody production for the screening of hybridoma. Initially, polyclonal antiserum produced in rabbit is used to trap the antigen. After addition of antigen, the hybridoma secreting antibodies (obtained by fusing spleen cells of the mouse injected with the virus and cancerous cells) are used as detecting antibody. The antigen–antibody reaction is visualized by adding anti-mouse IgG-alkaline phosphatase (Fig. 30.9) (Koenig 1981; Koenig and Paul 1982; Clark et al. 1986).

30.3.3.1 Materials

1. As listed in DAS-ELISA.
2. Polyclonal antiserum (or IgG) produced in rabbit for coating the wells.

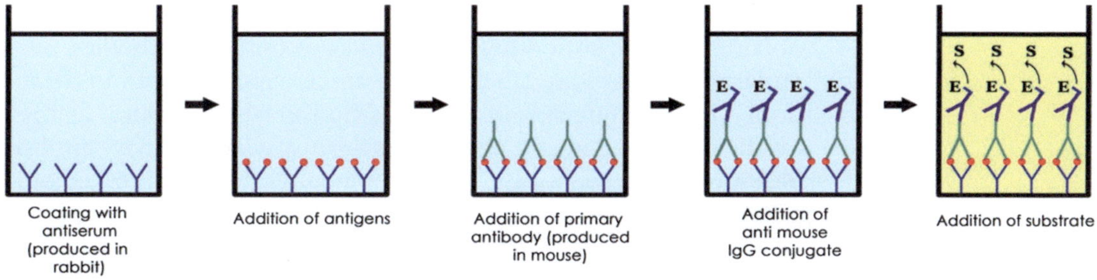

Fig. 30.9 Schematic flowchart of triple sandwich or plate trapped ELISA

3. Polyclonal or monoclonal antibodies produced in mouse against the same antigen.
4. Anti-mouse IgG-alkaline phosphatase conjugate: Commercially available.

30.3.3.2 Method

1. Add 200 μL/well of polyclonal antiserum (or IgG) produced in rabbit to each well of the ELISA plate that has been appropriately diluted in coating buffer and incubate for 2 h at 37 °C or overnight at 4 °C.
2. Wash the plate with PBS-T as in **step 2** of DAS-ELISA procedure.
3. Add test antigen as in **step 3** of DAS-ELISA.
4. Wash the plate with PBS-T as described earlier.
5. Add 200 μL of polyclonal or monoclonal antibodies produced in mouse against the same antigen appropriately diluted in antibody buffer and incubate at 37 °C for 2 h.
6. Wash the plate as described earlier.
7. Add 200 μL per well of commercial anti-mouse conjugated with alkaline phosphatase conjugate appropriately diluted in antibody buffer. Incubate at 37 °C for 2 h or overnight at 4 °C.
8. Wash the plate as described earlier.
9. Add 200 μL of substrate (p-nitrophenyl phosphate, 0.6 mg/mL of substrate buffer) to each well and incubate at room temperature.
10. Stop reaction if desired by adding 50 μL per well of 3 N NaOH.
11. Measure the absorbance at 405 nm in an ELISA reader.

30.3.4 Direct Antigen-Coated ELISA (DAC-ELISA)

DAS-ELISA requires preparation of conjugates specific to each of the viruses to be tested, thus making it cumbersome and expensive. This problem is overcome in DAC-ELISA as in DAC-ELISA, the presence of the antigen–primary antibody complex is detected by antibody produced against the immunoglobulin fraction of blood of animal species from which the primary antibodies had been obtained. These antibodies are called as secondary antibodies. Secondary antibodies are labelled with an enzyme and used in detection. These conjugates are called as 'universal conjugates' as they can be used for detecting all viruses against which primary antibodies are produced in one kind of animal (such as rabbit). Commonly used conjugates of this type are alkaline phosphatase-labelled goat anti-rabbit IgG (for polyclonal antibodies as many polyclonal antibodies are produced in rabbits) or alkaline phosphatase-labelled goat anti-mouse IgG (for monoclonal antibodies). The sensitivity of DAC-ELISA is usually lower than that of DAS-ELISA. However, it is better suited to study serological relationships among different viruses (Clark and Adams 1977; Clark et al. 1986).

The necessary steps in DAC-ELISA include coating the wells of microtitre plate with antigens, followed by blocking the remaining reactive sites of the wells by the addition of non-specific heterologous proteins such as bovine serum albumin or nonfat spray dried milk. Wells are then added with primary antibody (not labelled with the enzyme). The detection of the reaction is done by addition of enzyme-labelled second animal antibody (against antibodies of the first animal) and its substrate (Fig. 30.10).

30.3.4.1 Materials

1. As in DAS-ELISA.
2. Antigen extraction buffer: Coating buffer containing 2% polyvinyl pyrolidone.
3. Anti-rabbit IgG-alkaline phosphatase conjugate (commercially available).

30.3.4.2 Method

1. Add test antigens ground in coating buffer containing 2% polyvinyl pyrrolidone (PVP MW 40,000) to ELISA plate at 200 μL per well. Incubate 1–2 h at 37 °C or overnight at 4 °C.

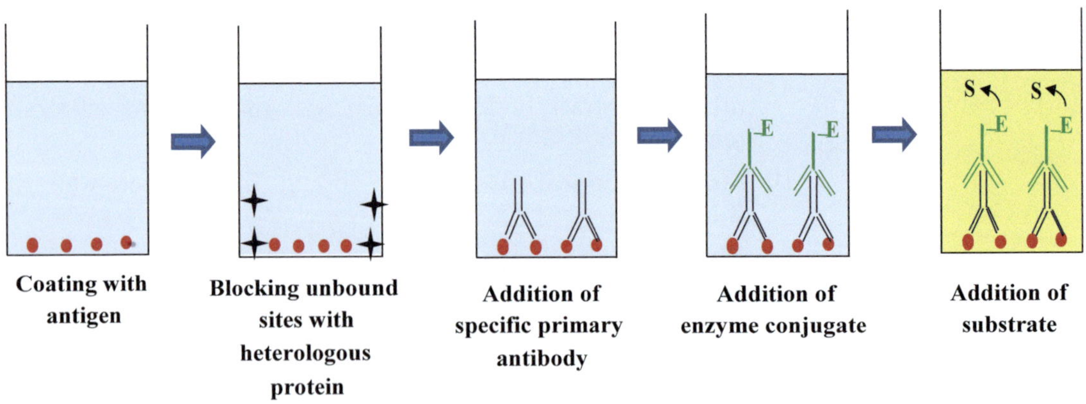

Fig. 30.10 Basic steps involved in DAC-ELISA

2. Wash wells with PBS-T three times as described in **step 2** of DAS-ELISA.
3. Add 200 μL per well of blocking solution (1% ovalbumin in PBS-T) and incubate at 37 °C for 1 h.
4. Wash the plate three times with PBS-T as described earlier.
5. Add primary antiserum or IgG, diluted in antibody buffer at the rate of 200 μL per well and incubate at 37 °C for 2 h.
6. Wash the plate as described earlier.
7. Add conjugated second animal species (anti-first animal immunoglobulin) antibody (anti-rabbit IgG-alkaline phosphatase conjugate) diluted in antibody buffer and incubate for 2 h at 37 °C or 4 °C overnight.
8. Wash the wells three times with PBS-T.
9. Add p-nitro phenyl phosphate (PNPP) at 0.6 mg/mL of substrate buffer, 200 μL per well and incubate at room temperature until colour develops.
10. Measure the intensity of colour in each well at 405 nm by using ELISA reader after 15, 30 and 60 min. Stop the reaction by adding 50 μL of 3 M NaOH (stop solution) to the wells.
11. Compare the absorbance values of the test samples with healthy control. Consider samples showing absorbance (OD_{405}) values more than two times of healthy control as positive.

30.4 Dot Immunobinding Assay (DIBA)

The basic principle of dot immunobinding assay is similar to that of the ELISA. In a dot immunobinding assay (DIBA), plant extract is directly dotted onto a nitrocellulose membrane (NCM) by hand or with a manifold usually in a grid pattern. A non-reactive heterologous protein to prevent non-specific adsorption of antibody to the membrane then blocks the remaining membrane surface. This step is done using bovine serum albumin, gelatin or non-fat spray-dried milk. After incubation with the primary antibody, the unreacted antibody is rinsed away, and the membrane is incubated with the second antibody conjugated with an enzyme. The enzyme is later reacted with a soluble substrate to form an insoluble coloured product at the site of reaction indicating which samples contained antigen of interest (Fig. 30.11). The intensity of colour can be read visually or through a densitometer. It has also been referred to as "dot ELISA" and found to be at least twofold more sensitive compared to ELISA on microtitre plates. Several procedures (like direct or indirect ELISA) utilizing various enzymes, enzyme conjugates, extraction buffers and membranes are available. The other advantages of DIBA include detection in subtle volumes of test samples, it is more suited in disease surveys because the membranes

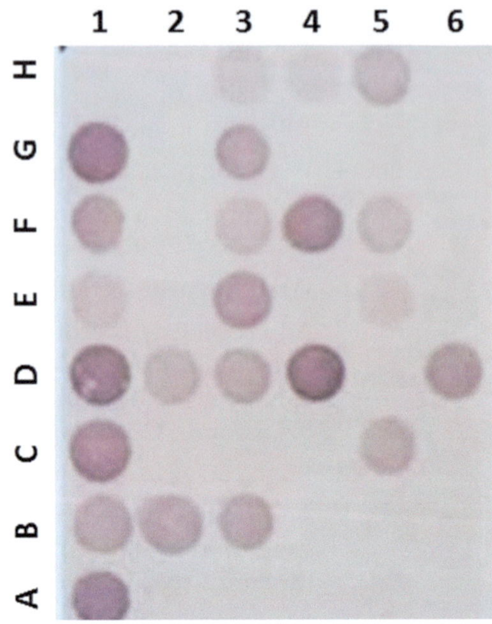

Fig. 30.11 Dot immunobinding assay (DIBA) for the detection of the virus. High intensity of the spots indicates the high amount of antigen (virus)

can be transported more easily than ELISA and the membranes are easier to process than ELISA plates (Banttari and Goodwin 1985; Hampton et al. 1990).

30.4.1 Materials

1. Tris-buffered saline (TBS) (0.02 M Tris and 0.15 M NaCl) (pH 7.5): Dissolve 2.42 g of Tris base and 8.76 g of NaCl in about 900 mL of water, adjust the pH to 7.5 using HCl and make up the volume to 1000 mL.
2. Blocking solution (TBS-SDM) (5% spray dried milk (SDM) in TBS): Add and dissolve 5 g of SDM in 100 mL of TBS.
3. Antibody/conjugate buffer: Same as TBS-SDM.
4. Antigen extraction buffer (TBS containing 50 mM diethyl dithiocarbamate): Dissolve 8.5 g of sodium diethyl dithiocarbamate in 1 L of TBS.
5. Chloroform.
6. Enzyme-labelled anti-rabbit IgG (secondary antibody or universal conjugate): Commercially available.
7. Fixing solution (10 mM Tris, 1 mM EDTA) (pH 7.5): Dissolve 1.21 g of Tris base and 0.1336 g of EDTA disodium salt in about 800 mL of water, adjust the pH to 7.5 using HCl and make up the volume to 1 L.
8. Glass dish or disposable plastic Petri dishes for processing membrane (do not use polystyrene as it can bind antibody).
9. Nitrocellulose membrane (NCM) or nylon or PVDF membrane.
10. Pencil.

11. Shaker.
12. Substrate buffer (0.1 M Tris, 0.1 M NaCl, 5 mM $MgCl_2$) (pH 9.5): Dissolve 12.1 g of Tris base, 5.84 g of sodium chloride and 0.475 g of magnesium chloride in about 900 mL of water, adjust the pH to 9.5 with HCl and make up the volume to 1 L.
13. Substrate solution: Prepare solution A by dissolving 75 mg of nitro blue tetrazolium (NBT) in 1 mL of dimethyl formamide (DMF) and solution B by dissolving 50 mg of 5-bromo-3-chloroindolyl phosphate (BCIP) in 1 mL of DMF. Add solution A and solution B to give a final concentration of 0.33 mg and 0.175 mg, respectively, per mL of the substrate buffer just before use. For example, to prepare 10 mL of the substrate solution, dissolve 44 μL of solution A and 37 μL of solution B.
14. Virus-infected and healthy plant samples.
15. Virus-specific antiserum (primary antibody).

30.4.2 Method

1. Extract tissue 1:10 (w/v) in antigen extraction buffer by grinding in a mortor and pestle and filter through a muslin cloth.
2. Take 0.8 mL of filtrate into a microfuge tube and add 0.4 mL of chloroform, vortex and centrifuge at 12,000 × *g* for 2 min.
3. Take 200 μL of the clarified sap into 800 μL of antigen extraction buffer and vortex.
4. Cut the desired size of nitrocellulose membrane (NCM) and make squares of 1 cm × 1 cm with a pencil (handle NCM with gloves/forceps) (alternatively nylon or PVDF membrane can also be used).
5. Wet NCM by floating it on TBS and then air dry.
6. Spot the test and control samples (5 or 10 μL each) as per the layout prepared.
7. Allow the NCM to air dry and immerse in blocking solution with gentle shaking for 1 h at room temperature.
8. Rinse the membrane once in TBS for 10 min.
9. Incubate the NCM with primary antibody (antiviral antibody produced in rabbit) diluted in TBS-SDM for 1 h at room temperature.
10. Wash the NCM three times 10 min each with TBS.
11. Transfer the NCM into the second antibody (anti-rabbit IgG conjugated with alkaline phosphatase) diluted in TBS-SDM for 1 h at room temperature.
12. Wash the NCM three times as mentioned above.
13. Incubate the NCM in substrate solution at room temperature and watch for colour development (good colour development

take 5–10 min). The positive samples show deep purple colour (*see* Fig. 30.11).

14. Rinse the NCM twice (10 min each) in fixing solution and then air dry. Record the result visually or through a densitometer. Photograph the NCM (if required) and store protected from light.

30.5 Tissue Blotting Immunoassay (TBIA)

It is a form of DIBA where tissue is directly applied to membranes without any extraction. Tissue blots are made by pressing the freshly cut tissue surface directly on nitrocellulose or nylon membranes. Antigens in tissue blots on membranes are detected by enzyme-labelled immunological probes essentially as described in DIBA. This technique has been applied to viruses belonging to the genera *Cucumovirus, Luteovirus, Potexvirus, Potyvirus* and *Tospovirus*. In a few cases, background due to pigments from plant tissue may interfere. Use of detergents like Triton X-100 or sodium hypochlorite has been recommended to remove such pigments. Like DIBA, once the tissue is applied to membranes, it can be stored for several weeks, thus suitable for surveys (Lin et al. 1990; Hsu 1991; Samson et al. 1993).

30.5.1 Materials

1. As listed in DIBA.
2. Razor blade.

30.5.2 Method

1. Collect infected and healthy tissues (leaves, petioles, stem, flowers, tubers, etc.).
2. For thin tissues prepare them to form a tight roll.
3. Cut the tissues with 'sharp razor blade' to obtain a single plane cut surface.
4. Immediately press the cut surface onto a nitrocellulose/nylon membrane to form a visible print (tissues containing high concentration of latex, the cut surface may be first drained on tissue paper to remove excess exudates before blotting onto membrane). Air-dry the membrane.
5. Process the membrane as described in DIBA. The positive samples show deep purple colour (*see* Fig. 30.12).

30.6 Electro-Blot Immunoassay (EBIA)/Western Blotting

Electro-blot immunoassay (EBIA)/Western blotting combines the protein fractionating power of gel electrophoresis with the sensitivity and specificity of solid-phase immunoassay. The fundamental principle is the transfer of electrophoresed proteins from the gel

Fig. 30.12 Detection of virus through tissue blot immunoassay (TBIA)

matrix onto a membrane and rendering them accessible for further analysis by specific immunoprobes. The advantage of the technique over dot immunobinding assay is that it identifies molecular weight of the protein and can be used to detect impurities in antigens or antisera besides its use in the detection, identification, quantification and comparison of plant viruses.

The basic steps involved in Western blotting are: (1) separation of proteins according to their molecular weight by sodium dodecyl sulphate-poly acrylamide gel electrophoresis (SDS-PAGE), (2) transfer of proteins from gel matrix onto a membrane (electroblotting), (3) blocking of free binding sites on the membrane with heterologous proteins (BSA or skimmed milk), (4) incubation with specific (primary) antibody, (5) incubation with enzyme-labelled anti-immunoglobulin molecule (secondary antibody) and (6) incubation with enzyme substrate and determination of enzyme-mediated colour detection (Fig. 30.13). The substrate used are chromogenic, and the colour and insolubility of the product are ideal for the detection of proteins on the nitrocellulose membrane. The intensity of colour produced is proportional to the amount of target protein on the membrane (O'Donell et al. 1982; Rybicki and Von Wechmar 1982).

The technique can be employed in the detection of viral proteins (either structural or non-structural), to determine serological relationships among viruses, epitope mapping as well as identification of virus-encoded proteins and their precursors under in vitro translation systems.

30.6.1 Materials

1. As in SDS-PGE (*see* Chapter 24)
2. As in DIBA.
3. Filter paper (Whatman No 3).
4. Forceps.

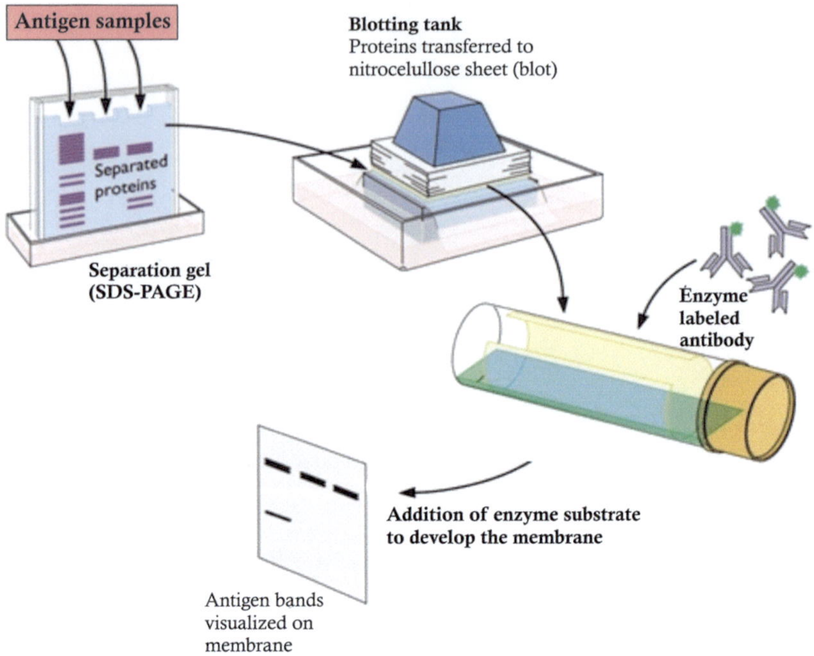

Fig. 30.13 Basic steps involved in electro-blot immunoassay (EBIA)

5. Glass tray or Petri plates for processing membrane.
6. Pre-stained protein molecular standards: Commercially available.
7. Purified virus or clarified extracts from infected plants.
8. Scissors.
9. Transblot (western transfer) apparatus: For transferring proteins from the gel to membrane.
10. Vertical gel apparatus with power pack.
11. Blotting/transfer buffer (TBS) (39 mM Tris base, 48 mM glycine in 20% methanol) (pH 8.3): Dissolve 4.719 g of Tris base and 3.6 g of gylcine in about 700 mL of water, adjust the pH to 8.3 using HCl, add 200 mL of methanol and finally make up the volume with distilled water to 1 L.

30.6.2 Method

SDS-PAGE

1. Carry out SDS-PAGE essentially as described earlier (*see* Chapter 24). SDS-PAGE is done in duplicates, and one gel will go for staining and de-staining after SDS-PAGE while the other will be carried forward through electro-blotting. Use the gel loaded with pre-stained marker proteins (in place of ordinary marker proteins) for electro-blotting.

Electro-blotting

2. When SDS-PAGE run time is going to be over, wear gloves and cut Whatman 3 filter paper (four strips), and nitrocellulose membrane (one strip) of the exact size of SDS-polyacrylamide gel and mark left the corner of the filter.
3. Remove gel from the apparatus and soak in transfer buffer. Also, float Whatman 3 filter paper and membrane strips on the surface of transfer buffer for 5 min.
4. Assemble 'transfer-sandwich' under buffer by placing SDS-PAGE gel on the nitrocellulose membrane and sandwiching these between filter paper strips, sponge pads and the sandwich holder (Fig. 30.14).
5. Transfer the 'transfer sandwich' to the blotting apparatus containing transfer buffer with nitrocellulose membrane on the anode side of the gel.
6. Run at a constant voltage of 20 V for 2–4 h.

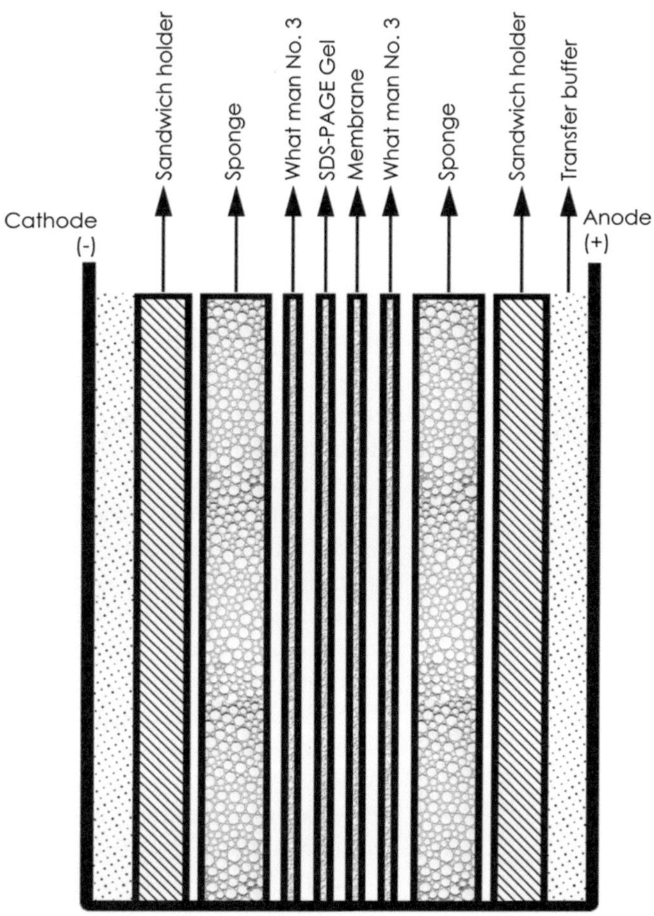

Fig. 30.14 Assembly of western transfer sandwich for transfer of proteins from the gel to membrane through electro-blotting

Immunological Probing

7. After electro-blotting, disassemble the transfer sandwich layer one by one and transfer the blot to a small tray containing blocking solution (gel can be stained with Coomassie blue to check whether the transfer is complete). Leave the blot in blocking solution with gentle shaking for 1 h at room temperature or 4 °C overnight.
8. Wash the blot three times in TBS for 10 min each at room temperature.
9. Incubate the blot in the primary antibody (against target protein) diluted appropriately based on titre of the antiserum in TBS for 1–4 h at room temperature with gentle shaking.
10. Wash the blot in TBS as above.
11. Transfer and incubate the blot in the second antibody conjugated with alkaline phosphatase diluted appropriately in TBS with gentle shaking for 1–4 h at room temperature.
12. Wash the blot as mentioned earlier.
13. Submerge and incubate the blot in substrate prepared in substrate buffer. Substrate consists of nitro blue tetrazolium (NBT) and bromochloro-indolyl phosphate (BCIP) at 0.33 mg and 0.175 mg per mL of substrate buffer. Avoid light.
14. After 20 min, stop the reaction by transferring the blot to stop solution before background colour becomes significant. Rinse twice in stop solution.
15. Air-dry the membrane and store under folds of blotting paper at room temperature protected from light. The blot may be photographed either wet or dry.
16. The time for adequate colour development will vary with concentration of reactants, the amount of antigen present and desired sensitivity. Typically right colour will develop within 5–10 min. However, if no significant reaction is observed within 20 min, it is unlikely to occur. Binding between antigen in the membrane and enzymatically labelled antibody leads to the production of a coloured product that appears as a distinct band at the site of antigen (*see* Fig. 30.15). The molecular weight of the band can be estimated by comparing the mobility of pre-stained marker protein bands as explained in SDS-PAGE. Presence of more than one band could be due to many reasons: (1) degraded product of the protein of interest (when a band or a series of bands seen lower than the size expected), (2) dimer formation (if molecular weight of the additional band is nearly twice that of the expected), (3) presence of more than one polypeptide species in the virus coat protein. Also, if primary antibodies used for detection contains antibodies against host proteins, the protein bands corresponding to this will also get visualized in the reaction.

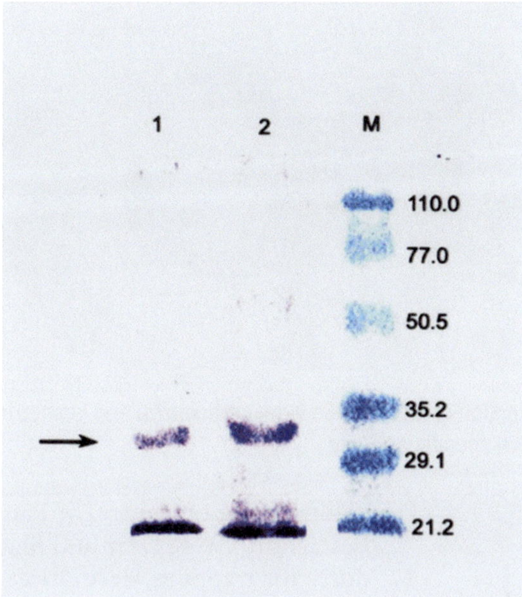

Fig. 30.15 Detection of virus through electro-blot immunoassay (EBIA). Lane M shows pre-stained protein markers. Lanes 1 and 2 were loaded with purified virus preparation

30.7 Immunosorbent Electron Microscopy (ISEM)

Immunosorbent electron microscopy (ISEM) or electron microscope serology (Derrick 1973) is an important diagnostic method for plant viruses. Two techniques, namely trapping and decoration, are commonly used. In 'trapping' the antibody-coated grid is covered with a sap homogenate, so that the antibodies trap virus particles from the solution, with the result that a higher concentration of particles becomes attached to the grid than is normally possible by usual quick leaf dip method. It is particularly useful when a virus occurs at low concentration. In 'decoration' antibodies are added to a grid that already has virus particles attached so that particles become coated or decorated with the antibodies (Fig. 30.16). It is particularly useful to detect individual viruses from a mixture of viruses. A third technique, namely gold labelling is useful when virus particles with the membrane (e.g. orthotospoviruses, rhabdoviruses) are difficult to decorate. Gold labelling is performed by adsorbing immunoglobulins or protein A to negatively charged colloidal gold particles. A protein A–gold complex binds specifically to the antigen–antibody complex, and electron-dense gold label is visible in the electron microscope as black dots (Fig. 30.17). Gold labelling is also useful for in situ identification of virus particles in thin sections of infected cells

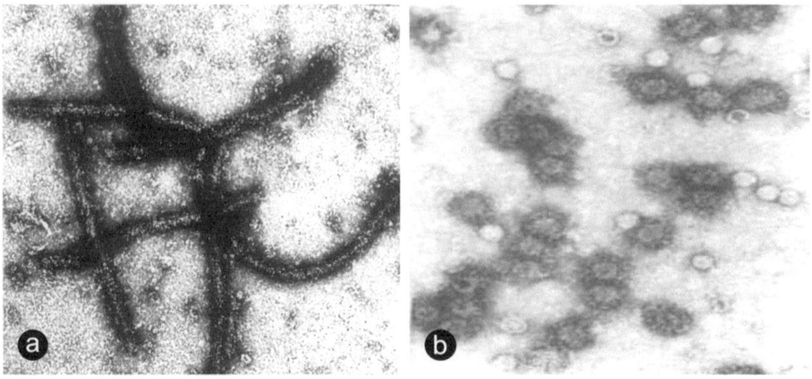

Fig. 30.16 Detection of filamentous virus particles (**a**) and isometric particles (**b**) through decoration technique using respective specific antisera

(Milne and Luisoni 1975; Pares and Whitecross 1982; Louro and Lesemann 1984; Draz and Shafiee 2018). In the following, procedures for trapping, decoration and gold labelling are given.

30.7.1 Trapping

30.7.1.1 Materials

1. Clean glass Petri dish.
2. Copper grids 400 mesh size with support film of carbon.
3. Electron microscope.
4. Filter paper.
5. Forceps.
6. Parafilm.
7. Pestle and mortar.
8. Distilled water.
9. Negative stain: Uranyl acetate 2%: Dissolve 0.2 g of uranyl acetate in 10 mL of water, filter solution with a 0.22 μm filter using a syringe, aliquot into eppendorf tubes and wrap with aluminium foil and store in dark at room temperature.
10. Phosphate buffer (0.07 M pH 6.5): Dissolve 1.14 g of Na_2HPO_4 in water and make up the volume to 100 mL. Similarly prepare 0.07 M of sodium dihydrogen phosphate (NaH_2PO_4) by dissolving 0.42 g of NaH_2PO_4 in water and make up the volume to 50 mL. To Na_2HPO_4 solution, add NaH_2PO_4 solution slowly while stirring till desired pH of 6.5 is obtained.
11. Virus-infected tissue.
12. Virus-specific antiserum.

30.7.1.2 Method

1. Float the carbon-coated grids (film side downward) over a drop of 10 μL of antiserum diluted in phosphate buffer. Ensure that grid surface becomes wet. Leave the grids for half an hour at 37 °C.

2. Pick up the grid by its edge with fine forceps and wash the grids for 10–15 min in phosphate buffer.
3. Drain the grids briefly and then place them over a drop of 10 μL of the extract containing the virus. Leave them for 45 min at room temperature. Partially purified preparation or extract prepared by macerating infected tissue in phosphate buffer is used.
4. Remove the grids, wash with approximately 20 drops of distilled water.
5. Stain with 2–4 drops of 2% freshly prepared uranyl acetate stain.
6. Drain the grid to dry, and examine in an electron microscope at an appropriate magnification. Count the number of virus particles from different areas of the grid and compare with a grid prepared by leaf dip preparation.

30.7.2 Decoration

30.7.2.1 Materials

1. As listed in 'Trapping.'

30.7.2.2 Method

1. Follow **steps 1–4** as in 'trapping' procedure except that the washing at fourth step will be in phosphate buffer.
2. Place the grid over specific antibodies diluted in phosphate buffer for 30 min at room temperature.
3. Wash the grid as earlier with distilled water and stain with 2% uranyl acetate.
4. Drain the grid to dry and examine in an electron microscope. Observe coating or decoration of virus particles due to binding of antibodies on the particle surface (Fig. 30.16).

30.7.3 Immunogold Labelling

This technique was introduced by Pares and Whitecross (1982) and later modified by Louro and Lesemann (1984) and Draz and Shafiee (2018). Gold labelling is performed by adsorbing immunoglobulins or protein A to negatively charged colloidal gold particles. A protein A–gold complex binds specifically to the antigen–antibody complex, and the electron-dense gold label is visible in electron microscope as black dots (Fig. 30.17). This can be done either on fresh tissue or fixed tissue. It is used for fresh tissues when the fragment of virus particles especially with the membrane is difficult to decorate such as orthotospoviruses and rhabdoviruses. In such labelling conditions, an antibody with gold can offer definite advantages. The technique is useful for fixed tissue as this will help to find out the in situ location of virus particles. Immunogold labelling with fixed tissue can be done in two ways: pre-embedding and post-embedding. In the pre-embedding, labelling will be done before embedding the tissues while in the post-embedding,

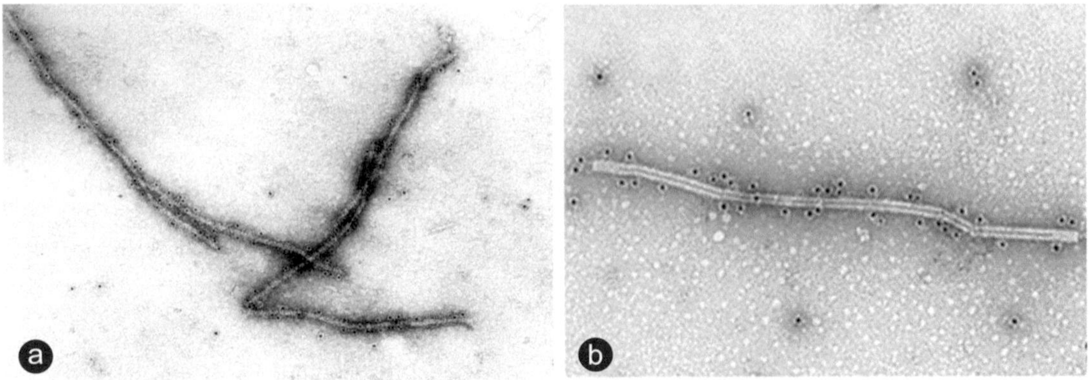

Fig. 30.17 Immunogold labelling of plant virus particles in ISEM. The dark pin head-sized dots indicate the location of gold (**a**, **b**)

sections taken out after embedding are labelled and stained. In the following protocol for all the three methods are given.

30.7.3.1 With Fresh Tissues

Materials

1. As listed in Trapping.
2. Secondary antibody or protein A conjugated with gold 10 nm size.
3. Bovine serum albumin.

Method

1. Grind the infected leaf about 2–3 mm diameter over a clean glass slide in 0.07 M phosphate buffer, pH 6.5.
2. Place 10 μL of sap homogenate on parafilm.
3. Place a carbon-coated grid (film side downward) on the surface of the droplet, ensuring that the grid surface becomes wet. Leave for 2–10 min.
4. Pick up the grid by its edge with fine forceps and wash the grid from the filmed surface using a continuous flow of 10–15 drops of double-distilled water to remove the sap.
5. Carefully add 2–4 drops of 2% uranyl acetate on the filmed surface of the grid.
6. Remove excess stain with a strip of filter paper and rinse carefully with 0.1 M phosphate buffer.
7. Place the grid for 1 h at room temperature on 1% bovine serum albumin (BSA) in 0.1 M phosphate buffer pH 7.0 (BSA-PB) for blocking remaining surface of the grid.
8. Incubate the grid for 15 min on specific antibody diluted in BSA-PB.
9. Rinse with BSA-PB and transfer the grid to a second antibody or protein A conjugated with gold (commercially available), diluted in BSA-PB for 30 min to 1 h.

10. Rinse with phosphate buffer and stain with 1% uranyl acetate.
11. Drain the grid to dry and examine under electron microscope. Virus particles show decoration with antibody and gold particles along the virus (*see* Fig. 30.17).

30.7.3.2 With Pre-embedded Tissues

Materials

1. As listed in Chapter 23 on electron microscopy with ultrathin sections.
2. Secondary antibody (anti-rabbit IgG) or protein A conjugated with gold 10 nm size.
3. As listed in Sub heading 30.7.1.1.

Method

1. Fix washed and cut tissue of 1 × 1 mm size in 1% GA diluted in 0.1 M PO_4 buffer, pH 7.0 for 1 h. Vacuum infiltrate for 10 min and keep in GA for 24 h.
2. Wash the tissue with PO_4 buffer for 15 min.
3. Incubate the tissue with 1% antibody diluted in PO_4 buffer containing 1% BSA and 0.1% Tween 20 (PB-BT) at room temperature for 2–4 h.
4. Remove liquid with Pasteur pipette, rinse with PB-BT for 20 min. Incubate with anti-rabbit IgG-10 nm gold or protein A-10 nm gold diluted 1/50 in PB-BT at room temperature for 24 h.
5. Rinse in PB, fix in 0.2% osmium tetroxide for 1 h.
6. Complete the embedding in epoxy resin and sectioning as described earlier (Chapter 23).
7. Stain in uranyl acetate and lead citrate as described earlier (Chapter 23) before being observed under electron microscope.

30.7.3.3 With Post-embedded Tissue

Materials

1. Same as in pre-embedded tissue.

Method

1. Carry out fixing, embedding and sectioning as described earlier (Chapter 23).
2. If osmium tetroxide is used, treat sections for 5–6 min, with a saturated solution of sodium metaperiodate (NaIO4). Wash in distilled water for 5 min.
3. Treat sections with primary antibody diluted in phosphate buffer containing 0.1–1% BSA, 0.1% Tween 20 and 0.05% sodium azide (PB-BT) for 2 h. Around 10–100 μg/mL IgG is optimal concentration for primary antibody.
4. Rinse in PB-BT, incubate with gold-labelled secondary antibody for 30–60 min.
5. Rinse in water.

6. Stain first with 5% uranyl acetate and then in 5% lead acetate as explained in the Chapter 23 and observe under an electron microscope.

30.8 Immunofluorescence

Immunofluorescence provides sensitive assays for the detection of antigens in frozen or fixed tissue sections or viable cells. By staining with specific antibody conjugated to a fluorescent chromophore and illuminating with ultraviolet light, the location of specific antigens in a tissue or cell preparation may be studied. Fluorescein emits green light and rhodamine orange light, whereas the weak natural fluorescence of some biological materials occurs in the blue region of the spectrum. In plant virology, it is used for the detection of the pathogen (both viruses and fastidious prokaryotes) and in situ location of the pathogen in the plant tissue. To perform the test, sections or sap from the infected plant is incubated with primary antibody. After washing to remove unbound antibodies, samples are incubated with secondary antibodies conjugated with fluorescent dyes such as fluorescein isothiocyanate (FITC). After washing, tissues are observed under fluorescent microscope.

30.8.1 Materials

1. Freeze microtome.
2. Binocular microscope.
3. Fluorescent microscope.
4. Cavity slide.
5. Glass slide.
6. Water bath.
7. Evans blue (1% w/v): Dissolve 0.5 g of Evans blue dye powder in 50 mL of normal saline, the solution is filter sterilized using a syringe and 0.45 μm filter into sterile 1.5 mL tubes.
8. Glycerine.
9. PBS-600 (pH 7.4): Dissolve 9.828 g NaCl, 0.09526 g KH_2PO_4, 1.4325 g $Na_2HPO_412H_2O$, 0.096915 g KCl in 500 mL distilled water.
10. PBS-T (PBS containing 0.05% Tween 20): Add 500 μL of Tween 20 in 1 L of 1× PBS and use.
11. Secondary antibodies conjugated with FITC (commercially available).
12. Virus infected and healthy tissues.
13. Virus-specific polyclonal or monoclonal antibodies.

30.8.2 Method

1. Make 3–4 mm long pieces of midribs of the test sample (leaves).
2. Cut 10 mm thin sections in a freeze microtome in ice blocks.
3. Select the appropriate sections having intact vascular bundles under a binocular microscope and place them in cavity slides.
4. Incubate slides 55 °C for 90 min.
5. Cover each section with a drop of appropriately diluted specific antiserum.
6. Incubate at room temperature for 1 h.
7. Wash thrice in PBS-T using a Pasteur pipette.
8. Drop 15 mL conjugate (1 mL PBS + 10 mL secondary antibodies conjugated with FITC +10 mL Evan's blue) over each section.
9. Incubate in dark for 30 min.
10. Mount in glycerine 1:1 (v/v) in PBS.
11. Observe under a fluorescent microscope at 40× magnification. Apple green fluorescence will be observed in affected tissue of the section, especially in the phloem. Healthy tissues show dark red colour (Fig. 30.18).

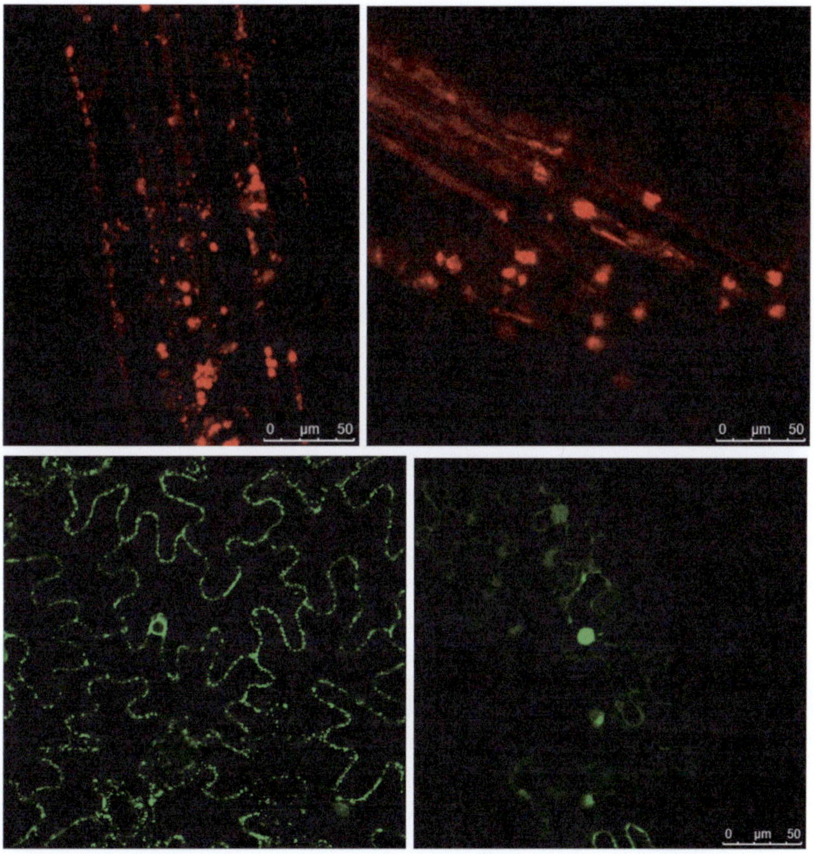

Fig. 30.18 Immunofluorescence reaction in virus-infected plant tissues

30.9 Lateral Flow Immunoassay Assay (LFIA)

Lateral flow immunoassay (LFIA) is an immunochemical-based method in which the fluid from the sample runs through a porous membrane (often nitrocellulose) by capillary force. LFIAs are based on the principle of immunochromatography where the specific interaction of antigen and antibody is responsible for colour development. LFAs are prefabricated strips of a carrier material containing dry reagents that are activated by applying the fluid sample. LFIAs are especially designed for single use at point of care/need, i.e. outside the laboratory, and are important for user-friendly diagnostic purposes. It is available for diagnosis of several viruses, bacteria and fungal pathogens and also used in medical and veterinary practice aimed to indicate the cause of an infection, ascertain pregnancy or failure of internal organs (e.g. heart attack, renal failure or diabetes). A lateral flow test incorporates a sample pad that is intimately associated with a conjugate pad. This, in turn, touches a membrane onto which test and control reagents have been immobilized. An absorbent pad wicks fluid away from the membrane. In these kits, the pathogen-specific antibodies (capture antibodies) and species-specific antibodies are pre-immobilized onto membranes at the test and control lines, respectively (Fig. 30.19). To perform the test, the sample is ground and the extracted sap from plant tissue is mixed with pathogen-specific antibody conjugated with coloured latex beads or gold particles. The pathogen-specific antibody within it will form a complex with tagged antibodies. The sample pad region of the strip is then immersed in this solution. The bound antigen/antibody complex will continue to move by lateral flow through the membrane until it reaches the test line (where pathogen-specific capture antibody is immobilized). At the test line, the movement of pathogen-specific

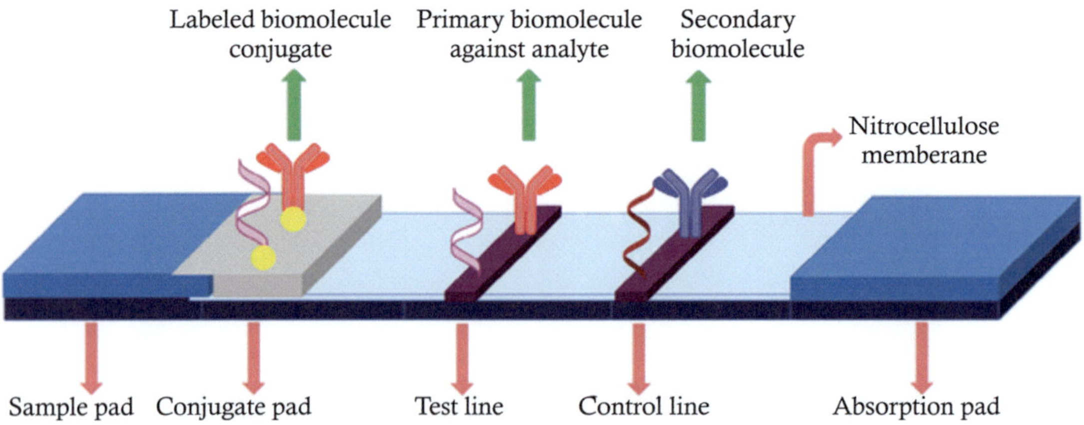

Fig. 30.19 Schematic representations showing the lateral-flow immune-assay (LFIA) strip assembly

antigen–antibody complex is arrested as they get bound the pathogen-specific capture antibody. As more and more antigen–antibody complex captured at the test line, the line becomes visible as the coloured latex or gold accumulates. The second capture line is the control line, containing species-specific antibodies designed to capture all unbound tagged antibodies that would also become visible to confirm that the assay has run correctly (Koets et al. 2006; Nielsen et al. 2008; Posthuma-Trumpie and van Amerongen 2013; Kulabhusan et al. 2017).

Generally, the LFIA technique is used in the form of an immunostrip, simply dipped into the fluid sample, or mounted in a device for easier handling. On this membrane at least one test line and a control line have been sprayed. The test line can be made up of anti-analyte antibody (primary antibody) or antigen, which in case of a hapten is conjugated to a protein. The control line is made up of immunoglobulin (IgG) raised against the animal species of the labelled antibody.

30.9.1 Preparation and Assembly of the Strip

To develop LFAs the minimal requirements in a research laboratory are a sprayer and a cutter to cut the strips that come as a sheet or as a roll. The preparation starts with the immobilization of the capture agent on the strip. Strips can be ordered with a backing of a polymer film; otherwise one has to stick the strip material to a backing first. The material is cut to an appropriate size, e.g. 20 × 5 cm, and the capture reagent for the test line and the control line (if applicable) is dissolved in a low salt buffer of high pH (e.g. 5 mM borate buffer pH 8.5 to an appropriate concentration between 100 and 1000 μg/mL). The solutions are sprayed on the strip at appropriate distances from the bottom by using piezoelectric dispensing using a Biodot or Kinematic sprayer. Another option is to pipet 1 μL of the solution for initial research. After spraying, the strip is dried at 37 °C for at least 2 h, or overnight. Separately, the sample and conjugate pads are being prepared. Together with the absorbent pad these pads are placed in particular positions on the polymeric membrane. Subsequently, a backing card is applied to the strip and the resulting assemblage is cut into sticks, for example 5 × 0.5 cm using a guillotine cutter.

30.9.1.1 Test and Control Lines

These lines are typically an enzyme, antibody, antigen, nucleic acid, etc. These are immobilized on a membrane, usually made from nitrocellulose. The decision on whether to opt for contact versus non-contact is taken as per membrane manufacturers' recommendations, and on the kinetics with which the immobilization occurs in the membrane matrix. Both FrontLine and BioJet (fully quantitative dispensers) are used to apply the lines. BioDot offers contact (FrontLine™) and non-contact (BioJet™) technologies. The FrontLine meters liquid using a high-resolution pump to drive a low volume syringe connected to a dragging tip. The BioJet drop-on-demand technology dispenses individual drops from 20 nL to 4 μL. The BioJet systems produce clearly defined, fine lines by overlapping individual drops at a tight pitch. The ability to control drop size, line volume and energy of the drop gives the researcher the control needed in the critical step of marrying reagents to materials.

30.9.1.2 Conjugate

LFIAs that use metal nanoparticles for colorimetric detection of targets are simple, rapid, low-cost and convenient for field deployment. Gold nanoparticles (AuNPs) are the most commonly used visual readout detection labels in LFIAs developed for a variety of proteins, metal ions, hormones, bacteria, viruses, etc. Colloidal gold, monodisperse latex or paramagnetic particle conjugates are applied to a variety of pad materials using commercial AirJet dispenser. These conjugate solutions typically contain high concentrations of sugars and polymers for stabilization and release.

30.9.1.3 Pad and Membrane Treatments

Conjugate pads, sample pads and membranes all typically require either pre-treatment or post-treatment in order to make them hydrophilic, to stabilize proteins and the membrane structure, and/or to prevent non-specific binding. These treatments can be performed either by quantitative dispensing using an BioDotAir Jet Quantito evenly and quantitatively apply the conjugate to the pad in a surface coating, which improves the efficiency of drying and conjugate release over immersion application methods followed by drying in high-efficiency forced air ovens.

30.9.1.4 Membrane Materials

The membrane used as strip material is a very important part of the test system. Most often the strip material is made of nitrocellulose of size 5 × 0.5 cm (Whatman, Sartorius, Pierce, or MilliPore). This material comes in different pore sizes, usually indicated by the time it takes for a fluid front to travel a distance of 4 cm/min called the capillary rise time or wicking rate (min/4 cm).

30.9.1.5 Material of the Sample Pad, Conjugate Release Pad and Absorbent Pad

The sample pad can be used to perform multiple tasks, of which the promotion and even distribution of the sample to the conjugate pad is most important. The sample pad can also be impregnated with proteins, detergents, viscosity enhancers and buffer salts to influence the ionic conditions and flow rate of the sample. The conjugate release pad, made of cross-linked silica, is in close contact with the strip material and the sample application pad. Labelled analyte(s) or recognition element(s) (depending on the application) are dried on this pad, and after addition of the (buffer-diluted) sample, this material dissolves in the fluid; the formation of specific interactions will be initiated here and will continue during the chromatographic process. The primary function of the absorbent pad is to drain off the fluid from the membrane.

30.9.1.6 Labels

Now a days many labels are used, often these are made of coloured nanoparticles based on gold, but enzyme labels are being applied as well. In that case there is no strict single-step assay, as the substrate of the enzyme has to be added in a second step.

30.9.2 Steps in Lateral Flow Immunoassay

The assembly of LFIA strips is composed of three steps.

30.9.2.1 Production of Polyclonal Antibody Against Plant Virus

Polyclonal antiserum against plant virus need to be raised in New Zealand white rabbits using purified virus or recombinant coat protein of the virus (see details of polyclonal antibody production in Chapter 28 and recombinant coat protein production in Chapter 27). After 6 weeks, antiserum is collected from the immunized animals, which is then subjected to standard purification steps such as ammonium sulphate precipitation (50%), overnight dialysis, and protein A affinity column chromatography to obtain IgG fractions. SDS-PAGE and ELISA are performed to determine the purity and titre of the antibody. The virus polyclonal antibody (pAb) thus obtained is stored at −80 °C and used in subsequent experiments.

30.9.2.2 Synthesis of Colloidal Gold Nanoparticles–Antibody Conjugate

Gold nanoparticles (AuNPs) used as detection label in the LFIA are synthesized by standard citrate reduction method.

Materials

1. Polyclonal antibodies.
2. Gold nanoparticles.
3. Potassium carbonate (K_2CO_3) (0.2 M): Dissolve 2.764 g of potassium carbonate in 100 mL of water.
4. BSA (1%): Dissolve 100 mg of BSA in 10 mL of water.
5. Phosphate-buffered saline (PBS): For preparation of this buffer, *see* Subheading 30.2.1.1.
6. Spectrophotometer.

Method

1. Add polyclonal antibodies (pAb) (25 μg/mL) to 500 μL of gold nanoparticles (AuNPs) (pH 9 using 0.2 M K_2CO_3) and allow to form AuNPs-pAb conjugate at ambient temperature with gentle stirring for 10 min.
2. Centrifuge the AuNPs–pAb conjugate at 15,000 × *g* for 15 min.
3. Add 10 μL of 1% BSA to the soft pellets, then resuspend in 0.1 M phosphate-buffered saline (PBS).
4. Confirm the pAb–AuNPs conjugate formation by UV-Vis spectroscopy, gel retardation assay and ELISA before storage at 4 °C for further use.

30.9.2.3 Assembly of LFIA Strips in Polypropylene Cassettes

The LFIA strips are assembled in polypropylene cassettes (5 cm × 1 cm) with sample and observation windows. The cassettes encased four overlaying components, viz. sample application pad, conjugate release matrix, high protein binding nitrocellulose membrane (0.8 μ) and absorbent pad. The polypropylene cassettes, all types of overlaying components and the necessary hardware/machinery (antibody printer, strip cutter and reagent dispenser) are procured. The LFIA devices are assembled as follows:

Materials

1. Polypropylene cassettes (5 cm × 1 cm).
2. Nitrocellulose membrane.
3. Absorbent pad.
4. Antibody printing machine.
5. Antivirus–AuNP conjugate.
6. Sample pad.
7. Conjugate release matrix.
8. Absorbent pad.
9. Strip cutter.

Method

1. Print the test (T) and control (C) lines of anti-pAb (1 mg/mL) and protein A (1 mg/mL), respectively, on nitrocellulose membrane (NC) using automated antibody printing machine.
2. Soak the conjugate release matrix with anti-virus pAb–AuNPs and place at one end of the NC membrane.
3. Place the sample pad in such a way that it overlaps the conjugate release matrix.
4. Place the absorbent pad at the opposite end of the sample pad to collect excess reagents.
5. After fabrication of these components, cut the assay strips to a size of 4 × 0.4 cm (l × b) by using a programmable strip cutter and encase.
6. The entire LFIA cassette fabrication procedure is carried out under dust-free conditions in a laminar airflow chamber.

30.9.3 Standardization of LFIA Procedure

30.9.3.1 Materials

1. Virus-infected leaf.
2. NTE buffer [0.2 M NaCl, 0.02 M Tris–HCl and 0.02 M EDTA, pH 7.4]: Dissolve 2.42 g of Tris base, 11.62 g of NaCl and 6.72 g of EDTA Na salt in about 900 mL of water, adjust the pH to 7.4 with the help of HCl and make the volume to 1 L.
3. Assembled lateral flow immunoassay cassette.
4. Phosphate-buffered saline (PBS): For preparation, *see* Subheading 30.2.1.1.

5. Syringe filter.
6. Running buffer (0.1% SDS and 0.1% of Tween-20 in PBS): Add 1 mL of Tween 20 and 1 g of SDS in 1× PBS and make up the volume to 1 L with distilled water.
7. Pestle and mortar.

30.9.3.2 Method

1. Collect the plant virus-infected sample and homogenize in a mortar and pestle.
2. The tissue homogenate is made with NTE buffer [0.2 M NaCl, 0.02 M Tris–HCl and 0.02 M EDTA, pH 7.4].
3. Centrifuge the suspension at 3000 *g* for 20 min at 4 °C.
4. Re-centrifuge the supernatant at 8000 *g* for 30 min at 4 °C and the filter the final supernatant through a 0.4 μm filter.
5. PBS buffer without virus will serve as negative control.
6. Pipette out 40 μL virus suspensions on the sample pad.
7. Allow to flow through the nitrocellulose membrane to the other end.
8. Add 100 μL of running buffer (0.1% SDS in PBS and 0.1% of Tween-20) to the sample window after 5 min. The appearance of red lines in the test and control zones indicating positive reaction (Fig. 30.20). Red lines in the test and control zones could be visualized easily within 20 min. The presence of two red lines indicates a positive reaction confirming the presence of the virus in the sample. The appearance of a red line only in control zone suggests the absence of virus, whereas formation of no line in the control and test zones indicates an invalid assay.

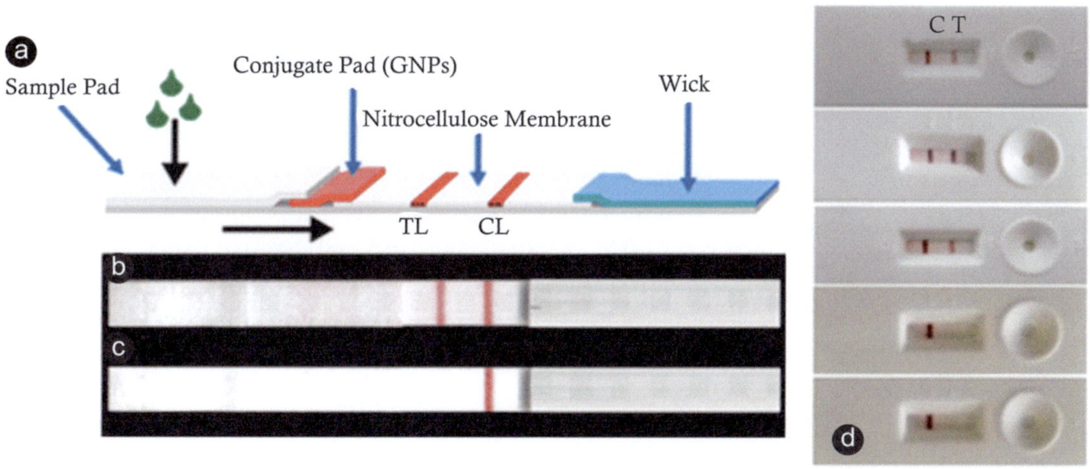

Fig. 30.20 Lateral flow immunoassay (LFIA). Application of sample (**a**), detection of the virus positive reaction (**b**), healthy or negative reaction (**c**), results of testing of different samples, of the five samples tested, three showed positive reaction for the virus; C is control and T is treatment (**d**)

30.10 Notes (Clark and Adams 1977; Clark et al. 1986; Hill 1984; Banttari and Goodwin 1985; Hampton et al. 1990; Koenig and Paul 1982; Lin et al. 1990; Milne and Luisoni 1975; Kulabhusan et al. 2017; O'Donell et al. 1982)

1. The tube precipitin, microprecipitin and ELISA tests can be used to determine antiserum titre or antigen dilution end point and also as a means of detecting the virus.
2. The agar gel diffusion test can also be done using glass slides instead of Petri dish.
3. Use formvar-coated Petri dish or glass slides for the agar gel diffusion experiment to prevent movement of reactants under the gel.
4. In agar gel diffusion test, isometric viruses can be tested without any problems as they can diffuse in the agar gel easily. But if elongate viruses are used, prior treatment with SDS is required to disrupt the particles. This can be done by boiling the virus (antigen) for 3–5 min in SDS solution. Also, SDS may also be added to gel (during melting of agar).
5. For ELISA, use the plates as supplied by manufacturers. Do not wash or rinse them before use. Do not reuse the plates.
6. Use separate beakers for making antigen, antiserum (or IgG), conjugate and substrate dilutions.
7. Use appropriate dilutions of antigens, antisera and conjugates. Usually, the antigen is used at 1:5 or 1:10 dilution while antisera are diluted according to their titre value. Commercial conjugates are to be used as per the manufacturer's directions.
8. Sap constituents of certain host plants may have an inhibitory effect on the virus. In such cases, the addition of polyvinyl pyrrolidone (PVP) or reducing agents such as mercaptoethanol, diethyl dithiocarbamate or thioglycolic acid to the extraction buffer is recommended.
9. For ELISA, do all incubations in a humid chamber. A small rectangular plastic box with moist paper towels is adequate.
10. In all kinds of serological tests, keep appropriate controls, such as the buffer, negative (healthy) and positive (infected).
11. Include negative (healthy) and positive control to validate the serological assays. Use tissue from a non-infected plant (tested negative for the test virus) of the same species as healthy control.
12. Include at least three replications for each of the samples.
13. For better results, use freshly prepared substrate buffer in the case of ELISA, DIBA, TBIA and EBIA.
14. PNPP is photo-degradable. Hence, avoid exposure to light. Cover the beaker with aluminium foil to prevent direct contact

with sunlight. Add as quickly as possible once PNPP solution is prepared.

15. Plan of ELISA will vary depending on the number of samples to be tested. It is suggested to keep the border wells of the plate empty.
16. In place of nitrocellulose, nylon or PVDF membranes may also be used in DIBA, TBIA and EBIA.
17. Wear gloves while handling nitrocellulose or nylon membranes.
18. Membranes are fragile and should be handled with blunt-tipped forceps.
19. The antibody and conjugate solution can be re-used for 3–5 times in the case of DIBA, TBIA and EBIA.
20. In case of DIBA, TBIA and EBIA if the antigen concentration is right, colour development starts immediately after the addition of substrate solution. Membranes can be kept in substrate solution for a maximum duration of 20 min as non-specific background colour starts developing after 20 min.
21. Antiserum used should be free from antibodies against any host proteins to avoid any false-positive reactions. If so, cross-absorb the antiserum with sap extracted from healthy plant (by incubating the required quantity of antiserum in about 1:50 dilution of healthy plant sap extracted in PBS). Incubate at 37 °C for 1 h and centrifuge at 7000 × *g* for 5 min. The supernatant is directly used for treating the membrane or loading onto ELISA plate wells.
22. The membrane once developed may be photographed or preserved between folds of filter paper for several years. The membrane can also be sent to other distant places.
23. Extra care should be taken while spotting the samples in the case of DIBA. See that sample diameter is as less as possible to ensure concentration of antigen at the same area. Spotting can be done manually or also through the use of dot-blot apparatus. If done manually, do not use more than 2.5 μL for dotting each time. If planning to dot 10 μL of the sample, take 2.5 μL of the sample and slowly release it on the centre of the grid area without touching the membrane. Allow it air dry. Repeat this process for another three times.
24. If one does not have antiserum or facilities for processing the membrane, the dotted membrane can be sent anywhere in the world where such facility exists. After processing, the membrane can also be sent back to the original place.
25. In the case of TBIA, use fresh blade to cut each sample to avoid cross contamination.

26. Plants that gave negative results should be further confirmed for the virus-free nature through other sensitive tests.
27. In all serological assays, samples must be very carefully washed between the incubation with each reagent.
28. While setting up electro-blotting apparatus to transfer protein from the gel to the membrane, remove any trapped air bubbles between gel and membrane by gently rolling a pipette over the set up.
29. During electro-blotting, care should be taken to see that membrane is facing towards anode and gel towards to cathode.
30. Keep the membranes floating in the solution during all the steps of membrane development in DIBA, TBIA and EBIA.
31. In the case of immunofluorescence, samples must be very carefully washed between the incubation with each reagent.

References

Avrameas S (1969) Coupling of enzymes to proteins with glutaraldehyde and use of conjugates for the detection of antigens and antibodies. Immunochemistry 5:43–52

Banttari EE, Goodwin PH (1985) Detection of potato viruses S, X and Y by enzyme-linked immunosorbent assay on nitrocellulose membrane (Dot-ELISA). Plant Dis 69:202–205

Barbara DJ, Clark MF (1982) A simple indirect ELISA using F(ab')$_2$ fragments of immunoglobulin. J Gen Virol 58:315–322

Clark MF, Adams AN (1977) Characteristics of the microtitre plate method of enzyme-linked immunosorbent assay for the detection of plant viruses. J Gen Virol 34:475–483

Clark MF, Lister RM, Bar-Joseph M (1986) ELISA techniques. Methods Enzymol 118:742–766

Derrick KS (1973) Quantitative assay for plant viruses using serologically specific electron microscopy. Virology 56:652–653

Draz MS, Shafiee H (2018) Applications of gold nanoparticles in virus detection. Theranostics 8(7):1985–2017

Gella FJ, Serra J, Gener J (1991) Latex agglutination procedures in immunodiagnosis. Pure Appl Chem 63:1131–1134

Hampton R, Ball E, De Boer S (1990) Serological methods for detection and identification of viral and bacterial plant pathogens. APS Press, St. Paul, MN

Hill SA (1984) Methods in plant virology. Blackwell Scientific, Cambridge

Hsu HT (1991) Direct tissue blotting for detection of tomato spotted wilt virus in *Impatiens*. Plant Dis 75:292–295

Koenig R (1981) Indirect ELISA methods for the broad specific detection of plant viruses. J Gen Virol 55:53–62

Koenig R, Paul HL (1982) Variants of ELISA in plant virus diagnosis. J Virol Methods 5:113–125

Koets M, Sander I, Bogdanovic J, Doekes G, van Amerongen A (2006) A rapid lateral flow immunoassay for the detection of fungal alpha-amylase at the workplace. J Environ Monit 8:942–946

Kulabhusan PK, Rajwade JM, SugumarV TG, Hameed AS, Paknikar KM (2017) Field-usable lateral flow immunoassay for the rapid detection of white spot syndrome virus (WSSV). PLoS One 3:2017. https://doi.org/10.1371/journal.pone.0169012

Lin NS, Hsu YH, Hsu HT (1990) Immunological detection of plant viruses and a mycoplasma-like organism by direct tissue blotting on nitrocellulose membranes. Phytopathology 80:824–828

Louro D, Lesemann D-E (1984) Use of protein A-gold complex for specific labeling of antibodies bound to plant viruses. I. Viral antigens in suspensions. J Virol Methods 9:107–122

Milne RG, Luisoni E (1975) Rapid, high-resolution immune electron microscopy of virus preparations. Virology 68:270–274

Nielsen K, Yu WL, Kelly L, Bermudez R, Renteria T, Dajer A et al (2008) Development of a lateral flow assay for rapid detection of bovine antibody to *Anaplasma marginale*. J Immunoassay Immunochem 29:10–18

Noordam D (1975) Identification of plant viruses: methods and experiments. Cen Agric Pub Doc, Wageningen

O'Donell IJ, Shukla DD, Gough KH (1982) Electroblot immunoassay of virus infected plant sap—a powerful technique for detecting plant viruses. J Virol Methods 4:19–26

Ouchterlony O (1968) Handbook of immunodiffusion and immunoelectrophoresis. Ann Arbor Scientific Publications, Michigan

Pares RD, Whitecross WI (1982) Gold-labeled antibody decoration (GLAD) in the diagnosis of plant viruses by immuno-electron microscopy. J Immunol Methods 51:23–28

Posthuma-Trumpie GA, van Amerongen A (2013) Lateral flow assays. In: Eline P (ed) Antibodies applications and new developments. Bentham Science, Meulenberg

Rybicki EP, Von Wechmar MB (1982) Enzyme-assisted immune detection of plant virus proteins electroblotted onto nitrocellulose paper. J Virol Methods 5:267–278

Samson RG, Allen TC, Whitworth JL (1993) Evaluation of direct tissue blotting to detect potato viruses. Am Potato J 70:257–263

Talley J, Warren FHJB, Torrance L, Jones RAC (1980) A simple kit for detection of plant viruses by the latex serological test. Plant Pathol 29:77–79

Van Slogteren DHM (1955) Serological microreactions with plant viruses under paraffin oil. Proceedings of the Conference on Potato virus diseases

Walkey DGA (1985) Applied plant virology. Heinemann, London

Chapter 31

Isolation of Total DNA from Plants

Abstract

In order to amplify plant DNA viruses, either viral DNA isolated from purified viral preparations or total DNAs isolated from virus-infected plants are used as template. The quantity and quality of DNA extracted are critical for successful amplification. Isolating quality DNA from plant tissues/cells involves several challenges, and certain plant tissues pose difficulty in processing due to the presence of polysaccharides, phenols, pigments and various secondary plant metabolites. A protocol that works with one plant group may fail with others. Consequently, a number of DNA isolation methods have been developed for different target groups. Many protocols for the isolation of total DNA from different plant species have been reported, but no single method that is applicable for all plant species is available. In this chapter, two widely used methods of DNA isolation is discussed with suitable protocols for obtaining a good quality DNA.

Key words Nucleic acid, Virus-infected plant, Extraction, Isolation, Yield, Quality, Quantity

31.1 Introduction

DNA isolation is a routine and basic procedure to collect DNA from a virus-infected plant for subsequent molecular analysis. In order to amplify DNA viruses, either viral DNA isolated from purified viral preparations or total DNAs isolated from virus infected plants are used as template. The purity and integrity of extracted DNA are critical for successful amplification. Wherever possible, use of viral nucleic acid as template is the best option to avoid non-specific amplification. However, this requires prior purification of the virus particles from diseased plants. Purification of viruses from plants is dealt in Chapter 21 while isolation of nucleic acids from purified virus preparation is dealt in Chapter 25. Certain plant viruses are known to occur in very low concentration in plants and are difficult to purify. Hence in such cases, total DNA isolated from infected plant can serve as template for PCR assay. In this procedure, along with the DNA of the host, if the plant is infected with a virus, its genome also gets isolated.

Certain plant tissues pose difficulty in processing due to the presence of various secondary plant metabolites. A protocol that works with one plant group may not be equally efficient with others. Consequently, a number of DNA isolation methods have

Alangar Ishwara Bhat and Govind Pratap Rao, *Characterization of Plant Viruses*, Springer Protocols Handbooks, https://doi.org/10.1007/978-1-0716-0334-5_31, © Springer Science+Business Media, LLC, part of Springer Nature 2020

been developed for different target groups (Murray and Thompson 1980; Dellaporta et al. 1983; Harper and Creamer 1985; Zhang et al. 1998; DeSilva et al. 2002). There are three main steps involved in DNA extraction. Cell disruption is the first step in the DNA isolation that is usually achieved by grinding or sonication of tissues. The second step is the isolation of DNA in extraction buffer which generally consists of detergents, buffering agents, salts and chelating agents. The third step is the main isolation step. Disruption of lipids is achieved by the addition of SDS which also binds to histones in DNA helping in the release of DNA into solution besides its action on nucleases. CTAB (cetyl trimethyl ammonium bromide) is a biological buffer and often is added to bind the cell wall debris, proteins and polysaccharides in the homogenate. During purification with phenol-chloroform after the CTAB step, the CTAB-debris-polysaccharide-protein complex gets precipitated and thus helps in the removal of these undesired materials. The extraction of DNA is done at pH 8.0, and to stabilize the pH, Tris–HCl buffer is used. The Na^+ ion of NaCl bonds with phosphates on the DNA that are negatively charged, thus neutralizing the charges of DNA. Chelating agent like EDTA is used to bind with metal ions such as Ca^{2+} and Mg^{2+} that are the co-factors of several enzymes such as DNA polymerase, DNases, etc. Neutralizing these ions allows DNA to be in their intact state. Sometimes reducing agents are also added to the extraction buffer. Polysaccharides from homogenates are removed by the addition of PVP (polyvinylpyrrolidone). Generally mercapto-ethanol is added to the extraction buffer as a reducing agent to prevent oxidation of phenols, tannins, etc. After initial extraction of DNA, the solution still contains protein and RNA in addition to DNA. In the last step, DNA is precipitated by the addition of isopropanol or ethanol. As DNA is insoluble in isopropanol and ethanol, it will aggregate collectively as pellet on centrifugation. A protocol that works well for one species of plant may not be suitable for another species. Hence large numbers of protocols with several modifications are available in published literature. Readers may refer published literature available that is suitable for the plant species they handle. In addition, many commercial kits are available for the isolation of DNA from plants (Dellaporta et al. 1983). Thus users do have many options to choose the best one that suits their requirement. In the following, two protocols used for isolation of total plant DNA from plants are described.

31.2 Protocol I

31.2.1 Materials

1. Disposable syringe (1 mL).
2. Eppendorf tubes.

3. Liquid nitrogen.
4. Microcentrifuge.
5. Perplex grinding shaft.
6. Pestle and mortor.
7. Vortex.
8. Water bath or temperature block.
9. Infected plant tissue.
10. Tris–HCl (1 M) (pH 8.0): Dissolve 12.1 g of Tris base in about 80 mL of water, set the pH to 8.0 by adding concentrated HCl and make up the volume to 100 mL.
11. EDTA (0.5 M) (pH 8.0): Dissolve 16.8 g of ethylene diamine tetra acetic acid disodium salt in about 80 mL of water, set the pH to 8.0 by adding NaOH, make up the volume to 100 mL.
12. NaCl (5 M): Dissolve 292.2 g of NaCl in 800 mL of distilled water, volume made up to 1 L and sterilize by autoclaving.
13. CTAB (cetyl trimethyl ammonium bromide) (10%): Dissolve 10 g of CTAB powder in water, dissolve by heating at 65 °C and make up the volume to 100 mL.
14. Extraction buffer [2.5% CTAB (cetyl trimethyl ammonium bromide), 100 mM Tris–HCl (pH 8.0), 20 mM EDTA (pH 8.0), 1.4 M NaCl, 1% polyvinyl pyrolidone (PVP), 0.1% β mercaptoethanol]: Prepare 100 mL of extraction buffer by adding 10 mL of 1 M Tris–HCl (pH 8.0), 25 mL of 10% CTAB, 4 mL of 0.5 M EDTA (pH 8.0), 1 g of polyvinyl pyrolidone (PVP) and 100 μL of β mercaptoethanol in a beaker and make up the volume to 100 mL.
15. High salt TE [10 mM Tris–HCl (pH 8.0), 1 mM EDTA (pH 8.0), 1 M NaCl]: Prepare 100 mL of high salt TE by adding 1 mL of 1 M Tris–HCl (pH 8.0), 200 μL of 0.5 M EDTA (pH 8.0), 20 mL of 5 M NaCl and make up the volume to 100 mL with water.
16. Chloroform: isoamyl alcohol (24:1): Mix 24 mL of chloroform with 1 mL of isoamylalcohol and store at 4 °C.
17. TE buffer [10 mM Tris–HCl (pH 8.0), 1 mM EDTA (pH 8.0)]: Prepare 100 mL of TE buffer by adding 1 mL of 1 M Tris–HCl (pH 8.0) and 200 μL of 0. 5 M EDTA (pH 8.0) and make up the volume to 100 mL with water.
18. Isopropanol.
19. 70% ethanol: Prepare by adding 30 mL of sterile water to 70 mL of absolute ethanol, store at 4 °C.
20. RNase A (10 mg/mL): Prepare by dissolving 10 mg of RNase A in 1 mL of Tris–HCl (10 mM, pH 7.5) in a vial, keep the vial

in boiling water for 5 min, cool at room temperature, make aliquots and store at −20 °C.

31.2.2 Method

1. Weigh and place 100 mg of tissue in sterile microcentrifuge tube and add sufficient quantity of liquid nitrogen to grind.
2. Quickly grind it to a fine powder with the help of perplex grinding shaft.
3. Add 400 μL of pre-warmed extraction buffer and mix gently.
4. Incubate at 60 °C in a water bath for 1 h (if dry sample is used incubate overnight at 37 °C).
5. To the homogenate, add equal volume of chloroform:isoamyl alcohol (24:1) and slowly invert the tube for 15 min for proper mixing.
6. Subject the contents to centrifugation at room temperature for 10 min at 7500 × *g*.
7. Collect the aqueous phase to a fresh microcentrifuge tube, add 150 μL of 5 M NaCl and 0.6 volumes of isopropanol and mix gently (no vortexing).
8. Incubate the tube at room temperature for 1 h followed by centrifugation at room temperature for 10 min at 7500 × *g*.
9. Throw away the supernatant and add about 500 μL of 70% ethanol pellet followed by centrifugation at 7500 × *g* for 2 min.
10. Throw away the supernatant, air dry the pellet and dissolve in 400 μL of high salt TE buffer.
11. Add 5 μL of RNase A and place the tube for 30 min at 37 °C (optional step).
12. Add equal volume to chloroform: isoamylalcohol (24:1), mix gently and centrifuge at 7500 × *g* for 10 min.
13. Collect the supernatant to a new microcentrifuge and add two volumes of chilled ethanol and incubate at −80 °C for 30 min or −20 °C for 2 h.
14. Centrifuge for 10 min at 7500 × *g* at room temperature.
15. Throw away the supernatant and wash the pellet with 80% ethanol.
16. Air dry the pellet and dissolve in 200 μL of sterile distilled water.
17. Measure the DNA concentration by taking absorbance at 260 nm or by running an aliquot on 0.8% agarose gel. About 25 ng of DNA may be used for PCR amplification.

31.3 Protocol II

31.3.1 Materials

1. Same as in method I.
2. Extraction buffer: 10% CTAB (hexadecyl trimethyl ammonium bromide) containing 0.2% β mercaptoethanol.
3. Sodium acetate (3 M) (pH 5.2): Dissolve 24.6 g sodium acetate in 70 mL of distilled water, set the pH to 5.2 by adding glacial acetic acid, make up the total volume to 100 mL with water.

31.3.2 Method

1. Grind fresh infected tissue (100 mg) using liquid nitrogen in a mortor and pestle.
2. Collect the ground powder in an eppendorf tube containing 500 μL extraction buffer and incubate the tube for 30 min at 65 °C.
3. Allow cooling to room temperature and then add an equal volume of chloroform: isoamyl alcohol (24:1) to the tube and invert the tube gently to mix.
4. Centrifuge the contents at room temperature for 10 min at 2500 × *g*.
5. Collect the aqueous phase and add equal volume of 10% CTAB.
6. To the above, add equal volume of chloroform: isoamyl alcohol (24:1), mix gently and centrifuge the contents as in **step 4**.
7. Collect the supernatant in a new eppendorf tube, add 0.1 volumes of sodium acetate and 1 volume of isopropanol.
8. Subject the contents to centrifugation for 15 min at 13,000 × *g* and discard the supernatant.
9. Add 1 mL of 70% ethanol to the pellet, centrifuge for 5 min at 13,000 × *g* and discard the supernatant.
10. Air dry and resuspend the DNA pellet in about 50 μL of TE buffer.
11. Quantify the DNA by taking spectral reading at 260 nm and use about 25 ng of DNA as template for PCR.

31.4 Notes (Dellaporta et al. 1983; Murray and Thompson 1980; Zhang et al. 1998)

1. Many commercial kits are available for the isolation of total DNA from the wide range of plant species.
2. Before starting, ensure that the tissue to be used is infected with virus. ELISA or any other sensitive methods can be used to confirm the presence of the virus.

3. For better yield of DNA, use young growing region of the plant as virus multiplies at a faster rate in this region.
4. Always fresh sample gives better result than the dried sample. If dried sample, use only 1/5th by weight of fresh sample.
5. After the sample is ground finely, immediately add extraction buffer (do not allow to thaw).
6. Care should be taken while collecting aqueous phase (avoid organic phase).
7. Change in extraction buffer or protocol may be required if isolated DNA fails to get amplified in PCR.
8. The DNA pellet should not be over dried. A hard overdried DNA pellet takes very long time to dissolve completely.
9. Only A_{260} readings between 0.1 and 1.0 are reliable enough for accurate DNA quantification.
10. Many plants such as bhindi, jute, mesta, sida, members of Euphorbiaceae, which serve as natural hosts for different viruses contain mucilage and other secondary metabolites that makes the isolation of DNA difficult and also interferes with further downstream applications such as the PCR and other enzymatic reactions. Following modifications in the CTAB method can overcome the problem of mucilage (Ghosh et al. 2009): (a) take less leaf tissue and increase in volume of DNA isolation buffer: Take 100 mg leaf tissue and use 10 mL of buffer (b) instead of dissolving crude nucleic acid pellet in Tris–EDTA (TE) buffer 1 mL, dissolve it in 1 M NaCl, (c) give RNase treatment to crude nucleic acid resuspended in 1 M NaCl at 37 °C for 30 min, (d) after purification, dissolve the DNA in TE buffer. The above steps would reduce the viscosity of the mucilage and yield mucilage-free DNA.

References

Dellaporta SL, Wood J, Hicks JB (1983) A plant DNA minipreparation: version II. Plant Mol Biol Rep 1:19–21

DeSilva DPP, Jones P, Shaw MW (2002) Identification and transmission of *Piper yellow mottle virus* and *Cucumber mosaic virus* infecting black pepper (*Piper nigrum*) in Sri Lanka. Plant Pathol 51:537–545

Ghosh R, Paul S, Ghosh SK, Roy A (2009) An improved method of DNA isolation suitable for PCR-based detection of begomoviruses from jute and other mucilaginous plants. J Virol Methods 159:34–39

Harper K, Creamer R (1985) Hybridization detection of insect-transmitted plant viruses with digoxigenin-labeled probes. Plant Dis 79:563–567

Murray MG, Thompson WF (1980) Rapid isolation of high molecular weight plant DNA. Nucleic Acids Res 8:4321–4325

Zhang YP, Uyemoto J, Kirkpatrick BC (1998) A small scale procedure for extracting nucleic acids from woody plants infected with various phytopathogens for PCR assay. J Virol Methods 71:45–50

Chapter 32

Isolation of Total RNA from Plants

Abstract

About 90% of viruses-infected plants have ssRNA as their genome. Complementary DNA (cDNA) copy for the ssRNA is made using reverse transcriptase enzyme which can then be used as template for PCR assay. This procedure requires isolation of RNA from either purified virus particles or from virus-infected plants. A large number of protocols are available for the extraction of total RNA from different plant species depending on, the type of tissue used for extraction and the presence of inhibitors and secondary metabolites. Majority of the protocols are based on the use of trizol while phenol-based method is the best suitable for species with high amount of secondary metabolites, and use of polyethylene glycol (PEG) may be suitable to extract RNA from species with high amount of polysaccharides and polyphenols.

Key words Plant species, RNA isolation, RNA extraction, Yield, Quality, Quantity

32.1 Introduction

The use of RNA isolated from purified virus particles is always better as template as compared to the total RNA isolated from the plants. The procedure for purification of plant viruses and isolation of RNA from purified virus particles are provided in Chapters 21 and 25, respectively. For viruses that occur at very low concentration in plants and difficult to purify, total RNA extracted from such plants can be used as template for reverse transcription PCR (RT-PCR). The basic steps involved in the isolation of total RNA from plants consist of following steps: (1) disruption of tissue, (2) separation of RNA from DNA, protein, polysaccharides, carbohydrates, etc. and (3) precipitation of RNA (Vunsh et al. 1991). Ribosomal RNA (rRNA) accounts for 85% of total RNA followed by low molecular weight RNA species such as tRNA and siRNA while only about 1–3% consists of mRNA which includes viral RNA as well. Obtaining non-degraded RNA is important for all experiments.

Acid guanidium thiocyanate phenol chloroform is the largely used protocol for the isolation of total RNA from plants (Chomczynski and Sacchi 1987; Spiegel and Martin 1993; MacKenzie et al. 1997; Siju et al. 2007). Guanidium thiocyanate is a chaotropic agent that destroys the three-dimensional structure of proteins;

Alangar Ishwara Bhat and Govind Pratap Rao, *Characterization of Plant Viruses*, Springer Protocols Handbooks, https://doi.org/10.1007/978-1-0716-0334-5_32, © Springer Science+Business Media, LLC, part of Springer Nature 2020

reducing agents used in the extraction buffer breaks the protein disulphide bonds, and SDS disrupts hydrophobic interactions and inhibits RNase activity. Commercial extraction buffers such as TRI-zol and TRI reagents which use guanidium thiocyanate as the major component are available for the isolation of total RNA.

The separation of RNA from DNA is based on the pH. RNA is soluble under acidic pH while DNA is insoluble. At alkaline pH, RNA is insoluble while DNA is soluble. Following cell disruption, total RNA is isolated by removing the proteins by phenol/chloroform extraction followed by precipitation using salt and ethanol. Otherwise, RNA can also be isolated after proteinase and DNase treatment followed by purification through column. The isolated RNA must be analysed through agarose gel electrophoresis to determine the quantity and to ensure integrity and quality.

RNA isolation (compared to DNA isolation) needs additional care due to omnipresent occurrence of RNases that degrade ssRNA molecules. The key step in plant virus RNA isolation is to minimize the RNase activity during the initial steps of total RNA extraction from infected plant tissue. This is achieved by including chemicals to prevent RNase activity in the extraction buffer (Levy et al. 1994). The care to be taken while isolating RNA from plants is provided in Subheading 32.4 of this chapter. Like DNA, certain plant tissues pose difficulty in processing due to the presence of various secondary plant products. Large number of protocols have been reported by various workers for the extraction of total RNA from different plant species. In addition, ready to use reliable commercial kits that can be employed for the extraction of total RNA from large number of plant species are also available. Here we describe two protocols: (1) acid guanidium phenol thiocyanate and (2) extraction and purification of RNA by column.

32.2 Acid Guanidium Thiocyanate Phenol Chloroform (AGPC) Method (Chomczynski and Sacchi 1987)

32.2.1 Materials

1. −20 °C deep freezer.
2. −80 °C deep freezer.
3. Pestle and mortor.
4. Water bath or temperature block.
5. Infected plant samples.
6. Guanidium thiocyanate (10 M): Weigh and dissolve 118.16 g of guanidium thiocyanate in water and make up the volume to 100 mL.
7. Sodium citrate (100 mM) (pH 7.0): Dissolve 2.58 g of sodium citrate in about 80 mL of water, set the pH to 7.0 using 0.1 N HCl and make up the volume to 100 mL with water.

8. Sarcocyl: Prepare 10% stock solution by dissolving 10 g of sarcocyl in 100 mL of water.
9. Denaturing solution [4 M guanidium thiocyanate, 25 mM sodium citrate (pH 7.0) and 0.5% sacrcocyl]: Prepare 100 mL of denaturing solution by adding 40 mL of 1 M guanidium thiocyanate, 25 mL of 0.1 M sodium citrate (pH 7.0), 5 mL of 10% sarcocyl and make up the volume to 100 mL with water, autoclave and store at room temperature; add 0.5% β-mercaptoethanol for the requisite volume just before use.
10. Sodium acetate (2 M, pH 4.0): Dissolve 27.2 g of sodium acetate in about 80 mL of water, set the pH to 4.0 by adding glacial acetic acid, make up the volume to 1 L with water and sterilize by autoclaving.
11. Phenol (water saturated): Prepare this solution in Fume hood, wear gloves and eye protection. Take required quantity of distilled phenol and add equal volume of water at room temperature, stir for 15 min and allow it to stand for the separation of phases. Aspirate upper aqueous phase using a glass pipette (avoid mouth sucking) and discard. Add an equal volume of water to the phenol, stir for 15 min, turn off the stirrer and remove the upper aqueous phase. Repeat this process for 2–3 times. The solution should be stored in an amber-coloured bottle at 4 °C.
12. Chloroform : isoamylalcohol (49:1): 49 mL of chloroform mixed with 1 mL of isoamylalcohol.
13. Isopropanol.
14. Ethanol 75%: Add 25 mL sterile water to 75 mL of absolute ethanol, mixed and stored at 4 °C.

32.2.2 Method

1. Grind 50 mg tissue in a pestle and mortor using 500 μL of denaturing solution.
2. Collect the contents in a centrifuge tube and add the following solutions sequentially while mixing after each addition: 50 μL of 2 M sodium acetate pH 4.0, 500 μL phenol (water saturated), and 100 μL of chloroform: isoamyl alcohol (49:1).
3. Mix for 10 s vigorously, incubate for 15 min on ice followed by centrifugation at 4 °C for 20 min at 10,000 × *g*.
4. Collect the aqueous phase in a new microcentrifuge tube, add 500 μL of isopropanol and place for 1–2 h at −20 °C.
5. Centrifuge the contents for 20 min at 10,000 × *g*.
6. Throw away the supernatant and slowly add 500 μL of 70% ethanol to the pellet.

7. Centrifuge immediately for 10 min at 10,000 × *g*, throw away the supernatant and air dry the pellet.
8. Resuspend the pellet in 50 μL RNase free water and store at −70 °C.

32.3 RNA Extraction and Column Purification

32.3.1 Materials

1. Same as in AGPC method.
2. Toothpick.
3. Disposable syringe.
4. Glass wool.
5. Liquid nitrogen.
6. Tris–HCl (1 M) (pH 8.0): Prepare by dissolving 12.1 g of Tris base in about 80 mL of water, set the pH to 8.0 by adding HCl and make up the volume to 100 mL.
7. EDTA (0.5 M) (pH 8.0): Dissolve 16.8 g of ethylene diamine tetra acetic acid disodium salt in about 80 mL of water, set the pH to 8.0 by adding NaOH and make up the volume to 100 mL.
8. SDS (10%): Prepare by dissolving 10 g of SDS in water and make up the volume to 100 mL.
9. Extraction buffer [0.1 M Tris–HCL (pH 8.0), 2% SDS and 2 mM EDTA (pH 8.0)]: Prepare 100 mL of extraction buffer by adding 10 mL of 1 M Tris–HCl (pH 8.0), 20 mL of 10% SDS, 400 μL of 0.5 M EDTA (pH 8.0) and make up the volume to 100 mL with sterile water.
10. TE buffer [10 mM Tris–HCl (pH 8.0), 1 mM EDTA (pH 8.0)]: Mix 1 mL of 1 M Tris–HCl (pH 8.0) and 200 μL of 0. 5 M EDTA (pH 8.0) and make up the volume to 100 mL with water.
11. Phenol (TE saturated): Prepare this solution in Fume hood, wear gloves and eye protection. Take required quantity of distilled phenol and add equal volume of TE buffer at room temperature, stir for 15 min and allow it to stand for the separation of phases. Aspirate upper aqueous phase using a glass pipette (avoid mouth sucking) and discard. Add an equal volume of TE buffer to the phenol, stir for 15 min, turn off the stirrer and remove the upper aqueous phase. Repeat this process for 2–3 times. The solution should be stored under TE in an amber-coloured bottle at 4 °C.
12. Chloroform.
13. Sephadex G50 equilibrated with TE: Dissolve about 4–5 g of Sephadex G-50 in RNase-free water, wash 3–4 times by

removing the liquid and replacing with fresh RNase-free water, remove liquid and add an equal volume of TE buffer to the hydrated Sephadex and store at room temperature.

32.3.2 Method

32.3.2.1 Extraction of Total Nucleic Acids

1. Take 1 cm^2 leaf tissue in a microcentrifuge tube and freeze in liquid nitrogen.
2. Grind the tissue with sterile toothpick.
3. Add 300 μL extraction buffer, 150 μL of TE saturated phenol and 150 μL of chloroform.
4. Vortex and heat the contents for 5 min at 70 °C.
5. Centrifuge at room temperature for 10 min at 15,000 × *g*.
6. Collect aqueous phase in a new centrifuge tube and freeze in liquid nitrogen.
7. Store the extract at −80 °C until further use.

32.3.2.2 Column Purification

1. Block the bottom of a 1 mL disposable syringe to be used as column support with sterile glass wool.
2. Keep the syringe in a centrifuge tube.
3. Fill the syringe with equilibrated Sephadex G 50 and centrifuge the syringe at 1600 × *g* for 4 min.
4. Add 100 μL of TE buffer to the column and centrifuge as above.
5. Discard the effluent and place a decapped microcentrifuge tube in the centrifuge tube. Place the syringe with column back in the centrifuge tube.
6. Add 100 μL of tissue extract (from step 7 above) to the column and centrifuge as above and collect the effluent into the centrifuge tube.
7. Transfer the effluent to a fresh microcentrifuge tube and store at −80 °C.

32.4 Notes (Chomczynski and Sacchi 1987; Sambrook and Russel 2001)

1. Precautions while working with RNA: RNA is more susceptible to degradation than DNA, due to the ability of the two hydroxyl groups adjacent to the phosphodiester linkages in RNA to act as intramolecular nucleophiles in both base and enzyme-catalysed hydrolysis. Whereas deoxyribonucleases (DNases) require metal ions for activity and can therefore be inactivated with chelating agents (e.g. EDTA), many ribonucleases (RNases) bypass the need for metal ions by taking advantage of the two hydroxyl groups as a reactive species.

Since RNases are difficult to inactivate and even minute amounts are sufficient to destroy RNA. RNase A is single-strand specific endoribonuclease that is resistant to metal chelating agents and can survive prolonged boiling or autoclaving. However, RNase A-type enzymes rely on active site histidine residues for catalytic activity, and consequently, these enzymes can be inactivated by the histidine-specific alkylating agent diethyl pyrocarbonate (DEPC).

2. Source of RNase contamination: It is possible to reintroduce RNases during the course of experimentation via aqueous solutions or from the environment at large. Autoclaving a solution will kill contaminating bacteria, but RNases liberated from the dead bacteria will still be active. Additionally, ungloved fingers can introduce bacteria into solutions resulting in RNase contamination. Fortunately, the use of DEPC to remove ribonuclease activity from aqueous solutions, coupled with few laboratory precautions, is generally sufficient to prevent RNase contamination.
3. Always wear gloves when working with RNA.
4. It is a good idea to maintain a separate area for RNA work that has its own lab equipment. This is especially important if your work requires the use of RNase A (e.g. plasmid preps).
5. Sterile, disposable plastic ware can safely be considered RNase-free and should be used when possible.
6. Metal spatulas can be quickly decontaminated by holding in a burner flame for several seconds.
7. Contaminating RNases can be inactivated by baking glassware at 180 °C or higher for several hours.
8. Alternatively, glassware can be soaked in freshly prepared 0.1% (v/v) DEPC in water or ethanol for 1 h, drained and autoclaved (necessary to destroy any unreacted DEPC which can otherwise react with other proteins and RNA).
9. DEPC will destroy polycarbonate or polystyrene materials (e.g. electrophoresis tanks), which should instead be decontaminated by soaking in 3% hydrogen peroxide for 10 min. Remove peroxide by extensively rinsing with DEPC-treated and autoclaved water prior to use.
10. Submerge the microcentrifuge tubes in a 0.1% (v/v) DEPC-treated water solution. Leave the beaker of the submerged tubes in a fume hood overnight and then dispose of the DEPC-treated water. Autoclave the microcentrifuge tubes for at least 30 min and place the tubes in a drying oven overnight.
11. DEPC is extremely toxic and should be handled with care.

12. Purified RNA should be stored at −80 °C. Avoid frequent freezing and thawing of a sample containing RNA.
13. Addition of reducing agents such as sodium sulphite is known to enhance RNA yield in certain plant tissues.

References

Chomczynski P, Sacchi N (1987) Single step method of RNA isolation by acid guanidium thiocynate phenol chloroform extraction. Anal Biochem 162:150–159

Levy L, Lee IM, Hadidi A (1994) Simple and rapid preparation of infected plant tissue extracts for PCR amplification of virus, virioid, and MLO nucleic acids. J Virol Methods 49:295–304

MacKenzie DJ, McLean MA, Mukerji S, Green M (1997) Improved RNA extraction from woody plants for the detection of viral pathogens by reverse transcription polymerase chain reaction. Plant Dis 81:222–226

Sambrook J, Russell DW (2001) Molecular cloning: a laboratory manual, 3rd edn. Cold Spring Harbor Laboratory Press, New York

Siju S, Madhubala R, Bhat AI (2007) Sodium sulphite enhances RNA isolation and sensitivity of *Cucumber mosaic virus* detection by RT-PCR in black pepper. J Virol Methods 141:107–110

Spiegel S, Martin RR (1993) Improved detection of potato leaf roll virus in dormant potato tubers and microtubers by polymerase chain reaction and ELISA. Ann Appl Biol 122:493–500

Vunsh R, Rosner A, Stein A (1991) Detection of bean yellow mosaic virus in gladioli corms by the polymerase chain reaction. Ann Appl Biol 119:289–294

Chapter 33

Isolation of Double-Stranded (ds) RNA from Virus-Infected Plants

Abstract

All single-stranded RNA (ssRNA) virus-infected plants produce double-stranded RNA (dsRNA) during their replication cycle. dsRNAs are more stable than its counterpart ssRNA. Many protocols are available for the isolation of dsRNA from virus-infected plants that are used as template for cDNA synthesis followed by PCR assay. In this chapter, suitable protocol for the isolation of ds RNA from plant sources is discussed.

Key words Replicative form of virus, RNA template from plant, CF-11 column

33.1 Introduction

Except for the members of *Reovirus*, all other viruses infecting plants have single-stranded RNA (ssRNA) as their genome. All ssRNA viruses are known to form double-stranded RNA (dsRNA) during their replication in the plants. These ds forms of RNA are referred to as replicative forms (RFs) or replicative intermediates (RI). Thus ssRNA virus-infected plants also produce dsRNA molecules (Dodds et al. 1984). As dsRNAs are stable and resistant to RNase attack, they are relatively easy to isolate. dsRNAs are denatured and then used as template for cDNA synthesis or RT-PCR. Several protocols are available for the isolation of dsRNA from plants (Morris and Dodds 1979; Morris et al. 1983; Dodds et al. 1984). The procedure involves extraction of nucleic acid and purification of dsRNA through CF-11 column.

33.2 Materials

1. Glass wool.
2. Infected tissue.
3. Liquid nitrogen.
4. Pestle and mortor.
5. Syringe (12 mL).

Alangar Ishwara Bhat and Govind Pratap Rao, *Characterization of Plant Viruses*, Springer Protocols Handbooks, https://doi.org/10.1007/978-1-0716-0334-5_33, © Springer Science+Business Media, LLC, part of Springer Nature 2020

6. Whatman CF 11 cellulose.
7. NaCl (5 M): Dissolve 292.2 g of NaCl in 800 mL of distilled water, make up the volume up to 1 L and sterilize by autoclaving.
8. Tris–HCl (1 M, pH 8.0): Dissolve 12.1 g of Tris base in about 80 mL of water, adjust the pH to 8.0 by adding HCl and make up the volume to 100 mL.
9. EDTA (0.5 M pH 8.0): Dissolve 16. 8 g of ethylene diamine tetra acetic acid disodium salt in about 80 mL of water, adjust the pH to 8.0 by adding NaOH and make up the volume to 100 mL.
10. SDS (10%): Dissolve 10 g of SDS in water and make up the volume to 100 mL.
11. STE (10×) (pH 6.8) (100 mM NaCl, 10 mM Tris–Cl, pH 8.0 and 1 mM EDTA): Prepare 100 mL of 10× STE by adding 2 mL of 5 M NaCl, 1 mL of 1 M Tris–HCl pH 8.0, 200 μL of 0.5 M EDTA pH 8.0 and make up the volume to 100 mL with water.
12. Sodium acetate (3 M, pH 5.2): Dissolve 246.1 g of sodium acetate in 500 mL of deionized H_2O, adjust the pH to 5.2 with glacial acetic acid and make up the final volume to 1 L with deionized H_2O and sterilize by autoclaving.
13. Bentonite: It is used as an adsorbent for nucleases during the phenol degradation of the virus.
14. Ethanol.
15. Saturated phenol: Prepare this solution in Fume hood, wear gloves and eye protection. Take the required quantity of distilled phenol and add equal volume of water at room temperature, stir for 15 min and allow standing for the separation of phases. Aspirate upper aqueous phase using a glass pipette (avoid mouth sucking) and discard. Add an equal volume of water to the phenol, stir for 15 min, turn off the stirrer and remove the upper aqueous phase. Repeat this process for 2–3 times. Store the solution in an amber-coloured bottle at 4 °C.

33.3 Method

1. Grind up to 7 g of infected tissue in liquid nitrogen and transfer the powder to a beaker containing 14 mL 2× STE buffer, 20 mL saturated phenol, 2 mL 10% SDS and 14 g bentonite.
2. Stir the contents of the beaker for 30 min to 1 h at 4 °C.
3. Transfer the contents to a centrifuge tube and spin for 15 min at 5000 × *g*.

4. Collect the aqueous phase in a 50 mL beaker and adjust the volume to 20 mL with 1× STE, add 3.8 mL of ethanol.
5. Add 1.0 to 1.5 g Whatman CF-11 powder.
6. Stir for 1 h at 4 °C or just leave on ice and swirl intermittently.
7. Make a column using this slurry in a 10 mL syringe plugged at the bottom with glass wool.
8. Pass about 100 mL of 1× STE containing 15% ethanol solution through the column or until the cellulose becomes colourless.
9. Force all the liquid out of syringe using syringe piston.
10. Add 14 mL of 1× STE alone and collect the drops. To this add 0.1 volume of 3 M sodium acetate, pH 5.2 and 2.5 volumes of ethanol. Incubate at −20 °C overnight.
11. Next day spin the tubes at 7500 × *g* for 30 min, wash once with 70% ethanol, dry the pellet and dissolve the RNA in 50–100 μL of sterile water. Keep it frozen at −70 °C or alternatively dsRNA can be left under ethanol at −20 °C until required.
12. Determine the concentration spectrophotometrically and use appropriate amounts for gel electrophoresis. Include appropriate healthy controls.
13. Prepare 1% agarose gel.
14. Load the required quantity on the gel. Load suitable size marker in another well and run the gel at a constant 100 V for 1 h.
15. Observe the gel under transilluminator. Determine the size and quality of dsRNA molecules. Monopartite viruses will produce single bands, while bi- and tripartite viruses will produce two and three bands, respectively.
16. dsRNA from the gel can be eluted and purified if run on a low melting point agarose gel (details of gel elution procedure is given in Chapter 42).

33.4 Notes (Morris et al. 1983; Dodds et al. 1984)

1. For better yield of dsRNA, use fresh young growing region of plant, as the rate of virus multiplication is faster at this region.
2. Confirm the presence of virus in the tissue used through ELISA or other sensitive methods before subjecting to dsRNA isolation.
3. Do not use plants that are infected with more than one type of viruses.

4. dsRNA is also used as a diagnostic tool for the identification of viruses based on the number and size of RNA molecules on the gel.
5. Purified dsRNA from the gel can be used for cDNA synthesis followed by amplification by PCR, cloning, etc.

References

Dodds JA, Morris TJ, Jordan RL (1984) Plant viral double stranded RNA. Annu Rev Phytopathol 22:151–168

Morris TJ, Dodds JA (1979) Isolation and analysis of double stranded RNA from virus infected plant and fungal tissue. Phytopathology 69:854–858

Morris TJ, Dodds JA, Hillman B, Jordon R, Lommel SA, Tamaki SJ (1983) Viral specific dsRNA. Diagnostic value for plant virus disease identification. Plant Mol Biol Rep 1:27–30

Chapter 34

Dot-Blot Hybridization Technique

Abstract

The dot-blot hybridization is a nucleic acid hybridization technique where complementary single-stranded sequences of the probe (either RNA or DNA) hybridizes with single-stranded sequences of the test samples (either RNA or DNA) under suitable conditions of temperature and salt concentration. It is based on the homology between two strands of nucleic acid (DNA:DNA, DNA:RNA or RNA:RNA). In this assay, a probe which is a single-stranded nucleic acid (either DNA or RNA) is 'labelled' with a reporter molecule. The labelled probe is then allowed to form hybrid with nucleic acid isolated from the test plant. The double-stranded hybrid molecule is then detected using appropriate method depending on the reporter molecule used. The process involves isolation of nucleic acid from test plant; it's spotting on a membrane, pre-hybridization, hybridization using labelled probe and their detection. The DNA probes may be labelled using different methods, namely nick translation, random primed labelling or by polymerase chain reaction. The RNA probes are labelled by in vitro transcription. The labels can be either radioactive or non-radioactive. If radioactive probes are used, then the detection is done through autoradiography. If non-radioactive labels are used, then suitable detection methods through either chromogenic or chemiluminescence is used.

Key words Spot-blot hybridization, Radioactive probe, Non-radioactive probe, Nucleic acid hybridization, Virus detection

34.1 Introduction

The affinity of one strand of DNA for its complementary sequences is the basic principle of nucleic acid hybridization assays. It is based on the homology between two strands of nucleic acid or in hybrid molecules that consist of strands of RNA and DNA. Dot-blot hybridization is one form of nucleic acid hybridization that involves mixing single strands of two sources of nucleic acids, a labelled probe which typically consists of homogenous population of identical molecules (e.g. cloned DNA) and a target (test sample) which usually consists of a heterologous population of nucleic acid molecules. Both the probe and the target are made single stranded by heating or by alkaline treatment. Single-stranded probe and target DNA are mixed and allowed to re-associate. Complementary probe strands may bind with its complementary target DNA strand (test DNA) and form a probe–target heteroduplex which can be

Alangar Ishwara Bhat and Govind Pratap Rao, *Characterization of Plant Viruses*, Springer Protocols Handbooks,
https://doi.org/10.1007/978-1-0716-0334-5_34,

detected by different techniques. The complementary strands of nucleic acid may be allowed to bind in solution (solution hybridization), membrane or filter paper (filter hybridization) or on the tissue itself (in situ hybridization) (Britten and Kohne 1968; Gillespie and Spiegelman 1965). Among these, dot-blot hybridization uses mixed phase (filter) hybridization. In this assay, a single-stranded complementary nucleic acid labelled with a reporter molecule (either radioactive or non-radioactive) will be used as the probe. It forms a hybrid with the target nucleic acid in the test samples which can be later detected by a variety of methods depending on the reporter molecule used. The important steps in the process include: (1) the target nucleic acid is isolated, spotted and immobilized onto a support matrix such as nitrocellulose or nylon membrane; (2) remaining free binding sites on the membrane are then blocked with non-heterologous DNA (usually salmon sperm or calf thymus DNA) or a protein source (like bovine serum albumin or non-fat dried milk); (3) the membrane is incubated with denatured probe which is complementary to the target of interest in annealing condition; (4) unhybridized probe is removed from hybridized material in a series of washing steps and (5) the presence of target sequences is assayed by the detection of the reporter molecule in the hybridized probe sequences (Fig. 34.1) (Brandsma and Miller 1980; Owens and Diener 1981). The simplest format of nucleic acid hybridization test is the dot blot or spot blot (Garger et al. 1983; Maule et al. 1983). The dot-blot assay does not provide any information on the size and number of target nucleic acid molecules. This information can be obtained by electrophoresing the nucleic acid in a gel and then transferring it to a membrane followed by hybridization with the probe. If the sample is DNA, then it is known as a Southern blot (Southern 1975), and if the same is RNA, the method is known as the northern blot (Alwine et al. 1977).

The probe nucleic acid can be DNA or RNA, ss or ds, and should have sequences complementary to the target nucleic acid. The common type of probe used for the detection of plant viruses is cDNA as more than 80% of plant viruses contain ssRNA as their genome. cDNA can be either made directly for each experiment or be cloned into a bacterial plasmid or phage vector. The cloning approach has the advantage that they can be prepared in large amounts when needed and provide very high purity and specificity of the probe. Single-stranded RNA probes (also called riboprobes) generated by in vitro transcription of a DNA template cloned into a transcription vector have also been used in detecting plant viruses. Several methods are available to label nucleic acid probes. Important ones include nick translation, random primer labelling, synthesis of double- or single-stranded DNA probes using polymerase chain reaction and in vitro synthesis of RNA probe from cloned cDNA. Nick translation is characterized by the exchange of

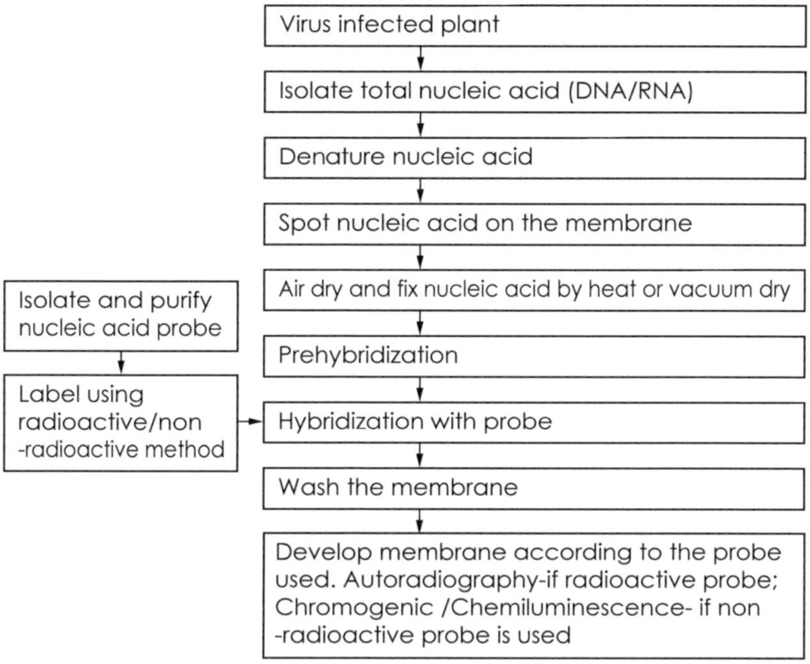

Fig. 34.1 Different stages involved in dot-blot hybridization test

unlabelled nucleotides with labelled nucleotides (replacement synthesis), while the other two methods result in the synthesis of new DNA/RNA sequences (net synthesis). RNA probes are prepared by synthesizing 'run-off' transcripts from cloned cDNA fragment in a transcription vector (Rigby et al. 1977).

Most of the hybridization techniques were developed using radioactively labelled nucleotides as reporter molecules. The most common labels are ^{32}P and ^{35}S. However, there are major limitations of using radioactive probes such as (1) short half-life of the radio isotype, (2) potential health hazardous during preparation and use of the labelled nucleic acid, (3) the need for separate facilities and trained personnel for handling radio isotopes and (4) the need for special disposal facilities for radioactive waste. In recent years, a number of different systems have been developed which use non-radioactive reporter molecules for the detection of viruses in nucleic acid hybridization assays. These systems fall into three basic types, those which directly modify bases in probe DNA and those which attach precursors to DNA or RNA. The most popular are biotin-streptavidin and the digoxigenin-antidigoxigenin systems. In these systems, the reporter molecule (biotin or digoxigenin) attached to the nucleotide in the probe is detected by using ligands like streptavidin (for biotin) or anti-digoxignen antibody (for digoxigenin) covalently conjugated to an enzyme such as alkaline phosphatase or horseradish peroxidase. The location of the hybridized target sequence is then visualized by incubating the membrane with a signal generating substrate which gives a coloured product or a luminescent compound (Roy et al. 1989; Smith et al. 1994; Blok et al. 1995).

34.2 Production of Radiolabelled Probe by Random Priming Method

In this method, short oligonucleotides (which are heterologous in sequence) are used as primers for the initiation of DNA synthesis on single-stranded templates by DNA polymerase. The products of DNA synthesis are radiolabelled by using one ($\alpha^{32}P$) dNTP and three unlabelled dNTPs as precursors to generate probes with high specific activities (2×10^9 dpm/μg). The probes thus produced are homogenous in size and behave more reproducibly in hybridization reactions (Feinberg and Vogelstein 1983; Sambrook and Russell 2001).

34.2.1 Materials

1. Facility for radioactivity work and disposal.
2. Ice flakes.
3. Microcentrifuge.
4. Water bath.
5. DNA polymerase (*E. coli* DNA polymerase I or klenow fragment).
6. Ethanol.
7. Radiolabel (one radiolabelled [$\alpha^{32}P$] dNTP (sp. act. 3000 Ci/mmol): obtained commercially.
8. Random primer buffer and random primers: Heterologous oligonucleotides of 6–7 nucleotides in length: commercially available.
9. Stop or storage buffer: 50 mM Tris-HCl (pH 7.5); 50 mM NaCl; 5 mM EDTA (pH 8.0); 0.5% SDS.
10. Template DNA (linear double-stranded DNA templates up to 1 kb in length).
11. Unlabelled dNTPs (other three dNTPs).

34.2.2 Method

1. Add the following to a microcentrifuge tube:

DNA template	25 ng (1–23 μL)
Distilled water	0–23 μL
Random oligonucleotide primer (125 ng)	1 μL

2. Heat the reaction tube in boiling water for 5 min.
3. Remove the tube and place it on ice for 1 min, briefly centrifuge and return the tube to the ice bath.

4. Add the following:

dNTPs (5 mM/ μL)	1 μL
Random priming buffer (5×)	10 μL
(α^{32}P) dNTP (10 mCi/mL) (sp. act. 3000 Ci/mmole)	5 μL
Water to make	50 μL

5. Add 5 units (1 μL) of the Klenow fragment. Mix the contents and centrifuge briefly.
6. Incubate the reaction at room temperature for 60 min.
7. Add 10 μL of stop/storage buffer to the reaction and store the probe at −20 °C until needed for use.

34.2.3 Notes

1. All institutes working with radiolabels should have a 'Radiation Safety Officer' who will train the personnel in the safe handling and disposal of radioisotopes, as well as the standard operating procedures for record keeping and surveying for contamination. Researchers also need training in methods of measuring the amount of radioactivity in samples. If you are going to be the first person at your institution to use radioisotopes, you should find out what are the legal requirements in your area. Then visit another institution to learn the procedures first-hand.
2. Store radioactive materials only in identified and marked areas in the laboratory. The room should be labelled with signs (Caution—Radiation Area and In Case of Emergency, Call...) and kept under lock when not in use. Eating, drinking and smoking should be totally avoided in the laboratory.
3. It is mandatory to wear appropriate personal protective equipment to avoid direct contact with radioactive materials. Do not pipette solutions by mouth.
4. It is important to contain radioactive materials at all stages of handling to avoid any spillage. Use a fume hood whenever possible for work with open sources. It is a must to use a fume hood for work that requires boiling.
5. It is mandatory to use radioactive shield when working with radioactive materials.
6. Store radioactive isotopes only at designated areas in refrigerators or freezers. Do not use glassware and other equipment reserved for radioactive work for other purposes.
7. Containers and solutions of radioactive materials should be clearly marked with warning symbols indicating the nature and amount or radioactivity. Similarly mark other equipment that is contaminated with radioisotopes and store it in appropriately shielded locations in the radioisotope lab.

8. All radioactive waste should be stored in a safe and secure place under shielded area to ensure that the surface radiation does not exceed 2.5 Sv/h (0.25 mR/h).
9. Clean and isolate contaminated supplies and equipment after the work. Check and decontaminate glassware, wipe floors and working surfaces. Wash hands, monitor clothes, shoes and hands before leaving the laboratory. Log on to https://www.tru.ca/safety/workinglearningsafely/work/radiation/radiationrules.htmL and https://www.bu.edu/ehs/ehs-topics/radiation-safety/isotopes/radioisotope-safety/for more details on the safe handling of radioactive materials.
10. Reaction volumes can be doubled if more probe is needed.
11. Linear template of a plasmid may be prepared by either PCR amplification or restriction digestion.
12. Efficiency of label incorporation can be determined through a radioactive counter.

34.3 Production of Non-radioactive Probe by Random Primed Labelling

The method is based on the same principle as that of production of radioactive probe by random primed labelling except that in place of a radiolabelled dNTP; in this method a non-radioactive dNTP is used. Two methods of labelling systems are available. Biotin-streptavidin system uses biotin (a naturally occurring vitamin) as the reporter and the streptavidin (bacterial protein) as the affinity molecule. Streptavidin conjugated with alkaline phosphatase or horseradish peroxidase is used for detection with appropriate substrates. In digoxigenin (DIG) system, a plant steroid obtained from *Digitalis purpurea* plant is used as reporter and anti-DIG antibody coupled with enzymes (alkaline phosphatase or horseradish peroxidase) as conjugates. Detection method involves two steps. (1) Membrane is exposed to conjugate consisting of reporter enzyme and a ligand that binds with labelling moiety (biotin-streptavidin-alkaline phosphatase and digoxigenin-anti-DIG antibody-alkaline phosphatase). (2) Reporter enzyme is assayed with colorimetric, fluorogenic or chemiluminescent substrates. Chemiluminescence is the fastest and more sensitive. Colorimetric assays are two or more orders of magnitude less sensitive, and their corresponding substrates are used in detection. Labelling using DIG is given below.

34.3.1 Materials

1. Same as in Subheading 34.2.1.
2. Freezer (−20 or −80 °C).
3. DIG-dUTP: commercially available.
4. dNTP mix: Commercially available.

5. EDTA (0.2 M, pH 8.0): Weigh and dissolve 7.444 g of disodium EDTA (Na_2EDTA) in about 80 mL of distilled water, adjust the pH to 8.0 by the addition of NaOH and make up the volume to 100 mL.
6. Ethanol.
7. LiCl (4 M): Weigh 16.96 g of LiCl, dissolve in water and make up the volume to 100 mL, sterilize the solution by autoclaving.
8. Random primer (hexanucleotide primer mix).
9. TE buffer (pH 8.0): 10 mM Tris-HCl, 1 mM EDTA.
10. Template DNA.

34.3.2 Method

1. Denature the template DNA (1–3 μg) in a microcentrifuge tube by heating for 10 min in a boiling water bath and quickly place in ice.
2. Add to the denatured DNA the following on ice:
   2 μL random hexanucleotide primer mix (5×).
   2 μL dNTP mix (125 ng).
   2 μL DIG-dUTP (25 ng).
   5 μL denatured template (1–3 μg).
   10 μL sterile water.
   1 μL klenow enzyme (5 units).
3. Centrifuge briefly and incubate for 60 min at 37 °C.
4. Stop the reaction by adding 2 μL of 0.2 M EDTA, pH 8.0.
5. Precipitate the labelled DNA by adding 2.5 μL of 4 M LiCl and 75 μL cold ethanol, mix well and leave for 30 min at −70 °C or 2 h at −20 °C overnight.
6. Centrifuge at 7500 × *g* for 15 min to pellet the DNA, wash the pellet with 50 μL of cold 70% ethanol.
7. Dry the pellet and dissolve in 50 μL of TE buffer.

34.3.3 Note

1. Instead of DIG-dUTP, biotin-labelled nucleotides may also be used.

34.4 Production of Radiolabelled DNA Probes by Polymerase Chain Reaction

Polymerase chain reaction (PCR) is another method used to produce DNA probes with high specific activity. The principle involved in the PCR is described elsewhere (*see* Chapter 35). The main advantage of the method is that the defined segment of the target DNA can be amplified and labelled independently. Only a small amount of template DNA is required, and specific activity of the probe can exceed 10^9 dpm/μg. Both double- and single-stranded probes can be prepared using PCR. The following protocol describes the generation of double-stranded radiolabelled probe.

34.4.1 Materials

1. Same as in Subheading 34.2.1.
2. Syringe (1 mL).
3. Thermal cycler.
4. Ammonium acetate (10 M): Dissolve 770 g of ammonium acetate in H_2O, make up the volume to 1 L and sterilize by filtration.
5. 10× PCR amplification buffer.
6. Chloroform.
7. dNTP mix containing dATP, dGTP, dTTP, dCTP.
8. Ethanol.
9. Forward and reverse primers: Custom synthesized.
10. Radiolabel (α^{32}P dCTP).
11. Sephadex G-75.
12. *Taq* DNA polymerase.
13. TE buffer (pH 8.0): 10 mM Tris-HCl and 1 mM EDTA.
14. Template DNA.
15. NaCl (5 M): Dissolve 292.2 g of NaCl in 800 mL of distilled water, make the volume to 1 L and sterilize by autoclaving.
16. Tris-HCl (1 M, pH 8.0): Dissolve 12.1 g of Tris base in about 80 mL of water, adjust the pH to 8.0 by adding HCl and make up the volume 100 mL.
17. EDTA (ethylene diamine tetra acetic acid) (0.5 M pH 8.0): Dissolve 16. 8 g of ethylene diamine tetra acetic acid disodium salt in about 80 mL of water adjust the pH to 8.0 by adding NaOH, make up the volume to 100 mL.
18. TEN buffer (40 mM Tris-HCl, pH 8.0, 1 mM EDTA, pH 8.0 and 150 mM NaCl): Prepare 100 mL of TEN buffer by adding 4 mL of 1 M Tris-HCl pH 8.0, 200 μL of 1 mM EDTA, 3 mL of 5 M NaCl and make up the volume to 100 mL.

34.4.2 Method

1. In a PCR tube add the following:

10× PCR amplification buffer	5.0 μL
10 mM dNTP solution	1.0 μL
0.1 mM dCTP	1.0 μL
20 μM forward primer	2.5 μL
20 μM reverse primer	2.5 μL

(continued)

Template DNA (2–10 ng)	5.0 μL
10 mCi/mL ((α^{32}P dCTP) (sp. act 3000 Ci/mmole)	5.0 μL
Water to make	48.0 μL
Taq DNA polymerase (5 units/μL)	2.0 μL

2. Mix and briefly centrifuge.
3. Subject the tubes to amplification for at least 30 cycles in a thermal cycler using appropriate denaturation, annealing and extension times and temperatures followed by a final extension of 10 min at 72 °C.
4. Remove the tubes from thermal cycler and add an equal volume of 4 M ammonium acetate and 2.5 volumes of ethanol. Store the tubes for 1–2 h at −20 °C or for 30 min at −70 °C.
5. Centrifuge at maximum speed for 10 min at 4 °C and discard the supernatant.
6. Add 1 mL of 70% ethanol to the pellet and centrifuge at maximum speed for 5 min.
7. Aspirate and discard the supernatant. Dry and dissolve the pellet in 20 μL TE buffer.
8. Subject the contents to spun column chromatography through Sephadex G-75 as described in the Chapter 43 to remove unincorporated dNTPs and primers. Measure the amount of radioactivity in 1.0 μL of the void volume of the spun column in a liquid scintillation counter and store the remainder of radiolabelled DNA at −20 °C until needed.

34.4.3 Notes

1. Take all necessary precautions while handling radioactive material as indicated in Subheading 34.2.3. All the radioactivity work should be carried out within the radioactive shield. Have a dedicated microcentrifuge, micropipette set and tips for radioactive work. Disposal of radioactive waste should be done as per the guidelines.
2. Single-stranded probe can be prepared using the same procedure except that only one primer (either forward or reverse) need to be used in PCR.

34.5 Production of Non-radioactive DNA Probe by PCR

34.5.1 Materials

1. Same as in Subheading 34.3.1 and 34.4.1
2. PCR DIG labelling mix.

34.5.2 Method

1. In a PCR tube add the following:

10× PCR amplification buffer	10.0 μL
10× PCR DIG labelling mix	10.0 μL
2 mM dNTP solution	10.0 μL
20 μM forward primer	1.0 μL
20 μM reverse primer	1.0 μL
Template DNA (2–10 ng)	1.0 μL
Water	To make 99.0 μL
Taq DNA polymerase (5 units/μL)	1.0 μL

2. Mix and briefly centrifuge.
3. Carry out PCR amplification for 35 cycles (each cycle consisting of a denaturation step at 94 °C for 30 s, annealing at 55 °C for 1 min and extension at 72 °C for 1 min) followed by a final extension at 72 °C for 10 min (annealing temperature may vary depending on the Tm of primer and the duration of extension may vary depending on the length of insert to be amplified).
4. Subject the PCR product to agarose gel electrophoresis to check the size and amount of product.
5. Analyse PCR amplification product labelling efficiency by hybridization to a dot blot
6. Store the probe at −20 °C.

34.6 Dot-Blot Hybridization

It is one of the easy and quick methods for the detection of virus in a crude plant sap. The technique is more sensitive compared to ELISA, can detect up to 1–10 pg of virus. The method is suitable for the detection of large number of samples. In this technique, nucleic acid from the plant material is irreversibly bound to a nitrocellulose (NCM) or nylon membrane and is allowed to hybridize with the probe either radiolabelled or non-radiolabelled to a virus genome suspected to be present in the sample. Positive hybridization indicates the presence of the suspected virus in test samples.

The important steps include (1) isolation of total nucleic acid from plants, (2) denaturing and spotting nucleic acid onto membrane in a grid pattern, (3) blocking remaining reactive sites of membrane by dipping the membrane in a heterologous protein or nucleic acid samples, (4) hybridization of membrane with denatured probe, (5) removal of excess probe by washing of membrane and (6) detection of hybridization by suitable method depending on the reporter molecule used in the probe. If radioactive probe is used, auto radiography is done. If non-radioactive probe is used, the detection method will vary depending on the reporter molecule used.

34.6.1 Materials

1. Dark room.
2. Dot-blot apparatus.
3. Hybridization oven.
4. Nylon or nitrocellulose membrane.
5. Pencil.
6. Pestle and mortor.
7. Petri plate.
8. Saran wrap.
9. Scale.
10. Vacuum drier.
11. X-ray cassette.
12. X-ray film.
13. Infected and healthy plant tissues.
14. TE buffer (10 mM Tris-HCl, 1 mM EDTA).
15. Sodium acetate (3 M, pH 5.0): Dissolve 24.6 g sodium acetate in 70 mL of distilled water, adjust the pH to 5.0 by adding glacial acetic acid and make up the volume to 100 mL by adding water.
16. Sodium hydroxide (0.05 N): Dissolve 200 mg of sodium hydroxide in 100 mL of water.
17. Sodium dodecyl sulphate (SDS): Prepare a 10% stock solution by dissolving 10 g of SDS in 100 mL of water.
18. Developer: For developing X-ray film.
19. Fixer: For developing X-ray film.
20. Saline sodium citrate (SSC) (20×): Dissolve NaCl (175.3 g) and sodium citrate (88.2 g) in water and make up the volume to 1000 mL and autoclave before use.
21. Denhard's solution (50×): Prepare 50× solution by dissolving 5 g each of BSA (bovine serum albumin), Ficoll and PVP (polyvinyl pyrrolidone) in 500 mL of water.
22. Pre-hybridization buffer/hybridization buffer (for radioactive probes): Prepare by dissolving 6× solution of SSC, 5× of Denhardt's solution, 0.5% SDS and 100 μg/mL of salmon sperm DNA.
23. Probe: Labelled with radioactive material (details of the preparation is provided in Subheadings 34.2.2 and 34.4.2)

34.6.2 Method

34.6.2.1 Sample Preparation

Sample to be spotted can be plant sap, virion particles or nucleic acid. Crude extract is made in appropriate buffer and denatured by alkali or heat before application. After spotting the samples, the membrane is baked at 80 °C in a vacuum drier for 2 h to immobilize the samples to the membrane. If nylon membrane is used, baking may not be needed.

1. Grind 1 g of leaf in 1 mL of TE buffer in a pestle and mortor. Filter through muslin.
2. Take 200 μL of this extract in an eppendorf tube and add 50 μL of 0.5 N NaOH. Leave at room temperature for 10 min.
3. Add 25 μL of 3 M sodium acetate, pH 5.0.
4. Make dilution of 1:10 and 1:100 in TE buffer.

34.6.2.2 Preparation of Membrane and Dotting Samples

5. Cut the required size of membrane, either nylon or nitrocellulose (handle membrane with gloves). Mark 1 cm squares using soft pencil. Give a small cut at one edge (left top) to indicate orientation.
6. Soak the membrane in sterile water, then in 20× SSC for 5 min (nylon membrane should be soaked only in water).
7. Drain off the fluid and spread the membrane on a clean blotting paper. Dot the samples as quickly as possible as per the pre-decided plan.
8. Carefully dot 2.5–5 μL of sample in each square of the membrane by keeping the dot as small as possible. Alternatively dot-blot apparatus (available from Bio-Rad or Schleicher and Schruel or other manufacturers) may also be used. A negative (healthy) and a positive (plant inoculated with the virus for which test is done) should always be included in the tests for comparison.
9. Air dry the membrane and bake in a vacuum drier at 80 °C for 2 h (baking is not required if nylon membrane is used). At this step the membrane can be stored between the folders of Whatman paper at room temperature until proceeding further for hybridization.

34.6.2.3 Processing and Development of Membrane (When Radioactive Probe Is Used)

Pre-hybridization and Hybridization

The membranes are either suspended in heat sealable plastic bags or in bottles at a temperature of 65 °C. Before hybridization, the membranes are soaked in pre-hybridization solution to block any non-specific binding of the probes to the unoccupied surface in the membrane. Pre-hybridizing solution contains a protein (usually bovine serum albumin), a large molecular weight DNA (calf thymus or salmon sperm that is sonicated to give small fragments), Ficoll and polyvinyl pyrrolidone. After 1–3 h of pre-hybridization, the probe is added to the solution and hybridization continued overnight.

10. Add 100 μL/cm^2 (area of membrane) pre-hybridization solution (for radioactive probes) in to the hybridization tube or bag containing the membrane.
11. Seal the bag or tighten the tube carefully preventing any leakage.
12. Incubate at 65 °C in a water bath (for bags) or hybridization oven (for tubes) with gentle shaking for 1–3 h.
13. Denature the radiolabelled probe at 100 °C for 10 min and immediately chill on ice.
14. Take out the bags, cut at one end or open the lid of the tube and add probe carefully to the buffer and seal the bag or tighten the tube. Incubate at 65 °C in water bath (for bags) or in hybridization oven (for tubes) with gentle shaking overnight.

Washing

After 18 h of hybridization, the membrane is washed in salt solutions to remove unhybridized radiolabelled probes. A heterologous duplex of nucleic acid is less stable than the homologous one and has a lower Tm. Both temperature and salt concentration affect the melting of double-stranded nucleic acids of different degrees of complementarity: low stringency will allow hybridization with less complementary sequences. Hence, a selection of right temperature and salt concentration is important to obtain successful result. Initial experiments with low to high stringencies will have to be done to arrive at ideal combination of temperature and salt concentration to give consistent results.

15. Pour off the hybridization solution and wash the membrane as follows: First, wash with 1× SSC and 0.1% SDS at 65 °C for 15 min; second, wash with 0.5× SSC and 0.1% SDS at 65 °C for 15 min and third, wash with 0.1× SSC and 0.1% SDS at 65 °C. Wrap the blot with Saran wrap, check the count and perform autoradiography.

Autoradiography

The membrane after washing are dried and exposed to X-ray film to visually locate positive hybridization. The intensity of autoradiographic signal depends on the intensity of the emitted radiation and duration of exposure. In general, ^{32}P is used in dot-blot hybridization as it emits high-energy β-particles that increase the sensitivity of detection.

16. Place the wrapped membrane in an exposure X-ray cassette. Use adhesive tape at corners if needed to fix the membrane.
17. Take the cassette to the dark room, place X-ray film on the membrane and close the cassette. Give a small cut at left top corner on X-ray film for easy identification of orientation.

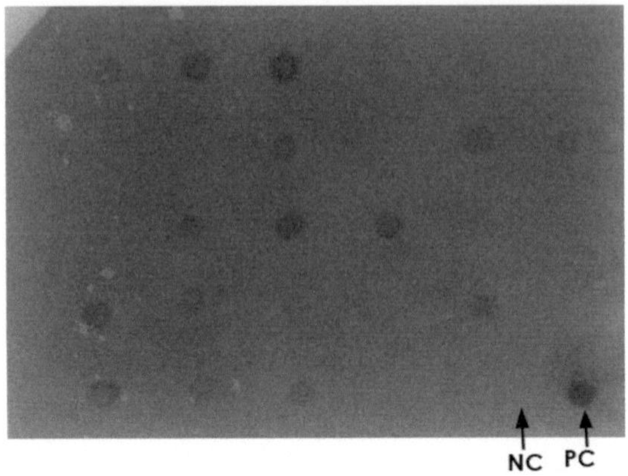

Fig. 34.2 Detection of virus using dot blot hybridization. Spots NC and PC correspond to known healthy (negative control) and infected plants (positive control) while the rest of the spots are from test plants

18. Cover the cassette with black cloth and keep at −70 °C for 1 h to 3 days depending upon the counts.
19. Remove the X-ray film from the cassette and place it in a glass tray containing sufficient quantity of X-ray developer solution in the dark room for about 5 min with slow agitation.
20. Remove the X-ray film from the developer solution, drain off excess developer and briefly rinse the film in distilled water before transferring on to the next tray containing fixer solution. Agitate the tray slowly 5 min under dark.
21. Wash the X-ray film thoroughly in running tap water and air dry.
22. Check for positive hybridization: Presence of black dot indicates the positive reaction with the corresponding sample. The intensity of dark dot is proportional to the concentration of the virus in the sample used. Presence and absence of black dot at positions where nucleic acid from known positive and negative samples, respectively, indicate that the test performed was OK. Thus among the various test samples used, the presence of the black dot indicates the presence of virus in that sample (Fig. 34.2).

34.6.2.4 Processing and Development of Membrane (When a Non-radioactive Probe Labelled with DIG Is Used)

Digoxigenin, a steroid hapten is used to label DNA or RNA for hybridization and subsequent colour luminescence detection. During hybridization, the probe having the DIG nucleotides binds to the homologous nucleic acid molecule, and the remaining gets washed away in the washing step. During detection, the anti-DIG-alkaline phosphatase conjugate antibody binds to the DIG nucleotides on the hybridized bands. After adding the chemiluminescence substrate, the alkaline phosphatase enzyme reacts emitting light, which can be detected by X-ray film. Alternatively, a calorimetric substrate (NBT and BCIP) can be used for colour development.

Materials

1. As in Subheading 34.6.1.
2. Anti-DIG antibody coupled with alkaline phosphatase (DIG-AP): Commercially available.
3. DIG-labelled probe: Details of preparation provided in Subheadings 34.3.2 and 34.5.2.
4. Colour substrate (NBT and BCIP): Prepare solution A by dissolving 75 mg of nitro blue tetrazolium (NBT) in 1 mL of dimethyl formamide (DMF) and solution B by dissolving 50 mg of 5-bromo-3-chloroindolyl phosphate (BCIP) in 1 mL of DMF. Add solution A and solution B to give a final concentration of 0.33 mg and 0.175 mg per mL, respectively, of the substrate buffer just before use. For example to prepare 10 mL of the substrate solution, dissolve 44 μL of solution A and 37 μL of solution B.
5. Pre-hybridization/hybridization solution (5× SSC, 0.1% w/v *N*-lauroylsarcosine, 0.02% w/v SDS): Prepare 100 mL of the buffer by mixing 25 mL of 20× SSC, 1 mL of 10% *N*-lauroylsarcosine, 200 μL of 10% SDS and make up the volume to 100 mL with distilled water.
6. Maleic acid (1 M, pH 7.5): Dissolve 116.1 g of maleic acid in water, adjust the pH to 7.5 and make up the volume to 1 L.
7. Maleic acid buffer (pH 7.5) (0.1 M maleic acid and 0.15 M NaCl): Prepare 100 mL of maleic acid buffer by mixing 10 mL of 1 M maleic acid solution (pH 7.5), 3 mL of 5 M NaCl and make up to 100 mL with water.
8. Washing buffer (maleic acid buffer containing 3% Tween 20): Add 3 mL of Tween 20 in 100 mL of maleic acid buffer.
9. Blocking solution: 1% blocking agent (BSA or ovalbumin) in maleic acid buffer: Add 1 g of BSA or ovalbumin in 100 mL of maleic acid buffer.
10. Detection buffer (pH 9.5) (0.1 M Tris-HCl, 0.1 M NaCl and 50 mM $MgCl_2$): Dissolve 12.1 g of Tris base, 5.84 g of sodium chloride and 0.475 g of magnesium chloride in about 900 mL of water, adjust pH to 9.5 with HCl and make up the volume to 1 L.
11. CSPD: It is a chemiluminescent substrate for alkaline phosphatase that provides fast and sensitive detection of the target molecule by producing visible light that can be captured on X-ray film or on luminescence imager systems.

Pre-hybridization and Hybridization

1. Follow **steps 1–9** as indicated in Subheadings 34.6.2.1 and 34.6.2.2 for sample preparation, membrane preparation and dotting of samples.
2. Preheat the hybridization buffer to 60 °C. The required volumes are 20 mL/100 cm^2 of membrane in bags or hybridization ovens.
3. Pre-hybridize the blots with buffer at least for 30 min at 60 °C with constant gentle shaking.
4. Place the tube containing probe in a boiling water for 5 min to denature followed by snap cooling on ice.
5. For hybridization, take 2.5 mL per 100 cm^2 (area of membrane) of pre-warmed hybridization buffer and mix with required amount of probe (5–25 ng/mL). The probe concentration may be reduced for longer duration of hybridization. Avoid putting hybridization buffer directly on the membrane.
6. Hybridize the blots overnight at 60 °C with constant gentle agitation.
7. Wash the blots as follows: Wash twice with 2× SSC containing 0.1% SDS at room temperature for 5 min each followed by two more washes with 0.1× SSC containing 0.1% SDS at 60 °C for 15 min each.
8. Do not allow the blot to dry. If at all desired for storage, it should be wrapped with Saran wrap and stored at 2–8 °C in the dark.

Detection of Hybridized Bands by Colorimetric Method

9. Rinse the membrane briefly in washing buffer for 1–2 min.
10. Place for 30 min in 1% blocking solution.
11. Add anti-DIG–AP conjugate to the blocking solution as per the dilution suggested by the manufacturer. Incubate the membrane for 30 min in 50 mL of antibody solution.
12. Wash the membrane two times with 100 mL of washing buffer for 15 min each and equilibrate 2–5 min in 20 mL of detection buffer.
13. Place the membrane in 10 mL of freshly prepared colour substrate solution for 5 min in dark. Do not shake during colour development.
14. The formation of colour precipitate is seen within a few minutes, and the reaction may sometimes extend up to 1 h. During the process, do not expose substrate solution to light except for a short time to monitor the colour development.
15. When the desired band or spot intensities are achieved, stop the reaction by washing the membrane for 5 min with 50 mL of distilled water.

Detection of Hybridized Bands by Chemiluminescence Method

1. Follow **steps 1–8** as indicated above.
2. Rinse the membrane briefly (1–5 min) in washing buffer.
3. Place the membrane in blocking solution for 30 min.
4. Add anti-DIG–AP conjugate to the blocking solution as per the dilution suggested by the manufacturer. Keep the membrane for 30 min in 50 mL of antibody solution.
5. Give two washes with 100 mL of washing buffer for 15 min each and equilibrate 2–5 min in 20 mL of detection buffer.
6. Place the membrane on a development folder or hybridization bag in such a way that DNA-spotted side facing up and add 1 mL of CSPD.
7. Immediately use another sheet of the folder to spread the substrate uniformly without air bubbles and cover the membrane.
8. Wait for 5 min at room temperature.
9. Remove the excess substrate and seal the corners of development folder.
10. Place the membrane in an exposure cassette. Use adhesive tape at corners if needed to fix the membrane.
11. Take the cassette to the dark room, place X-ray film over the membrane and close the cassette. Give a small cut at left top corner on X-ray film for easy identification of orientation.
12. Cover the cassette with black cloth and keep at room temperature for 15–20 min.
13. Remove the X-ray film from the cassette and develop using developer (5 min) in the dark room and rinse in cold distilled water.
14. Transfer the developed film to the fixer (5 min) in the dark room.
15. Wash the film thoroughly in running tap water and air dry.
16. Check for positive hybridization: Presence of black dot indicates the positive reaction with the corresponding sample. The intensity of dot is proportional to the quantity of the virus in the sample used. Presence and absence of dot at positions where nucleic acid from known positive and negative samples respectively indicate that the test performed was OK.

34.6.3 Notes (When Radioactive Probe is Used)

1. Autoradiography is used to detect the location of radioactivity on membranes or in gels.
2. If X-ray exposure is made at −70 °C, allow cassette to warm to room temperature before opening, to prevent condensation.

3. When ^{32}P is used, cover the membrane with clear Saran wrap in order to protect the cassette, intensifying screen from contamination with penetrating radioactivity.
4. The clear Saran wrap may block a significant amount of the ^{35}S signal. Hence if ^{35}S is used as the radioactive source, do not cover with Saran wrap.
5. The wash temperature will determine the signal-to-noise ratio. More stringency (less background) will be gained with a higher temperature. The homology of probe to the test DNA, probe length, and salt concentration can also affect the signal-to-noise ratio.
6. If more or less exposure is desired, repeat autoradiography for an appropriate period.
7. If the background is too high, go back to **step 15**, and re-wash at a higher stringency (either low salt or high temperature or both).

34.6.4 Notes (When Non-radioactive Probe is Used)

1. Do not allow drying of membrane during exposure as it will result in dark background.
2. Luminescence continues for 48 h. The signal increases in first few hours after the initiation of detection reaction until it will reach a plateau where signal intensity remains almost constant during the next 24–48 h. Multiple exposures can be taken to achieve the desired signal strength.
3. Instead of anti-DIG-alkaline phosphatase (DIG-AP), anti-DIG-horseradish peroxidase (DIG-HRP) conjugate with appropriate substrates can also be used for detection.
4. Like DIG, biotin can also be used as non-radioactive labelling moiety.
5. If biotin is used as probe, streptavidin-alkaline phosphatase or streptavidin-horseradish peroxidase conjugate can be used in detection either colorimetrically or through chemiluminescence.

References

Alwine JC, Kemp DJ, Stark GR (1977) Method for detection of specific RNAs in agarose gels by transfer to diazobenzyloxymethyl paper and hybridization with DNA probes. Proc Natl Acad Sci U S A 74:5350–5354

Blok VC, Ziegler A, Scott K, Dangora DB, Robinson DJ, Murant AF (1995) Detection of groundnut rosette umbravirus infections with radioactive and non-radioactive probes to its satellite RNA. Ann Appl Biol 127:321–328

Brandsma J, Miller B (1980) Nucleic acid spot hybridization: rapid quantitative screening of lymphoid cell lines for Epstein-Barr viral RNA. Proc Natl Acad Sci U S A 77:6851–6855

Britten RJ, Kohne DE (1968) Repeated sequences in DNA. Science 161:529–540

Feinberg AP, Vogelstein B (1983) A technique of radiolabeling DNA restriction fragments to high specific activity. Ann Biochem 132:6–15

Garger SJ, Turpen T, Carrington JC, Morris TJ, Jordan RL, Dodds JA, Grill U (1983) Rapid detection of plant RNA viruses by dot-blot hybridization. Plant Mol Biol Rep 1:21–25

Gillespie D, Spiegelman S (1965) A quantitative assay for DNA/RNA hybrids with DNA immobilized on a membrane. J Mol Biol 12:829

Maule AJ, Hull R, Donson J (1983) The application of spot hybridization to the detection of DNA and RNA viruses in plant tissues. J Virol Methods 6:215–224

Owens RA, Diener TO (1981) Sensitive and rapid diagnosis of potato spindle tuber viroid disease by nucleic acid hybridization. Science 213:670–672

Rigby PWJ, Dieckmann M, Rhodes C, Berg P (1977) Labeling deoxyribonucleic acid to high specificity by nick translation with DNA polymerase I. J Mol Biol 113:237–257

Roy BP, Abou Haidar MG, Alexander A (1989) Biotinylated probes for the detection of potato spindle tuber viroid (PSTV) in plants. J Virol Methods 23:149–156

Sambrook J, Russell DW (2001) Molecular cloning: a laboratory manual, 3rd edn. Cold Spring Harbor Laboratory Press, New York

Smith GR, Clarke ML, de Velde RV, Dale JL (1994) Chemiluminescent detection of Fiji disease virus with biotinylated DNA probes. Arch Virol 136:325–334

Southern EM (1975) Detection of specific sequences among DNA fragments separated by gel electrophoresis. J Mol Biol 98:503–508

Chapter 35

Polymerase Chain Reaction

Abstract

Polymerase chain reaction (PCR) is the most widely used nucleo-based method for the detection of plant viruses. It is a primer-mediated in vitro reaction involving amplification of target nucleic acid sequences. A standard PCR is a three-step procedure: (1) denaturation at a high temperature (90–95 °C), (2) annealing of target specific primers (45–60 °C) and (3) primer extension by a thermostable DNA polymerase at 72 °C. Different variants of PCR have been developed ranging from conventional PCR to real-time PCR to improve the sensitivity and specificity for the detection of plant viruses. Some virus detection methods are: DNA-based PCR, nested PCR (nPCR), immunocapture PCR (IC-PCR), multiplex PCR (M-PCR), real-time PCR and DNA finger printing. Others are RNA-based reverse transcription (RT) PCR, real-time RT-PCR, IC-RT-PCR and AmpliDet RNA. All these methods enable a rapid and accurate detection and quantification of plant viruses. Protocols of PCR assays for the detection of plant viruses are discussed in this chapter.

Key words PCR, Immunocapture PCR, Nested PCR, RT-PCR, IC-RT-PCR

35.1 Introduction

Polymerase chain reaction (PCR) is a technique used to amplify a segment of DNA that lies between two regions of known sequences. Kary Mullis invented the PCR in 1983 (Mullis et al. 1986; Saiki et al. 1988). It is a method by which any number of copies (several million) of any specific piece of DNA can be made within a few hours in the laboratory. During cell division in the enzyme, polymerase makes a copy of the DNA in the cell. The process involves separation of two strands of DNA after which polymerase synthesizes copies of the DNA using each of the separated strands as template, dNTPs (four nucleotide bases, namely dATP, dGTP, dCTP and dTTP) and two primers (forward and reverse). The enzyme primase in the cell makes the primer. Once the primer anneals to the template strand, polymerase adds nucleotides to the 3′ end of the primer and extends synthesis of the new strand.

PCR mimics the above process in a small vial in the laboratory. A PCR requires the following components such as template DNA, four nucleotides (dATP, dGTP, dCTP and dTTP), DNA

Alangar Ishwara Bhat and Govind Pratap Rao, *Characterization of Plant Viruses*, Springer Protocols Handbooks,
https://doi.org/10.1007/978-1-0716-0334-5_35,

polymerase and buffer. The automation of the PCR is possible due to the use of *Taq* polymerase (DNA polymerase isolated from the bacterium, *Thermus aquaticus* that lives in hot springs) that remains active even at high temperatures. Unlike other polymerases, the polymerase extracted from *T. aquaticus* (*Taq*) is stable in the fluctuating temperatures. The uses of thermostable *Taq* polymerase and automated machine (thermocycler) that can be programmed to a particular temperature for specified number of cycles have significantly enhanced the utilization of PCR. With the advent of PCR, many PCR-based diagnostic methods for the identification of plant pathogens have been developed (Pappu et al. 1993; Mumford et al. 1996; Ward et al. 2005; Jain et al. 2008; Vincelli and Tisserat 2008). Furthermore, the amplification of pathogen-specific sequences and the coupling of PCR with other techniques have been described in order to improve specificity and sensitivity (Merighi et al. 2000; Brasileiro et al. 2004; Carrasco-Ballesteros et al. 2007). RT-PCR is a variant of PCR, wherein normal PCR is preceded by reverse transcription converting RNA to complementary DNA (cDNA) using the enzyme reverse transcriptase. Since majority of the plant viruses have RNA genome, RT-PCR is used to amplify a segment of RNA that is present between two regions of known sequences. IC-PCR, another variant of PCR, is a highly specific and sensitive method for the detection of viruses, in which actual viral particles are trapped with virus-specific antibodies. It is particularly useful when the presence of inhibitory plant compounds or low virus titres pose a problem in the detection by conventional PCR. Multiplex PCR allows the simultaneous and sensitive detection of different DNA or RNA targets in a single reaction. It is a useful tool since crops are infected by more than one virus, and their individual detection by PCR is not only expensive but also time consuming (Henson and French 1993; Hadidi et al. 1995).

35.2 Steps in PCR

There are three basic steps in PCR. The first step separates the two DNA chains in the double helix called as melting of DNA (denaturation). Denaturation is done by heating the vial containing PCR components to 90–95 °C for about 30 s (Fig. 35.1). The second step is hybridization or annealing of the primers to their complementary bases on the new single-stranded DNA (primer annealing). This is achieved by cooling the vial to 45–65 °C (depending on the melting temperature of the primers) for about 30 s. The final step is to make a complete copy of the template called as the extension. During this, the temperature in the vial is raised to 72 °C as *Taq* polymerase works best at this temperature. The *Taq* polymerase extends the chain by adding nucleotides to the 3′ end of

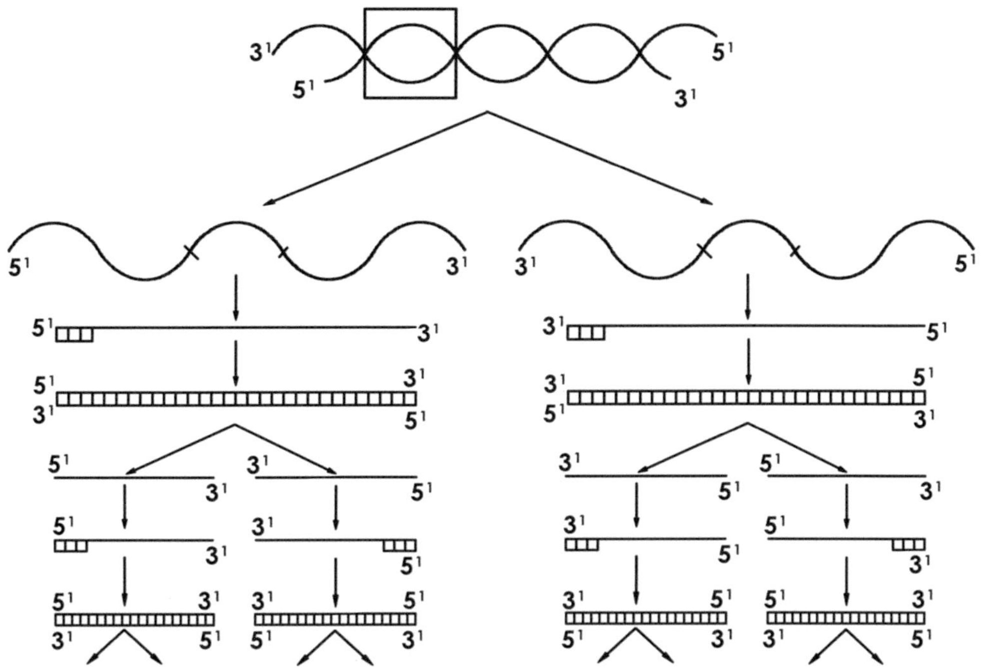

Fig. 35.1 Schematic representation of the process of DNA amplification by polymerase chain reaction (PCR)

the primer and makes a complimentary copy of the template. With this, one cycle of PCR is over, during which each piece of DNA in the vial gets duplicated (Fig. 35.1). Each newly synthesized DNA part can now act as a new template for the subsequent cycles, thus after 30 cycles, several millions of copies of a particular piece of DNA can be formed. The instrument called as thermocycler (PCR machine) is used to control automatically the required temperature for each of the three steps. Determination of the optimum concentration of reaction components (dNTPs, $MgCl_2$, primers, polymerase), as well as temperature and time, is necessary to ensure the reproducibility and efficiency of PCR (Sambrook and Russell 2001).

35.3 Components of PCR

To perform PCR, the following components are essential.

1. *Oligonucleotide primer pair*

   To synthesize primers, the sequence of at least a portion of the DNA molecule that one wishes to amplify should be known. Primers consist of short oligonucleotides (containing about two dozens of nucleotides) that are precisely complementary to the sequence at the 3′ end of each strand of the DNA to be amplified. Primers can be designed manually or by using software packages. While designing primers, care should be taken to avoid complementarity in the sequence at the 3′ ends of forward and reverse primers. Also, self-complementarity within the primer should be avoided. Ideally primers should have G + C ratio of about 40–60%. Specificity of the primers can be checked through BLAST (basic local alignment search tool) suite of NCBI (National Centre for Biotechnology Information) (https://blast.ncbi.nlm.nih.gov/Blast.cgi). Primers are synthesized on an automated DNA synthesizer. Careful design of primers is required to obtain the desired products in high yield. As far as possible, design both forward and reverse primers with same melting temperature (Tm). Primer annealing temperature depends on its Tm value that is calculated from the following equation:

Tm (°C) = 4 (G + C) + 2 (A + T), where A, T, G and C stand for the number of corresponding nucleotides in the primer. Primer annealing may fail at a temperature much higher than Tm, whereas temperatures much below Tm lead to non-specific amplification.

2. *Thermostable DNA polymerase*

   Taq DNA polymerase is the most commonly used enzyme, although, many other thermostable enzymes are now available. One unit of *Taq* DNA polymerase is the amount of enzyme which incorporates 10 n moles of total dNTPs into acid-perceptible DNA within 30 min at 74 °C.

3. *Deoxynucleoside triphosphates (dNTPs)*

   This contains equimolar amounts of dATP, dTTP, dCTP and dGTP. Generally a concentration of 200–250 μM of each dNTP is recommended for each PCR.

4. *Divalent cations*

   All thermostable DNA polymerases require free divalent cations such as Mg^{+2} for activity. The concentration of Mg^{+2} required for optimal reaction needs to be determined for each of the primers and template. In general, the concentration would vary from 1.5 to 4.5 mM depending on the primers and template.

5. *Buffer*

   A 10 mM Tris-HCl buffer at a pH between 8.3 and 8.8 is required for the standard PCR.

6. *Template DNA*

   Template DNA can be in single- or double-stranded form. Closed circular DNA can also be used as templates though they are slightly less efficient than linear DNAs. One microgram of 1000 bp DNA fragment corresponds to 1.52 pico moles. The following equation depicts the relation between nanogram and picomole quantities of primer:

   Nanogram of oligonucleotide primer = picomole of the primer × number of nucleotides × 0.33.

35.4 Running PCR

After the addition of all the components in a 0.2 or 0.5 mL thin-walled PCR tube, it is placed in a thermocycler. An initial denaturation step (3–5 min at 95 °C) is given at the beginning of the reaction to completely denature the template DNA. This will be followed by amplification cycles consisting of denaturation (30–60 s at 95 °C), annealing (1 min at Tm of the primers or Tm − 2 °C) and extension (72 °C for 1–4 min depending on the length of the PCR product to be amplified). The most efficient synthesis of product in PCR is in the range of 300–600 bp in length. The reaction is cycled 25–35 times with a final extension of 7–10 min at 72 °C to complete any partially extended product. The reaction tube is then removed from the machine and placed at 4 °C. This will then be analysed by agarose gel electrophoresis to visualize the reaction products.

35.5 PCR for the Amplification of DNA Viruses

Plant viruses whose genome is made up of either single- or double-stranded DNA in the genera such as *Babuvirus*, *Badnavirus*, *Begomovirus*, *Caulimovirus*, *Curtovirus*, *Mastrevirus*, *Nanovirus*, and *Tungrovirus* can be used as template for amplifying either full or partial genomes of these viruses. As sequences of many species of viruses infecting different crops of the above virus genera are already available, primers can be designed to any region of genome for their amplification through PCR (Hadidi et al. 1995; Atzmom et al. 1998; Henson and French 1993).

The basic steps in the PCR for DNA plant viruses include:

- Isolation of DNA from purified virus preparations or total DNA from virus-infected plant.
- Design and synthesis of two virus-specific oligonucleotide primers (forward and reverse).
- Set up the PCR in a tube by adding template nucleic acid, oligonucleotide primers, dNTPs, magnesium chloride and *Taq* polymerase. Set up a positive and a negative control reaction using nucleic acid isolated from known infected and healthy plants, respectively.
- Keep the tubes in the PCR (thermocycler) machine and start the run as per the pre-decided program.
- Run the contents of the tube in an agarose gel and observe for the presence of expected size band (presence of band at the expected size indicates that the sample under test is positive) (Fig. 35.2).

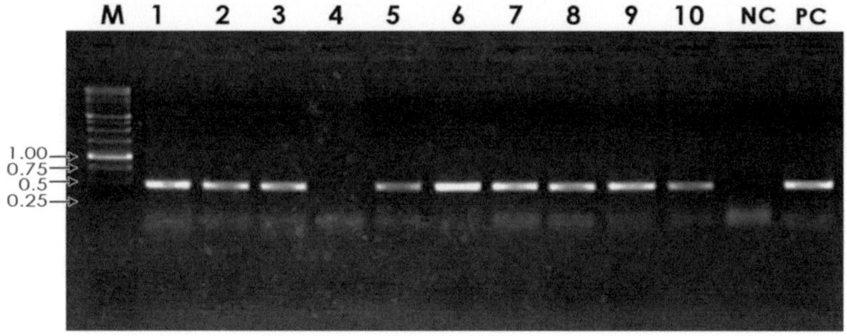

Fig. 35.2 Detection of a DNA of virus-infected plants through PCR. Presence of the 450 bp band indicates positive reaction for the virus. Lane M: DNA marker ladder; Lane 1–10: test plants; Lane NC: healthy (negative) control; Lane PC: a known virus-infected (positive) control

35.5.1 Materials

1. Agarose: It is used for the separation of DNA and RNA, by electrophoresis. Generally agarose gels of 0.7–2.0% is used for the separation of nucleic acid based on size of the expected PCR product.
2. dNTP mix: It contains equimolar concentrations of nucleotides, dATP, dCTP, dGTP and dTTP, each at a final concentration of 10 mM. Usually in most PCRs, 2.5 mM dNTP mix is used.
3. Ethidium bromide: Prepare a stock solution of 10 mg/mL by dissolving 10 mg of ethidium bromide in 1 mL of water, make aliquots and store at room temperature in amber-coloured bottles. As ethidium bromide is a potent mutagen, recently safe alternatives for this (ethidium bromide safe stain) from different firms are available. These stains have same sensitivity like ethidium bromide, and DNA in the gels can be viewed and documented using same gel documentation unit.
4. Gel documentation unit: This equipment is required for the imaging and documentation of nucleic acid bands within poly acrylamide or agarose gels. The gels are usually stained with ethidium bromide or ethidium bromide is added with the sample to be loaded in the gel. The unit includes an ultraviolet (UV) light transilluminator, a hood or a darkroom to guard external light sources and protect the user from UV exposure, and a CCD or CMOS camera for image capturing. If you are interested only in viewing the DNA bands present in the gel, a transilluminator is enough.
5. Gel loading dye: Prepare by adding 200 μL of 1% bromophenol blue, 200 μL of glycerol, 60 μL of 10% SDS, 50 μL of 0.5 M EDTA, 60 μL of 10× TAE and 30 μL of H_2O.
6. Horizontal gel apparatus (submarine unit) with power pack: It is used to run agarose gels submerged in running buffer. Samples are introduced to an electric field and will migrate to the anode or cathode depending on their intrinsic charge. Systems

can be used to separate DNA and RNA for applications such as sample quantification, size determination or PCR amplification detection. Systems typically come with submarine tank, casting tray, combs, electrodes and power supply. Mini, midi and maxi gel sizes are available for small, medium and large gels, respectively. System size choice will depend on sample type, size, running time and electrophoresis conditions.

7. Ice machine: To produce ice flakes needed to keep all PCR components under cool conditions while setting up of PCR.
8. $MgCl_2$: Commercially available in different molarities.
9. PCR buffer (10×): This comes with *Taq* polymerase.
10. PCR work station: This includes a laminar air flow for contamination-free work area for setting up PCR mix. Keep autoclaved microtips and micro pipettes exclusively for PCR work. It is good to have a vortex, a microfuge dedicated for PCR work in the workstation.
11. Primers (forward and reverse): This can be custom synthesized commercially by sending the nucleotide sequence of the required primers. The thumb rules explained in Subheading 35.3 should be strictly followed while designing primers as success of PCR mainly depend on the primers.
12. TAE buffer (10×): Prepare by mixing 48.4 g of Tris base, 11.42 mL of glacial acetic acid, 20 mL of 0.5 M EDTA (pH 8.0) and water to make 1000 mL. In order to prepare 1× TAE, dilute ten times with water, for example, 10 mL of 10× made to 100 mL by adding sterile water.
13. *Taq* DNA polymerase: It is a thermostable DNA polymerase (can withstand high temperatures) isolated from thermophilic bacterium *Thermus aquaticus*. This is available commercially from different firms.
14. Template DNA: Either DNA isolated from purified virus particles or total DNA isolated from infected plants (detailed method for isolation of DNA from purified virus preparation and total DNA from plants are provided elsewhere, *see* Chapters 25 and 31).
15. Thermal cycler (PCR machine): Many commercial companies are manufacturing PCR machines. One can choose the best that suits his/her work depending on the features available.
16. Thin-walled PCR tubes: Available commercially, they are specially prepared thin-walled tubes for conducting PCRs.
17. DNA molecular weight markers: This is available in different ranges starting from 50 bp (base pairs) to 20 kbp (kilo base pairs). Depending on the expected product size of the PCR, one can choose the right type needed.

Table 35.1
Reagents of a PCR

Reagent	Stock concentration	Working concentration	Volume in reaction mix (μL)
PCR buffer	10×	1×	5.0
dNTP mix	10 mM/μL	0.2 mM	1.0
$MgCl_2$	25 mM/μL	1.5 mM	3.0
Forward primer	100 μM/μL	50 μM	0.5
Reverse primer	100 μM/μL	50 μM	0.5
Taq DNA polymerase	3 U/μL	1.5 U	0.5
Template DNA	20 ng/μL	100 ng	5.0
Sterile water	–	–	34.5
Total volume			*50.0*

35.5.2 Method

1. Thaw all components of PCR on ice and briefly centrifuge.
2. Add the following reagents to a 0.2 mL PCR tube on ice (Table 35.1).
3. Mix the contents and briefly centrifuge.
4. Set up similar reaction using template DNA isolated from a known virus-infected plant (to serve as positive control) and virus-free plant (to serve as negative control). Set up another negative control without any template (use water instead of DNA) to check for any non-specific amplification.
5. Place the tube in a thermal cycler (PCR machine) and carry out PCR as per pre-decided programme using the temperature and time required for denaturation, annealing and extension as per the guidelines provided in Table 35.2. Generally thumb rule for the time required for extension is 1 min for every 1000 bp fragment to be amplified. Denaturing and extension are usually carried out at 94 °C and 72 °C, respectively. Temperature for annealing of primers should be either Tm or Tm − 2 °C of the primers (Table 35.2).
6. After the PCR, analyse a portion (about 10–20 μL) of the reaction on a 1% agarose gel electrophoresis along with DNA markers of appropriate size (for details of agarose gel electrophoresis, refer Chapter 26).
7. Observe gel for the presence of DNA using gel documentation unit. The presence of band at the expected size in positive control and absence of the same in negative controls (healthy plant and no template) indicate the success of PCR. The test

Table 35.2
Temperatures for different steps of PCR

Step[a]	Stage	Temperature (°C)	Time (min)
1	Initial denaturation	90–94	3–5
2	Denaturation	90–94	0.50
3	Primer annealing	45–60 (depends on Tm of primers)	1
4	Extension	72	1–4 (depends on the length of the fragment to be amplified)
5	Final extension	72	5–10

[a]Repeat steps (2–4), 30–40 times

samples depending on the presence or absence of virus may show the presence or absence of the band, respectively (Fig. 35.2).

35.6 Reverse Transcription (RT) PCR for Detection of RNA Viruses

Reverse transcription polymerase chain reaction (RT-PCR) is a method to amplify complementary DNA (cDNA) copies of RNA. In this, the first step is the enzymatic conversion of RNA to a single-stranded cDNA template using the enzyme, reverse transcriptase. An oligo deoxynucleotide primer is hybridized to the mRNA and is then extended by an RNA-dependent DNA polymerase (reverse transcriptase) to create cDNA copy that can be later amplified by PCR. The enzyme reverse transcriptase (RT) catalyses the synthesis of cDNA from RNA template. Either RNA isolated from purified viral preparations (*see* Chapter 25 for details) or total RNA isolated from infected plant (*see* Chapter 32 for details) or dsRNA isolated from infected plant (*see* Chapter 33 for details) can be used as template to initiate cDNA synthesis. The primer for first strand cDNA synthesis can be specifically designed to hybridize to a particular target gene, or it can be an oligo (dT) [for viruses containing poly (A) tail at their 3′ ends]. Amplification of the desired portion of cDNA can be achieved in PCRs primed by sense (forward) and antisense (reverse) oligonucleotide primers corresponding to specific sequences in particular cDNAs. For maximum specificity, the antisense primer should be located upstream of the oligonucleotide used to prime cDNA synthesis. Positive and negative controls should always be included when setting up RT-PCRs. In this chapter, we have described two protocols. In the first one, a single-tube RT-PCR in which both RT and PCR are done in the same tube. In the second protocol, cDNA synthesis

Table 35.3
Reaction mixture of RT-PCR assay

Component	Volume to be taken
PCR buffer (10×)	10.0 μL
DTT (0.1 M)	10.0 μL
$MgCl_2$ (25 mM)	5.0 μL
dNTP mix (10 mM)	1.0 μL
Forward primer (10 μM)	1.0 μL
Reverse primer (10 μM)	1.0 μL
RNasin (32 units/μL)	0.5 μL
AMV RT (20 units/μL)	1.0 μL
Taq DNA polymerase (3 units/μL)	1.0 μL
Template RNA[a]	40.0 μL
Water (to make)	100.0 μL

[a]Denature the RNA by heating for 10 min at 80 °C and snap cooling on ice for at least 5–10 min before adding into RT-PCR mix

is carried out first using antisense primer. A portion of cDNA synthesized is then used for PCR using both sense and antisense primers (Singh et al. 1997; Bhat et al. 2002).

35.6.1 Single-Tube RT-PCR

35.6.1.1 Materials

1. As in PCR for the amplification of DNA viruses.
2. Dithiothreitol (DTT): Dissolve 1.54 g of dithiothreitol (DTT) in water and make up the volume to 10 mL with sterile water, filter sterilize through a 0.2 μm syringe filter, aliquot into 2 mL tubes and store at −20 °C.
3. Reverse transcriptase: Available commercially, it is thermosensitive, always store at −20 °C.
4. RNA template: Either RNA isolated from purified virus (*see* Chapter 25) or total RNA isolated from infected plants (*see* Chapter 32) may be used.
5. RNase free water.
6. RNase inhibitor (RNasin).

35.6.1.2 Method

1. Thaw all components (except reverse transcriptase) of RT-PCR on ice and briefly centrifuge. Add the following reagents to a 0.2 mL PCR tube on ice (Table 35.3).
2. Mix the components, place the tube on the thermal cycler and carry out RT-PCR as per pre-decided programme. Guidelines for the temperature and time for the run are provided in Table 35.4.

Table 35.4
Temperature profile for performing reverse transcription (RT)-PCR

Step	Temp (°C)	Duration
Reverse transcription	42	45 min
Denaturation	90–94	30–60 s
Annealing[a]	37–55	30–60 s
Extension[b]	72	1–3 min
Final extension	72	10 min

Repeat steps of denaturation, annealing and extension for 30–40 times
[a]Depend on the Tm of primers
[b]Provide 1 min for every 1000 bp to be amplified

3. Analyse about 10–20 μL of the RT-PCR product on 1% agarose gel electrophoresis and visualize the DNA with the help of gel documentation unit (for details of agarose gel electrophoresis, refer Chapter 26).

35.6.2 Two-Step RT-PCR (cDNA Synthesis Followed by PCR)

In addition to the materials listed in one-step RT-PCR, an oligonucleotide primer for synthesis of cDNA is also required.

35.6.2.1 Materials

1. As in 35.5.1, 35.6.1.1.
2. Primer for CDNA synthesis.

35.6.2.2 Method

1. Take 10 pg of viral RNA (isolated from purified virus) or up to 1 μg of total RNA (isolated from infected plant) in a microfuge tube and make up the volume to 10 μL by using RNase free water. Heat the tube at 80 °C for 5 min to denature the RNA and place it on ice.
2. Add the following reagents on ice under a PCR work station.

Amplification buffer (10×)	2 μL
dNTP mix (20 mM)	1 μL
Primer (10 pmoles) (for cDNA synthesis)	1 μL
RNase inhibitor (20 units/μL)	1 μL
$MgCl_2$ (50 mM/μL)	1 μL
Reverse transcriptase (100 units/μL)	1 μL
Sterile RNase-free water (to make)	30 μL

3. Incubate the reaction for 60 min at 42 °C for cDNA synthesis.
4. Place the tube in a water bath maintained at 95 °C for 5 min to inactivate the reverse transcriptase and denature the template cDNA.

5. Place the tube on ice and add the following reagent under PCR work station:

Sense (forward) primer (5 pmole/μL)	1 μL
Antisense (reverse) primer (5 pmole/μL)	1 μL
Amplification buffer (10×)	8 μL
Taq DNA polymerase (3 units/μL)	1 μL
Water (to make)	100 μL

6. Mix and briefly centrifuge the tube. Place the tube in the thermal cycler and carry out PCR using required temperature and time profile (Table 35.5).
7. After the PCR reaction, analyse about 10–20 μL of the reaction product by electrophoresis through an agarose gel and document the image as explained earlier (for details of agarose gel electrophoresis refer Chapter 26) (Fig. 35.3).

35.7 Immunocapture (IC) PCR and IC-RT-PCR

In this method, prior to PCR, virus particles are captured, immobilized and concentrated on a sterile solid surface previously coated with virus-specific antiserum. It is a technique that combines PCR assay and serology. The procedure has been successfully used for the amplification of several plant viruses. The method can also be used for the amplification of RNA virus, and then it is called as IC-RT-PCR. Thin-walled tubes or microtitre plates are coated with 100 μL of appropriately diluted virus-specific antiserum and incubated overnight at 4 °C. The tubes are then washed and added with extracted plant sap from infected plant and incubated for 3–4 h at room temperature. Samples are removed and tubes are washed before adding the required quantity of water for the PCR. To release viral RNA, 2–3 cycles of freezing (−80 °C for 10 min) and thawing (70 °C for 5 min) are given. Alternatively, viral RNA can be released by the addition of Tris-HCl and 1% Triton X-100 and by incubation at 65 °C for 10 min with vortexing. This is followed by either PCR or RT-PCR after the addition of required components. The method results in greater sensitivity, and problems encountered with RNA extraction are minimized and inhibitors of RT-PCR washed away prior to amplification. Thus IC-PCR is a very useful alternative for RT-PCR in virus detection from plant material and insect vectors (James et al. 1997; Latvala et al. 1997; Mumford and Seal 1997; Candresse et al. 1998; Jain et al. 1998).

Table 35.5
Temperature profile for different steps of PCR

Step	Stage	Temperature (°C)	Time (min)
1	Denaturation	94	0.50
2	Annealing	50–65[a]	1
3	Extension	72	1–4[b]
4	Final extension	72	5–10

[a]Will depend on the melting temperature (Tm) of sense and antisense primers
[b]Extension should be carried out for 1 min for every 1000 bp length of the target DNA

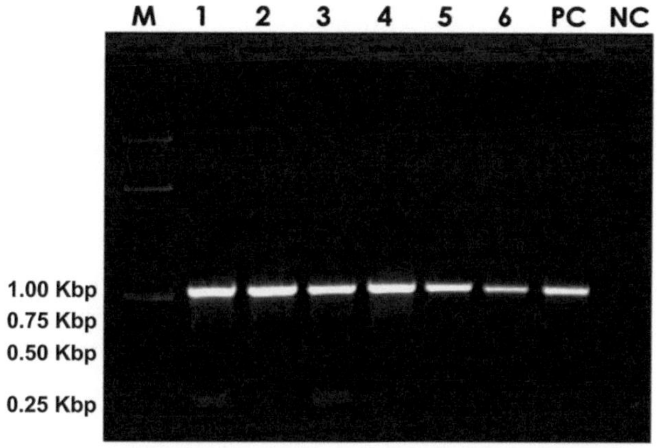

Fig. 35.3 Detection of an RNA virus-infected plant through RT-PCR assay. Lane M: DNA molecular size markers; Lanes 1–6: RT-PCR products from test plants; Lane PC: positive (infected) control and Lanes NC: negative (healthy) control

35.7.1 Materials

1. As listed in RT-PCR.
2. Water bath or temperature block.
3. Pestle and mortar.
4. Freezer −80 °C.
5. Phosphate-buffered saline (PBS) (10×): Dissolve 8.00 g NaCl, 0.20 g KH_2PO_4, 2.90 g Na_2HPO_4 (or 1.15 g anhydrous), 0.20 g KCl, 0.20 g NaN_3 in distilled water and make up the volume to 1000 mL. For working solution of 1× PBS, mix 100 mL of 10× PBS with 900 mL of distilled water.
6. PBS-T (phosphate-buffered saline with 0.05% Tween 20): Add 50 μL Tween 20 in 1 L of 1× PBS.
7. ELISA coating buffer (carbonate buffer): Dissolve 1.59 g of Na_2CO_3, 2.93 g of $NaHCO_3$ and 0.20 g of NaN_3 in 1 L of distilled water.
8. ELISA antigen extraction buffer (1× PBS containing 2% PVP and 0.2% albumin): Dissolve 2 g of polyvinyl pyrrolidone (PVP, MW 40,000) and 200 mg of albumin in 100 mL of 1× PBS.

9. Infected tissue.
10. Virus-specific antiserum.

35.7.2 Method

1. Add 100 μL of appropriately diluted (in ELISA coating buffer) antiserum to the 0.2 mL PCR tube and incubate at 4 °C overnight (or 37 °C for 4 h).
2. Decant the antiserum and wash the tube 3–5 times with sterile distilled water or PBST. Remove the traces of water by tapping the tube over a blotting paper.
3. Extract the antigen by grinding 100 mg of infected plant tissue in 1 mL of ELISA extraction buffer using a pestle and mortar.
4. Collect the extract in an eppendorf tube and centrifuge for 30 s.
5. Take 100 μL of the supernatant and dispense into the pre-treated PCR tube.
6. Incubate the tube at room temperature for 3 h.
7. Keep a water bath or temperature block ready at 70 °C.
8. Decant the tissue extract from the PCR tube and wash the tube for 3–4 times with sterile water. Remove as much of the water as possible. Add 74.5 μL of sterile water to the tube.
9. Freeze the tube at −80 °C for 10 min followed by snap thawing at 70 °C for 5 min. Repeat the process 2–3 times.
10. Keep the PCR tube on ice and add the following RT-PCR mix.

PCR buffer (10×)	10 μL
DTT (0.1 M)	10 μL
$MgCl_2$ (25 mM)	5 μL
dNTP s (10 mM)	2 μL
Forward primer (20 μM)	1 μL
Reverse primer (20 μM)	1 μL
RNasin (10 units/μL)	0.5 μL
AMV-RT (20 units/μL)	1.0 μL
Taq DNA polymerase (3 units/μL)	1.0 μL

11. Run the RT-PCR using temperature profile for cDNA synthesis and amplification as follows:

Step	Temperature (°C)	Duration
Reverse transcription	42	45 min
Denaturation	90–94	30–60 s
Annealing[a]	37–55	30–60 s

(continued)

Step	Temperature (°C)	Duration
Extension[b]	72	1–3 min
Final extension	72	10 min

Repeat steps denaturation, annealing and extension for 30–40 times
[a]Depend on the Tm of primers
[b]Provide 1 min for every 1000 bp to be amplified

12. Analyse about 10–20 μL of the reaction product on 1% agarose gel and visualize the DNA with the help of gel document unit as explained earlier in the Chapter 26 (Fig. 35.4).

35.8 RT-PCR Using dsRNA Template

Certain viruses are difficult to purify from infected plants. All ssRNA virus-infected plants are known to produce virus-specific dsRNA molecules as replicative intermediates during their replication process. Since dsRNA is quite stable compared to ssRNA, it will be relatively easy to handle them. Protocols are available for the isolation of dsRNA from a variety of infected plants (Morris and Dodds 1979; Dodds et al. 1984). RT-PCR can be performed using dsRNA as templates. In this method, dsRNA is first denatured to separate the strands. This can be achieved through heating or by using chemicals such as methoxy mercury hydroxyl. The denatured RNA is then used as template for the RT-PCR. The procedure involves isolation of dsRNA from plants to use as template. The dsRNA template is first denatured to separate the strands. Then, the components of the RT-PCR are added to the tube and placed in the thermocycler for cDNA (first-strand cDNA) synthesis followed by amplification using specific primers. After the RT-PCR assay, contents are run on an agarose gel to visualize the amplicons for an expected size.

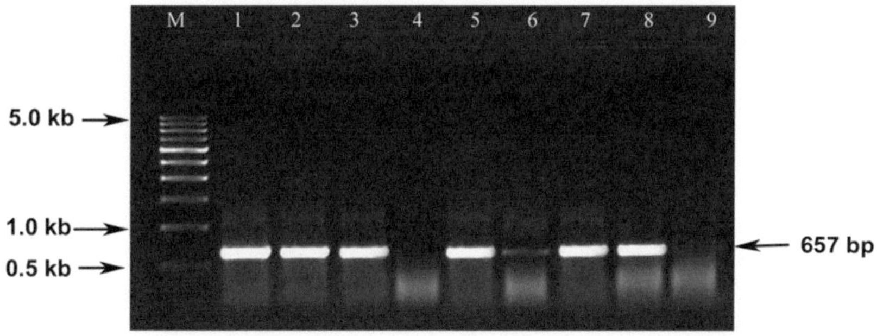

Fig. 35.4 Analysis of immunocapture reverse transcription PCR (IC-RT-PCR) products through agarose gel electrophoresis. Positive reaction is identified by the presence of 657 bp band. Lane M: Molecular size markers; Lanes 1–9 are loaded with IC-RT-PCR products from test plants

35.8.1 Materials

1. As in RT-PCR.
2. dsRNA from infected plant: For isolation of dsRNA, *see* Chapter 33.
3. LMP (low melting point) agarose.
4. Methyl mercury hydroxyl (CH_3Hg_2) (100 mM): Dissolve 232 mg of methyl mercury hydroxyl in sterile water and make up the volume to 10 mL. Note that methyl mercury is a very poisonous form of mercury. Hence extra care is required in its handling and disposal as it is one of the top ten chemicals of major public health concern.

35.8.2 Method

1. Isolate and purify the dsRNA from of virus-infected plant through LMP agarose as explained earlier in Chapter 33.
2. Dissolve purified dsRNA in 10 μL of RNase-free water.
3. Denature the ds RNA by adding the following in a sterile microcentrifuge tube.

dsRNA	3 μL
Forward primer (10 pM)	3 μL
Reverse primer (10 pM)	3 μL
100 mM CH_3 Hg_2	1 μL
Total	10 μL

4. Mix gently, centrifuge briefly and incubate for 8 min at room temperature. Quickly freeze the tube in liquid nitrogen and remove to −20 °C until ready to use for first-strand synthesis.
5. Transfer the tube from −20 °C onto ice and add the following for first-strand cDNA synthesis.

RT buffer (5×)	5 μL
DTT (0.1 M)	2.5 μL
dNTPs (10 mM)	2 μL
RNasin (10 units/μL)	0.5 μL
RNase-free water	4.0 μL
AMV-RT (20 units/μL)	1.0 μL
Total	25 μL

Keep at 42 °C for 1 h and transfer to ice.

6. Set up the PCR reaction in a thin-walled PCR tube by adding the following components

PCR buffer (10×)	10 μL
25 mM $MgCl_2$ (25 mM)	5 μL
dNTP mix (10 mM/μL)	2 μL
Forward primer (100 ng/μL)	1 μL
Reverse primer (100 ng/μL)	1 μL
Taq DNA polymerase (5 units/μL)	1 μL
Template (from step 5)	10 μL
Sterile water	70 μL
Total	100 μL

7. Amplify for 40 cycles using appropriate PCR profile as given below

Step	Stage	Temperature (°C)	Time (min)
1	Initial denaturation	90–94	3–5
2	Denaturation	90–94	0.50
3	Primer annealing	45–60 (depends on Tm of primers)	1
4	Extension	72	1–4 (depends on the length of fragment to be amplified)
5	Final extension	72	5–10

Repeat steps 2–4, 35 times

8. Analyse 10 μL aliquot of PCR product by electrophoresing on 0.8% agarose gel.
9. Examine the gel under UV transilluminator and record the image using gel documentation unit. Presence of band at expected size indicates positive reaction.

35.9 Multiplex-PCR

Multiplex-PCR is a method used for the detection of more than one virus in a single PCR. The method requires design of primers unique to each of the viruses that do not show complementarity within and between primers especially at their 3′ ends and produce amplicons of different lengths for easy differentiation of viruses. It is useful for indexing plants that are infected by more than one virus, thus saving time and reagents (López et al. 2009). It can be used for simultaneous and sensitive detection of different DNA and RNA viruses in a single reaction (Bhat and Siju 2007). In the following protocol for combined isolation of DNA and RNA and detection of an RNA and a DNA virus-infected black pepper is given (Bhat and Siju 2007).

35.9.1 Materials

1. As in total RNA isolation by AGPC method (*see* Chapter 32).
2. As in single step RT-PCR method (Subheading 35.6.1.1).
3. Tris-HCl (0.1 M, pH 10.5): Dissolve 1.21 g of Tris base in about 80 mL of sterile water, adjust the pH to 10.5 using HCl and make up the volume to 100 mL.
4. Primers (one pair specific for each of the test viruses).

35.9.2 Method

35.9.2.1 Isolation of Total RNA and DNA from Plants

1. Grind 50 mg of tender leaf tissue from infected plants in 500 μL of denaturing solution [4 M guanidium thiocyanate, 25 mM sodium citrate (pH 7.0), 0.5% (w/v) sodium *N*-lauryl sarcosine, 0.1% (v/v) β mercaptoethanol, 0.5% (v/v) sodium sulphite)] and collect in a microcentrifuge tube.
2. To the homogenate sequentially add 50 μL of 2 M sodium acetate (pH 4.0), 500 μL of water saturated phenol and 100 μL of chloroform: isoamyl alcohol (49:1) with mixing at each addition.
3. Incubate on ice for 15 min, centrifuge at 4 °C for 15 min at 12,000 × *g* for the separation three phases—aqueous phase, organic phase and interphase.
4. Collect the aqueous phase in a fresh tube, add equal volume of isopropanol and incubate at −80 °C (this contains RNA). While RNA is being precipitated, isolate the DNA from the interface and organic phase as described below.
5. To the organic phase and interphase from **step 4**, add 0.5 volumes of 0.1 M Tris-HCl (pH 10.5), vortex for 15 s and incubate on ice for 30 min followed by centrifugation for 15 min at 2000 ×*g* at 4 °C.
6. Re-extract the aqueous phase twice with equal volume of chloroform:isoamyl alcohol (24:1) and centrifugation for 10 min at 12,000 × *g* at 4 °C.
7. To the final aqueous phase, add 0.1 volumes of 3 M sodium acetate (pH 4) and 1 volume of isopropanol and mix well by gentle pipetting.
8. The above entire sample (from **step** 7) is then added to the tube containing RNA kept at −80 °C (**step 4**) and incubate for a further 1 h for the precipitation of total nucleic acids.
9. Centrifuge the contents at 4 °C for 10 min at 12,000 × *g* followed by washing the pellet with 75% ethanol and air dry the pellet.

10. Dissolve the final pellet that contains both DNA and RNA in 80 μL of sterile HPLC-grade water.

35.9.2.2 Single-Tube Multiplex RT-PCR (mRT-PCR)

1. Design and synthesize one pair of compatible oligonucleotide primers for each of the viruses to be tested in the single-tube mRT-PCR based on published sequence data of viruses to be tested. While designing primers care should be taken such that all primers have almost same melting temperature, they do not form primer dimers and lack 3′ complementarity between primers. Also, care should be taken such that the expected PCR product of each virus should differ in their size by at least 200 bp for easy differentiation of viruses.
2. Set up 50 μL single-tube mRT-PCR that contain 1× PCR buffer, 0.1 M DTT, 10 U RNasin, 15 pM each of the primers, 1.5 U *Taq* DNA polymerase and 10 U MuMLV RT and 1 μL of template volume. Make up the final volume to 50 μL by adding sterile HPLC-grade water.
3. Also set up separate PCR (for DNA virus) and RT-PCR (for RNA virus) containing respective individual sets of specific primers to serve as controls.
4. Keep the tubes including the controls in the PCR machine using the following temperature profile: cDNA synthesis at 42 °C for 45 min and amplification for 40 cycles where each cycle consist of denaturation at 94 °C (30 s), primer annealing at 50 °C (this temperature may vary depending on the Tm of primers) (1 min) and extension at 72 °C (1 min) (this duration can vary depending upon the maximum expected product size) followed by a final extension at 72 °C (10 min).
5. Run amplified products (25 μL of 50 μL reaction mix) along with appropriate DNA ladder on 1% agarose gel containing ethidium bromide.
6. Visualize the bands and photograph using an UV transilluminator and a gel documentation apparatus. Look for the expected single band in the control reactions while the mRT-PCR may show either single band (indicating infection of only virus) or two bands (indicating infection by two viruses) (Fig. 35.5).

35.10 Nested PCR Assay

Nested PCR assay is used to increase both specificity and sensitivity of the PCR. It is ideally suitable for plants in which viruses usually present in low titres or inhibitors are present that may interfere in the PCR efficacy. The method involves carrying out two PCRs, the first reaction to increase the amount of template for the second. Degenerate primers are generally used in the first round of PCR. Then, an aliquot of the first reaction is used for the second PCR

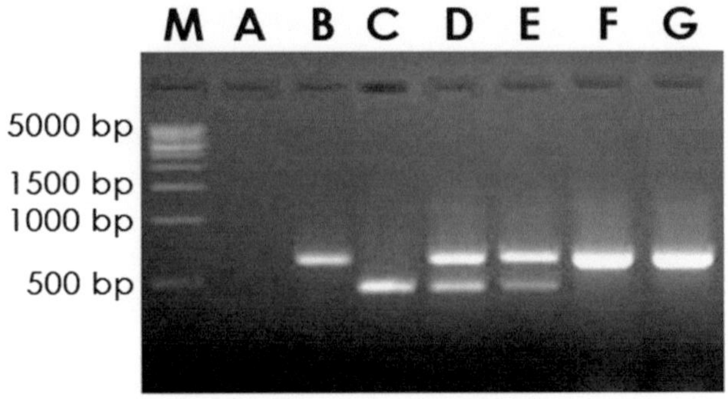

Fig. 35.5 Detection of an RNA virus-infected (cucumber mosaic virus (CMV)) and a DNA virus-infected (piper yellow mottle virus (PYMoV)) black pepper using duplex RT-PCR. Lane M: DNA ladder, Lane A: healthy plant; Lane B: a known plant infected with CMV alone, Lane C: a known plant infected with PYMoV alone and Lanes D–G: test plants

with primers that anneal within the first amplicon. This increases the template molecule and dilutes inhibitors. It is normally used for the detection of viruses that occur in low concentration in plants (Henson and French 1993; Hadidi et al. 1995; Simmonds et al. 1990; Olmos et al. 1997).

35.11 Notes (Hadidi et al. 1995; Candresse et al. 1998; López et al. 2009; Mumford and Seal 1997; Sambrook and Russell 2001)

1. PCR is a versatile technique that is more sensitive than serological assays like ELISA. It can be used for the specific detection and differentiation of viruses and their strains. It can be used to index plants or seeds to avoid introduction of viruses into field.
2. Always maintain high cleanliness during nucleic acid isolation and PCR set up to avoid any false-positive result. It is a good idea to keep no template control while setting up PCR to check false positives if any.
3. PCR workstation containing a dedicated laminar air flow and micropipette and tips would reduce the contamination to a great extent.
4. Proper thawing of all components of PCR is important before addition.
5. Always keep a negative control (without template) for checking false amplification due to contamination in PCR.
6. When developing PCR assay for the detection of a virus, use nucleic acid isolated from a known virus-infected plant (positive control) and virus-free plant (negative control). Also, keep another control without any template (use water instead of

nucleic acid). Agarose gel electrophoresis of the PCR products obtained using above templates should give band at expected size only in the positive control, no band should be seen in the virus-free plant and no template control. Fine tuning of the reaction may be done by varying the concentrations of PCR components such as template, $MgCl_2$, dNTPs and temperature profile especially the annealing temperature.

7. Once a PCR assay is developed for the detection of a virus, it should be validated using large number of field samples collected from different agro-ecological regions in order to ensure the ability of the developed assay to detect all isolates of the virus.
8. Whenever a PCR assay is being developed for the detection of a virus, design primers to the conserved region present in all isolates/strains of the virus species. It is always a good idea to design at least 3–4 pairs of primers and select the best one based on the performance in PCR test.
9. Immunocapture PCR assay can be used for all kinds of viruses with both DNA and RNA genomes. Since PCR tubes of different make differ in their binding capacity, tubes can be pre-treated to increase the efficiency of viral antigen binding as follows: (a) wash tube with 0.1 N HCl for 15 min, (b) rinse with ELISA PBS-T buffer, (c) treat with 4 N NaOH for 15 min, (d) rinse with ELISA PBS-T buffer, (e) treat with 95% ethanol for 15 min, (f) air dry tubes.
10. If virus-specific antiserum is not available, direct binding (DB) PCR can be performed. In this, pre-treated PCR tubes are added with extract from infected plant and incubated in ice for 30–60 min. During incubation, virus particles will get adsorbed to the walls of the PCR tubes. Rest of the procedure remains same as in IC-PCR.
11. In the case of IC-PCR, to trap more intact virus particles, certain virus and host may require the addition of additives such as reducing agents to the antigen extraction buffer.
12. Distribution of virus concentration in a plant may vary. Hence selection of right kind of tissue having more virus concentration is important for all kinds of PCR assays.
13. Template DNA should be of good quality. Check quality of the DNA by running on the agarose gel. It should be free from other contaminants without any shearing.
14. To avoid contaminations in PCR: (a) store all PCR reagents separately at the designated place in the freezer; (b) thaw and briefly centrifuge all reagents before opening the tubes; (c) carry out assembly of PCR in a laminar hood equipped with UV lights. Keep in the hood, disposable glove, microfuge,

eppendorf tubes, thin-walled PCR tubes, micro tips and a set of the micropipette to handle only reagents for PCR.

15. The following points to be remembered while designing primers: (a) The primers can be of 18–25 nucleotides long. (b) G + C contents in a primer should lie between 40 and 60% with an even distribution of all four bases along the length of the primer. (c) The 3′ terminal sequences of one primer should not be able to bind to any site on the other primer to avoid formation primer dimers. (d) For effective base pairing, 3′ base of each primer should be G or C. (e) Tm values of primer pairs used in a PCR should not differ by >5 °C. (f) Sequences of restriction sites not complementary to the target DNA can be added to the 5′ end of primer if planning to clone the amplified product at a particular restriction site in the vector.

16. Most thermal cyclers have an option where after the PCR programme, the tube can be kept on hold at 4 °C. This facility may be used especially when you are not able to collect the tube (vial) from the thermocycler immediately after cycling programme ends due to any reasons.

References

Atzmom G, van Oss H, Czosnek H (1998) PCR amplification of tomato yellow leaf curl virus (TYLCV) DNA from squashes of plants and whitefly vectors: application to the study of TYLCV acquisition and transmission. Eur J Plant Pathol 104:189–194

Bhat AI, Siju S (2007) Development of a single tube multiplex RT-PCR for the simultaneous detection of *Cucumber mosaic virus* and *Piper yellow mottle virus* associated with stunt disease of black pepper. Curr Sci 93:973–976

Bhat AI, Jain RK, Ramiah M (2002) Detection of *Tobacco streak virus* from sunflower and other crops by reverse transcription polymerase chain reaction. Indian Phytopathol 55:216–218

Brasileiro BTRV, Coimbra MRM, De Morais MA, De Oliveira NT (2004) Genetic variability within Fusarium solani specie as revealed by PCR-fingerprinting based on PCR markers. Braz J Microbiol 35:205–210

Candresse T, Hammond RW, Hadidi A (1998) Detection and identification of plant viruses and viroids using polymerase chain reaction (PCR). In: Hadidi A, Khetarpal RK, Koganezawa K (eds) Control of plant virus diseases. APS Press, St. Paul, pp 399–416

Carrasco-Ballesteros S, Castillo P, Adams BJ, Perez-Artes E (2007) Identification of *Pratylenchus thornei*, the cereal and legume root-lesion nematode, based on SCAR-PCR and satellite DNA. Eur J Plant Pathol 118:115–125

Dodds JA, Morris TJ, Jordan RL (1984) Plant viral double stranded RNA. Annu Rev Phytopathol 22:151–168

Hadidi A, Levy L, Podleckles EV (1995) Polymerase chain reaction technology in plant pathology. In: Singh RP, Singh US (eds) Molecular methods in plant pathology. CRC Press, Boca Raton

Henson JM, French R (1993) The polymerase chain reaction and plant disease diagnosis. Annu Rev Phytopathol 31:81–109

Jain RK, Pappu SS, Pappu HR, Culbreath AK, Todd JW (1998) Molecular diagnosis of tomato spotted wilt tospovirus infection of peanut and other field and greenhouse crops. Plant Dis 82:900–904

Jain RK, Pappu HR, Pappu SS, Krishnareddy M, Vani A (2008) Watermelon bud necrosis tospovirus is a distinct virus species belonging to serogroup IV. Arch Virol 143:1637–1644

James D, Trytten PA, Mackenzie DJ, Towers GHN, French CJ (1997) Elimination of apple stem grooving virus by chemotherapy and development of an immunocapture RT-PCR for rapid sensitive screening. Ann Appl Biol 131:459–470

Latvala S, Susi P, Lemmetty A, Cox S, Jones AT, Lehto K (1997) Ribes host range and erratic distribution with in plants of blackcurrant reversion associated virus provide further evidence for its role as the causal agent of reversion disease. Ann Appl Biol 131:283–295

López MM, Llop P, Olmos A, Marco-Noales E, Cambra M, Bertolini E (2009) Are molecular tools solving the challenges posed by detection of plant pathogenic bacteria and viruses? Mol Biol 11:13–46

Merighi M, Sandrini A, Landini S, Ghini S, Girotti S, Malaguti S (2000) Chemiluminescent and colorimetric detection of *Erwinia amylovora* by immunoenzymatic determination of PCR amplicons from plasmid pEA29. Plant Dis 84:49–54

Morris TJ, Dodds JA (1979) Isolation and analysis of double stranded RNA from virus infected plant and fungal tissue. Phytopathology 69:854–858

Mullis KB, Faloona F, Scharf S, Saiki R, Horn G, Erlich H (1986) Specific enzymatic amplification of DNA *in vitro*: the polymerase chain reaction. Cold Spring Harbor Symp Quart Biol 51:263

Mumford RA, Seal SE (1997) Rapid single-tube immunocapture RT-PCR for the detection of two yam potyviruses. J Virol Methods 69:73–79

Mumford RA, Barker I, Wood KR (1996) An improved method for the detection of *Tospovirus* using the polymerase chain reaction. J Virol Methods 57:109–115

Olmos A, Cambra M, Dasi MA, Candresse T, Esteban O, Gorris MT, Asensio M (1997) Simultaneous detection and typing of plum pox potyvirus (PPV) isolates by hemi-nested-PCR and PCR-ELISA. J Virol Methods 68:127–137

Pappu SS, Brand R, Pappu HR, Rybicki E, Gough KH (1993) A polymerase chain reaction method adapted for selective cloning of 3′ non translated regions of potyviruses: application to dasheen mosaic virus. J Virol Methods 41:9–20

Saiki RK, Gelfand DH, Stoffel S, Scharf SJ, Higuchi R, Horn GT, Mullis KB, Erlich HA (1988) Primer-directed enzymatic amplification of DNA with a thermostable DNA polymerase. Science 239:487–491

Sambrook J, Russell DW (2001) Molecular cloning, vol I–III, 3rd edn. Cold Spring Harbor Laboratory Press, New York

Simmonds P, Zhang LQ, Watson HG, Rebus S, Ferguson ED, Balfe P et al (1990) Hepatitis C quantification and sequencing in blood products, haemophiliacs, and drug users. Lancet 336:1469–1472

Singh RP, Kurz J, Boiteau G, Moore LM (1997) Potato leafroll virus detection by RT-PCR in field collected aphids. Am Potato J 74:305–313

Vincelli P, Tisserat N (2008) Nucleic acid-based pathogen detection *in* applied *plant pathology*. Plant Dis 92:660–669

Ward E, Foster SJ, Fraaije BA, Mccartney HA (2005) Plant pathogen diagnostics: immunological and nucleic acid-based approaches. Ann Appl Biol 145:1–16

Chapter 36

Real-Time Polymerase Chain Reaction

Abstract

The real-time PCR, also known as quantitative PCR assay, detects the presence of the target pathogen (virus) in real time during PCR amplification without the need for post-PCR gel electrophoresis. The real-time detection of the accumulating amplicons is made possible by labelling of primers, probes or amplicons with fluorogenic molecules. Quantification of the amplicon in real-time PCR is achieved by the detection of the fluorescence signal produced proportionally during the amplification. Dyes such as SYBR-Green I or ethidium bromide that bind non-specifically to any double-stranded DNA or fluorescent probes that specifically bind to the targeted region are used for real-time detection of the amplicon during amplification. The real-time PCR instrument consists of optical sensor to record the transmission emitted by the fluorogenic agent during amplification. The results of real-time PCR are displayed in the form of a curve that has four different phases which are used to determine the cycle threshold (Ct) and amplification efficiency. In order to quantify or determine copy number of the target DNA molecule produced, a standard curve is initially plotted by using Ct values obtained for different known copy numbers of the targeted amplicon region. Once the graph is done, copy number of the same target amplicon in a test sample can be determined using the standard curve.

Key words Quantitative PCR assay, DNA intercalating dye, *Taq*Man PCR, Cycle threshold, Quantification, Florescent probes, FRET probes

36.1 Introduction

Real-time PCR assay can be used for the detection and quantification of the target viruses in the sample. Besides its sensitivity and specificity, it avoids post-PCR gel analysis to view results. However, the assay requires special equipment and reagents compared to conventional PCR technology. Detection and quantification of the target virus is possible based on the amount of fluorogenic molecule accumulated during the process of amplification (Dietzgen et al. 1999; Eun et al. 2000; Roberts et al. 2000; Mumford et al. 2000; Lopez et al. 2003). Following two methods are generally used for the quantitative detection of amplicons:

1. Dyes binding non-specifically to the double-stranded DNA.
2. Fluorescent probes which have specificity for binding to targeted DNA sequence.

Alangar Ishwara Bhat and Govind Pratap Rao, *Characterization of Plant Viruses*, Springer Protocols Handbooks, https://doi.org/10.1007/978-1-0716-0334-5_36, © Springer Science+Business Media, LLC, part of Springer Nature 2020

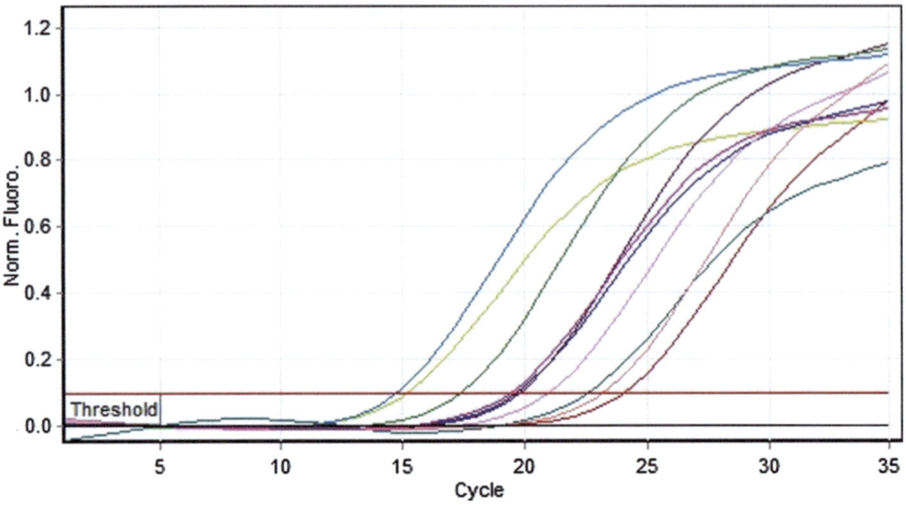

Fig. 36.1 Real-time PCR showing amplification curve

36.1.1 Dyes Binding to the Double-Stranded DNA

In this, fluorogenic molecules bind non-specifically to the double-stranded DNA. Dyes such as ethidium bromide (EtBr), Yo-PRO-1 and SYBR-Green I bound to the dsDNA molecule fluoresce under UV light of specific wavelength (Mackay et al. 2002). The most popular and easiest among all these is SYBR-Green I dye.

The free SYBR-Green exhibits little fluorescence at the time of the denaturation. SYBR-Green dye intercalates into the minor grooves of dsDNA and produces a fluorescent signal, the intensity of which is proportional quantity of dsDNA in the PCR reaction. During polymerization step, more and more of SYBR-Green dye molecules bind to the nascent strand and the increase in fluorescence can be followed in real time (Fig. 36.1). Thus, the fluorescence intensity increases proportionally to dsDNA concentration (Wittwer et al. 1998). The only disadvantage with SYBR-Green dye is that it binds non-specifically with double-stranded DNA, i.e. primer-dimer or non-specific amplification products. However, it can be confirmed by melting curve assay.

36.1.2 Melt Curve and Detection of Non-specific Amplification

In addition to monitoring the DNA synthesis during PCR, the real-time PCR instrument is also used to find out the melting temperature (Tm) or melting point of the PCR product at the end of the PCR run. The Tm of the amplicon varies depending upon its base composition. Each amplicons obtained for a pair of primers will have a single melting temperature. Presence of more than one peak indicates that there are either contamination, mis-priming or primer-dimer artefacts, or some other problem (Fig. 36.2).

*Taq*Man probe is the most common fluorescent probe which is used in real-time PCR for more specificity and sensitivity. In 5′-nuclease PCR technology (*Taq*Man PCR), an oligonucleotide

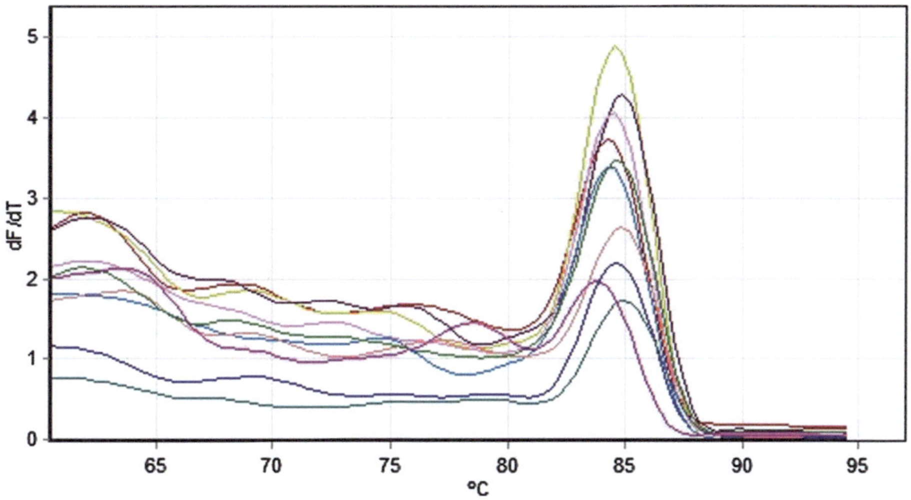

Fig. 36.2 Dissociation curve showing specificity of primers in the detection of the virus in a real-time PCR assay. The peak at about 84 °C represents specific amplification

36.1.3 Fluorescent Probes Which Have Specificity for Binding to Targeted DNA

probe labelled at both ends with a reporter and a quencher dye that has exact sequence complementary to a region internal to the PCR primers is used. When the oligonucleotide probe is intact and bound to the template nucleic acid, fluorescence emitted by the reporter dye is absorbed by the quencher dye present on the other end of the probe (Mackay et al. 2002). During amplification, the probe is cleaved by the 5′-nuclease activity of *Taq* polymerase that would separate the reporter and quencher dyes from the probe resulting in an increase in fluorescence, the amount of which is directly related to the amount of PCR product amplified. During amplification, the increase in reporter fluorescence is observed in real time using a combined thermal cycler-fluorescence reader system. There are various dyes used as fluorescing agent like FAM, VIC, JOE, etc. Quencher (TAMRA, BHQ) is the agent which absorbs or quenches the fluorescence when present in close proximity with dye. During the denaturing step, the probe is free on solution. During the annealing step, probe hybridizes to its target sequence. The proximity of the quencher allows the inhibition of fluorescence. The polymerase moves and hydrolyses the probe. The fluorescing agent is released from the environment and thus emits fluorescence. Simultaneous detection of several viruses in a sample is possible by using probes with different fluorescent reporter dyes (Boonham et al. 2002).

36.1.4 Quantification

Quantification of the amplicon produced during real-time PCR can be performed by using standard curve method, and a serial dilution of cloned DNA or purified DNA of pathogen is used for making standard curve. The standard curve method produces a linear plot for approximate copy number for the above standard samples versus cycle threshold (Ct) value. The Ct values obtained in the reaction of unknown samples are plotted on standard curve for quantification. The concentration of the nucleic acid is determined by taking OD values at 260 nm through a spectrophotometer. The approx. copy number of the nucleic acid is calculated by using the following formula:

$$\text{copies}/\mu\text{L} = \left(\text{concentration in ng} \times 6.023 \times 10^{23}\right)/\left(\text{genome length} \times 1 \times 10^{9} \times 650\right)$$

36.1.5 Real-Time PCR Instrument

Real-time PCR is just an improvement version of normal PCR machine which consists of optical sensor which catches the transmission emitted during PCR having fluorescing agent. In order to run the machine in real-time mode, it is connected to a computer and software. Software analyses the data by observing fluorescence released during PCR reaction (Patel et al. 2016).

36.1.6 Amplification Curve and its Phases

The real-time PCR amplification curve shows the quantity of fluorescent emission at each reaction cycle. When large quantity of target DNA is used in the sample, the real-time PCR reaction will be faster and enters the exponential phase of amplification quickly. The quantity of PCR amplicon produced at each cycle is measured using fluorescent dyes or probes and cycle threshold (Ct) is calculated for each sample. Ct is the cycle number at which a statistically significant increase in fluorescence is detected. Higher Ct value is observed when the quantity of target DNA in the sample is low and *vice versa*. In general, an amplification curve has four phases—the linear ground, early exponential, log-linear and plateau phases (Fig. 36.1). Data from these phases are used for determining background noise, Ct and amplification efficiency (Rn). Rn is the ratio of the intensity of fluorescent emission of the reporter dye to the intensity of fluorescent emission of the passive dye (a reference dye added into the real-time PCR reaction mix to control for differences in master mix volume). Rn represents the magnitude of signal generated during PCR and calculated as the difference in Rn values of a sample and no template control.

36.1.7 Primer and Probe Design

1. Many online software are available freely to design primers and probes. One such software is Beacon Designer software (www.premierbiosoft.com).
2. Ideally melting temperature of the primers should be between 58 and 60 °C while that of the probe should be 10 °C higher. The Tm of both forward and reverse primers should be equal. The length of primers should be of 15–30 bases in length.

3. Avoid continuous stretch of a single base in the primer. This is especially true for 'G', where four or more continuous 'G' is not preferable.
4. The ideal amplicon size is 50–150 bases (maximum amplicon size should not exceed 400 bases.

36.1.8 Application

Quantitative real-time PCR (qPCR) can be applied in many experiments, like comparison of gene expression pattern at mRNA level, mutational analysis, allelic determination (homozygous or heterozygous), and DNA copy number calculation in case of genome and virus or other pathogens. But most widely it is applied for the detection as it is a quick and sensitive method and works in closed system.

36.2 Performing Real-Time PCR/Real-Time RT-PCR Using SYBR-Green

36.2.1 Materials

1. Real-time PCR machine.
2. Primer pair: design primer pair to amplify target region of about 75–150 bp with a Tm in the range of 58–60 °C, G + C content in the range of 40–60%.
3. Template DNA/RNA: For preparation, please refer Chapters 31 and 32.
4. SYBR-Green master mix.
5. Reverse transcriptase (If RNA is used as template).

36.2.2 Method (When RNA is Used as Template)

1. Under the PCR workstation, set up the real-time RT-PCR reaction in a thin walled PCR vial by adding 12.5 μL of 2× SYBR-Green PCR master mix, 1.0 μL (1 μM/μL) each of both primers and 1 μL template RNA (about 100 ng), 50 U of RevertAid reverse transcriptase and make up the volume to 25 μL with RNase-free water.
2. Set up two more control reactions as indicated above using total RNA isolated from healthy plant and water as template.
3. Keep the tubes in a real-time PCR instrument and set up the following programme: 42 °C for 45 min for cDNA synthesis followed by 95 °C for 10 min for initial denaturation. This should be followed by 35 cycles of 95 °C for 15 s and 60 °C (this may vary depending on the Tm of primers) for 45 s.
4. Analyse output data with the software available in the real-time PCR instrument using preset parameters and identify positive samples based on the Ct values (Fig. 36.3a).
5. After the run, subject the amplicons to melt curve analysis as follows: 95 °C for 1 min, then at 55 °C for 45 s followed by

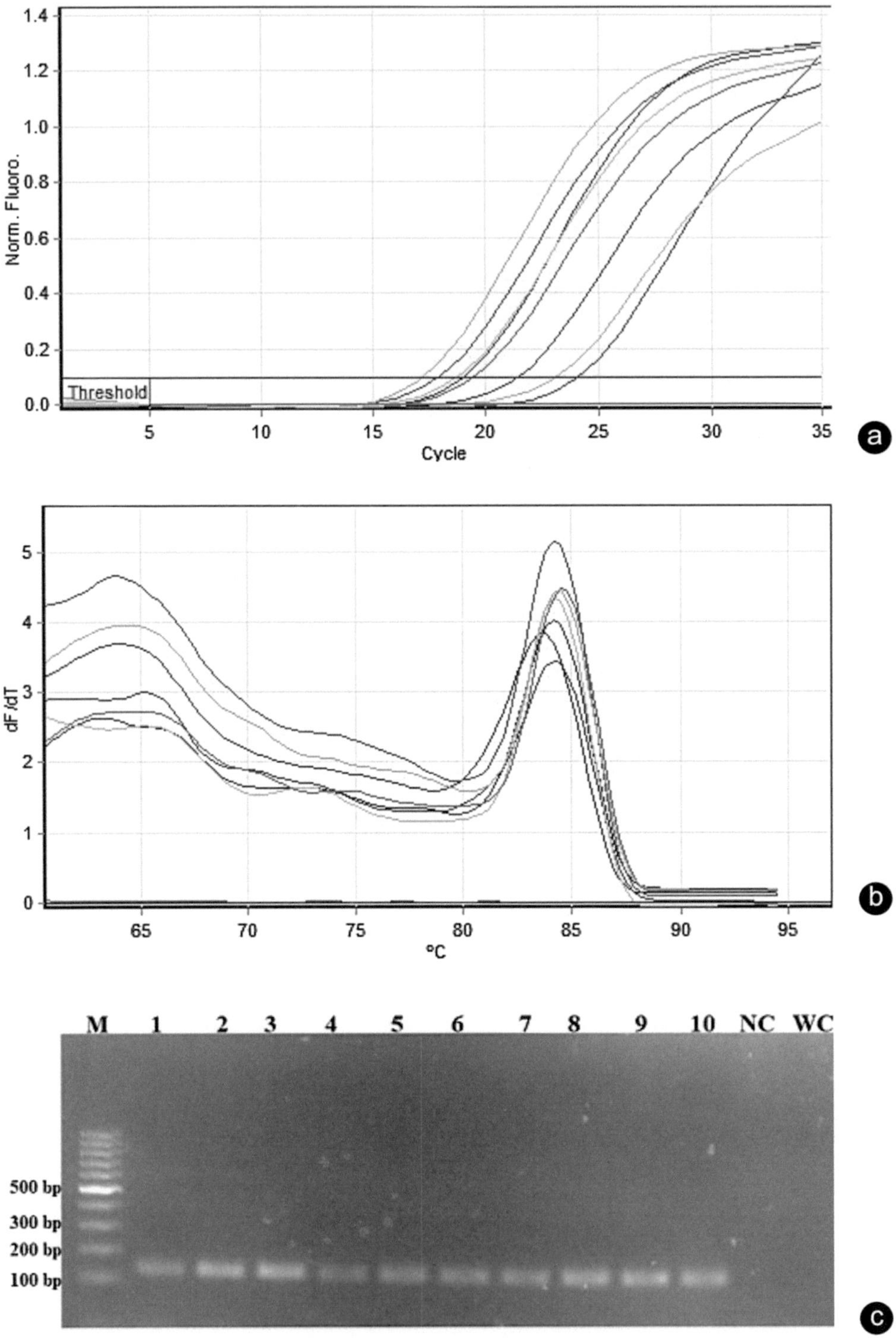

Fig. 36.3 Real-time RT-PCR for the detection of virus: (**a**) amplification curves, (**b**) melt curve analysis and (**c**) agarose gel electrophoresis of real-time RT-PCR products; lane M, 100 bp DNA ladder; lanes 1–10, infected plant samples; lane NC, negative (healthy control); lane WC, water (no template control)

heating back to 95 °C at a rate of 0.5 °C increments (Fig. 36.3b).

6. Apart from this, verify the RT-qPCR products by running the product on a 1.5% agarose gel electrophoresis (Fig. 36.3c).

36.2.3 Method (When DNA is Used as Template)

1. Under the PCR workstation, set up the real-time PCR in a vial by adding 12.5 μL of 2× SYBR-Green PCR master mix, 1.0 μL (1 μM/μL) each of both primers and 1 μL template DNA (about 100 ng) and make up the volume to 25 μL with sterile water.
2. Set up two more control reactions as indicated above using total DNA isolated from healthy plant and water as template.
3. Keep the tubes in a real-time PCR instrument and set up the following programme: Denaturation step at 95 °C for 10 min followed by 35 cycles of 95 °C for 15 s and 60 °C (this may vary depending on the Tm of primers) for 45 s.
4. Analyse output data with the software available in the real-time PCR instrument using preset parameters and identify positive samples based on the Ct values.
5. After the run, subject the amplicons to melt curve analysis from 60 to 95 °C to check the specificity of the real-time PCR product, In melting curve analysis, the product is heated to 95 °C for 1 min, then decreased to 55 °C for 45 s (pre melt conditioning) followed by heating back to 95 °C at a rate of 0.5 °C increments (Fig. 36.3b).
6. Apart from this, verify the qPCR products by running the product on a 1.5% agarose gel electrophoresis (Fig. 36.3c).

36.3 Performing Real-Time RT-PCR Using *Taq*Man Assay

36.3.1 Materials

1. As in Subheading 36.2.1.
2. Probe: Design probe of about 30 bp in length to align within the target region (between forward and reverse primers). It should have a Tm of around 65 °C, G + C content up to a maximum of 80%. The reporter dyes (such as 6-FAM, ROX, JOE, Cy5, Vic, Tet, Ned and Hex) and quencher dye (such as tamra dye, minor groove binding molecule, MGB; black hole quencher, BHQ) have to be linked to the 5′ and 3′ ends of the probe, respectively.

36.3.2 Method

1. Under the PCR workstation, set up the *Taq*Man real-time RT-PCR reaction using *Taq*Man one step RT-PCR master mix kit (Applied Biosystems) in a total volume of 25 μL containing 12.5 μL of 2× master mix, 1.0 μL (1 μM/μL) of each forward and reverse primer, 1 μL (20 μM) of probe, 0.625 μL

of 40× enzyme mixture, template RNA (about 100 ng) and 50 U of RevertAid reverse transcriptase and sterile water to make the final volume to 25 μL.

2. Set up two more control reactions as indicated above using total RNA from healthy plant and a water to serve as negative controls.
3. Keep the tubes in a real-time PCR instrument and set up the following programme: 45 min at 42 °C for cDNA synthesis and at 95 °C for 10 min for denaturation. This should be followed by 35 cycles of 95 °C for 15 s and 60 °C (this may vary depending on the Tm of primers) for 45 s.
4. Analyse output data with the software available in the real-time PCR instrument using preset parameters and identify positive samples based on the Ct values (Fig. 36.3a).
5. After the run, subject the amplicons to melt curve analysis by heating the product at 95 °C for 1 min, then at 55 °C for 45 s. Heat back to 95 °C at a rate of 0.5 °C increments.
6. Apart from this, verify the RT-qPCR products by running the product on an 8% acrylamide gel.

36.4 Developing Standard Curve for Quantification

36.4.1 Materials

1. As in Subheadings 36.2.1 and 36.3.1.
2. Target DNA/RNA region.
3. Spectrophotometer.

36.4.2 Method

1. Determine yield of template RNA/DNA using a spectrophotometer (*see* Chapter 25 for details).
2. Take a known quantity of RNA transcript or DNA and determine the copy number using the formula: copies/μL = (concentration in ng × 6.023 × 10^{23})/(genome length × 1 × 10^{9} × 650).
3. Dilute the template DNA/RNA serially tenfold (from 1 ng to 10^{-8}) using total RNA/DNA isolated from corresponding healthy plant as diluents. Subject each of the dilutions to real-time RT-PCR (if RNA is the template) or real-time PCR (if DNA is the template) as explained above in triplicates.
4. Analyse and record all output data with the software available in the instrument using preset parameters.
5. Construct standard curve for the virus by plotting Ct values obtained for each of the dilutions on the '*Y*' axis and logarithm of the copy number on '*X*' axis and fitting a straight line to these data by simple linear regression. Determine the efficiency (*E*) of the PCR using slope of the standard curve. $E = [10^{-1/}$

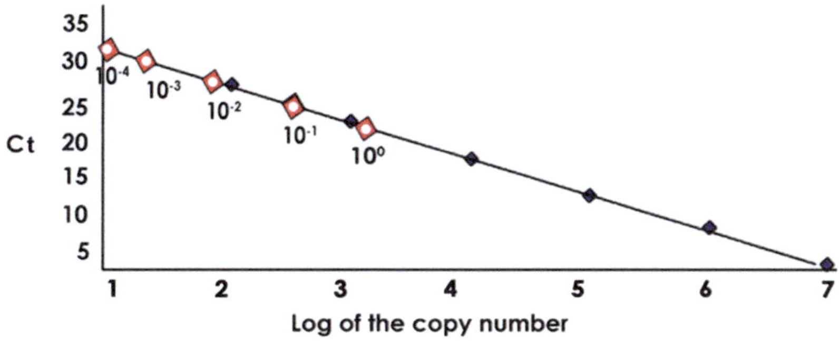

Fig. 36.4 Development of standard curve for detection and quantification of real-time RT-qPCR. Standard curve was prepared by plotting cycle threshold (Ct) in '*Y*' axis and the log of copy number of target-specific amplicon in the '*X*' axis. (•) represent Ct value obtained to corresponding copy number of the template (◇) shows Ct value obtained for different dilutions (10^0–10^{-5}) of total RNA from infected plant. (Reproduced from Siljo et al. 2014 with permission from Springer)

$^{slope}]$ − 1 considering 100% efficiency for a value of two. Optimal PCR efficiency is achieved when a slope of −3.32 is reached (Fig. 36.4).

6. Calculate the coefficient (R^2) to determine the validity of the linear regression. For detection purposes, a quantification cycle of 35 is established as the cutoff for distinguishing positive from negative samples.

36.5 Notes (Mackay et al. 2002; Lopez et al. 2003; Patel et al. 2016)

1. Real-time PCR collects data during exponential growth phase while conventional PCR is measured at end point (plateau).
2. In general, real-time PCR is 10^3–10^5 times more sensitive compared to conventional PCR.
3. Real-time PCR requires expensive equipment and reagent and needs more of optimization.

References

Boonham N, Smith P, Walsh K (2002) The detection of tomato spotted wilt virus (TSWV) in individual thrips using real time fluorescent RT-PCR (TaqMan). J Virol Methods 101:37–48

Dietzgen RG, Thomas JE, Smith GR, Maclean DJ (1999) PCR-based detection of viruses in banana and sugarcane. Curr Top Virol 1:105–118

Eun AJ-C, Seoh ML, Wong SM (2000) Simultaneous quantitation of two orchid viruses by the TaqMan real time RT-PCR. J Virol Methods 87:151–160

Lopez MM, Bertolini E, Olmos A, Caruso P, Gorris MT, Llop P, Penyalver R, Cambra M, Penyalver R, Cambra M (2003) Innovative tools for detection of plant pathogenic viruses and bacteria. Int Microbiol 6:233–243

Mackay IM, Arden KE, Nitsche A (2002) Real-time PCR in virology. Nucleic Acids Res 30:1292–1305

Mumford RA, Walsh K, Barker I, Boonham N (2000) Detection of potato mop top virus and tobacco rattle virus using a multiplex real-time fluorescent reverse-transcription polymerase chain reaction assay. Phytopathology 90:448–453

Patel RN, Patel DR, Bhandari R, Silawat N, Homkar U, Sharma J (2016) Upcoming plant pathological techniques in the disease diagnosis. J Microbiol Exp 3:87

Roberts CA, Dietzgen RG, Heelan LA (2000) Real-time PCR fluorescent detection of tomato spotted wilt virus. J Virol Methods 88:1–8

Siljo A, Bhat AI, Biju CN (2014) Detection of *Cardamom mosaic virus* and *Banana bract mosaic virus* in cardamom using SYBR Green based reverse transcription-quantitative PCR. Virusdisease 25:137–141

Wittwer C, Ririe K, Rasmussen R (1998) Fluorescence monitoring of rapid cycle PCR for quantification. In: Ferré F (ed) Gene quantification. Advanced Biomedical Technologies, Birkhäuser Boston, Cambridge, MA

Chapter 37

DNA Microarray for Detection of Plant Viruses

Abstract

Microarray technique is used for the simultaneous detection of multiple pathogens infecting plants in a single reaction. This method uses virus-specific oligos immobilized on a membrane or glass slide as probe. The total RNA isolated from infected plant is converted into cDNA and amplified through PCR using pathogen-specific primers, labelled using suitable molecules for detection. The amplified and labelled products are then applied to the array and allowed for hybridization. After washing, array will be developed depending on the label used and result visualized using CCD with suitable software.

Key words Biochip, DNA chip, Virus detection, Simultaneous detection

37.1 Introduction

In diagnosis of plant viruses, the microarray offers a new approach for detection of multiple viruses in a single plant species. DNA arrays have created a revolution in nucleic acid detection of plant viruses in early 1990s. The resulting ‘microarrays’ or DNA chips allows the simultaneous detection of thousands of cDNAs probes arrayed on a small surface of a microscope slide or chip approximately of 1 cm^2 in size. Oligo DNA microarrays have been used extensively for detection of known and unknown plant viruses. The different steps involved in the technique are: nucleic acid extraction, cDNA synthesis followed by PCR amplification, labelling, hybridization, washing and scanning the array for automated detection of nucleic acids directly on the array (van Doorn et al. 2007; Liu et al. 2007; Jeong et al. 2014; Martinelli et al. 2015).

Each cDNA is located at a specific spot on the array surface and the detection is carried out by reverse mixed-phase hybridization format. The cDNAs are labelled with a specific fluorophore and the probes are 50–75 bp long cDNAs of the targeted pathogens to be detected. The detection system uses one or more fluorophores, which are detected with laser technology. The microarray can be used to detect all known viruses whose sequences are known and unknown, but at the same time the technique is too expensive. During the experiment, ssDNA probes are immobilized or spotted

Alangar Ishwara Bhat and Govind Pratap Rao, *Characterization of Plant Viruses*, Springer Protocols Handbooks, https://doi.org/10.1007/978-1-0716-0334-5_37,

on a fixed matrix and exposed to different DNA fragments. The strong hybridization of complementary fragments for the probe spots is detected and quantified using appropriate scanners. DNA microarrays have been successfully used to diagnose plant viruses and viroids in potatoes, cucumbers, tomatoes, apples, apricots and grapevines (Barba and Hadidi 2007, 2012; Hadidi and Barba 2008; Hadidi et al. 2004; Lenz et al. 2008).

DNA microarrays or biochips are made of a surface capable of linking and capturing multiple probes, each one being specific for a DNA or RNA sequence of the target pathogen. Their purpose is to allow the expression of different target genes to be investigated simultaneously in a highly parallel fashion. Various supports like nylon, glass and different polymers are used for the elaboration of microarrays and up to thousands of DNA probes (up to 30,000 gene sequences) can be arrayed onto a single chip. The RNA/DNA sample to be detected is fluorescently labelled and hybridized to the array. The fluorescence allows the hybridization events on the solid support to be identified, and the identity of the target virus is deduced from its position on the array.

In a DNA microarray, each DNA spot contains a specific DNA sequence (probe) of short (15–25 nucleotides) or long oligonucleotides (50–120 nucleotides). Probes are short section of a gene or other DNA elements that are used to hybridize a cDNA sample (called target). The oligonucleotide synthesis in situ on a solid support involves the use of photolithography to build up each element of the array, nucleotide by nucleotide up to 25 bases. Alternatively, longer nucleotides and cDNAs can be spotted directly onto glass slides or membranes (Bates et al. 2005). Thousands of probes can be spotted mechanically by robotic instrumentation onto a glass slide. DNA microarray can also be used to measure the expression levels of large numbers of genes simultaneously. The process of microarray technology is illustrated in Fig. 37.1. Probe-target hybridization is detected and quantified by detection of fluorescence or chemiluminescence (Fig. 37.2).

The DNA microarray method is fast developing sensitive technology with broad spectrum of detectable viruses and viroids in a single assay. Microarray allows the simultaneous detection of thousands of diverse viruses and viroids (Boonham et al. 2003; Lee et al. 2003; Bystricka et al. 2005; Deyong et al. 2005; Agindotan and Perry 2007; Pasquini et al. 2008; Tiberini et al. 2010; Martinelli et al. 2015). This technology may be used in different aspects of plant virology such as detection of viruses, virus–host interactions, virus certification and quarantine evaluation programmes. The major steps involved in microarray are as follows (Fig. 37.1):

- Synthesis of microarray slides
- Total RNA isolation from tissue samples

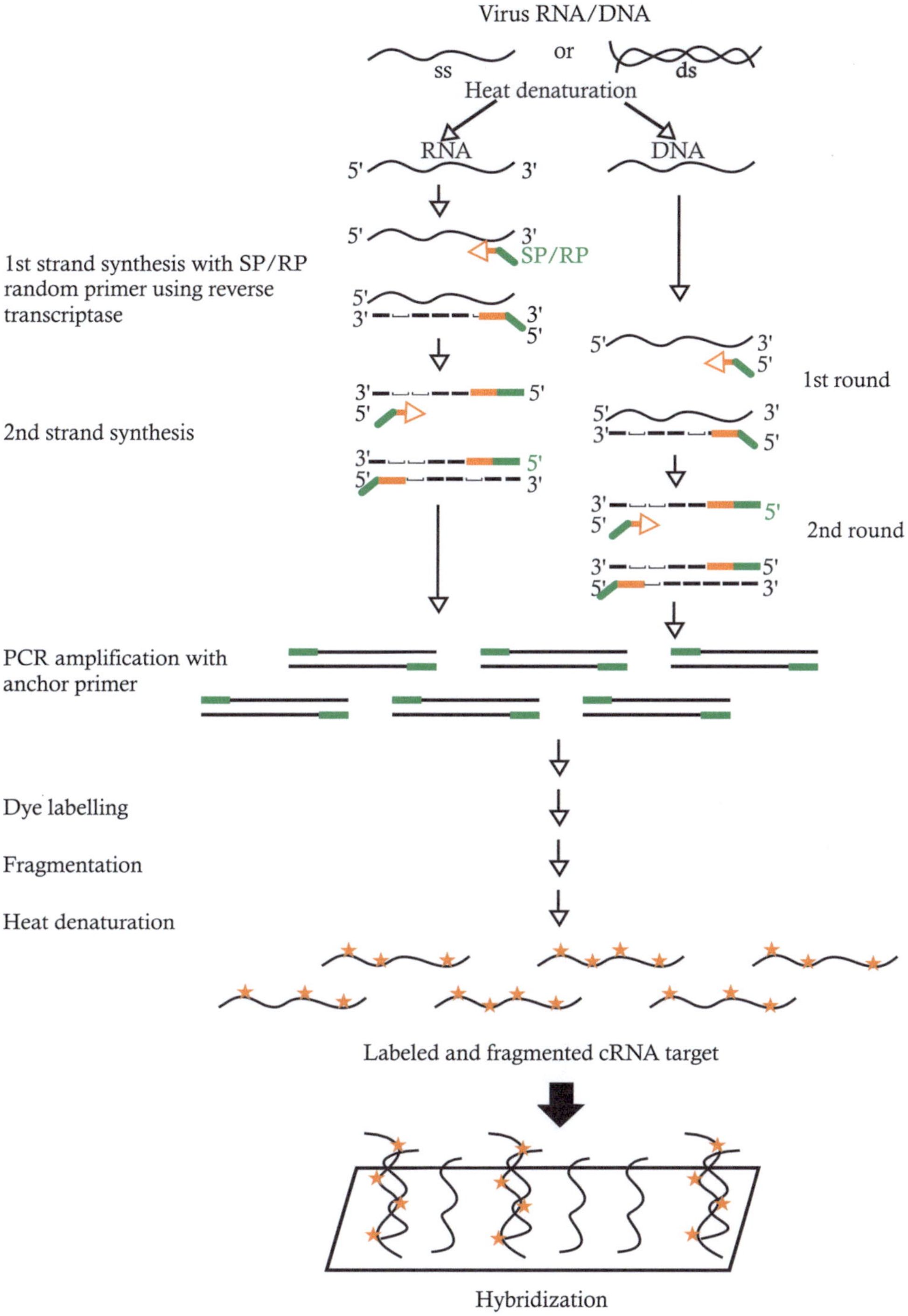

Fig. 37.1 Schematic outline of the strategy used to convert viral RNA and DNA into labelled cDNA for the microarray detection of viral sequences

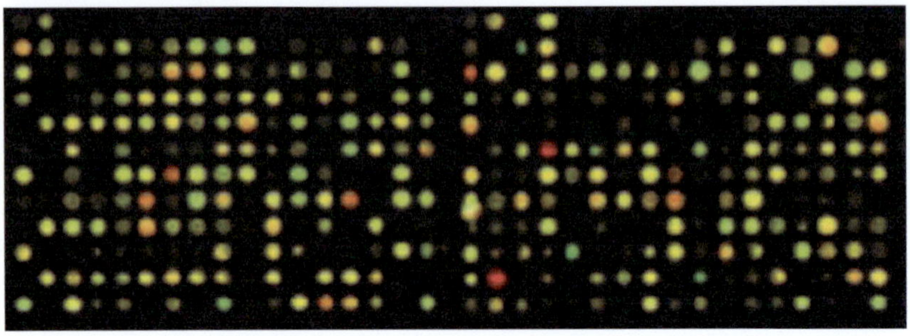

Fig. 37.2 A view of the glass microarray used for the detection of viruses. Green spots indicate positive reaction

- cDNA preparation by using oligo dT primers or random primers
- PCR amplification of the cDNA and label incorporation
- Hybridization on microarray followed by washing to remove un-hybridized cDNA
- Scanning hybridized slides using array scanner
- Fluorescent intensities quantification
- Identification and localization of the DNA spots
- Measurement of fluorescence and local background fluorescence for each DNA spot
- Consideration of positive signals (at least fivefold intensity above the local background)

37.2 Materials

1. Microarray slide: This is custom synthesized based the oligonucleotide sequence.
2. CTAB extraction buffer: Mix 2% cetyl trimethylammonium bromide (CTAB), 1% polyvinyl pyrrolidone (PVP), 100 mM Tris-HCl, 1.4 M NaCl, 20 mM EDTA. Always use freshly prepared extraction buffer. To prepare 100 mL of extraction buffer, add 2 g of CTAB, 1 g of PVP, 10 mL of 100 mM Tris-HCl (pH 8.0), 28 mL of 5 M NaCl, 4 mL of 0.5 M EDTA (pH 8.0) and make up the volume to 100 mL with water.
3. Lithium chloride (LiCl)(4 M): Weigh and dissolve 16.96 g of lithium chloride in 90 mL of H_2O, adjust the volume to 100 mL with H_2O. Sterilize the solution by passing it through a 0.22 μM filter or by autoclaving for 15 min at 15 psi (1.05 kg/cm^2) on liquid cycle.
4. RLT buffer: Commercially available from Qiagen.
5. RPE buffer: Commercially available from Qiagen.

6. Chloroform: isoamyl alcohol (24:1): Mix 24 parts of chloroform and 1 part of isoamyl alcohol.
7. Infected and healthy plant tissues.
8. Chemicals for cDNA synthesis: Please *see* Chapter 43.
9. Cy3 or Cy5 dye: Commercially available.
10. Na $(HCO_3)_2$ buffer (0.1 M, pH 9.0): Prepare 0.1 M of sodium carbonate (106 mg in 10 mL of water) and 0.1 M of sodium bicarbonate (8.4 mg in 10 mL) solution. To the solution of sodium carbonate, add solution of sodium bicarbonate till you get the desired pH of 9.0.
11. CyScribe purification kit: Commercially available, used for the purification of Cy3 and Cy5-labelled cDNA.
12. Microarray high speed centrifuge.
13. Cover slip for the slide: Provided by the array manufacturer.
14. Formamide.
15. Sodium chloride.
16. Sodium phosphate.
17. Ethylene diamine tetra acetic acid (EDTA).
18. Tween 20.
19. Denhardt's solution (50×): Prepare 50× solution by dissolving 5 g each of BSA (bovine serum albumin), Ficoll and PVP (Polyvinyl pyrrolidone) in 500 mL of water.
20. Salmon sperm DNA (100 mg/mL).
21. Phosphate buffer saline (PBS) (10×): Dissolve 8.00 g NaCl, 0.20 g KH_2PO_4, 2.90 g Na_2HPO_4 (or 1.15 g anhydrous), 0.20 g KCl, 0.20 g NaN_3, in distilled water and make up the volume to 1000 mL. For working solution of 1× PBS, mix 100 mL of 10× PBS with 900 mL of distilled water.
22. Tris-acetate buffer (200 mM pH 8.1): Dissolve 2.422 g of Tris base in about 80 mL of water, adjust the pH to 8.1 with glacial acetic acid, make up the volume to 100 mL.
23. Potassium acetate (KOAc) (500 mM): Dissolve 490 mg of KOAc in water and make up the volume to 10 mL with water.
24. Magnesium acetate (MgOAc) (150 mM): Dissolve 322 mg of MgOAc in water and make up the volume to 10 mL with water.
25. Sodium dodecyl sulphate (SDS): Prepare a 10% stock solution by dissolving 10 g of SDS in 100 mL of water.
26. SSPE buffer (6% sodium chloride-sodium phosphate-EDTA): Prepare by adding 175.3 g of NaCl, 27.6 g NaH_2PO_4, 9.4 g EDTA in about 800 mL of water, set the pH to 7.4 using NaOH and make up the volume to 1 L with water, autoclave for 20 min.

27. Blocking buffer [6% sodium chloride-sodium phosphate-EDTA (SSPE), 0.05% Tween 20, 20 mM EDTA (pH 8.0), 5× Denhardt's solution, 100 ng/mL denatured salmon sperm DNA, 0.05% SDS]: Prepare 100 mL of blocking buffer by adding 6 mL of SSPE, 50 μL of Tween 20, 4 mL of 0.5 M EDTA (pH 8.0), 10 mL of 5× Denhardt's solution, 100 μL of denatured salmon sperm DNA, 50 μL of 10% SDS and make up the volume to 100 mL with sterile water.
28. Hybridization buffer: 6% SSPE, 0.05% Tween 20, 20 mM EDTA, 25% formamide, 100 ng/mL denatured salmon sperm DNA and 0.04% SDS.
29. Hybridization chamber.
30. Hydroxylamine (4 M): Dissolve 1.39 g of solid hydroxylamine in 1.5 mL of H_2O in a glass test tube by shaking under hot tap water and make up the volume to 2 mL.
31. Dimethyl sulphoxide (DMSO).
32. Nanodrop spectrophotometer.
33. Water bath.
34. Hybridization oven/thermo mixer.
35. Microarray scanner and data analysis software.

37.3 Method

37.3.1 Design and Synthesis of Microarray Slide

1. Retrieve all available sequences of each of the identified viruses from NCBI GenBank.
2. Align the sequences using clustalW or MEGA software.
3. Identify most conserved regions and use these regions as target region for oligo probe design.
4. Design at least 5–7 oligo probes for each of the viruses using the oligodesign software tool available at www.oligo.inatools.com using the following format: oligo length: 40–70 bases with melting temperatures in the range of 82–88 °C; secondary structure threshold: 70 °C; cross hybridization temperature threshold: 65 °C and GC content: 40–60%.
5. Confirm specificity of the software generated oligos/probes through BLAST search against NCBI database.
6. Align all the designed oligos with respective virus sequences to confirm their proper alignment.
7. Assess specificity of the probes by nucleotide identity score and BLAST N alignment.

8. Test all designed probes for their secondary folding pattern using 'M fold' software and melting temperature using 'bio-tools' software.
9. Send all the designed probes for custom array synthesis to appropriate manufacturer. Each of the oligoprobes has to be synthesized and spotted in 2–4 replications.
10. Design 3–5 oligos/probes to plant genes to serve as positive controls.

37.3.2 RNA Isolation

1. Weigh about 150 mg of leaf sample in a polybag and dip in liquid nitrogen for about 10 min. Remove the bag from liquid nitrogen and grind finely with a wooden roller and add about 2 mL CTAB buffer to sample, continue grinding.
2. Transfer about 1 mL of homogenate to a 2 mL tube and incubate at 65 °C for 15 min.
3. Add 1 mL chloroform: isoamyl alcohol (24:1), mix to emulsion by inverting the tube and centrifuge at 11,000 × *g* for 10 min.
4. Transfer 800 μL of aqueous layer to a new 2 mL tube and add 800 μL of chloroform: isoamyl alcohol (24:1), mix by inverting and centrifuge at 11,000 × *g* for 10 min.
5. Transfer 600 μL of aqueous layer and add with an equal volume of 4 M LiCl, incubate at 4 °C overnight or 20 °C for 2 h.
6. Centrifuge at 13,000 × *g* for 25 min, discard supernatant, dissolve pellet in 50 μL nuclease-free water.
7. Add 350 μL of RLT buffer (Qiagen) and 250 μL of ethanol, mix well.
8. Transfer the contents to the RNeasy column (Qiagen) and centrifuge at 8000 × *g* for 30 s.
9. Wash the column by adding 500 μL of RPE buffer (Qiagen) and spin for 30 s at 8000 × *g*.
10. Wash with RPE buffer (Qiagen) once again, discard flow through and centrifuge at maximum speed for 1 min.
11. Transfer column to a fresh collection tube, add 50 μL of nuclease-free water and spin at 8000 × *g* for 30 s.
12. Repeat elution step and store RNA at −30 °C.
13. Determine RNA yield by a spectrophotometer/nanophotometer.

37.3.3 cDNA Synthesis and Fluorescent Labelling of cDNA

1. To synthesize first strand cDNA, mix the following in a sterile microcentrifuge tube on ice.

Component	Volume (μL)
10× first strand buffer	3.0
Random hexamer (1 μM/μL)	2.0
Oligo d (T) primer (3 μM/μL)	2.0
dNTP mix (10 mM/μL)	5.0
Dithiothreitol (0.1 M)	10.0
RNasin (10 units/μL)	0.5
RNA template[a] (1 μg/μL)	1.0
AMV-RT (20 units/μL)	1.0
Water to make	30.0

[a]denatured by heating at 80 °C for 5 min and snap cooling on ice for 2 min before addition

2. Mix the reagents by gentle vortexing followed by quick spin.
3. Incubate the reaction for 1 h at 42 °C.
4. Heat the reaction to 70 °C for 10 min and then transfer to ice.
5. Purify the cDNA using any commercial purification kit using manufacturer's protocol.
6. Dry the purified cDNA and re-suspended it in a 50 μL of 0.1 M Na $(HCO_3)_2$ buffer, pH 9.0.
7. Add 10 μL of Cy5 or Cy3 dye suspend in dimethyl sulphoxide buffer.
8. Incubate cDNA and labelling mixture for 18 h at room temperature in a dark place.
9. Deactivate the unincorporated dyes by adding 15 μL of 4 M hydroxylamine.
10. Clean the labelled cDNA using commercial kit such as Cyscribe cleaning kit.
11. Check the efficiency of labelling by reading an aliquot of the Cy5 or Cy3-labelled cDNA in a nanodrop spectrophotometer. Determine the labelling efficiency using the formula:

$$\frac{\text{dye molecules}}{1000\,\text{nt}}\ \frac{A\,\text{dye}}{A\,260} \times \frac{9010}{\text{dye extinction coefficient}} \times 1000$$

37.3.4 Hybridization and Scanning of Microarray Slide

1. Insert the microarray slide into the hybridization chamber.
2. Pre-hybridize the slide for 30 min at 42 °C with the blocking buffer (30 μL) consisting of 6% sodium chloride-sodium phosphate-EDTA (SSPE), 0.055 Tween 20, 20 mM EDTA, 5×

Denhardt's solution, 100 ng/μL denatured salmon sperm DNA and 0.05% SDS.

3. While pre-hybridization is on, prepare 1 μg of labelled target cDNA by fragmenting with a solution of 200 mM Tris-acetate pH 8.1 and 500 mM KOAc (or 150 mM of MgOAc) followed by incubation at 95 °C for 20 min. Denature the labelled target at 95 °C for 3 min in hybridization buffer (6% SSPE, 0.05% Tween 20, 20 mM EDTA, 25% formamide, 100 ng/mL of denatured salmon sperm DNA, 0.04% SDS), cooled on ice at least for 1 min, centrifuge briefly and retain it on ice until it is ready to add onto the slide in **step 4**.
4. After pre-hybridization, wash the slide and cover it with 30 μL of fragmented and denatured labelled target in hybridization buffer (from **step 3** above) and incubate for 16 h at 42 °C.
5. Wash the slide with preheated solution as follows:
   - 6× SSPE, 0.05% Tween 20 for 5 min at 42 °C
   - 3× SSPE, 0.05% Tween 20 at room temperature for 5 min
   - 0.5× SSPE, 0.05% Tween 20 at room temperature for 5 min
   - 2× Phosphate buffer saline (PBS), 0.1% Tween 20 at room temperature for 5 min
   - 2× PBS for 5 min at room temperature
6. Remove the slide from hybridization chamber and immediately cover its semiconductor surface with imaging solution and cover slips provided by the manufacturer.
7. Scan the hybridized slide using array scanner at 532 nm laser (for Cy3) and at 635 nm laser (for Cy5) for fluorescence measurements. Fix laser power at 50% of its potential for both lasers and photomultiplier tube (PMT) from 45 to 65% depending on signal intensity. Select level of PMT to balance the background with the level of spot detection on a slide. Quantify the fluorescence intensity using suitable software and analyse the images by Combimatrix microarray imaging software. The DNA-cDNA hybrid spots are then located on the array using Gal file as feature indicator.

37.4 Notes (Boonham et al. 2003; Agindotan and Perry 2007; Bystricka et al. 2005; Hadidi et al. 2004)

1. RNA must be intact and of good quality. Degraded RNA will give poor results. All precautions listed in Chapter 32 for RNA isolation should be strictly followed.
2. Designing oligo probes is very crucial and determines the success of detection.

3. Glass, nylon and different polymers are used as support system for the elaboration of the microarrays (Barba and Hadidi 2007).
4. Another possibility of detection of viruses is based on nanochip technology developed by Nanogen (San Diego, USA) which involves the combination of microelectronics with microarray technology on a solid support covered with streptavidin to increase the power of union with biotin-labelled DNA (Miller and Tang 2009).
5. A DNA microarray can also be used for studying the pathogen–host gene interaction which may lead to the development of plants resistant or tolerant to the pathogen infection.
6. Microarray slide can be re-used after stripping. The stripping protocol is based on denaturation of DNA-cDNA hybrids that results in the removal of the labelled targets while retaining the original oligonucleotide probes on the array. The procedure involves washing and incubating the slide in a stripping solution at 65 °C for 15 min. The stripped array is scanned again to ascertain the absence of any fluorescence signal before re-hybridization with new target samples (Zhang et al. 2009).

References

Agindotan B, Perry KL (2007) Macroarray detection of plant RNA viruses using randomly primed and amplified complementary DNAs from infected plants. Phytopathology 97:119–127

Barba M, Hadidi A (2007) DNA microarrays: technology, applications and potential applications for the detection of plant viruses and virus-like pathogens. In: Rao GP, Valverdi RA, Dovas CI (eds) Techniques in diagnosis of plant viruses. LLC Press, Houston, pp 288–293

Barba M, Hadidi A (2012) Microarray-based detection and genotyping of plum pox virus. In: Rao GP, Baranwal VK, Mandal B, Rishi N (eds) Recent trends in plant virology. Studium Press LLC, Houston

Bates SR, Baldwin DA, Channing A, Gifford LK, Hsu A, Lu P (2005) Cooperativity of paired oligonucleotide probes for microarray hybridizationassays. Anal Biochem 342:59–68

Boonham N, Walsh K, Smith P, Madagan K, Grahamand I, Barker I (2003) Detection of potato viruses using microarray technology: towards a generic method for plant viral disease diagnosis. J Virol Methods 108:181–187

Bystricka D, Lenza O, Mraza I, Piherovad L, Kmochd S, Sipc M (2005) Oligonucleotide-based microarray: a new improvement in microarray detection of plant viruses. J Virol Methods 128:176–182

Deyong Z, Willingmann P, Heinze C, Adam G, Pfunder M, Frey B, Frey JE (2005) Differentiation of Cucumber mosaic virus isolates by hybridization to oligonucleotides in a microarray format. J Virol Methods 123:101–108

Hadidi A, Barba M (2008) DNA microarrays: 21st century pathogen detection. Acta Hortic (781):331–339

Hadidi A, Czosnek H, Barba M (2004) DNA microarrays and their potential applications for the detection of plant viruses, viroids and phytoplasmas. J Plant Pathol 86:97–104

Jeong JJ, Ju HJ, Noh J (2014) A review of detection methods for the plant viruses. Res Plant Dis 20:173–181

Lee GP, Min BE, Kim CS, Choi SH, Harn CH, Kim SU, Ryu KH (2003) Plant virus cDNA chip hybridization for detection and differentiation of four cucurbit infecting Tobamoviruses. J Virol Methods 110:19–24

Lenz O, Petrzik K, Spak J (2008) Investigating the sensitivity of a fluorescence-based microarray for the detection for the detection of fruit tree viruses. J Virol Methods 148:96–105

Liu CX, Lagae L, Borghs G (2007) Manipulation of magnetic particles on chip by magnetophoretic actuation and dielectrophoretic levitation. Appl Phys Lett 90:184109

Martinelli F, Scalenghe R, Davino S, Panno S, Scuderi G, Ruisi P, Villa P, Stroppiana D, Boschetti M, Goulart LR, Davis CE, Dandekar AM (2015) Advanced methods of plant disease detection—a review. Agron Sustain Dev 35:1–25

Miller MB, Tang YW (2009) Basic concepts of micro arrays and potential applications in clinical microbiology. Clin Microbial Res 22:611–613

Pasquini G, Barba M, Hadidi A, Faggioli F, Negri R, Sobol I, Tiberini A, Caglayan K, Mazyad H (2008) Oligonucleotide microarray based detection and genotyping of *Plum pox virus*. J Virol Methods 147:118–126

Tiberini A, Tomassoli L, Barbaa M, Hadidi A (2010) Oligonucleotide microarray-based detection and identification of 10 major tomato viruses. J Virol Methods 168:133–140

van Doorn R, Szemes M, Bonants P, Kowalchuk GA, Salles JF, Ortenberg E, Schoen CD (2007) Quantitative multiplex detection of plant pathogens using a novel ligation probe-based system coupled with universal, high-throughput real-time PCR on open arrays. BMC Genomics 8:276

Zhang X, Xu W, Tam J, Zeng Y (2009) Stripping custom micro RNA microarrays and the lessons learned about probe: slide interactions. Anal Biochem 386:222–227

Chapter 38

Loop-Mediated Isothermal Amplification (LAMP)

Abstract

Loop-mediated isothermal amplification (LAMP) can amplify nucleic acid with high specificity, sensitivity and speed under isothermal conditions. The LAMP method uses four to six primers and a DNA polymerase with strand-displacing activity to generate amplification products. The products can be detected by agarose gel electrophoresis or visually by turbidity or colour changes. The amplified products of LAMP are also visible as ladder-like patterns on a gel. LAMP methods have been used successfully for detecting both the RNA and DNA plant viruses.

Key words Reverse transcription (RT)-LAMP, Virus detection, Real-time LAMP, Turbidity, isothermal reaction

38.1 Introduction

Loop-mediated isothermal amplification (LAMP) is a powerful molecular diagnostic tool, which can amplify nucleic acid with high specificity, sensitivity and speed under isothermal conditions (Notomi et al. 2000) (Fig. 38.1). It can be carried out in a water bath at a constant temperature of 60–65 °C. The technique relies on the design of a set of primers that generate stem looped (hairpin) structures during early stage of DNA synthesis. The most vital step in any amplification reaction is the primer design. LAMP employs four to six primers specially designed to recognize six to eight distinct regions of a target gene, hence LAMP amplifies gene with high efficiency and precision. Another most important feature of LAMP is the amplification of non-denatured DNA. This feat is achieved with the aid of special LAMP polymerase enzyme. There are two polymerase enzymes suitable for LAMP reaction, the most common is the *Bst* DNA polymerase, and the less commonly used enzyme is the *Bsm* DNA polymerase. These enzymes are sourced from *Bacillus stearothermophilus* and *Bacillus smithii*, respectively. The enzymes possess strand displacement activity and catalyse 5′–3′ as well as the 3′–5′ DNA polymerization. The *Bst* and the *Bsm* enzyme retain their enzymatic activity at 66 °C and 63 °C, respectively, and exhibit strand displacement activity as it extends

Alangar Ishwara Bhat and Govind Pratap Rao, *Characterization of Plant Viruses*, Springer Protocols Handbooks, https://doi.org/10.1007/978-1-0716-0334-5_38, © Springer Science+Business Media, LLC, part of Springer Nature 2020

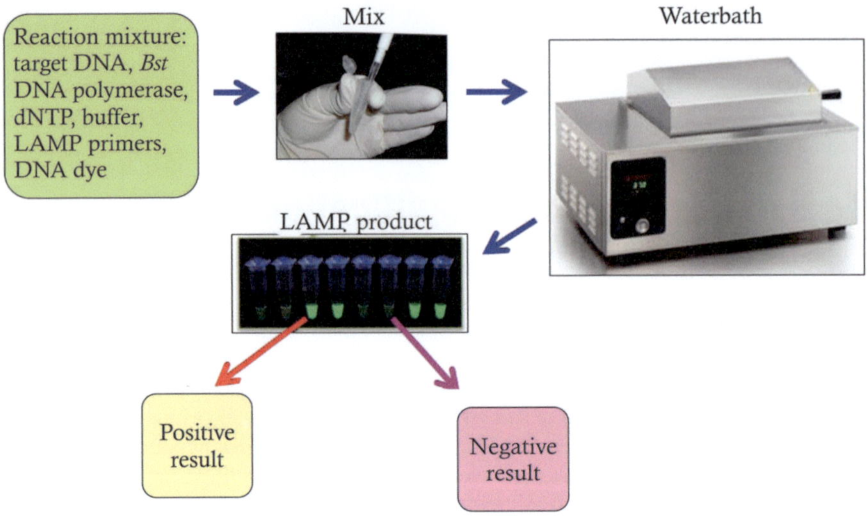

Fig. 38.1 Steps involved in LAMP assay

annealed primer sequence. This strand displacement activity is uniquely elicited by the LAMP polymerase enzyme at the 5′–3′ end without exonuclease. Outstanding quality of this polymerase enzyme also leads to the production of single-stranded DNA. Reaction occurs under constant temperature, with high stability (Nagamine et al. 2002; Abdullahi et al. 2015; Keizerweerda et al. 2015).

LAMP as a nucleic acid amplification technique operates under a unique amplification principle which involves two basic steps. They are the non-cyclical and cyclical phase. The non-cyclical precedes the cyclical phase of the amplification process. All the four primers as well as the *Bst* DNA polymerase enzyme with a strand displacement activity play a role in this first stage of LAMP reaction. However, the cyclical step builds upon the product of the non-cyclical which only involves the two primers and the *Bst* DNA polymerase enzyme. Loop primers might be involved in the cyclical step when six primers are used.

A chemical called betaine is used in stabilization of the AT and GC content, to ensure stability of the reaction. Other components include deoxynucleoside triphosphate (dNTP), which provides the required nucleotides; magnesium sulphate ($MgSO_4$) used in the reaction combines with pyrophosphate molecules released during DNA synthesis forming an insoluble Mg-pyrophosphate complex. As more and more Mg-pyrophosphate molecules accumulate, they give turbidity (white precipitate) to the reaction. Thus presence of turbidity in the tube itself is an indication that LAMP is happening (Fig. 38.2). Fluorescent DNA dye can be added to ease the visualization of the turbidity resulting from precipitate of pyrophosphate ions. Example of DNA dyes used in LAMP includes SYBR Green,

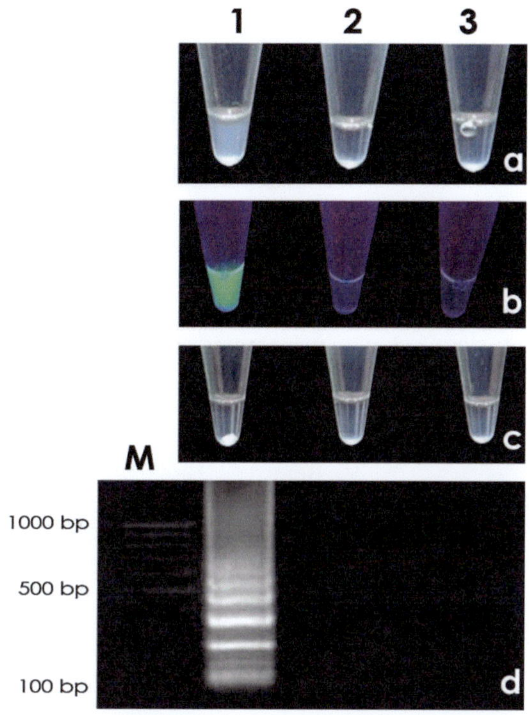

Fig. 38.2 Reverse transcription loop-mediated isothermal amplification (RT-LAMP) for detecting cucumber mosaic virus infecting black pepper. Lane 1: Infected pepper, Lane 2: healthy pepper, Lane 3: control (with water instead of nucleic acids). Making the products visible (**a**) under normal light, (**b**) under UV light, (**c**) centrifugation and (**d**) gel electrophoresis. Lane M shows a 100 bp DNA ladder. (Reproduced from Bhat et al. 2013 with permission from Elsevier)

HNB, Picogreen and a Calcein metal ion indicator. Lastly buffer solution containing $(NH_4)_2 SO_4$, Tris-HCl (pH 8.8), $MgSO_4$ and KCl are used as part of the LAMP reaction mixture. LAMP technique is more sensitive than the PCR, with a 10- to 100-fold higher sensitivity than PCR (Fukuta et al. 2004; Nie 2005; Bhat et al. 2013; Siljo and Bhat 2014).

38.2 Materials

1. Pestle and mortar.
2. Incubator.
3. Transilluminator.
4. Gel documentation unit.
5. Horizontal gel apparatus with power pack.
6. Primers: LAMP requires two outer primers (F3 and B3) and two long inner primers (FIP and BIP) that recognize six

specific sequences in the target DNA. The first inner primer containing sense and antisense sequences in the DNA hybridizes with the target sequence and initiates DNA synthesis. Six primers are specially designed to target six to eight regions in a gene of interest. They are Forward Inner Primer (FIP), Forward Outer Primer (F3), Backward Inner Primer (BIP), Backward Outer Primer (B3) and two optional loop primers—Forward loop (FL) and Backward loop (BL). The primers are designed manually or by the use of LAMP primer design software (Primer Explorer) freely available online (https://primerexplorer.jp/e/).

7. Healthy and infected plant material.
8. $MgSO_4$.
9. Betaine.
10. $MnCl_2$.
11. *Bst* polymerase.
12. dNTP mix: Use 25 mM stock.
13. Thermoscript RT: This reverse transcriptase enzyme can withstand temperature up to 65 °C and thus suitable for RT-LAMP.
14. Calcein.

38.3 Method

38.3.1 Performing LAMP and RT-LAMP

1. Isolate total DNA (if testing for DNA viruses) or RNA (if testing for RNA viruses) from infected plants. The detailed procedure of total DNA and RNA extraction have been explained in Chapters 31 and 32, respectively.
2. Isolate total DNA or RNA from corresponding virus-free plants to serve as negative control.
3. Set up LAMP reaction by adding the following in a 0. 2 mL tube: The LAMP reaction mixture (25 μL) contains 1 μL (about 60 ng) of the template nucleic acids, 2× thermopol buffer, 1.4 mM each of dNTPs, 0–14 mM $MgSO_4$ (optimum concentration need to be worked out for each set of primers and target DNA) and 0.4–2.4 M betaine (optimum concentration need to be worked out for each set of primers and target DNA), 200 nM each of the external primers F3 and B3, 2 μM each of the internal primers FIP and BIP, 1 μM each of the loop primers (B-loop and F-loop), 1 mM $MnCl_2$ and 50 μM calcein, and 8 U of *Bst* DNA polymerase (see Table 38.1). If performing reverse transcription (RT)-LAMP, in addition to the above reagents, add thermoscript reverse transcriptase enzyme (1.5 U). Incubate the tube at 65 °C for 60 min (temperature and time can be optimized for each target and primer

Table 38.1
Components of the loop-mediated isothermal amplification (LAMP) and reverse transcription (RT)-LAMP reaction mixture for the detection of DNA and RNA plant viruses (Reproduced from Bhat et al. 2013 with permission from Elsevier)

		Required concentration		Required volume per reaction (μL)	
Component	Stock conc.	LAMP	RT-LAMP	LAMP	RT-LAMP
Thermopol reaction buffer	10×	2×	2×	5.0	5.0
$MgSO_4$[a]	50 mM	–	–	–	–
dNTPs	10 mM	1.4 mM	1.4 mM	3.5	3.5
F3 primer	10 μM	0.2 μM	0.2 μM	0.5	0.5
B3 primer	10 μM	0.2 μM	0.2 μM	0.5	0.5
FIP primer	100 μM	2 μM	2 μM	0.5	0.5
BIP primer	100 μM	2 μM	2 μM	0.5	0.5
BL primer/FL primer	100 μM	1 μM	1 μM	0.25	0.25
Betaine[a]	5 M	–	–	–	–
Bst polymerase	8 U/μL	8 U	8 U	1.0	1.0
Thermoscript RT	1.5 U/μL	–	1.5 U	–	1.0
$MnCl_2$	20 mM	1 mM	1 mM	1.25	1.25
Calcein	1 mM	50 μM	50 μM	1.25	1.25
Sterile water[b]	–			–	–
Template nucleic acid	60 ng/μL			1.0	1.0
Total reaction volume	–			25.0	25.0

[a]Optimum concentration need to be worked out as indicated in the procedure
[b]To make final reaction volume to 25 μL

combinations by performing LAMP/RT-LAMP for different temperatures starting from 58 to 68 °C and time from 30 to 75 min) followed by 80 °C for 5 min to inactivate the *Bst* polymerase (Fig. 38.2).

4. For determining optimal temperatures, the LAMP reaction can be carried out from 58 to 68 °C for 75 min. Similarly to determine the optimum time, LAMP reaction can be carried out for 15 min, 30 min, 45 min, 60 min and 75 min at 65 °C.
5. For determining the optimum concentration of $MgSO_4$, betaine may be maintained at 0.8 M (for both LAMP and RT-LAMP), whereas in determining the optimum concentration of betaine, $MgSO_4$ may be maintained at 6 mM (for LAMP) and 4 mM (for RT-LAMP).

38.3.2 Visual Detection of LAMP and RT-LAMP Products

LAMP products can be visualized in agarose gel electrophoresis that provides a ladder-like pattern, by visual observation of turbidity or colour changes (by addition of SYBR Green, manganese chloride or fluroxin dye). In both LAMP and RT-LAMP reactions, DNA is synthesized in large quantities. The magnesium pyrophosphate is also produced in large quantities as an insoluble by-product. Because the by-product is insoluble, the liquid in the reaction tube becomes turbid. The degree of turbidity is measured by different methods (Mori et al. 2001; Tomita et al. 2008; Zhang et al. 2014).

1. The first and simplest method is by visual examination of the tube in normal daylight for the presence or absence of turbidity. The presence of turbidity indicates positive reaction indicting the presence of the test pathogen (Fig. 38.2).
2. The second method induces fluorescence by adding manganese in ionic form and calcein to the reaction solution. The reaction tube is then observed for the presence or absence of green fluorescence under UV light. The presence of green fluorescence indicates positive reaction indicting the presence of the test pathogen (Fig. 38.2).
3. In the third method instead of Mn and calcein, fluorexon dye can be added. The positive reaction is detected from colour changing from light orange to green.
4. In the fourth method, the reaction tubes are centrifuged at $5000 \times g$ for 3 min and observed for the presence or absence of a white pellet. The presence of pellet indicates positive reaction indicting the presence of the test pathogen (Fig. 38.2).
5. In the fifth method, amplification products are run on 2% agarose gel through electrophoresis (Fig. 38.2). The presence of multiple ladder-like bands indicates positive reaction indicting the presence of the test pathogen.

38.4 Notes (Mori et al. 2001; Tomita et al. 2008; Zhang et al. 2014)

1. It is advisable to prepare master mix of primers and make aliquots to avoid frequent freezing and thawing.
2. As LAMP reactions are highly sensitive, the work area should be free from contamination. It is always advisable to set up LAMP reaction in a laminar air flow to avoid contaminations.
3. All the visual methods (turbidity, fluorescence, pellet and gel electrophoresis) would give similar results in terms of detection limits.
4. The LAMP reaction can be carried out at a constant temperature with a short reaction time, is ideal for point-of-care

detection of plant viruses in fields and has been used to detect many plant viruses in less than 2 h.

5. LAMP has very high amplification efficiency and sensitivity as it generates large amounts of amplification product with low amounts of input DNA.
6. The method is relatively cost-effective as it requires simple equipment to perform the assay.
7. One of the limiting factors of LAMP is the complicated nature of the LAMP primer designing. For a good LAMP reaction, the size of the target DNA should be within 200–300 bp; this condition makes it more difficult to design primers for all targets.
8. Real-time LAMP can be performed in real-time LAMP instrument developed by Optigene, UK. It requires less than 1 h for real-time LAMP as amplification in positive sample can be seen by eye inspection via on-site screen on real-time LAMP instrument. Real-time LAMP is more practical providing a quick primary on-site screening (detection within 7–15 min). The anneal curve obtained confirms the accuracy of the reaction.
9. Another version of LAMP is called lyophilized LAMP that aims to simplify and make the process of LAMP more rapid, through combination of all LAMP reagents into a single mixture called lyophilized LAMP mix. User of this version of LAMP is required to only add the template DNA or sample into the standby mix to carry out the amplification. Lyophilized LAMP kit is commercially available for rapid diagnosis of some diseases.
10. LAMP can be combined with Lateral Flow Assay (LFA). It employs an absorbent pad or strip, containing an antibody specific to a target analyte. The antibody coupled with colour moiety binds to the LAMP product in a positive reaction, this device can be designed to serve as an ideal tool for plant virus detection in fields (Abdullahi et al. 2015).

References

Abdullahi UF, Rochman N, Wan R, Wan T, Ahmadu S, Anas M, Aliyu S, Baig A (2015) Loop-mediated isothermal amplification (LAMP), an innovation in gene amplification: bridging the gap in molecular diagnostics: a review. Indian J Sci and Tech 8:1–12

Bhat AI, Siljo A, Deeshma KP (2013) Rapid detection of *Piper yellow mottle virus* and *Cucumber mosaic virus* infecting black pepper (*Piper nigrum*) by loop-mediated isothermal amplification (LAMP). J Virol Methods 193:190–196

Fukuta S, Ohishi K, Yoshida K, Mizukami Y, Ishida A, Kanbe M (2004) Development of immunocapture reverse transcription loop-mediated isothermal amplification for the detection of *Tomato spotted wilt virus* from chrysanthemum. J Virol Methods 121:49–55

Keizerweerda AT, Chandra A, Grisham MP (2015) Development of a reverse transcription loop-mediated isothermal amplification (RT-LAMP) assay for the detection of *Sugarcane mosaic virus* and *Sorghum mosaic virus* in sugarcane. J Virol Methods 212:23–29

Mori Y, Nagamine K, Tomita N, Notomi T (2001) Detection of loop-mediated isothermal amplification reaction by turbidity derived from magnesium pyrophosphate formation. Biochem Biophys Res Commun 289:150–154

Nagamine K, Hase T, Notomi T (2002) Accelerated reaction by loop-mediated isothermal amplification using loop primers. Mol Cell Probes 16:223–229

Nie X (2005) Reverse transcription loop-mediated isothermal amplification of DNA for detection of *Potato virus Y*. Plant Dis 89:605–610

Notomi T, Okayama H, Masubuchi H, Yonekawa T, Watanabe K, Amino N, Hase T (2000) Loop-mediated isothermal amplification of DNA. Nucleic Acids Res 28:e63

Siljo A, Bhat AI (2014) Reverse transcription loop-mediated isothermal amplification assay for rapid and sensitive detection of *Banana bract mosaic virus* in cardamom (*Elettaria cardamomum*). Eur J Plant Pathol 138:209–214

Tomita N, Mori Y, Kanda H, Notomi T (2008) Loop-mediated isothermal amplification (LAMP) of gene sequences and simple visual detection of products. Nat Protoc 3:877–882

Zhang X, Lowe SB, Gooding JJ (2014) Brief review of monitoring methods for loop-mediated isothermal amplification (LAMP). Biosens Bioelectron 61:491–499

Chapter 39

Rolling Circle Amplification (RCA)

Abstract

Rolling circle amplification (RCA) is a method of isothermal amplification of circular DNA molecules. RCA assay involves DNA amplification using a DNA ϕ polymerase with strand displacement activity to extend a single or multiple primers annealed to a circular DNA template. The strand displacement activity allows the newly synthesized DNA template to displace the previously generated DNA molecule releasing single-stranded DNA (ssDNA). This enzymatic process of primer extension combined with DNA strand displacement generates a long single-stranded DNA containing a repeated sequence complementary to the circular DNA template. The method is successfully used for the detection of many circular single- and double-stranded DNA viruses (begomovirus, badnavirus) infecting different crops. In this chapter, different steps involved in the RCA of a begomovirus virus genome are discussed.

Key words Amplification, Circular single-stranded DNA virus, Circular double-stranded DNA virus, Isothermal amplification

39.1 Introduction

Rolling circle amplification (RCA) is an isothermal DNA synthesis reaction that can rapidly synthesize multiple copies of circular molecules of DNAs (Blanco et al. 1989; Fire and Xu 1995; Dean et al. 2001; Johne et al. 2009) (Fig. 39.1). It uses the ϕ 29 DNA polymerase that has a 5′- to -3′ polymerization activity and a 3′- to -5′ ssDNA exonucleolytic activity, which provides a proof-reading ability. The ϕ 29 DNA polymerase also has a strong strand displacement activity that allows for the complementary strand to be displaced during replication, consistently creating a new template for the amplification (Blanco et al. 1989; Dean et al. 2001; Johne et al. 2009). Thus, RCA is a new promising method for the detection of circular DNA viruses such as begomoviruses (Inoue-Nagata et al. 2004) and badnaviruses (Wambulwa et al. 2012). The use of random hexamers to prime the reaction offers the possibility to amplify circular templates without any prior knowledge of the virus sequence, which has the advantage in the diagnosis of new viruses.

The procedure involves isolation of total DNA, its denaturation and subjecting to RCA using either random hexamers or virus specific or combination of both hexamers and virus-specific

Alangar Ishwara Bhat and Govind Pratap Rao, *Characterization of Plant Viruses*, Springer Protocols Handbooks,
https://doi.org/10.1007/978-1-0716-0334-5_39,

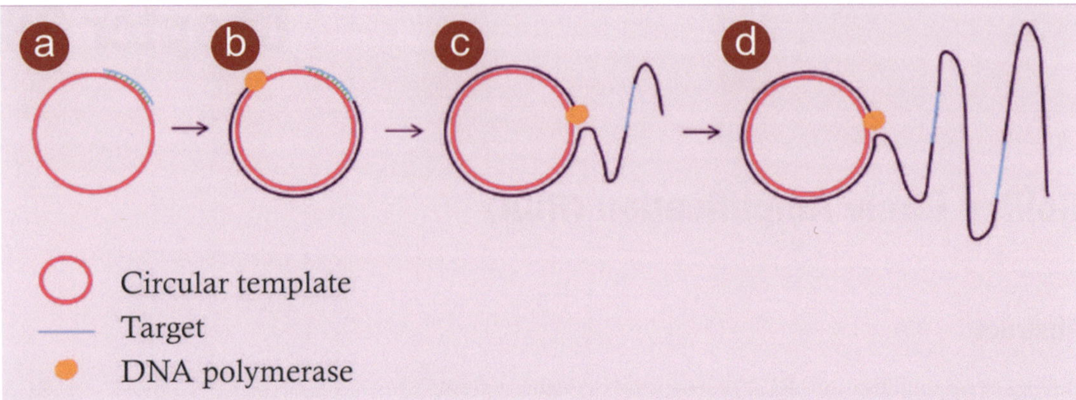

Fig. 39.1 Schematic diagram of rolling circle amplification. (**a**) A primer anneals to the circular template; (**b**) Initiation of DNA synthesis with DNA polymerase; (**c**) Continuation of DNA synthesis with strand displacement along the circular template; (**d**) DNA synthesis continues to generate ssDNA product. (Reproduced from Lau and Botella 2017)

primers. Following RCA, the product will be subjected to restriction digestion using unique restriction site for the target virus and analysed in the agarose gel. The expected size product can be cloned in a plasmid vector and sequenced to confirm specificity of the DNA band and identity of the virus. Alternatively, the specificity of the band can be confirmed through Southern hybridization using virus-specific probe. Rolling circle amplification offers the simplest available isothermal reaction mechanism (Lau and Botella 2017). With additional manipulation, linear DNA is also suitable as a template for RCA reaction.

Rolling circle amplification is widely used for plant pathogen detection since early 2000s (Inoue-Nagata et al. 2004). Several techniques have been used in combination with RCA such as RFLP and direct sequencing to characterize plant viruses efficiently with minimum effort and cost than the conventional methods (Schubert et al. 2007). The method is also used to differentiate circular (episomal) and linear forms (integrated or endogenous form) of badnaviruses (James et al. 2011; Wambulwa et al. 2012).

39.2 Materials

1. Agarose.
2. Virus-infected plant samples or DNA sample.
3. Random hexamer primer mix.
4. Custom-made virus-specific primers.
5. ϕ 29 DNA polymerase.
6. Bovine serum albumin (BSA).

7. dNTP mix.
8. Restriction enzyme.
9. Water bath.

39.3 Methods

1. Isolate total DNA from known infected and a healthy plant using the protocol available. A detail of the protocol is given in Chapter 31.
2. Check quality and quantity of DNA through gel electrophoresis and spectrophotometer, respectively. For detailed method, refer Chapters 25 and 26.
3. Take required quantity of DNA (1 μL, maximum 2 μL) and denature at 95 °C for 3 min.
4. To the above tube, add amplification solution containing custom-synthesized virus-specific or random primers or both (0.4 μM each), 1× ϕ 29 buffer, 2 ng/μL BSA, 15 mM dNTP mix, 2 U/μL of ϕ 29 DNA polymerase and water to make final volume of 20 μL.
5. Incubate the tube at 30 °C for 18–20 h in a water bath.
6. Stop reaction by keeping the tube at 65 °C for 10 min. Chill on ice.
7. About 10 μL of the product analysed by restriction digestion using restriction enzyme having unique restriction site in the expected product.
8. Analyse the restriction product in the agarose gel along with appropriate markers. Presence of single band at the expected size indicates positive result (Fig. 39.2).
9. Alternatively, the specificity of the RCA product can also be confirmed through PCR or real-time PCR using specific primers or by cloning and sequencing or through Southern hybridization using virus-specific probe.

39.4 Notes (Dean et al. 2001; Inoue-Nagata et al. 2004; Lau and Botella 2017)

1. Inclusion of a known plasmid in the reaction would serve as a method control in the experiment.
2. Thiophosphate modification between the last two nucleotides at the 3′ end of the primers will increase the stability of primers, thus contributing to the success of RCA.
3. In some cases, DNA denatured at 95 ° C for 3 min followed by cooling at room temperature in stages (50 °C for 15 s, 30 °C

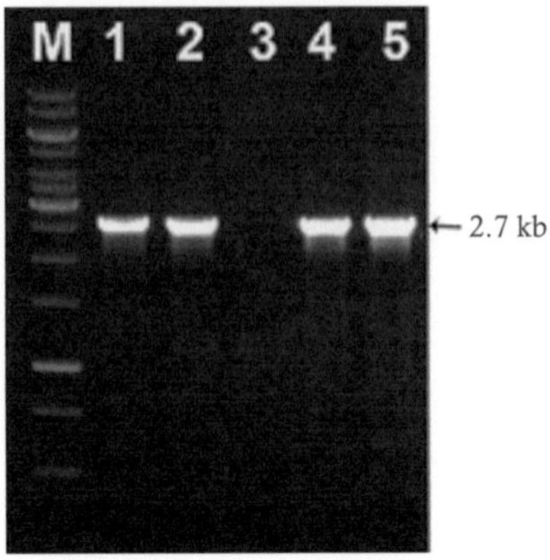

Fig. 39.2 Isolation of complete genome of a begomovirus (2.7 kb) through rolling circle amplification (RCA) followed by restriction digestion. Lane M: DNA size markers; lanes 1–5: digestion of RCA product with *Eco RI*

for 15 s and 20 °C for 1 min) before placing on ice was found better.

4. RCA can also be performed using commercial kit, Illustra TempliPhi amplification kit (GE Healthcare, UK).
5. Use of random hexamers to prime the reaction leads to non-specific amplification that reduces the assay's sensitivity to the specific targets. To mitigate the lack of specificity, enrichment of the specific template through linear DNA digestion prior to the RCA amplification would increase the specific amplification and sensitivity (Lau and Botella 2017).
6. RCA method allows generation of up to 0.5 MB (megabase) of DNA per probe (primer) in overnight incubation (Baner et al. 1998) and generates more number of copies (10^9) of each circle in 90 min (Lizardi et al. 1998).
7. RCA is least or less prone to carry-over contamination of the amplification products as there is no new 3′end ssDNA product generation throughout the RCA process.

References

Baner J, Nilsson M, Mendel-Hartvig M, Landegren U (1998) Signal amplification of padlock probes by rolling circle replication. Nucleic Acids Res 26:5073–5078

Blanco L, Bernad A, Lazaro JM, Martin G, Gar-mendia C, Salas M (1989) Highly efficient DNA synthesis by the phage Phi29 DNA polymerase. J Biol Chem 264:8935–8940

Dean F, Nelson J, Giesler T, Lasken R (2001) Rapid amplification of plasmid and phage DNA using Phi29 DNA polymerase and multiply primed rolling circle amplification. Genome Res 11:1095–1099

Fire A, Xu SQ (1995) Rolling replication of short DNA circles. Proc Natl Acad Sci U S A 92:4641–4645

Inoue-Nagata AK, Albuquerque LC, Rocha WB, Nagata T (2004) A simple method for cloning the complete begomovirus genome using the bacteriophage 29 DNA polymerase. J Virol Methods 116:209–211

James AP, Geijskes RJ, Dale JL, Harding RM (2011) Development of a novel rolling-circle amplification technique to detect *Banana streak virus* which also discriminates between integrated and episomal virus sequences. Plant Dis 95:57–62

Johne R, Müller H, Rector A, van Ranst M, Stevens H (2009) Rolling-circle amplification of viral DNA genomes using phi29 polymerase. Trends Microbiol 17:205–211

Lau HY, Botella JR (2017) Advanced DNA-based point of care diagnostic methods for plant diseases detection. Plant Sci 8:2016

Lizardi PM, Huang XH, Zhu ZR, Bray-Ward P, Thomas DC, Ward DC (1998) Mutation detection and single-molecule counting using isothermal rolling-circle amplification. Nat Genet 19:225–232

Schubert J, Habekuss A, Kazmaier K, Jaske H (2007) Surveying cereal infecting geminiviruses in Germany-diagnostics and direct sequencing using rolling circle amplification. Virus Res 127:61–70

Wambulwa MC, Wachira FN, Karanja LS, Muturi SM (2012) Rolling circle amplification is more sensitive than PCR and serology-based methods in detection of *Banana streak virus* in *Musa* germplasm. Am J Plant Sci 3:1581–1587

Chapter 40

Recombinase Polymerase Amplification

Abstract

Recombinase polymerase amplification (RPA) assay is a single tube, amplification method that is carried out at a constant single temperature used for detection of DNA. It can also detect RNA by addition of reverse transcriptase enzyme to the RPA reaction without any additional step to produce cDNA. The amplification process is very quick and can be completed in 10–15 min. Because it is performed at a constant temperature, it requires just an equipment to maintain the set temperature like an incubator. In most cases, the optimum temperature for performing RPA lies between 37 and 42 °C. Thus RPA is a low-cost, rapid, point-of-care diagnostic assay. It is being successfully used for the detection of many plant viruses.

Key words Isothermal amplification, Virus detection

40.1 Introduction

Recombinase polymerase amplification (RPA) is a fast amplification method that is carried out at a constant temperature. RPA was developed and launched by TwistDx Ltd. based in Cambridge, UK. It has high specificity and sensitivity and does not require denaturing of the template DNA by heating. Denaturing is done by an enzymatic activity that will help in annealing of primers to the complementary sequences in the target region (Yan et al. 2014; Londono et al. 2016; Lobato and O'Sullivan 2018). In RPA, amplification of the target region of the template is done by the use of enzymes and proteins that includes the recombinase, single-stranded DNA binding proteins (SSB) and strand displacing polymerase. The primer pair forms the primer-recombinase complex which will facilitate denaturing of strands. The SSB stabilizes the denatured strand allowing primers to anneal to the specific sequences in the template (Fig. 40.1) (Zhang et al. 2014). After the annealing of primers, the recombinase dissociates from the primers and allows the DNA polymerase to initiate the amplification. The ideal temperature for RPA reaction is between 37 and 42 °C for a maximum duration of 60 min although in most cases a duration of just 10–15 min is enough (Piepenburg et al. 2006; Yan

Alangar Ishwara Bhat and Govind Pratap Rao, *Characterization of Plant Viruses*, Springer Protocols Handbooks,
https://doi.org/10.1007/978-1-0716-0334-5_40,

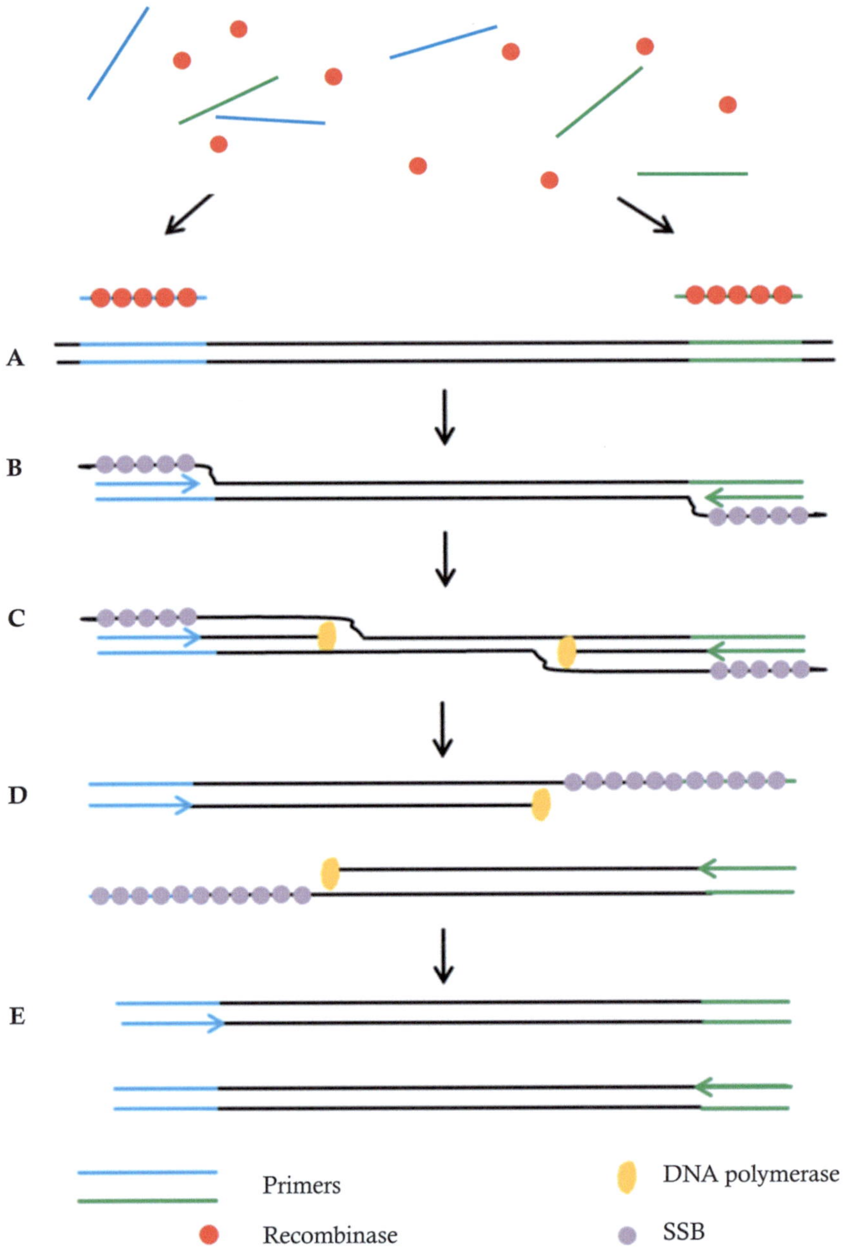

Fig. 40.1 Schematic diagram of the recombinase polymerase amplification (RPA). (**a**) Formation of recombinase-primer complexes. (**b**) Denaturing of template DNA strands and binding of single-stranded binding proteins (SSB) to the DNA to form a D-loop. (**c**) Initiation of DNA amplification. (**d**) Continuation of amplification. (**e**) Newly synthesized DNA will serve as template for the subsequent amplification cycles. (Reproduced from Lau and Botella 2017)

et al. 2014). After the run, reaction product can be analysed on a 1.5–2% agarose gel to visualize the result (Fig. 40.2).

RPA assay has been used with high sensitivity and specificity for the detection of many viruses infecting different plant species

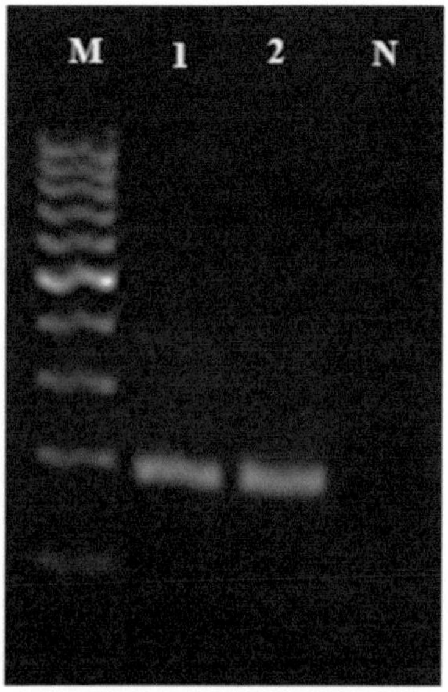

Fig. 40.2 Analysis of RPA amplicons generated from virus-specific primers on agarose gel electrophoresis. Lane M loaded with 100 bp ladder; Lanes 1–2: RPA product from infected plant samples; Lane N: known healthy plant

(Kapoor et al. 2017). The short reaction time (15–30 min) and low incubation temperature make RPA a suitable assay for quick and sensitive detection of viruses. Another advantage of RPA is in designing of primers where annealing temperature is not taken into account as primer forms a complex with the recombinase to target the homologous sequences. RPA is one of the most sensitive method with detection limit of 6.25 fg of template DNA input with a specificity >95% (Boyle et al. 2014). Like LAMP, RPA can be used for amplification of only small DNA fragments of <500 bp in length and not suitable for amplification of longer amplicons. The longer primers (30–35 bp) used in RPA occasionally produce non-specific amplification in healthy samples used as control owing to the formation of primer dimers that may affect sensitivity of the assay (Boyle et al. 2014; Crannell et al. 2014; Kersting et al. 2014).

40.2 Materials

1. Incubator.
2. Primers (*see* **Note 4** for details).
3. Rehydration buffer.
4. Magnesium acetate.
5. Reaction pellet: It consists of recombinase, single-stranded binding protein and polymerase.

40.3 Methods (Based on Protocol of TwistDx, Cambridge, UK)

1. Isolate DNA from 100 mg of symptomatic and asymptomatic samples as discussed earlier in Chapter 31.
2. The DNA is quantified and stored at −20 °C until further use.
3. Set up RPA reaction as follows.

Forward primer (10 μM)	2.4 μL
Reverse primer (10 μM)	2.4 μL
Rehydration buffer	29.5 μL
Template DNA and H_2O	13.2 μL
Total volume	47.5 μL

4. Mix contents, vortex and spin briefly.
5. Transfer the entire contents of the tube to the reaction pellet provided in the kit. Mix through pipette till the entire pellet is resuspended.
6. Add 2.5 μL of 280 mM magnesium acetate and mix thoroughly. Incubate the reaction at 37 °C for 30 min in an incubator.
7. After 4 min, get the sample away from the incubator, mix thoroughly, spin down and return the samples to the incubator block. Repeat this process at every 4 min (The time interval for agitation may be optimized by testing different time interval from 3 to 6 min to get better amplification).
8. Continue the incubation for a total incubation time of 20–40 min.
9. After the incubation time, to visualize RPA amplicons, incubate about 10 μL of the product at 65 °C for 10 min prior to loading on 1.8% agarose gel (Fig. 40.2). Alternatively RPA product can also be visualized after mixing with 5% SDS followed by agarose gel electrophoresis.
10. If using for some downstream applications, purify the amplification product using commercial PCR purification kit. Alternatively, the amplification product can be diluted in water and then phenol/chloroform extracted according to standard molecular biology protocols.

40.4 Notes

1. RPA also works well with crude sap prepared from fresh or frozen leaf samples in 0.5 M NaOH (Londono et al. 2016).

2. Periodic agitation of the reaction components during the incubation period is very important. It is done by manual mixing 4 min after initiation of the reaction.
3. For RPA assay optimization, reaction temperature (37–42 °C), the magnesium acetate concentration (12–20 mM), the agitation regime (3–6 min) and the primer concentrations (150 and 600 nM) need to be done (Boyle et al. 2014).
4. RPA primers should be about 30–35 nucleotides long. While designing primers avoid lengthy tracks of 'G' (first 5 nucleotides) at the 5′end, although 'C' (and pyrimidines) are favourable. Presence of 'G' and 'C' (last three nucleotides) at the 3′end of the primer is beneficial. Avoid lengthy tracks of one particular nucleotide or a too many small repeats in the primer. Keep GC contents of the primers between 30 and 70% (Yan et al. 2014).
5. The ideal amplicon length is between 100 and 200 bp and it should not exceed 500 bp.
6. It is important to designate separate work area and pipette for pre- and post-amplification steps to prevent contamination.
7. Among isothermal amplification methods, RPA is considered as superior as it requires only two primers (as opposed to LAMP which requires 4–6 primers), lower incubation temperature of 37–42 °C (as opposed to LAMP which requires 60–65 °C) and lesser, 10–15 min of run time (as opposed to LAMP which requires 45–60 min).

References

Boyle DS, Mcnerney R, Low HT, Leader BT, Perez-Osorio AC, Meyer JC (2014) Rapid detection of *Mycobacterium tuberculosis* by recombinase polymerase amplification. PLoS One 9:e103091

Crannell ZA, Rohrman B, Richards-Kortum R (2014) Quantification of HIV-1 DNA using real-time recombinase polymerase amplification. Anal Chem 86:5615–5619

Kapoor R, Srivasatava N, Kumar S, Saritha RK, Sharma SK, Jain RK, Branwal VK (2017) Development of a recombinant polymerase amplification assay for the diagnosis of banana bunchy top virus in different banana cultivars. Arch Virol 162:2791–2796

Kersting S, Rausch V, Bier FF, Von Nickisch-Rosenegk M (2014) Rapid detection of *Plasmodium falciparum* with isothermal recombinase polymerase amplification and lateral flow analysis. Malar J 13:99

Lau HY, Botella JR (2017) Advanced DNA based point of care diagnostic methods for plant disease detection. Plant Sci 8:2016

Lobato IM, O'Sullivan CK (2018) Recombinase polymerase amplification: basics, applications and recent advances. Trends Anal Chem 98:19e35

Londono MA, Harmon CL, Polston JE (2016) Evaluation of recombinase polymerase amplification for detection of begomoviruses by plant diagnostic clinics. Virol J 13:48

Piepenburg O, Williams CH, Stemple DL, Armes NA (2006) DNA detection using recombination proteins. PLoS Biol 4:e204

Yan L, Zhou J, Zheng Y, Gamson AS, Roembke BT, Nakayama S, Sintim HO (2014) Isothermal amplified detection of DNA and RNA. Mol Biosyst 10:970–1003

Zhang S, Ravelonandro M, Russell P, McOwen N, Briard P, Bohannon S, Vrient A (2014) Rapid diagnostic detection of plum pox virus in *Prunus* plants by isothermal Amplify RP using reverse transcription-recombinase polymerase amplification. J Virol Methods 207:114–120

Chapter 41

Next-Generation Sequencing for Diagnosis of Viruses

Abstract

Next-generation sequencing (NGS) is an unbiased approach to plant viral disease diagnosis that requires no prior knowledge of the host or the pathogen. The method involves extraction of total DNA or RNA or small RNA from the infected plant, the production of cDNA and, finally, sequencing, which produce sequences from a large range of potential pathogens. Analysis of the sequence data, development of contigs, and blasting contigs sequence against plant virus database aids in the identification of potential viruses. The results of the NGS can then be verified by designing virus-specific primers (based on the sequence of the contigs that hit viruses) and subjecting them to RT-PCR using total RNA from the infected plant, cloning and sequencing of the RT-PCR product by Sanger sequencing method. Once verified, more routine assays based on nucleic acid such as RT-PCR, real-time RT-PCR, loop-mediated isothermal amplification and recombinase polymerase amplification can be developed for quick and reliable detection of the virus in plants. The sequence data analysis is mainly based on bioinformatics software that can distinguish between host plant and viral sequences, and provides new opportunities in the areas of virus diagnosis and viral ecology.

Key words Second-generation sequencing, Third-generation sequencing, Virus identification, Small RNA sequencing, RNA sequencing, dsRNA sequencing, Sequencing by synthesis, Contigs, Sequence data analysis

41.1 Introduction

Next-generation sequencing (NGS) consists of numerous sequencing reactions which take place simultaneously, resulting in generation of large amounts of sequencing data. All NGS technologies regulate the sequential addition of nucleotides to immobilized and spatially arrayed DNA templates. Different sequencing technology involves sequencing by synthesis (Illumina), sequencing by ligation (SOLID, Applied Biosystems), single-molecule real-time sequencing (Pacific Biosciences), pyrosequencing (Roche 454) and ion semiconductor sequencing (Ion Torrent, Applied Biosystems) (Nyren 2007; Egan et al. 2012; Vives et al. 2013; Pecman et al. 2017).

Alangar Ishwara Bhat and Govind Pratap Rao, *Characterization of Plant Viruses*, Springer Protocols Handbooks, https://doi.org/10.1007/978-1-0716-0334-5_41,

41.2 Chemistry Used by Different Platforms

NGS techniques are categorized into two types: sequencing by synthesis or by ligation. In synthesis-based sequencing, the complementary DNA strand is synthesized by DNA polymerase and the fluorescent signal is detected resulting from the nucleotide incorporation to identify the sequence. The major synthesis-based sequencing platforms are Illumina, Roche 454 pyrosequencing and Ion Torrent (Egan et al. 2012).

Roche 454 pyrosequencing is based on adhesion of single or primed DNA template to a microbead and amplification using emulsion PCR is done that each bead is individually placed within a well, incubated with DNA polymerase, luciferase, substrates luciferin, ATP sulfurylase and adenosine 5′-phosphosulfate (ASP). Nucleotide is then incorporated into the DNA strand which produces ATP required for converting luciferin to oxyluciferin and reflected by emitting light. After the first round of nucleotide addition, all nucleotides and ATP are washed and the process is repeated several times until the DNA template is elongated to the desired length (Nyren 2007).

In the Illumina-based system, solid-phase bridge amplification is used where each end of a DNA template is ligated with adapters (Guo et al. 2008). The one end of adapter-conjugated DNA fragment is attached to a substrate; however, the other end generates clusters of identical template in order to enhance the flourescent signal. This process continues in the presence of a mixture of four nucleotides in a continuous cycle (each nucleotide is labelled with a different fluorophore) followed by image capture. This cycle is repeated until the DNA fragment is synthesized to its target length.

In the Ion Torrent system, the incorporated nucleotide is detected by measuring pH change due to the release of H+ ions in the sequential addition of nucleotides (Rothberg et al. 2011). The Ion Torrent system is economical and appealing option for genome sequencing of known and unknown viruses.

NGS technologies have a great potential in the diagnosis of several pathogens in a single analysis without any previous knowledge (Massart et al. 2017; Withers et al. 2016; Rott et al. 2017). NGS equipment is quite sophisticated and the sample preparation for analysis is quite complex. The NGS sequencing platforms are too expensive for routine diagnostic analyses. NGS technologies allow a generic approach that does not require any prior knowledge on the targeted pathogens but can deliver identification of a species/strain specific result (Quan et al. 2008; Adams et al. 2009, 2012; AI Rwahnih et al. 2009, 2011, 2012; Pecman et al. 2017). NGS was first utilized for plant virus detection in 2009 (Adams et al. 2009; AI Rwahnih et al. 2009; Kreuze et al. 2009) and since then, several developments have occurred both in the sample

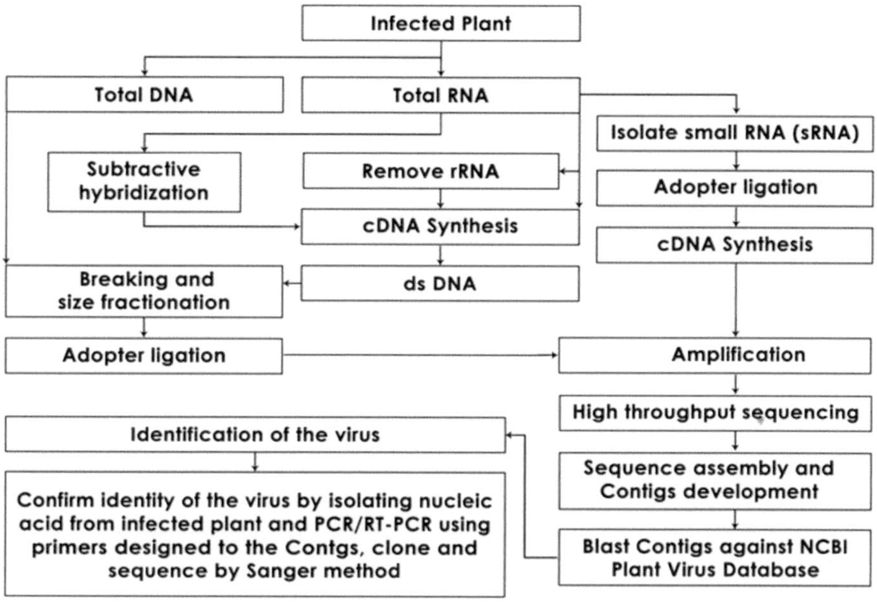

Fig. 41.1 Next-generation sequencing strategies used for the identification of viruses infecting plants

preparation and analysis of data. Researchers have used various kinds of nucleic acid inputs such as total RNA (tot RNA), ribosomal RNA depleted total RNA (rRNA depleted totRNA); double-stranded RNA (dsRNA), virus-derived small interfering RNA (sRNA) and RNA after subtractive hybridization with healthy plant RNA for the identification of unknown viruses infecting plants (Fig. 41.1) (Coetzee et al. 2010; Monger et al. 2010; Pallett et al. 2010; Yan et al. 2010; Silva et al. 2011; Zhang et al. 2011; Loconsole et al. 2012; Hany et al. 2014).

41.3 General Workflow of NGS

1. The starting material for all NGS experiments is double-stranded DNA; however, the source may vary, e.g. total DNA if suspected virus has DNA genome and reverse-transcribed RNA or cDNA if suspected virus has RNA genome (Fig. 41.1).
2. Library preparation (sequencing library): Fragmentation and size selection to break the DNA templates into smaller sequenceable fragments and addition of adapters, which serve as primers for downstream amplification and/or sequencing reactions.
3. Immobilization of libraries on beads/solid surface.
4. Depending on the NGS technology used, a library is either amplified then sequenced (clonally amplified templates) or

sequenced directly (single-molecule templates, this technology is called next-next generation or third-generation sequencing).

5. Clonal amplification: Amplification of immobilized libraries is required to produce signal for detection of nucleotide addition by the system. Amplification strategies may vary between platforms for example emulsion PCR (followed by Roche 454 platform) or bridging amplification method (followed by Illumina platform).

   Sequencing reaction: Each sequencing platform employs a series of chemical reactions in repeated cycles which are carried out and detected automatically. Reactions typically use a flow cell that consists of the immobilized templates and allows standardized addition and detection of nucleotides and washing/removal of reagents, and repetition of this cyclic process.

6. Data analysis: The read length (the actual number of continuous sequenced bases) for NGS is much shorter (50–500 continuous base pair reads), in comparison to Sanger sequencing. These short reads are a major limitation of this technology, but have the potential to read several continuous kilo base pairs (kb). NGS technology produces short reads for which, the genome coverage is a very important issue. Coverage is defined as the number of short reads of genome that overlap each other within a specific region. For example, a 50-fold coverage for coat protein gene means that every nucleotide is represented at least 50 distinct and overlapping short reads. Sufficient genome coverage is essential for accurate genomic sequence assembly. With the advent of third-generation sequencing, that mainly uses single-molecule sequencing approach, most of the above problems would be overcome as they provide accurate sequencing data with potentially long-reads.

41.4 Application in Plant Virology

NGS technologies have been mainly used for the identification of causal viruses associated with diseases of unknown aetiology affecting diverse plant species. Re-assembling significant portions of the virus sequence from overlapping siRNA sequences is another successful method to identify the viruses through massive sequencing of siRNAs called as small RNA (sRNA) sequencing. Vector-enabled metagenomics using NGS is another approach to study the diversity of viruses transmitted through vectors. NGS provides insights on the important role of host RNA dependent RNA polymerases in viral siRNA biogenesis and viral siRNA-mediated virus–host interactions in viral pathogenicity and host specificity.

Briefly, the method involved in the identification of virus through NGS includes: isolation of total nucleic acid or total double-stranded RNA (dsRNA) from virus-infected tissue and subjecting to NGS (Fig. 41.1). However, in other cases, nucleic acid of the host was partially eliminated by hybridization to enrich virus sequences in the infected plant material prior to subjecting them for sequencing through NGS. Use of nucleic acid isolated from partially purified virus preparation is another strategy to enrich virus sequences. Another NGS-based method for the identification of viruses is based on the sequencing of small interfering RNA (siRNA), which is also referred as small RNA sequencing (sRNA) isolated from diseased plant (Kreuze et al. 2009) (Fig. 41.1). In response to infection by RNA/DNA viruses, the host plant generates specific RNA molecules of 21–24 nucleotide length, called short interfering RNAs (siRNAs). RNA silencing (RNAi) is a cytoplasmic cell surveillance system to recognize dsRNA and specifically destroy ss and ds RNA molecules homologous to the inducer, using small interfering guide RNAs. NGS of siRNAs offers good opportunities to identify plant viruses at extremely low titers, and in symptomless host carriers, including previously unknown viruses. NGS can provide thousands to millions of siRNA sequences from virus-infected plant material. When virus-derived siRNAs are 21–24 nt in length, their sequences can be assembled.

Use of NGS technologies revealed that some diseases of unknown aetiology are caused by either new or known viruses. sRNA sequencing of diseased sweet potato showed the presence of novel viruses such as two new badnaviruses (named as sweet potato badnavirus A and sweet potato badnavirus B) and a mastrevirus (Kreuze et al. 2009). Similarly, total RNA isolated after subtractive hybridization when subjected to NGS revealed the occurrence of an unknown virus in *Gomphrena globasa* for which the name gayfeather mild mottle virus was proposed (Adams et al. 2009); complete sequence of Tanzanian isolate of cassava brown streak virusin cassava (Monger et al. 2010) and maize chlorotic mottle virus and sugarcane mosaic virus in maize (Adams et al. 2012) were obtained through NGS. The sRNA sequencing revealed the occurrence of cereal yellow dwarf virus infecting wild cocksfoot grass (Pallett et al. 2010), rice stripe virus in rice (Yan et al. 2010), gayfeather mild mottle virus in *Liatris spicata* and cotton leafroll dwarf virus in cotton (Silva et al. 2011). Complete genome sequence of piper yellow mottle virus was determined using viral and plant DNA isolated from virus-enriched fraction in black pepper (Hany et al. 2014). sRNA sequencing showed the occurrence of a new *Begomovirus* (citrus chlorotic dwarf-associated virus) (Loconsole et al. 2012), a *Cilevirus* (citrus leprosies virus cytoplasmic type 2) (Roy et al. 2018) and a *Luteovirus* (citrus vein enation virus) (Vives et al. 2013) in citrus. Using either total RNA or dsRNA or sRNA sequencing approaches revealed the association of many new viruses with grapevine such as *Marafivirus* (grapevine Syrah 1 virus) (AI Rwahnih et al. 2009), grapevine virus E, grapevine leaf roll-associated virus-3, grapevine rupestris pitting-associated virus, grapevine virus A (Coetzee et al. 2010), grapevine vein clearing virus (Zhang et al. 2011), 26 mycoviruses (AI Rwahnih et al. 2011) and grapevine red blotch-associated virus (AI Rwahnih et al. 2012).

References

Adams I, Glover RH, Monger WA, Mumford R, Jackeviciene E, Navalinskiene M, Samuitiene M, Boonham N (2009) Next-generation sequencing and metagenomic analysis: a universal diagnostic tool in plant virology. Mol Plant Pathol 10:537–545

Adams P, Miano DW, Kinyua ZM, Wangai A, Kimani E, Phiri N, Reeder R, Harju V, Glover R, Hany U, Souza-Richards R, Nath PD, Nixon T, Fox A, Barnes A, Smith J, Skelton A, Thwaites R, Mumford R, Boonham N (2012) Use of next generation sequencing for the identification and characterization of *Maize chlorotic mottle virus* and *Sugarcane mosaic virus* causing lethal necrosis in Kenya. Plant Pathol 62:741–749

AI Rwahnih M, Daubert S, Golino D, Rowhani A (2009) Deep sequencing analysis of RNAs from a grapevine showing Syrah decline symptoms reveals a multiple virus infection that includes a novel virus. Virology 387:395–401

AI Rwahnih M, Daubert S, U'rbez-Torres JR, Cordero F, Rowhani A (2011) Deep sequencing evidence from single grapevine plants reveals a virome dominated by mycoviruses. Arch Virol 156:397–403

AI Rwahnih M, Dav A, Anderson M, Uyemoto JK, Sudarshana MR (2012) Association of a circular DNA virus in grapevine affected by red blotch disease in California. In: Proceedings of the 17th Congress of ICVG, Davis, CA, USA, 7-- 14 October 2012

Coetzee B, Freeborough MJ, Maree HJ, Celton JM, Rees DJ, Burger JT (2010) Deep sequencing analysis of viruses infecting grapevines: virome of a vineyard. Virology 400:157–163

Egan AN, Schlueter J, Spooner DM (2012) Applications of next-generation sequencing in plant biology. Am J Bot 99(2):175–185

Guo J, Xu N, Li Z, Zhang S, Wu J, Kim DH, Sano Marma M, Meng Q, Cao H, Li X (2008) Four-color DNA sequencing with 30-O-modifiednucleotide reversible terminators and chemically cleavable fluorescent dideoxy nucleotides. Proc Natl Acad Sci U S A 105:9145–9150

Hany U, Adams IP, Glover R, Bhat AI, Boonham N (2014) The complete genome sequence of *Piper yellow mottle virus* (PYMoV). Arch Virol 159:385–388

Kreuze JF, Perez A, Untiveros M, Quispe D, Fuentes S, Barker I, Simon R (2009) Complete viral genome sequence and discovery of novel viruses by deep sequencing of small RNAs: a generic method for diagnosis, discovery and sequencing of viruses. Virology 388:1–7

Loconsole G, Saldarelli P, Doddapaneni H, Savino V, Martelli GP, Saponari M (2012) Identification of a single-stranded DNA virus associated with citrus chlorotic dwarf disease, a new member of the family Geminiviridae. Virology 432:162–172

Massart S, Candresse T, Gil J, Lacomme C, Predajna L, Ravnikar M (2017) A framework for the evaluation of biosecurity, commercial, regulatory, and scientific impacts of plant viruses and viroids identified by NGS technologies. Front Microbiol 8:45

Monger WA, Alicai T, Ndunguru J, Kinyua ZM, Potts M, Reeder RH, Miano DW, Adams IP, Boonham N, Glover RH et al (2010) The complete genome sequence of the Tanzanian strain of *Cassava brown streak virus* and comparison with the Ugandan strain sequence. Arch Virol 155:429–433

Pallett DW, Ho T, Cooper I, Wang H (2010) Detection of *Cereal yellow dwarf virus* using small interfering RNAs and enhanced infection rate with *Cocksfoot streak virus* in wild cocksfoot grass (*Dactylis glomerata*). J Virol Methods 168:223–227

Nyren P (2007) The history of pyrosequencing. Methods Mol Biol 373:1–14

Pecman A, Kutnjak D, Gutiérrez-Aguirre I, Adams I, Fox A, Boonham N, Ravnikar M (2017) Next generation sequencing for detection and discovery of plant viruses and viroids: comparison of two approaches. Front Microbiol 8:1998

Roy S, Coldren C, Karunamurthy A, Kip NS, Klee EW, Lincoln SE, Leon A, Pullambhatla M,

Temple-Smolkin RL, Voelkerding KV, Wang C, Carter AB (2018) Standards and guidelines for validating next-generation sequencing bioinformatics pipelines: a joint recommendation of the Association for Molecular Pathology and the College of American Pathologists. J Mol Diagn 20(1):4–27

Quan P, Briese T, Palacios G, Lipkin WI (2008) Rapid sequence-based diagnosis of viral infection. Antivir Res 79:1–5

Rothberg JM, Hinz W, Rearick TM et al (2011) An integrated semiconductor device enabling non-optical genome sequencing. Nature 475:348–352

Rott M, Xiang Y, Boyes I, Belton M, Saeed H, Kesanakurti P (2017) Application of next generation sequencing for diagnostic testing of treefruit viruses and viroids. Plant Dis 101:1489–1499

Silva TF, Romanel EAC, Andrade RRS, Farinelli L, Osteras M, Deluen C, Correa RL, Schrago CEG, Vaslin MFS (2011) Profile of small interfering RNAs from cotton plants infected with the polerovirus, *Cotton leafroll dwarf virus*. BMC Mol Biol 12:40

Vives MC, Velazquez K, Pina JA, Moreno P, Guerri J, Navarro L (2013) Identification of a new enamovirus associated with citrus vein enation disease by deep sequencing of small RNAs. Phytopathology 103:1077–1086

Withers S, Gongora-Castillo E, Gent D, Thomas A, Ojiambo PS, Quesada-Ocampo LM (2016) Using next-generation sequencing to develop molecular diagnostics for *Pseudoperonospora cubensis*, the cucurbit downy mildew pathogen. Phytopathology 106:1105–1116

Yan F, Zhang H, Adams M, Yang J, Peng J, Antoniw J, Zhou Y, Chen J (2010) Characterization of siRNAs derived from *Rice stripe virus* in infected rice plants by deep sequencing. Arch Virol 155:935–940

Zhang Y, Singh K, Kaur R, Qiu W (2011) Association of a novel DNA virus with the grapevine vein-clearing and decline syndrome. Phytopathology 101:1081–1090

Chapter 42

Cloning of PCR Product

Abstract

Cloning refers to the production of identical copies of a DNA fragment. The process involves ligating the DNA fragment of interest into a vector (usually a plasmid) molecule. After ligation, the recombinant vector that harbours the DNA fragment is introduced into cells usually a bacterium such as *Escherichia coli*. There are different methods available through which ligated plasmid can be inserted into *E. coli*. The bacterial cells are then cultured and the recombinant bacteria (with vector along with the DNA fragment) are selected by a number of methods that usually include blue/white screening on X-gal medium and antibiotics. Further tests such as PCR, restriction analysis and DNA sequencing are required to confirm that cloning was successful.

Key words Cloning viral genome, Plasmid vector, Ligation, Transformation, Recombinant clone, Colony PCR, Plasmid isolation, Restriction analysis

42.1 Introduction

The validity of the amplification reaction can be assessed by cloning the product followed by sequencing. Cloning is a process where a DNA fragment is inserted into a vector (usually a plasmid). The recombinant vector is then introduced into a suitable bacterium through a process referred as transformation. Transformed bacteria are then selected using appropriate markers present on the vector. The essential steps of DNA cloning involve: (1) production of target DNA fragment, (2) choice of plasmid vector to be used for cloning, (3) ligating DNA fragment with the plasmid vector, (4) preparation of competent *E. coli* cells, (5) introduction of recombinant vector into the susceptible bacterium, (6) selection and identification of recombinant clones. A simple flow chart of cloning steps is given in Fig. 42.1.

There are three different methods by which a PCR product can be cloned. They include: (1) TA cloning: PCR carried out using *Taq* polymerase adds a single 'A' overhang at 3′ ends of PCR product. Hence this product can be ligated to a plasmid vector with a single 'T' overhang at 5′ ends. (2) Blunt end cloning: PCR performed with *Pwo* or *Pfu* polymerase produces amplicons with blunt ends that can be cloned into a plasmid restricted with blunt

Alangar Ishwara Bhat and Govind Pratap Rao, *Characterization of Plant Viruses*, Springer Protocols Handbooks, https://doi.org/10.1007/978-1-0716-0334-5_42,

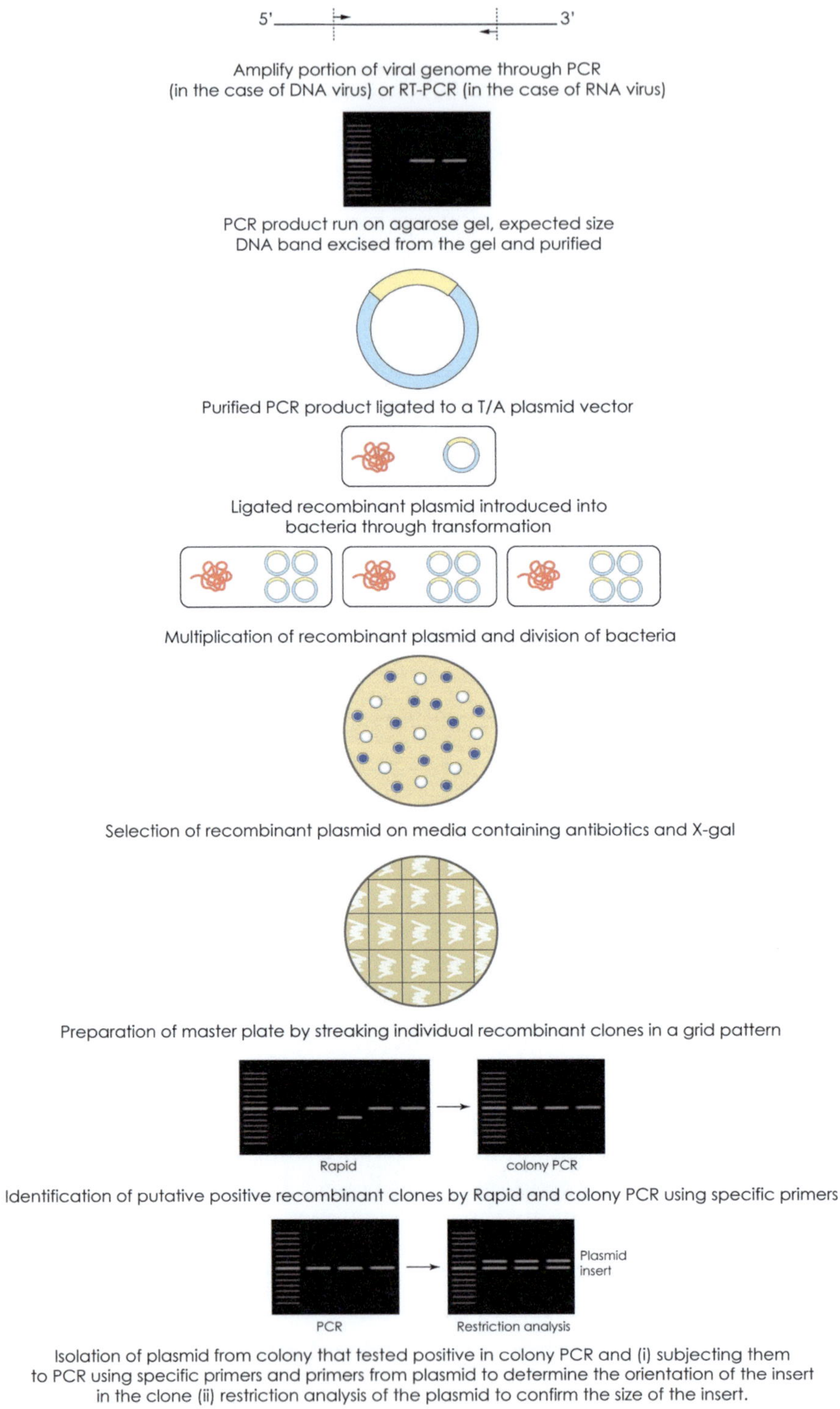

Fig. 42.1 Steps in PCR product cloning

end generating enzymes. (3) Directional cloning: PCR performed with primers containing restriction sites can be used for directional cloning in a plasmid vector. In such cases, PCR product is first restricted with appropriate restriction enzymes followed by ligating it to a plasmid restricted by same restriction enzymes.

In order to clone, PCR products are run on a low melting point agarose gel, the expected size of DNA band is excised from the gel and purified through phenol-chloroform method. Alternatively, PCR product can also be purified after running on normal high melting agarose gel; the band of interest containing target PCR product is excised and purified using commercial kits. The purified PCR product is quantified and ligated to a linearized dT or blunt end or cohesive end vector using *E. coli* T_4 DNA ligase. If planning for directional cloning, then restriction of PCR product is carried out with appropriate restriction enzymes before ligation with vector. The ligated vector is then introduced into competent *E. coli* cells and transformed colonies are selected using appropriate markers such as antibiotic or blue/white colony morphology (Davis et al. 1986; Ausubel et al. 1987; Sambrook et al. 1989; Sambrook and Russel 2001). Here we have described the protocol for PCR fragment cloning using TA vector.

42.2 Isolation of Target DNA by Purification of PCR Product Through Low Melting Point (LMP) Agarose Gel

The isolation of target DNA is done normally through PCR/RT-PCR.

42.2.1 Materials

1. Gel documentation unit.
2. Horizontal gel apparatus and power pack.
3. Microcentrifuge.
4. Razor blade.
5. Water bath.
6. Sodium acetate (3 M, pH 5.2): Add and dissolve 24.6 g sodium acetate in about 70 mL of distilled water, adjust the pH to 5.2 by adding glacial acetic acid and make up the volume to 100 mL by adding water.
7. Ethanol.
8. Ethanol (70%): Add 70 mL of ethanol and 30 mL of sterile distilled water, mix well and store at 4 °C.
9. Tris-saturated phenol: Melt liquified phenol stored at −20 °C at 65 °C and add hydroxyquinone to a final concentration of 0.1%. To this add equal volume of buffer (0.5 M Tris–HCl, pH 8.0) at room temperature, mix well for 15 min and allow standing for separation of phases. Aspirate upper aqueous

phase using a glass pipette (avoid mouth sucking) and discard. Add an equal volume of 0.1 M Tris–HCl pH 8.0 to the phenol, mix well for 15 min, turn off the stirrer and remove the upper aqueous phase. Repeat extraction until the pH of phenolic phase is more than 7.8. After final extractions, add 0.1 volume of 0.1 M Tris–HCl (pH 8.0) containing 0.2% β mercaptoethanol. Store the solution in an amber coloured bottle at 4 °C. Alternatively, commercially available Tris-saturated phenol can also be used.

10. Chloroform.
11. Ethidium bromide: Dissolve 1 g of ethidium bromide in water and make up the volume to 100 mL, store in dark or amber coloured bottle at room temperature. As ethidium bromide is a potent mutagen, recently safe alternatives for this (ethidium bromide safe stain) from different firms are available. These stains have same sensitivity like ethidium bromide and DNA in the gels can be viewed and documented using same gel documentation unit.
12. Low melting point (LMP) agarose: commercially available.
13. PCR product.
14. Phenol: chloroform (24:1): To 24 parts of phenol add 1 part of chloroform, mix well and use.
15. Sterile water.
16. TAE buffer: Prepare 10× TAE by dissolving Tris base (48.4 g), glacial acetic acid (11.42 mL), 0.5 M EDTA (pH 8.0) (20 mL) and water to make 1000 mL. Dilute 1: 10 with water to get 1× TAE.

42.2.2 Method

1. Prepare 0.8% LMP agarose in 1× TAE buffer and cool it to room temperature.
2. Cast the gel on a pre-cooled gel set and allow it to solidify at 4 °C.
3. Load PCR product with loading dye and run at 100 V for 60–90 min.
4. Examine the gel under UV transilluminator to locate the DNA band of interest.
5. Excise the band using a clean sterile razor blade or scalpel.
6. Trim off any excess agarose and transfer to a 1.5 mL sterile eppendorf tube.
7. Incubate the tube at 70 °C in a heating block for 7 min.
8. Place the tube at room temperature; add equal volume of buffer (Tris–HCl, pH 8.0) and saturated phenol and vortex for 30 s.

9. Centrifuge the tube at 7500 × *g* for 2 min and collect aqueous phase.
10. Re-extract the aqueous phase once with equal volume of phenol: chloroform (24:1) and then twice with chloroform alone.
11. To the aqueous phase add 0.1 volume of 3 M chilled sodium acetate and 2 volumes of ethanol and incubate at −80 °C for 1 h.
12. Centrifuge at 11,000 × *g* for 15 min at 4 °C to palletize the DNA.
13. Pour off the supernatant carefully and 1 mL of 70% cold ethanol to the pellet and re-centrifuge for 5 min.
14. Pour off the supernatant carefully, air dry the pellet and resuspend the pellet in a small volume (about 10–20 μL) of sterile water and store at −20 °C until ready to use.

42.3 Ligation of the Purified PCR Product into Vector

Plasmid vectors are generally used for cloning of PCR products (Fig. 42.2). Plasmids are small, circular, episomal, covalently closed, double-stranded DNA molecules that have the ability to replicate within a susceptible bacterium. Plasmids carry one or two antibiotic resistance genes that will be useful to select recombinant plasmids. The enzyme, DNA ligase helps to ligate two DNA molecules. It joins 5′ phosphate group to 3′hydroxyl group of DNA through the formation of phosphodiester bonds. T_4 DNA ligase and *E. coli* DNA ligases are the commercially available ligase and used for ligation process (Sambrook et al. 1989; Sambrook and Russel 2001).

42.3.1 Materials

1. Ice flakes.
2. Linearized TA vector (Commercially available).
3. T_4 DNA ligase buffer (10×).
4. T_4 DNA ligase.
5. Purified PCR product.

42.3.2 Method

In an eppendorf tube, add the following components:

- Linearized TA vector (20 ng/μL): 1.0 μL*.
- Purified PCR product (10 ng/μL): 7.0 μL*.
- 10× T_4 DNA ligase buffer: 1.0 μL.
- T4 DNA ligase (3 Weiss units/μL): 1.0 μL.

*A ratio of 1:4 (vector: insert) molecules should be maintained for successful cloning. The amount of DNA fragment to be used for

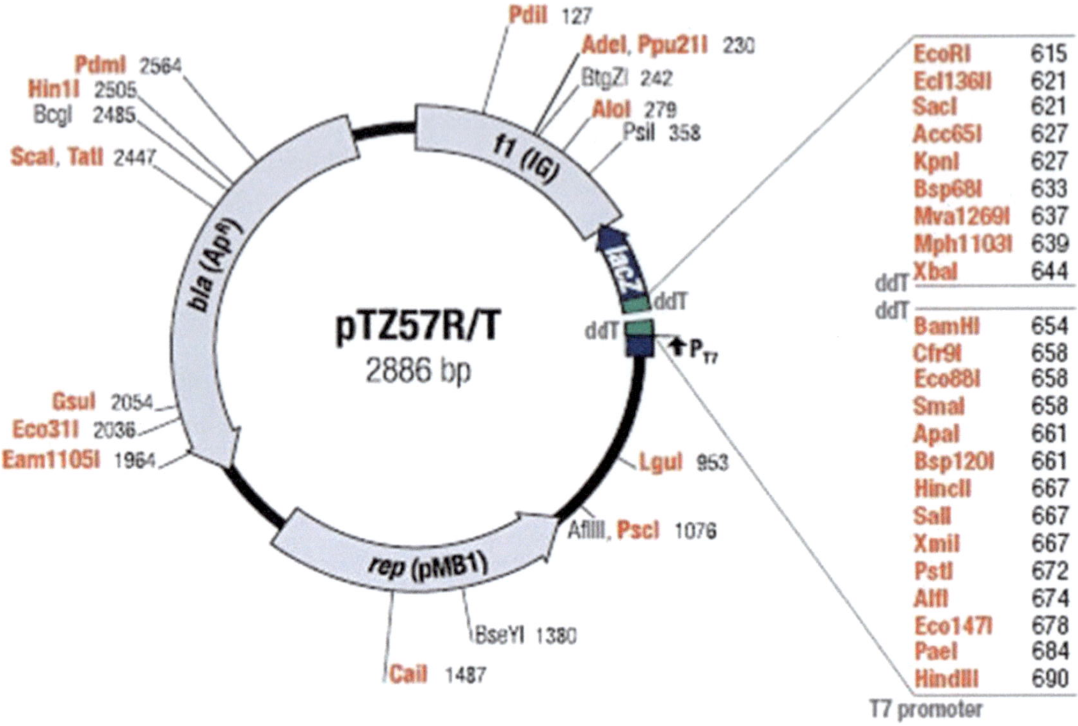

Fig. 42.2 Example of a TA plasmid vector used for cloning PCR product

Table 42.1
Conversion table for the amount of a PCR fragment required per ligation reaction

Length of DNA fragment (bp)	p moles of ends per 1 μg of DNA	Quantity of PCR fragments for ligation reaction in μg (0.54 pmol ends)
100	30	0.018
300	10	0.054
500	3.0	0.09
1000	3.0	0.18
2000	1.5	0.36
3000	1.0	0.54

ligation may be calculated using the following Table 42.1 (Source www.Thermofischer.com).

Mix the reactants by pipetting and incubate at 4 °C for overnight or 16 °C for 20 min to 1 h. Ligated plasmid will be ready for transformation of *E. coli* after the incubation time. If planning transformation at later date, store the ligation mix at −20 °C until you are ready for transformation.

42.3.3 Notes

1. Commercial kits available from different manufacturers may be used for gel purification of PCR products. Most of these kits can purify DNA from normal agarose (LMP agarose gel is not needed).
2. Commercial TA cloning kits are also available from different manufacturers. Such kits provide linearized 'T' vector, buffers and ligase needed for cloning.
3. Terminal transferase and dideoxy TTP are used to add a single protruding residue to the 3′ termini of a linearized vector. Alternatively *Taq* DNA polymerase may also be used to catalyse the addition of 'T' residues to the terminal 3′ hydroxyl groups of a linearized vector (Sambrook and Russel 2001).
4. In order to have directional cloning, restriction site sequences are added to the 5′ termini of primers. After PCR, purified product can be cleaved with appropriate restriction enzymes to generate amplified segments of DNA with cohesive termini which can be ligated to a plasmid vector restricted with same restriction enzymes. The restricted plasmid should be subjected to dephosphorylation using calf intestinal alkaline phosphatase to avoid self-ligation of plasmid.
5. Cohesive termini cloning is 10–100-fold efficient than blunt end cloning (Sambrook and Russel 2001).

42.4 Preparation of Competent *E. coli* Cells

The process of introducing the ligated plasmid into a bacterium is called as transformation. All bacteria cannot take up free-floating DNA; it does not pass through the bacterial cell's membrane. Naturally *E. coli* cells do not take up foreign DNA and hence plasmid cannot enter into *E. coli* cells naturally; however, they can be made competent to take up the DNA either by physical or chemical methods. Mendel and Higa (1970) developed a chemical process for making competent cells by suspending the bacterial cells in ice cold solutions of $CaCl_2$ and then briefly heated to 37 °C or 42 °C which could be transfected with bacteriophage lambda DNA. This method was later used to transform bacteria with plasmid DNA.

Due to rapid growth of bacterial cells, the cell wall possesses hundreds of pores called adhesion zone. These pores are large enough to admit the DNA double helix but the negatively charged phosphate group of the DNA is repelled by those of lipid molecules present on the cell membrane. The Ca^{+2} ion from $CaCl_2$ interacts with these negative charges of both the DNA and lipid and creates electrostatically stable environment. DNA is incorporated into the bacterial cells by incubating the cells and the DNA together in a tube on ice followed by heat shock at 42 °C for 1–2 min, and then putting them back on ice. This sudden heat shock causes transient opening of the cell wall pore through which the DNA that is adhering to the cell surface enters into the cells and then chilling on ice causes contraction of pores and thus not allow the DNA to come out from the cell. Bacterial strains which are used as competent cells must have no resistance gene and Dam/Dcm methylase mutation, should not produce any RE and must be able to multiply vigorously in any medium. The best examples are: DH5α, JM109, K-12 strains of *E. coli* (Davis et al. 1986; Ausubel et al. 1987; Sambrook et al. 1989; Sambrook and Russel 2001).

The physical method of making *E. coli* cells competent is through electroporation wherein bacterial cells are exposed to electrical charge to destabilize their cell membranes to induce the formation of transient membrane pores, thus making them competent to take up the plasmid DNA. The method is fast and efficient but requires expensive equipment to perform the transformation.

Here we describe preparation of competent *E. coli* cells using $CaCl_2$ method.

42.4.1 Materials

1. Centrifuge tubes.
2. High speed refrigerated centrifuge.
3. Ice flakes.
4. Incubator.
5. Inoculation needle.
6. Laminar air flow.
7. Shaker incubator.
8. *E. coli* bacterial strain (DH 5 α): Use 1–2-day-old streaked plate.
9. Luria broth: Add and dissolve 10 g of bacto-tryptone, 5 g bacto-yeast extract and 10 g NaCl in about 950 mL of water; adjust the pH to 7.4 with 1 N NaOH, make up the volume to 1 L and sterilize by autoclaving.
10. Luria agar: Add and dissolve 10 g of bacto-tryptone, 5 g bacto-yeast extract and 10 g NaCl in about 950 mL of water; adjust the pH to 7.4 with 1 N NaOH; add 15 g agar, make up the volume to 1 L and sterilize by autoclaving.
11. $MgCl_2$ (0.1 M): Add and dissolve 0.95 g of $MgCl_2.6H_2O$ in water and make up the volume to 100 mL, filter sterilize the contents and store at −20 °C.
12. $CaCl_2$ (0.1 M): Add and dissolve 1.47 g of $CaCl_2.2H_2O$ in water and make up the volume to 100 mL, filter sterilize the contents and store at −20 °C.

42.4.2 Method (Mendel and Higa 1970)

1. Pick a single bacterial colony (*Escherichia coli* strain DH 5α) from a recently prepared plate and inoculate into 5 mL Luria broth medium.
2. Subject the culture for shaking at 200 rpm at 37 °C overnight.
3. From the overnight culture take 1 mL and transfer it to a fresh 50 mL of LB and subject it shaking at 200 rpm at 37 °C for 1–2 h until the $O.D_{600\ nm}$ reach 0.6. Yellowish colour should disappear.
4. Collect the bacterial cells in a sterile ice cold 40 mL polypropylene centrifuge tubes and keep on ice for 10 min.
5. Pelletize the bacterial cells by centrifuging at 2500 × *g* for 5 min at 4 °C in a refrigerated centrifuge.
6. Resuspend the pellet (cells) by swirling in 4 mL of ice cold 0.1 M $CaCl_2$ solution followed by centrifugation at 4 °C for 10 min at 2000 × *g*.
7. Resuspend the pellet in 10 mL of ice cold 0.1 M $MgCl_2$ solution and incubate on ice for 1 h.
8. Recover the cells by centrifugation at 4 °C for 5 min at 2500 × *g*.
9. Resuspend the pellet in 5 mL of ice cold 0.1 M $CaCl_2$ containing 10% glycerol.
10. Dispense 200 μL aliquots in eppendorf tubes kept on ice and use directly for transformation. Cells can be stored at −80 °C for later use.

42.4.3 Notes

1. It is essential that all solutions to be made using the purest water possible, and the highest grade of reagents possible, otherwise the transforming efficiency will decrease considerably. Also, avoid using detergent residues at all costs (Mendel and Higa 1970).
2. Use plasticware whenever possible, or pre-rinse glassware with good water. When the solutions are filtered, nylon membranes are best because they have not been pretreated with wetting agents. If cellulose acetate or nitrate filter is to be used, pre-rinse it with good water.
3. Follow sterile procedure for all steps in this method.
4. Doubling time for *E. coli* is 20 min. If O.D (0.30) is overshot cell walls change and should not be used. However may dilute back 4× rather than start over fresh (Sambrook and Russel 2001).
5. Fresh competent cells may be used up to 12 h if kept on ice; this actually improves competence over time.

42.5 Transformation of *E. coli*

After preparation of competent cells, they are subjected to transformation using ligated recombinant plasmid. The method involves incubation of competent cells with recombinant plasmid DNA on ice followed by heat shock for 2 min at 42 °C in a water bath. The transformed bacterial cells are then allowed to multiply for a few cycles and subjected to selection using antibiotic and blue/white colour selection on appropriate media.

42.5.1 Materials

1. Same as in Subheading 42.4.1.
2. Appropriate antibiotics solution: Prepare stock solution (50 mg/mL) of the antibiotic by dissolving 50 mg in 1 mL of distilled water, filter sterilize through 0.22 μ filter and store at −20 °C.
3. Competent cells (from above).
4. IPTG (Isopropyl β-D-1-thiogalctopyranoside) (0.1 M): Dissolve 24 mg of IPTG in 1 mL of water, filter sterilize and store at −20 °C.
5. Recombinant ligated plasmid (from above).
6. X-gal (5 Bromo-4 Chloro-3 indolyl-ß-D-galactoside) (2%): Dissolve 20 mg of X-gal in 1 mL of dimethyl formamide (DMF), store at −20 °C, protect from light.

42.5.2 Method

1. Add 20 μL ligation mix to 200 μL of competent cells in a sterile chilled eppendorf tube.
2. Mix softly by flicking four to five times and incubate on ice for 30 min.
3. Place the tube in a water bath maintained at 42 °C for 2 min (to give heat shock to cells) without shaking the tube.
4. Immediately transfer the tube to an ice bath and let the cells to chill for 5–10 min.
5. Add 900 μL of Luria broth (LB) medium to the tube and keep at 37 °C for 1 h with shaking at 200 rpm.
6. Centrifuge the tube at maximum speed for 1 min, discard supernatant and dissolve the pellet in 200 μL of fresh LB medium. Plate cell suspension on the selection medium consisting of Luria agar (LA) with 0.1% of ampicillin, 0.2% of X-gal and 0.02% of IPTG.
7. Keep the plate inverted for 14–16 h at 37 °C and look for both blue and white colonies (Fig. 42.3).
8. Transformants are selected on the basis of blue/white colony colour (Fig. 42.3). If the ligation is successful, the bacterial colony should be white and carry target gene construct. Blue

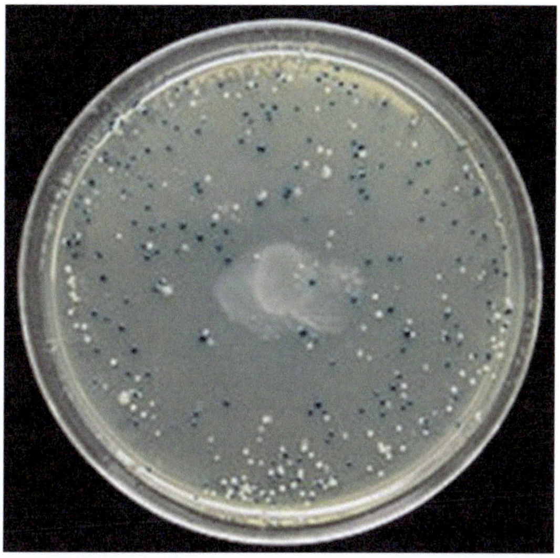

Fig. 42.3 A view of the transformed colonies in Luria agar showing blue and white colonies

colonies indicate self-ligated vector. This type of selection of recombinant bacteria is based on insertional inactivation of marker gene. The inactivation of a gene occurs by inserting a fragment of DNA into the coding sequence that lead to loss of a particular phenotype. When transformations are plated on the selection medium containing the substrate X-gal, colonies which contain vector with an insert appear white while those that contain vector without an insert are blue. Alpha complementation was first reported in 1967 after researchers found that co-expression of the N- and C-terminal coding sequences of the *lacZ* gene results in wild-type β-galactosidase activity. In most prokaryotic cloning vectors, the small α fragment is present on the vector with a multiple cloning site inserted into the open reading frame of the gene. The lacZ w fragment is expressed in the *E. coli* strain (encoded by lacZ M15). Insertion of a gene into the multiple cloning site disrupts the lacZ α open reading frame and results in an inactive, β-galactosidase protein. The substrate X-gal is hydrolyzed by functional β-galactosidase and produces a blue colour (Sambrook and Russel 2001).

9. Screen the blues: Many prokaryotic cloning vectors carry the fragment of the lacZ gene to allow blue/white screening of colonies. Sometimes small inserts, in general 1 kb or less will not disrupt the lacZ α reading frame or will cause re-initiation of translation therefore allowing expression of an intact, however, less active β-galactosidase protein. This results in colonies which contain inserts that are light blue or white with a blue centre. It is important, therefore, to screen these colonies for

the insert, particularly if the PCR product used for cloning is small.

10. Calculate transformation efficiency as follows:
    - Next morning, check plates.
    - Untransformed control should not develop colonies.
    - Count the number of colonies on the control plate. For example: 500 colonies/200 μL. i.e. 500 × 5 = 2500 colonies/1 mL total cells in the tube containing 2 ng vector.

      i.e. 1250 colonies/ng of plasmid DNA or 1250×10^3 or 1.25×10^6 colonies/ng of DNA.
    - Thus, the transformation efficiency of this batch of cells made is 1.25×10^6.

42.6 Selection of Transformants (Preparation of Master Plate)

42.6.1 Materials

1. As in Subheadings 42.4.1 and 42.5.1.
2. Scale.
3. Tooth pick.

42.6.2 Method

1. Melt the flask containing Luria agar and add required quantity of appropriate antibiotic when the temperature of the molten media is about 55 °C under aseptic condition. Pour the media onto Petri plates (about 20–25 mL/plate) and allow them to solidify.
2. After solidification, close the lid and place the plate upside down in a laminar flow. Make 1 cm^2 grid on the plate with the help of marker and scale and label each grid.
3. On the next day of transformation, observe the plate and select transformants on the basis of blue/white colour.
4. Pick up the white colonies (light blue/white colonies with a blue centre may also be used) and streak them on Luria agar (LA) plates with suitable antibiotics (ampicillin, X-gal and IPTG as mentioned earlier) on a grid with the help of tooth pick. Use separate tooth pick for each of the colonies.
5. Place the plate for 14–16 h at 37 °C.
6. Observe the plate for bacterial growth. This will serve as master plate (Fig. 42.4). Master plate can be stored for about 1 month at 4 °C. Each of the colonies in the master plate then should be subjected to colony PCR and/or restriction analysis to identify the positive recombinants.

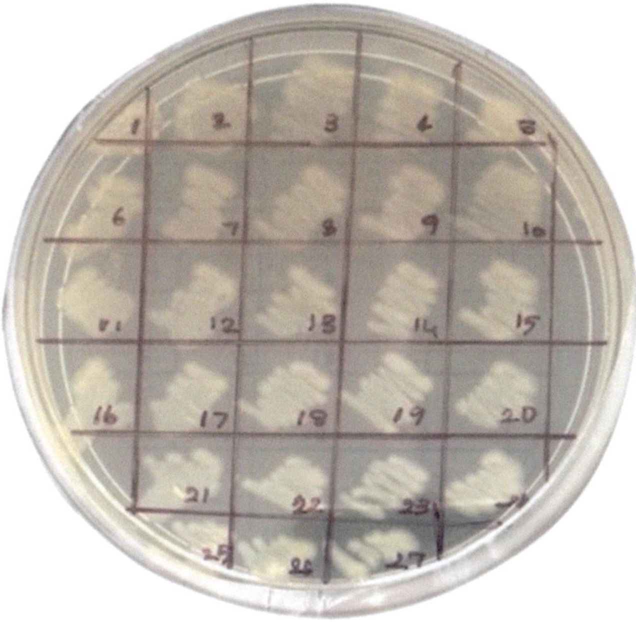

Fig. 42.4 An example of a master plate showing probable transformants in the selection medium

42.7 Screening and Identification of Positive (Recombinant) Clones

Subject the colonies in the master plate to any one or combination of methods given below to identify positive recombinants (Birnboim and Doly 1979; Davis et al. 1986; Ausubel et al. 1987; Sambrook et al. 1989; Sambrook and Russel 2001):

- Rapid disruption of bacterial colonies (Rapid) to test the size of plasmids.
- Colony PCR using insert-specific primers.
- Isolation of recombinant plasmid and subject them to restriction digestion.
- Isolation of recombinant plasmid followed by PCR using either insert-specific or combination of vector- and insert-specific primers.
- Nucleic acid hybridization using labelled probe (colony hybridization, dot blot hybridization).
- Sequencing of insert.

42.7.1 Rapid Disruption of Bacterial Colonies (Rapid) to Test the Size of Plasmids

In this method, a portion of bacterial growth from each of the colonies from the master plate is taken in a tube and added with disruption solution to disrupt the bacterial cells, followed by heating (Sambrook et al. 1989). After adding potassium chloride and loading dye, the tubes are subjected to a brief centrifugation. The supernatant is then run on an agarose gel and mobility of a control (non-recombinant colony) is compared with mobility of test colonies. The colonies that show less mobility compared to control are expected to have foreign DNA inserted.

42.7.1.1 Materials

1. Tooth pick (sterile).
2. Laminar air flow.
3. Water bath.
4. Vortex machine.
5. Horizontal gel apparatus with power pack.
6. Gel documentation unit.
7. EDTA (10 mM, pH 8.0): Dissolve 16.8 g of ethylene di amine tetra acetic acid disodium salt in about 80 mL of water, adjust the pH to 8.0 by adding NaOH and make up the volume to 100 mL.
8. Master plate containing potential recombinant clones.
9. Potassium chloride (KCl) (2 M): Dissolve 1.49 g of potassium chloride in water and make up the total volume to 10 mL.
10. Orange dye.
11. Disruption solution: 20 μL of 10 N NaOH, 50 μL of 10% SDS, 500 μL of 40% sucrose and 420 μL of water.

42.7.1.2 Method

1. Prepare 1.0% agarose gel.
2. Label the eppendorf tubes and put 50 μL of sterile 10 mM EDTA in each tube.
3. Using sterile tooth pick, pick off a little of the each colonies and swirl it in the respective tube containing EDTA solution (should be cloudy) under aseptic condition.
4. Add 50 μL of freshly made disruption solution to each of the tubes.
5. Vortex the mixture for at least 30 s.
6. Incubate the tubes for 5 min at 70 °C and then at room temperature.
7. Add 4 μL of 2 M KCl and 4 μL of orange dye. Vortex for 30 s.
8. Incubate on ice for 5 min.
9. Centrifuge for 3 min to pellet bacterial debris at 11,000 × *g* at 4 °C.
10. Process a bacterial colony having plasmid vector alone (blue colony) in the same way to serve as control.
11. Load 50 μL of the supernatant into the gel and run at a constant 100 V for 1 h. Load the control sample in the centre well for better comparison with test colonies.

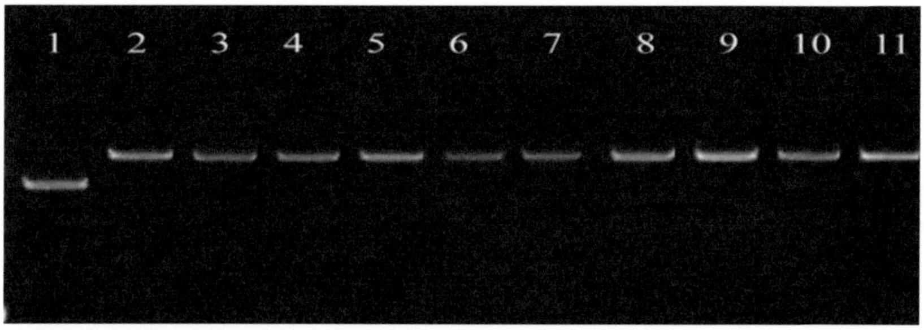

Fig. 42.5 An example of rapid disruption of bacterial colonies (Rapid) gel to test the size of plasmids. Lane 1: Control plasmid without insert; Lanes 2–11: putative transformants showing slow mobility indicating the presence of insert

12. Place the gel on the transilluminator and compare the mobility of both control and putative recombinants. The samples having less mobility compared to control may be expected to have foreign DNA inserted (Fig. 42.5).

42.7.2 Screening Putative Recombinants by Colony PCR

In this method, a portion of bacterial colony is boiled in tube to disrupt the contents. This is then subjected to centrifugation to pellet all cellular debris. An aliquot of the supernatant is taken and used as template for PCR. PCR can be done using primers specific for the insert or combination of primers from vector and insert which will help to know the orientation of the insert in the plasmid.

42.7.2.1 Materials

1. Same as in Subheadings 42.4.1, 42.5.1 and 42.6.1.
2. PCR work station.
3. Thermal cycler.
4. Triton X 100.
5. Master plate containing putative recombinant colonies.
6. PCR components: 10× PCR buffer, 25 mM $MgCl_2$, Primers, 10 mM dNTP mix, *Taq* DNA polymerase.

42.7.2.2 Method

1. Label the eppendorf tubes and add 50 μL of sterile water containing 0.1% Triton X 100 to each tube.
2. In the laminar air flow, carefully open the master plate containing colonies.
3. Pick up a portion of each of the colonies with the help of either sterile tooth pick or microtip and add into the corresponding eppendorf tube and mix well.
4. Keep the tubes in boiling water for 10–15 min.
5. Place the tube on ice for about 5 min and centrifuge for 5 min at 7500 × *g*.

6. Use 1 μL of the supernatant as template for setting up PCR reaction.
7. To set up PCR, mix together the following on ice (add enzyme at last). For multiple samples, make a required master mix (without template) and aliquot 49 μL in each PCR tube followed by addition of template (1 μL) to the respective PCR tubes.

Reactant	Volume (μL)
Sterile water	39.5
10× PCR buffer	5.0
25 mM $MgCl_2$	1.0
10 mM dNTPs	1.0
Forward primer (10 μM/μL)	1.0
Reverse primer (10 μM/μL)	1.0
Taq polymerase (3 U/μL)	0.5
Template (from **step 6** above)	1.0
Total	50

8. Mix the contents and give a short spin.
9. Place the tubes in thermocycler and start run after feeding the appropriate temperature profiles for denaturation, annealing and extension.
10. Analyse about 15–20 μL of PCR product along with DNA markers of known size in 0.8% agarose gel electrophoresis for 1 h as described earlier.
11. Place the gel on a transilluminator, observe for expected band, document the gel image using suitable gel documentation unit (Fig. 42.6).

42.8 Recombinant Plasmid DNA Isolation and Restriction Analysis

In general, isolation of recombinant plasmid from bacterial cells involves growing of bacterial cells overnight in a suitable medium supplemented with appropriate antibiotics. On the next day, bacterial cells are harvested by centrifugation, subjected to lysis of cell wall to release the plasmid into solution and purification of the plasmid by different methods. Lysis of cell wall is achieved by treatments such as alkali, organic solvents, nonionic detergents and heat (Birnboim and Doly 1979; Feliciello and Chinali 1993; Xiang et al. 1994; Sambrook and Russel 2001). The alkaline lysis

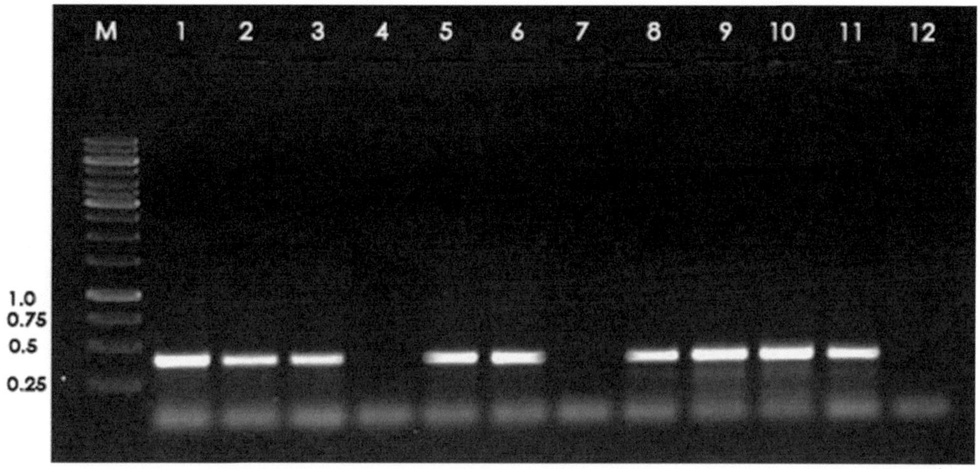

Fig. 42.6 An example of colony PCR of bacterial colonies for identification of transformants. Presence of band at the expected size indicates recombinant nature of the clone (note non-recombinant clones in the lane 4, 7 and 12)

technique was developed by Birnboim and Doly (1979). In this technique, high anionic detergent at high pH solution such as SDS and NaOH is used to break cell wall, denature chromosomal DNA and proteins and release plasmid DNA into the supernatant. Through centrifugation denatured chromosomal material is separated and native plasmid DNA present in the supernatant is purified through phenol-chloroform method followed by precipitation of the plasmid through ethanol or isopropanol. This technique is suitable for isolation of plasmids from all strains of *E. coli* with varying volume (1–500 mL). The isolated DNA is suitable for PCR, restriction digestion and nucleic acid hybridization test. With additional purification by polyethylene glycol (PEG), it is suitable for sequencing. However, extra care may be needed if isolating large sized plasmids (>15 kb) as they may be affected during cell lysis and subsequent purification. Recently, a variety of commercial kits are available for plasmid purification. These kits consist of disposable chromatography columns that are used for binding and elution of plasmid DNA. Different matrices such as glass, diatomaceous earth and anionic resins such as DEAE (diethyl aminoethyl) or QAE (diethyl [2-hydroxy propyl] aminoethyl) are used.

We describe here two methods: alkaline lysis method (Birnboim and Doly 1979) and modified alkaline lysis method (Xiang et al. 1994)

42.8.1 Plasmid Isolation by Alkaline Lysis Method (Birnboim and Doly 1979)

42.8.1.1 Materials

1. Master plate containing putative transformants.
2. Solution I (Resuspension buffer): Prepare by adding 2.25 mL of 20% glucose, 1 mL of 0.5 M EDTA, pH 8.0 and 45.5 mL of sterile distilled water.
3. Solution II (Lysis buffer): Prepare by mixing 0.4 mL of 10 N NaOH, 2 mL of 10% SDS and 18.6 mL of sterile distilled water (not to be autoclaved).
4. Solution III (Neutralization buffer) (3 M sodium acetate pH 4.8): Dissolve 40.81 g of sodium acetate.3H_2O in 60 mL water, bring the pH to 4.8 by adding NaOH and make up the volume to 100 mL.
5. Solution IV: Prepare by dissolving 2.5 mL of 1 M Tris pH 8.0, 1.65 mL of 3 M NaoAC pH 4.8 and 45.85 mL of sterile water.
6. RNase A: Dissolve 10 mg of RNase A in a 1 mL solution containing 10 mM Tris–HCl pH 7.9 and 15 mM NaCl; heat to 100 °C for 15 min, bring to room temperature. Make aliquots and store at −20 °C.
7. TE (pH 8.0): Dissolve 0.1211 g of Tris base and 0.0372 g of EDTA in about 80 mL of water, adjust pH to 8.0 by adding concentrated HCl and make up the volume to 100 mL.
8. Phenol: chloroform: isoamyl alcohol (25:24:1): Equal volume of equilibrated phenol and chloroform: isoamyl alcohol (24:1) are mixed and stored under 0.1 M Tris–HCl (pH 8.0) in amber coloured bottle at 4 °C.
9. Ethanol.

42.8.1.2 Method

1. Inoculate identified colonies separately in 5 mL LB broth with appropriate antibiotic at 37 °C for 14 h with shaking at 200 rpm.
2. Harvest the cells in a 1.5 mL eppendorf tube, centrifuge for 30 s and discard the supernatant.
3. Dissolve the pellet in 100 μL of solution I by vortexing. Place on ice for 5 min.
4. At room temperature add 200 μL of freshly prepared solution II, mix gently by inverting tubes (do not vortex) and place on ice for 5 min till SDS precipitates. Samples turn viscous and stringy when cap opened at the end of this step.
5. At room temperature add 150 μL of solution III, invert tubes gently several times and place the tube on ice for 5 min.
6. Centrifuge at 15,000 × *g* for 5 min at 4 °C and carefully collect the supernatant in a fresh eppendorf tube. Discard the pellet.
7. Add 0.9–1.0 mL cold ethanol, mix gently and place the tube at −70 °C for 15 min.

8. Centrifuge for 3 min at maximum speed and discard supernatant.
9. Add 100 μL of solution IV to the pellet and mix gently till pellet is dissolved.
10. Add 200 μL of cold ethanol and place at −70 °C for 15 min again.
11. Centrifuge for 3 min and discard the supernatant.
12. Add 1 mL of 70% ethanol to the pellet and centrifuge immediately for 1 min at maximum speed, discard supernatant and dry the pellet for 5 min under vacuum.
13. Add 20 μL of TE to the pellet and store at −20 °C until required.

42.8.2 Plasmid Isolation by Modified Alkaline Lysis Method (Xiang et al. 1994)

42.8.2.1 Materials

1. Same as listed in alkali lysis method.
2. Potassium acetate buffer (5 M): Mix 60 mL of 5 M potassium acetate, 11.5 mL glacial acetic acid and 28.5 mL water.
3. Antibiotic (appropriate) solution.
4. GTE (50 mM glucose, 25 mM Tris–HCl, pH 8.0, 10 mM EDTA, pH 8.0).
5. Lysis buffer (0.2 N NaOH, 1% SDS).
6. Terrific broth: Add 12 g of bacto-tryptone, 24 g of bacto-yeast extract and 4 mL glycerol in 900 mL of water and autoclave. Prepare separately 100 mL solution containing 0.17 M KH_2PO_4 and 0.72 M K_2HPO_4 and autoclave. Both autoclaved solutions are mixed to make 1 L of terrific broth.

42.8.2.2 Method

1. Grow *E. coli* carrying the plasmid in 4 mL of terrific broth medium containing appropriate antibiotics at 37 °C overnight with shaking at 250 rpm.
2. Harvest cell and centrifuge for 20 s at maximum speed.
3. Discard the supernatant and dissolve the pellet in 100 μL GTE by vortexing.
4. Immediately add 200 μL of lysis buffer and mix by inverting the tube.
5. Add 150 μL of 5 M potassium acetate buffer and mix quickly shaking by hand. Centrifuge immediately for 1 min and collect the supernatant into a fresh microcentrifuge tube.
6. Add 1 mL of 95% ethanol and invert the tube to mix.
7. Centrifuge for 1 min to harvest plasmid DNA, discard supernatant and vacuum dry the pellet.
8. Resuspend the plasmid DNA in 200 μL of TE or water containing RNase A.

42.8.3 Confirmation of Recombinant Clones by Restriction Analysis

Nucleases are the group of enzymes that cleave the nucleic acids. Exonucleases are known to cut DNA from ends while endonuclease cuts DNA at any point and S1 nuclease cuts ssDNA. But restriction endonucleases are known to cut DNA at a specific site known as recognition site. Each restriction enzymes have unique recognition site with varying size (4, 5, 6 nucleotides) and cleavage on each strand leaves a 3′-OH and a $5'\text{-}PO_4$. After cleavage both strands have same sequence in antiparallel orientation. Recognition site happens to be a specific palindrome. Different enzymes cut at different palindromes. DNA double helixes are cut at the axis of symmetry and both strands have same sequence in antiparallel orientation.

Depending upon the protein structure, recognition and cleavage site specificity, requirement of ATP, restriction enzymes are grouped into three types—Type I, Type II and Type III. Nomenclature of restriction enzymes consists of 3–4 parts (first part: initial of bacterial genus, second part: first two letters from species, third part: strain of the bacteria, fourth part: number of enzyme). For example in restriction enzyme, *Eco* R1, '*E*' stands of genus '*Escherichia*'; '*co*' for species '*coli*'; 'R' for strain and '1' for number of enzyme). Each restriction enzyme may produce either sticky (5′ or 3′ end) or blunt end product. Each DNA fragment will have its own unique restriction site. Map showing order and distances of restriction cut sites in a segment of DNA is called as restriction map. Restriction maps are useful in comparing DNA fragments to look for regions of identity.

All plasmid vectors used for cloning contain unique restriction sites for many restriction enzymes. Thus by restriction digestion of plasmid, it is possible to linearize or even release the insert (foreign DNA segment) from the plasmid. The size of recombinant clones will be higher (due to presence of insert) than their counterpart non-recombinant clones. Hence based on size, recombinants and non-recombinants can be differentiated. In this method, the plasmid DNA isolated from putative recombinants is subjected to restriction digestion using appropriate restriction enzymes (either to linearize or release the insert). The reaction mix will consist of appropriate restriction buffer and enzymes (buffers are specific for each enzymes) and plasmid DNA. The optimum temperature for most of the restriction enzymes is 37 °C and duration may vary from 1 to 3 h or even can be overnight. After the reaction, contents are run on an agarose gel along with suitable DNA size markers (Fig. 42.7).

42.8.3.1 Materials

1. Agarose.
2. Agarose gel apparatus with power pack.
3. Appropriate restriction enzymes and corresponding reaction buffers: Commercially available.

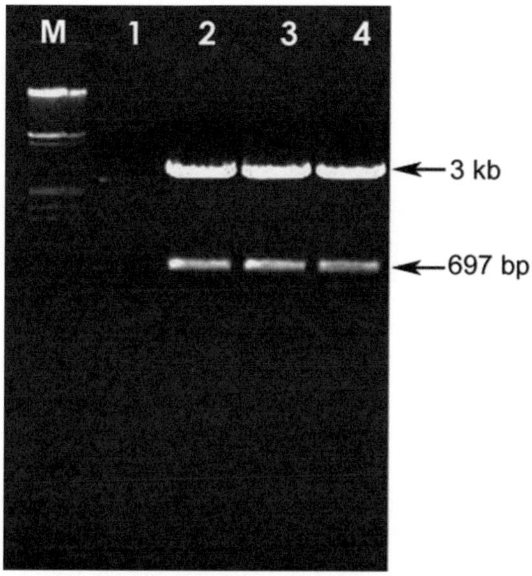

Fig. 42.7 Agarose gel electrophoresis of restricted recombinant plasmid showing release of insert. Lane M: DNA size marker bands; Lane 1: Restricted plasmid DNA (control); Lanes 2–4: Restricted plasmid DNA of putative transformants. The upper band corresponds to the vector (plasmid) while the lower band corresponds to the insert

4. DNA gel loading dye: Mix 10× TAE (60 μL), 0.5 M EDTA (50 μL), glycerol (200 μL), 10% SDS (60 μL), 1% bromophenol blue (200 μL) and H_2O (30 μL).
5. Gel documentation system.
6. Purified plasmid DNA sample.
7. Waterbath.

42.8.3.2 Method

1. Take about 1–17 μL (2.0 μg) of plasmid DNA.
2. Add sterile water to make up the volume to 17 μL.
3. Add 2 μL of 10× appropriate restriction enzyme buffer, mix well and add 1.0 μL (5–10 units) of restriction enzyme.
4. Gently mix the contents and spin briefly.
5. Incubate the samples at 37 °C for 1–3 h.
6. During incubation time, prepare 0.8% agarose gel containing ethidium bromide (refer Chapter 26 for details of agarose gel electrophoresis).
7. After restriction, spin down the samples and heat inactivate the restriction enzymes by placing the tube for 10 min at 80 °C in a waterbath.

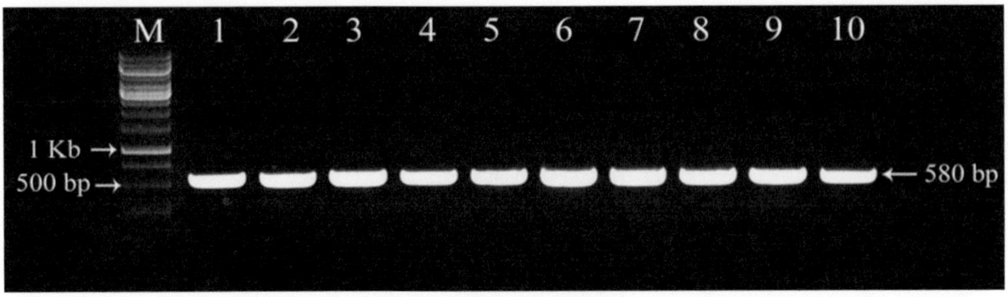

Fig. 42.8 PCR analysis of the plasmid isolated from putative transformants using insert-specific primers. Lane M: DNA size marker indicating size of each of the bands on the left side; Lane 1–10: PCR carried out using insert-specific forward and reverse primers

8. Add 2 μL of DNA loading dye to the restricted sample and load on the agarose gel. Use appropriate DNA size standards. Run at about 100 V for 60–90 min.
9. View the gel under transilluminator and check the size of the plasmid. Look for recombinant clones based on the size. Size of recombinant clone would be the size of plasmid + size of insert. In case restriction releases the insert, two bands will be seen in recombinant clones—one representing plasmid DNA while the other band represents the insert. Non-recombinants will have only one band corresponding to plasmid DNA (Fig. 42.7).

42.8.4 Confirmation of Recombinant Clones by PCR

After ascertaining the presence of the insert through restriction analysis, DNA PCR using recombinant plasmid is carried out to further confirm the presence of right insert. In this method, plasmid DNA isolated from putative recombinant clones are used as template. PCR is performed using primers specific for the insert (i.e. same primers used for initial amplification before cloning) to check for the presence and size of insert (Fig. 42.8). Further PCR can also be used to find out the orientation of the insert in the plasmid. For this, PCR is set using combination of primers from plasmid and insert such as (1) forward primer from plasmid and forward primer from insert, (2) forward primer from plasmid and reverse primer from insert, (3) reverse primer from plasmid and forward primer from insert, (4) reverse primer from plasmid and insert. Procedure involves setting up of PCR using above primer combinations and analysis by gel electrophoresis. Sense-oriented insert would give product in primer combination of (2) and (3) while antisense-oriented insert would give product in primer combinations of (1) and (4) above.

42.8.4.1 Materials

1. Same as in Subheading 42.8.3.1.
2. Template plasmid.
3. PCR amplification buffer (10×).

4. dNTP mix (10 mM).
5. $MgCl_2$ (25 mM).
6. *Taq* DNA polymerase.
7. RNase free water.
8. Oligonucleotide primers (both sense and antisense specific for insert and plasmid).

42.8.4.2 Method

1. Set up three different PCR reactions by adding components as shown in the table below.

Component	PCR reaction Forward primer	Reverse primer	Forward and reverse primer
10× PCR buffer	10.0 μL	10.0 μL	10.0 μL
25 mM $MgCl_2$	5.0 μL	5.0 μL	5.0 μL
10 mM dNTPs mix	2.0 μL	2.0 μL	2.0 μL
100 ng/μL Forward primer	1.0 μL	–	1.0 μL
100 ng/μL Reverse primer	–	1.0	1.0 μL
Taq DNA polymerase (3 U/μL)	0.5 μL	0.5 μL	0.5 μL
Template (plasmid)	1.0 μL	1.0 μL	1.0 μL
Sterile water to make	100.0 μL	100.0 μL	100.0 μL

2. After quick spin, place tubes in a PCR machine and run the program using required temperatures for denaturation, annealing and extension cycles.
3. Analyse PCR products by electrophoresing at 100 V on 0.8% agarose gel.
4. Formation of PCR product in a reaction where both the primers were used should confirm the appropriateness of the insert (Fig. 42.8).
5. If the results confirm the presence of insert, PCR may be set up using primers from plasmid and insert as indicated below in order to determine the orientation of the insert in the plasmid:
   - forward primer from plasmid and forward primer from insert
   - forward primer from plasmid and reverse primer from insert
   - reverse primer from plasmid and forward primer from insert
   - reverse primer from plasmid and reverse primer from insert

6. Analyse the PCR product in 0.8% agarose gel electrophoresis and observe for presence of bands under transilluminator. If the insert is oriented in sense orientation, band of expected size would be visible in PCR reaction set up with primer combinations (2) and (3) while antisense-oriented insert would give product in primer combinations of (1) and (4) above.

References

Ausubel FM, Bent R, Kingston RE, Moore DD, Seidman JG, Smith JA, Struthl K (1987) Current protocols in molecular biology. Greene Publishing Associates and Wiley Interscience

Birnboim HC, Doly J (1979) A rapid alkaline extraction procedure for screening recombinant plasmid DNA. Nucleic Acids Res 7:1513–1523

Davis LG, Dibner MD, Battey JF (1986) Basic methods in molecular biology. Elsevier, New York

Feliciello I, Chinali G (1993) A modified alkaline lysis method for the preparation of highly purified plasmid DNA from *Escherichia coli*. Anal Biochem 212:394–401

Mendel M, Higa A (1970) Calcium-dependent bacteriophage DNA infection. J Mol Biol 53:159–162

Sambrook J, Russel DW (2001) Molecular cloning (3rd edition): a laboratory manual, vol I–III. Cold Spring Harbor Laboratory Press, New York

Sambrook J, Fritsch EF, Maniatis T (1989) Molecular cloning: a laboratory manual, vol I–III, 2nd edn. Cold Spring Harbor Laboraratory Press, New York

Xiang C, Wang H, Shiel P, Berger P, Guerra DJ (1994) A modified alkaline lysis miniprep protocol using a single microcentrifuge tube. BioTechniques 17:30–32

Chapter 43

cDNA Synthesis and Cloning

Abstract

Majority of plant viruses contain single-stranded RNA as their genome. When genome sequence information of the virus is lacking, RT-PCR cannot be employed to amplify the viral genome. In such cases in order to characterize viral genomes, they have to be first converted into complementary DNA (cDNA) using reverse transcriptase enzyme and oligo d(T) or random primers. The cDNA is then made double stranded and ligated to a vector (usually a plasmid). The ligated vector is used to transform bacteria so that recombinant vector multiplies in the bacterium. The positive recombinant vector carrying viral insert is then identified and confirmed by different methods including sequencing. The identified recombinant clones can be stored in glycerol or long-term storage can be done through lyophilization.

Key words RNA virus, First strand cDNA synthesis, Second strand synthesis, Plasmid vector, Ligation, Transformation, Recombinant clones

43.1 Introduction

Whenever a new RNA virus is discovered, RT-PCR cannot be employed to amplify the genome of that virus due to lack of nucleotide sequence information. In such cases, cDNA synthesis and cloning of the test viral genome is done. In this procedure, viral RNA is enzymatically converted to complementary DNA (cDNA) followed by double-stranded DNA and inserted into prokaryotic vectors. Viral RNA is isolated from purified virus preparation or dsRNA or total RNA isolated from infected plant can be used as template for first strand cDNA synthesis (Fig. 43.1). The first strand of cDNA is synthesized by an RNA-dependent DNA polymerase (reverse transcriptase) using viral RNA as template and primed by either oligo (dT) [for those viruses having poly (A) tail at the 3′ end] or random oligonucleotide or virus-specific primers. Second strand synthesis is catalysed by *E. coli* DNA polymerase I and *E. coli* RNase H. The cloning of double-stranded cDNA into a vector is facilitated by the addition of various tails, linkers or adaptor sequences to the ends of cDNAs. Plasmids or the bacteriophage lambda are the commonly used vectors for cloning (Fig. 43.1).

The procedure involves addition of reverse transcriptase and primer (poly 'T' if virus has a poly 'A' tail or random primers) to the

Alangar Ishwara Bhat and Govind Pratap Rao, *Characterization of Plant Viruses*, Springer Protocols Handbooks, https://doi.org/10.1007/978-1-0716-0334-5_43,

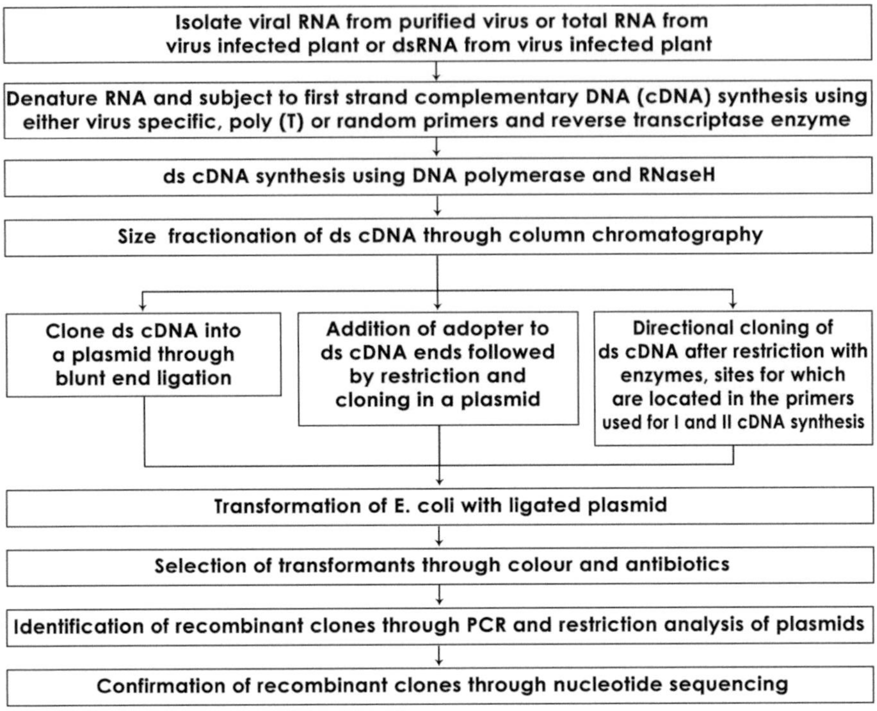

Fig. 43.1 Strategies used for cDNA synthesis and cloning of RNA viruses

denatured RNA and incubating for 1 h at 42 °C. During the process, first strand cDNA synthesis will take place. Using first strand cDNA as template, second strand synthesis is done by DNA polymerase in the presence of RNase H. This is then subjected to size fractionation through spun column chromatography to remove unincorporated nucleotides and very small sized cDNAs. The cDNA can then be cloned directly through blunt end ligation to a plasmid vector (restricted to produce blunt end) or after addition of suitable adapter to a plasmid vector (restricted with appropriate restriction enzyme to produce cohesive ends). The ligated recombinant plasmid vectors are then used to transform *E. coli* cells and transformants are selected on an agar plate containing appropriate selection markers such as X-gal and antibiotics. Master plate containing all putative transformants is then made and each of the colonies in the master plate is screened for the presence of insert through rapid disruption of bacterial colonies, isolation of plasmid DNA and subjecting them to restriction analysis and nucleotide sequencing of the insert DNA. If suitable probes are available, recombinant clones can be identified through colony hybridization, dot blot hybridization or Southern hybridization. Unlike PCR-based cloning, each of the clones obtained through cDNA synthesis will have varying insert lengths (as population of ds cDNA with different size are used for ligation). Hence size of insert cannot be criteria to select recombinants as two clones of similar insert size may represent two different regions of DNA or two clones of different size may represent overlapping regions of same DNA segment (Okayama and Berg 1982; Gubler and Hoffman 1983; Sambrook and Russel 2001).

43.2 First and Second Strand cDNA Synthesis

43.2.1 Materials

1. Freezers (−20 and −80 °C).
2. Glass wool.
3. Ice flakes.
4. Sterile syringe (1 mL).
5. Water bath at 16 and 70 °C.
6. EDTA (0.5 M, pH 8.0): Dissolve 16.8 g of ethylene diamine tetra acetic acid disodium salt in about 80 mL of water, adjust the pH to 8.0 by adding NaOH and make up the volume to 100 mL.
7. Ammonium sulphate (1 M).
8. Chloroform.
9. *E. coli* DNA ligase.
10. *E. coli* DNA polymerase I.
11. dNTP mix (dATP, dTTP, dCTP, dGTP) (5 mM or 10 mM).
12. Ethanol.
13. KCl (1 M).
14. Magnesium chloride (250 mM and 10 mM).
15. NaCl.
16. Phenol:chloroform (1:1): Mix equal volume of tris saturated phenol and chloroform, store in amber coloured bottle.
17. Primers.
18. Reverse transcriptase enzyme.
19. RNase H.
20. RNase inhibitor.
21. Dithiothreitol (0.1 M): Dithiothreitol (DTT) is a reducing agent capable of breaking disulphide bonds, perturbing the tertiary structure of THP and preventing polymerization. Weigh and dissolve 154 mg of dithiothreitol in 10 mL of sterile water, make aliquots and store at −20 °C.
22. Sephadex G 50.
23. Sephadex G-50 spun column: It separates molecules according to differences in size as they pass through the column.

23. Sodium acetate (3 M pH 5.2): Dissolve 24.6 g sodium acetate in 70 mL of distilled water, adjust the pH to 5.2 by adding glacial acetic acid and bring the total volume of solution to 100 mL with water.
24. β-NAD (β-Nicotinamide adenine dinucleotide) (50 mM).
25. T_4 DNA polymerase.
26. T_4 Polynucleotide kinase.
26. TE buffer: Dissolve 0.1211 g of Tris base and 0.0372 g of EDTA in about 80 mL of water, adjust pH to 8.0 with HCl and make up the volume to 100 mL.
27. Template RNA.
28. TEN buffer (1×): 40 mM Tris–HCl (pH 7.5), 1 mM EDTA (pH 8.0).
29. Tris–HCl (2 M, pH 7.4): Weigh and dissolve 24.22 g of Tris base in about 80 mL of water, adjust the pH to 7.4 with HCl and make up the volume to 100 mL.

43.2.2 Method

1. To synthesize first strand cDNA, mix the following in a sterile microcentrifuge tube on ice.

RNA template[a] (1 μg/μL)	10.0 μL
Primer (1 μg/μL)	1.0 μL
1 M Tris–HCl, pH 8.0	2.5 μL
1 M KCl	3.5 μL
250 mM $MgCl_2$	2.0 μL
5 mM dNTP mix	10.0 μL
0.1 M dithiothreitol	1.0 μL
Water to make	48.0 μL
Reverse transcriptase (10 units/μL)	2.0 μL

[a]Denatured by heating at 80 °C for 5 min and snap cooling on ice for 2 min before addition

2. Mix the reagents by gentle vortexing followed by quick spin.
3. Incubate the reaction for 1 h at 37 °C.
4. Heat the reaction to 70 °C for 10 min and then transfer to ice.
5. Add the following:

10 mM $MgCl_2$	70.0 μL
2 M Tris–HCl (pH 7.4)	5.0 μL
1 M $(NH_4)_2SO_4$	1.5 μL

(continued)

RNase H (1 unit/μL)	1.0 μL
E. coli DNA polymerase I (10 units/μL)	4.5 μL

6. Mix reagents by gentle vortexing and centrifuge briefly and incubate the reaction for 2–4 h at 16 °C.
7. Add the following reagents to the reaction mixture.
   β-NAD (50 mM) 1.0 μL.
   E. coli DNA ligase (4 units/μL) 1.0 μL.
   Incubate the reaction for 15 min at room temperature.
8. Add 1 μL of 10 mM dNTP mix and 2 μL (5 units) of bacteriophage T_4 DNA polymerase. Incubate the reaction for 15 min at room temperature. Remove an aliquot (3 μL) of the reaction for checking second strand synthesis.
9. To the remainder of the reaction, add 5 μL of 0.5 M EDTA (pH 8.0).
10. Add equal volume of phenol:chloroform (1:1), mix well and centrifuge at 11,000 × *g* for 10 min.
11. Collect aqueous phase and add equal volume of chloroform, mix well and centrifuge at 11,000 × *g* for 10 min.
12. Collect aqueous phase and add 0.1 vol. of 3 M sodium acetate (pH 5.2) and 2 volumes of chilled ethanol.
13. Incubate at −20 °C for 1 h and centrifuge at 17,000 × *g* for 15 min. Discard supernatant.
14. Add 1 mL of 70% ethanol and centrifuge at 17,000 × *g* for 5 min. Aspirate and discard supernatant, air dry the pellet and dissolve in 90 μL of TE.
15. To the DNA, add:
    - 10× T_4 polynucleotide kinase buffer: 10 μL.
    - T_4 Polynucleotide kinase (3 units/μL): 1 μL.
      Incubate the reaction at room temperature for 15 min.
16. Add equal volume of phenol: chloroform (1:1), mix well and centrifuge at 11,000 × *g* for 10 min.
17. Collect aqueous phase in a fresh eppendorf tube.
18. Separate the unincorporated dNTPs from the cDNA by spun column chromatography through Sephadex G-50 equilibrated in TE containing 10 mM NaCl (TEN).
    - Plug the bottom of a 1 mL syringe with a small amount of sterile glass wool.
    - Fill the syringe with sephadex G-50 equilibrated in 1× TEN buffer pH 8.0. Start the buffer flow by tapping the side of the syringe and keep adding more resin until the syringe is completely full.

- Insert the syringe into 15 mL disposable plastic tube. Centrifuge at 1600 × *g* for 4 min at room temperature. Discard the flow through.
- Add 100 μL of 1× TEN buffer to the column and re-centrifuge as above.
- Repeat the above step twice more.
- Apply the DNA sample to the column in a total volume of 0.1 mL. Place the spun column in a fresh disposable tube containing a decapped microfuge tube.
- Centrifuge again at 1600 × *g* for 4 min and collect the effluent from the bottom of the syringe into decapped microfuge tube.
- Remove the syringe. Using forceps, carefully recover the decapped microfuge tube, which contains the eluted DNA, and transfer its contents to a capped, labelled microfuge tube.

19. Precipitate the eluted cDNA by adding 0.1 volume of 3 M sodium acetate (pH 5.2) and two volumes of ethanol. Store the sample on ice for 30 min. Recover the precipitated DNA by centrifugation at maximum speed for 15 min at 4 °C in a microfuge.
20. Wash the pellet with 70% ethanol and centrifuge again. Gently aspirate all of the fluid and allow the pellet to dry in the air.
21. Dissolve the cDNA in a small quantity of TE or water.
22. The ds cDNA thus obtained can be directly ligated to a blunt ended vector and cloned. However, efficiency of blunt end ligation is low compared to cohesive end cloning. If cohesive end cloning is required, the dsDNA should be first ligated to a linker or an adapter and then ligated to a vector restricted with appropriate restriction enzyme. The following method describes blunt end ligation and transformation.

43.3 Ligation of DNA Fragment to the Vector DNA

43.3.1 Materials

1. As in Subheading 43.2.1.
2. Sodium acetate (3 M; pH 4.8).
3. Calf intestinal alkaline phosphatase (CIAP) enzyme and its buffer.
4. T_4 DNA ligase and its buffer.
5. Phenol: chloroform: isoamyl alcohol (25:24:1): Mix 25 mL of phenol, 24 mL of chloroform and 1 mL of isoamyl alcohol.
6. Plasmid DNA.

7. rATP (10 mM).
8. Restriction enzyme (*Sma* 1).

43.3.2 Method

43.3.2.1 Linearization of Vector DNA

1. Take about 2 μg of plasmid (vector) DNA in about 16 μL of sterile water.
2. Add 2 μL of restriction buffer and 2 μL of restriction enzyme (such as *Sma*I) that generates blunt end upon restriction.
3. Keep at 37 °C for about 1–3 h.
4. Prepare 0.8% agarose gel in 1× TAE.
5. After the restriction is over, load 1–2 μL of restricted sample along with undigested sample to check whether the restriction is complete.
6. Run for about 30 min at 50 V and check under UV light.
7. If restriction is complete, add about 80 μL of sterile distilled water to the remaining 18 μL of restricted sample (if not add 1 μL more of restriction enzyme and keep for some more time at 37 °C for complete restriction).
8. Add an equal volume of saturated phenol: chloroform: isoamyl alcohol (25:24:1) and mix the contents gently.
9. Centrifuge for about 2 min and collect the aqueous phase.
10. Add 0.1 vol. of 3 M sodium acetate pH 4.8, two volumes of ethanol and incubate the mixture at −70 °C for 1 h or −20 °C overnight.
11. Centrifuge at 13,000 × *g* for 15 min and discard the supernatant.
12. Wash the pellet with chilled 70% ethanol.
13. Air dry the pellet and dissolve in 10 μL of sterile water.
14. To check the concentration and quality of linearized vector DNA, load 1 μL on 0.7% agarose gel and run at 40 V for 1 h. View under transilluminator and assess the quality of DNA.

43.3.2.2 Dephosphorylation of Linearized Plasmid DNA

1. Set up the dephosphorylation reaction as follows:

Restricted plasmid DNA	17.5 μL (2 μg)
10× CIAP buffer	2.0 μL
CIAP	0.5 μL[a]

[a]1 unit/100 pmoles of 5′ termini (2 μg of linearized 5 kb DNA contain approx. 1.4 pmoles of 5′ termini)

2. Incubate at 37 °C for 30 min.
3. Inactivate CIAP by heating to 65 °C for 1 h in the presence of 5 mM EDTA (pH 8.0).

4. Cool the reaction to room temperature and add an equal volume of saturated phenol: chloroform: isoamyl alcohol (25:24:1) and mix the contents gently.
5. Centrifuge for about 2 min and collect the aqueous phase.
6. Add 0.1 vol. of 3 M sodium acetate pH 4.8, two volumes of ethanol and incubate the mixture at −70 °C for 1 h or −20 °C overnight.
7. Centrifuge at 13,000 × *g* for 15 min and discard the supernatant.
8. Wash the pellet with chilled 70% ethanol.
9. Air dry the pellet and dissolve in 10 μL of sterile water.

43.3.2.3 Ligation

1. Take 1 μL (100 ng) of the vector DNA (linearized) in a sterile eppendorf tube.
2. Add 1 μL of equimolar amount of insert DNA (eluted fragment) and 2 μL of 10× ligase buffer.
3. Briefly spin down the samples and add 1 μL of T_4 DNA ligase (3–4 U/μL). Make up the volume to 20 μL by adding sterile water.
4. Incubate at 12–15 °C overnight.
5. Set up two control reactions each containing the linearized vector alone and fragment DNA alone, separately.

43.4 Transformation

Ligated vector will be used for transformation of *E. coli* followed by selection of transformants on agar plates containing appropriate antibiotics and/or X-gal. A master plate containing all putative transformants is then made. In order to identify positive recombinants, each of the colonies in the master plate is subjected to rapid disruption of bacterial cells, colony PCR using primers from the vector, isolation and restriction analysis of plasmid DNA from colonies. The detailed procedures for all these methods are provided in Chapter 42.

References

Gubler U, Hoffman BJ (1983) A simple and very efficient method for generating cDNA libraries. Gene 25:263–269

Okayama H, Berg P (1982) High-efficiency cloning of full length cDNA. Mol Cell Biol 2:161–170

Sambrook J, Russel DW (2001) Molecular cloning, vol I–III, 3rd edn. Cold Spring Harbor Laboratory Press, New York

Chapter 44

DNA Sequencing

Abstract

DNA sequencing is the process of determining the order of nucleotides in a DNA fragment. Sequencing of viral genome refers to the determination of order of nucleotides in a viral genome. In order to sequence a viral genome, it needs to be first cloned in a plasmid vector. Sequence data helps to identify, characterize and develop DNA-based diagnostics for the diagnosis of the virus. It also helps to compare with other viral genomes and to establish phylogenetic relationships among distant and closely related virus strains. Chemical degradation and chain termination methods are the initial methods for DNA sequencing. However, at present the chain termination method developed by Sanger is commonly used for sequencing. Besides, next-generation sequencing and next-next generation sequencing methods are also being used at present, especially to diagnose and identify unknown viruses infecting plants.

Key words Viral genome sequence, Nucleotide, Amino acid, Open reading frame, Conserved sequence, Sanger sequencing, Next-generation sequencing, Automated sequencing

44.1 Introduction

The DNA sequencing determines the order of the arrangement of nucleotide bases, adenine, guanine, cytosine and thymine in a molecule of DNA. In order to determine the identities between virus isolates/species, their genome need to be sequenced. Cloned viral genomes in a plasmid vector can be subjected to sequencing. Sequence information of a virus isolate/strain is used to identify, compare and to determine phylogenetic relationships among virus isolates. Once complete genome of a virus isolate is known, it can be used to develop physical map and genome organization of the virus isolate. Sequence information is needed to design virus-specific primers to use in PCR and other nucleic acid based methods to detect and diagnose viruses in plants. Besides, sequence information is a pre-requisite for planning any manipulation of the DNA. DNA sequencing procedures were first discovered in the mid-1970s with two different procedures, (1) Chain termination method: In this method the sequence of a single-stranded DNA molecule is determined by enzymatic synthesis of complementary polynucleotide chains which are terminating at specific nucleotide positions (Sanger et al. 1977); (2) Chemical

Alangar Ishwara Bhat and Govind Pratap Rao, *Characterization of Plant Viruses*, Springer Protocols Handbooks, https://doi.org/10.1007/978-1-0716-0334-5_44,

degradation method: In this method the sequence of a double-stranded DNA molecule is determined by chemical treatment by cutting the DNA molecule at specific nucleotide positions (Maxum and Gilbert 1977). Out of these two methods, the chain termination procedure has gained better acceptance in recent years, especially for genome sequencing, because the chemicals used are less toxic and is easier to automate chain termination sequencing.

44.2 Maxam–Gilbert Method

This method is based on chemical modification of DNA and subsequent cleavage at specific bases. During initial years, the Maxam–Gilbert sequencing was more popular, since purified DNA could be used directly; however, the initial Sanger method required that each read start to be cloned for production of single-stranded DNA. However, with the improvement of the chain termination method, chemical degradation method got outdated due to its technical complexity. The method requires radioactive labelling and purification of the DNA fragment to be sequenced. Chemical treatment generates breaks at a small proportion of one or two of the four nucleotide bases and a series of labelled fragments are generated, from the radiolabelled end to the first 'cut' site in each molecule. The fragments in the four reactions are arranged side by side in gel electrophoresis for size separation. To visualize the DNA fragments, the gel is exposed to X-ray film for autoradiography, yielding a series of dark bands each corresponding to a radiolabelled DNA fragment, from which the sequence may be inferred.

44.3 Chain Termination DNA Sequencing

In chain termination DNA sequencing, a single-stranded DNA molecule with difference in length by a single nucleotide can be separated. For this sequencing, the preparation of identical single-stranded DNA molecules is a pre-requisite. The first step in this method is to anneal a short oligonucleotide at same position on each molecule. This oligonucleotide subsequently acts as the primer for synthesis of a new DNA strand, complementary to the original template. DNA polymerase enzyme along with the four deoxyribonucleotide triphosphates (dATP, dCTP, dGTP and dTTP, one or more of which is radioactively labelled) and a small amount of a dideoxynucleotide (e.g. ddNTP) are added. The polymerase enzyme does not discriminate between dNTPs and ddNTPs, hence dideoxynucleotide can be incorporated into the growing chain, but blocks further elongation because of lack of the 3′ hydroxyl group required to form a connection with the next nucleotide. For example, if ddATP is present, chain termination occurs at positions opposite to thymidines in the template. Because normal dATP is also present, the strand synthesis does not always terminate at first 'T' in the template, but in fact it may continue until several hundreds of nucleotides have been polymerized before a ddATP is incorporated. The result is therefore a set of new chains but each ending in ddATP but of different length. Similarly, the reaction terminating in ddCTP, ddGTP and ddTTP can be obtained. The reaction products generated in the presence of each of the ddNTPs are loaded into separate lanes of the denaturing polyacrylamide-urea gel. After electrophoresis, the DNA bands are visualized by autoradiography and the DNA sequence can be visualized directly from the band position on gel. The band that has moved the farthest represents the smallest piece of DNA, this being the strand that terminated by incorporation of a ddNTP at the first position in the template. The sequence reading can be continued up to the region of the gel where individual bands are well separated. With the advent of automated DNA sequencing method, the conventional method of using radioactive label and denaturing polyacrylamide gel electrophoresis to resolve the bands and autoradiography to visualize the bands is no more used.

44.4 Automated DNA Sequencing

Although chain termination technique was initially very slow and tedious, the process was sped up using fluorescent dye labels (ddA Green, ddT Red, ddG Yellow, ddC Blue) which allow the reaction to be carried out in a single tube. This process was further modified and improved with the use of a thermostable DNA polymerase for extension. Hence entire sequencing reaction can now be carried out using the PCR thermocycler. For automated sequencing, the primer or the ddNTPs are labelled with a fluorescent dye (Fig. 44.1a). Thus, rather than running the gel, the machine uses a laser to read the fluorescence of the dye as the bands pass a fixed point (Fig. 44.1b). Much longer sequences can be read from each track in this way. A further advantage is that the sequence read by the machine is fed automatically into a computer. This is not only much quicker than reading a gel manually and typing the resulting sequence into a computer, but also avoids the errors that would happen with manual data entry (Smith et al. 1985, 1986; Brown 2010).

Automated DNA sequencing instruments has capacity for sequencing up to 384 DNA samples in a single run and up to 24 runs a day. A DNA sequencer consists of capillary electrophore sis facility for size separation of DNA fragments, detection and recording of the dye fluorescence and data output as fluorescent peak based chromatograms (Fig. 44.2). Amplification of the template DNA by PCR, purification of PCR product and re-suspension

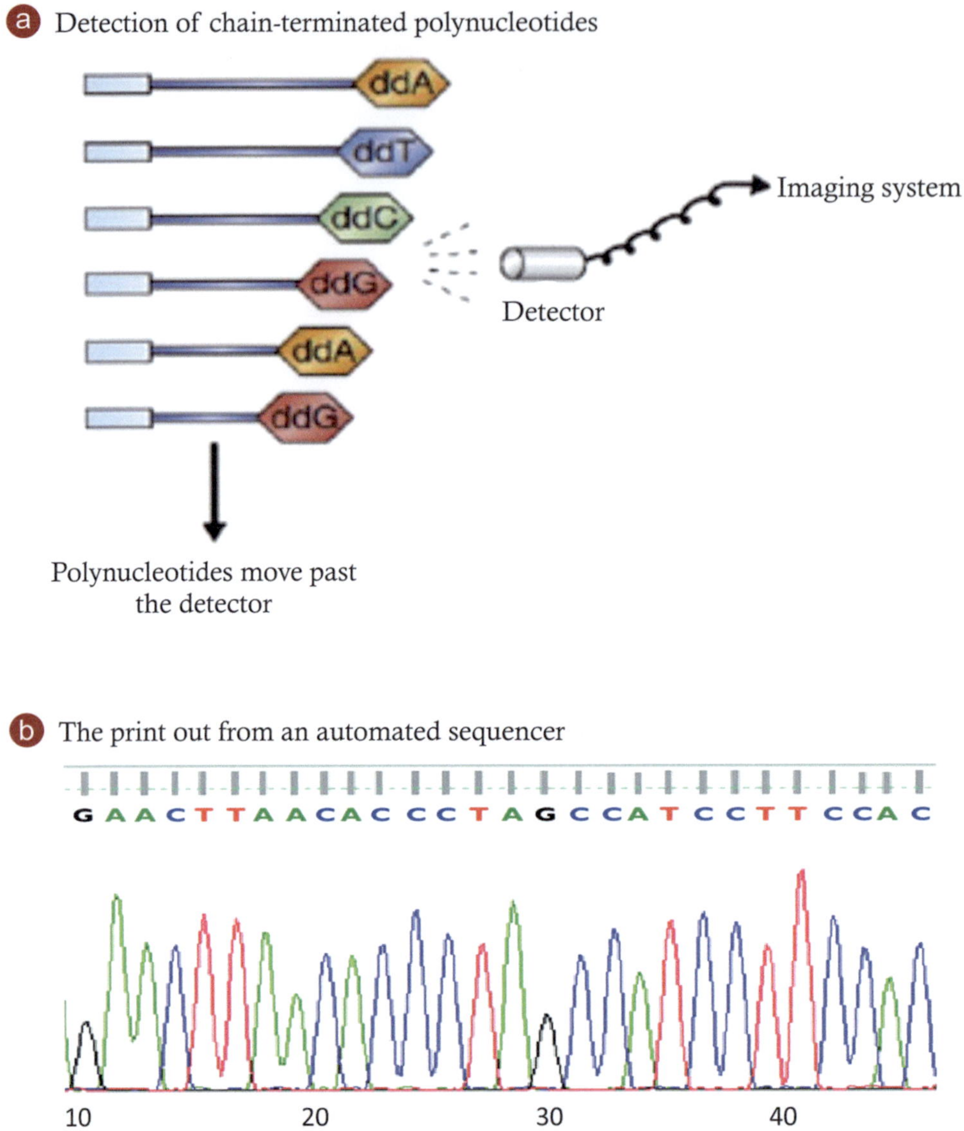

Fig. 44.1 Reading the DNA sequence generated by a chain termination method. (**a**) Each dideoxynucleotide is labelled with a different fluorochrome, so the chain-terminated polynucleotides are distinguished through the detector. (**b**) The print out example of a sequence. (Reproduced from El-Bondkly 2014)

in a buffer solution is done before loading the samples onto the sequencer. Samples are subjected to capillary electrophoresis and the presence or absence of a DNA strand is then recorded in a detector (Fig. 44.2). The shorter strands of DNA move faster through the gel matrix and hence are detected sooner and there is a direct correlation between the length of the DNA strand and time at the detector. This relationship is used to determine the actual length of the DNA sequence. Now software packages are available which can trim low-quality DNA traces automatically and score only the quality peaks. Using M13 forward primer and M13 reverse

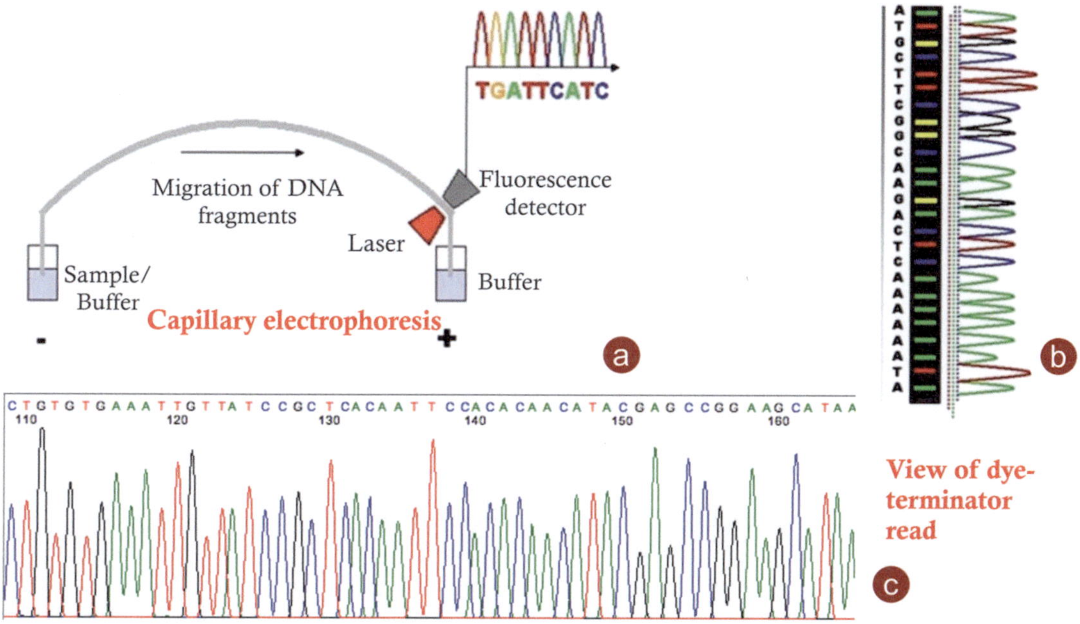

Fig. 44.2 Capillary electrophoresis for the separation of DNA fragments generated for sequencing (**a**), view of the chromatogram obtained during automated dye-terminator sequencing (**b**, **c**)

primers one can generate sequence for a clone from either ends. In order to bring the fragment to be sequenced nearer to the primer, the clones can be further subcloned or nested deletions can be generated by using exonuclease III. Alternatively, if related sequences are known sequencing grade primers can be made. Using this primer, sequence of 300–400 nucleotides can be generated. Based on sequence generated, next set of primers are made to sequence the region following it. This method of completing the sequencing of entire genome is called primer walking.

44.5 Whole-Genome Sequencing

Whole-genome sequencing has become important criteria in fine and authentic taxonomic classification of virus genus. The common approaches consist of cutting (with restriction enzymes) or shearing (with mechanical forces) large DNA fragments into shorter DNA fragments (Fig. 44.3). The fragmented DNA is cloned into a vector, and then amplified in *Escherichia coli*. Short DNA fragments purified from individual bacterial colonies are individually sequenced and assembled into one long and contiguous sequence electronically. This method does not require any pre-existing information about the DNA sequence and known as *de novo* sequencing. Gaps in the assembled sequence can be filled by primer walking. Shotgun methods are more commonly used for the sequencing of

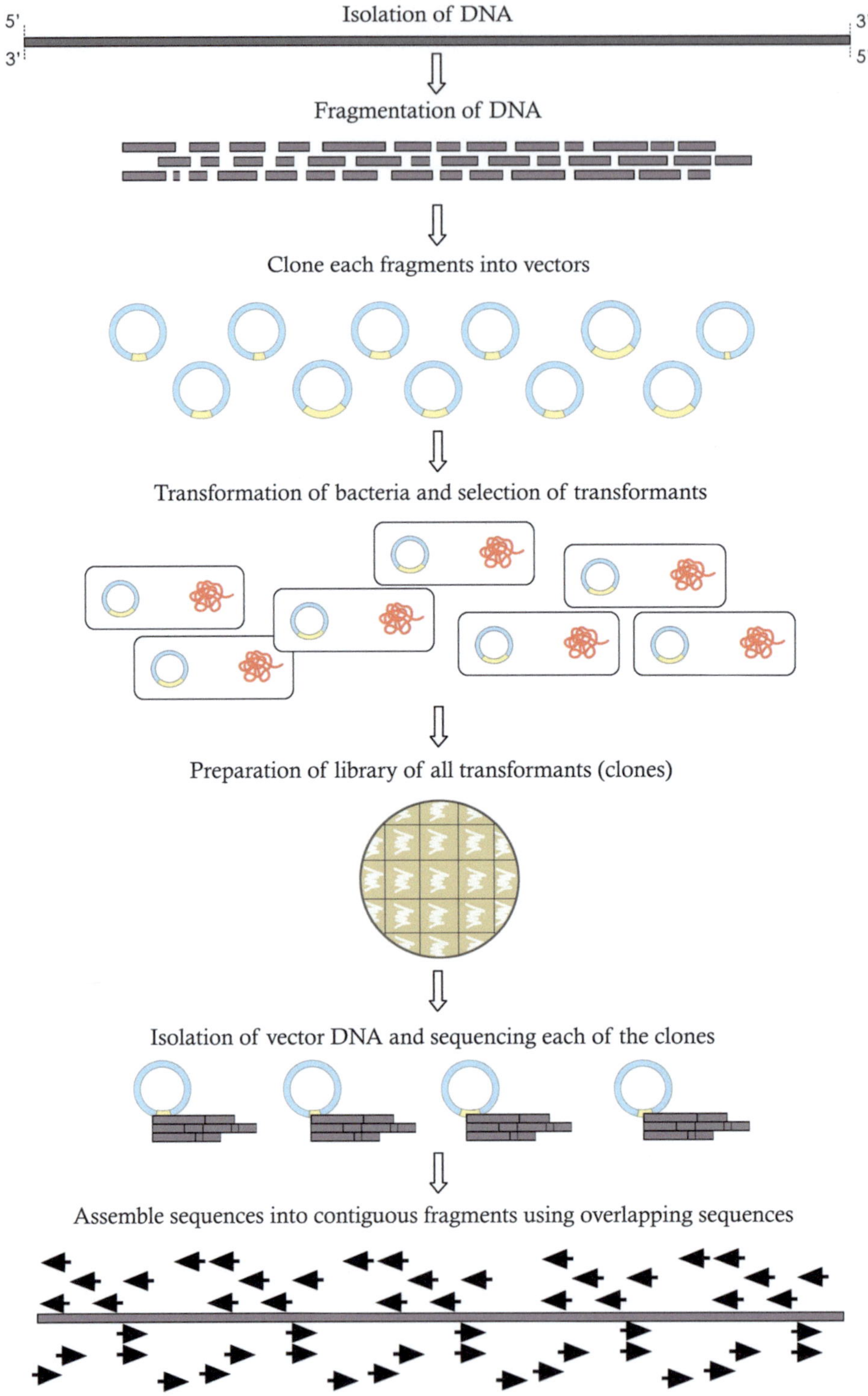

Fig. 44.3 Strategy used for whole-genome sequencing. DNA is fragmented into pieces and cloned as bacterial library. DNA from the individual bacterial clones is sequenced and sequence is assembled with overlapping DNA regions

large genomes, but during its assembly sometimes problems may arise as assembly process is complex and difficult, with sequence repeats often causing gaps in genome assembly.

44.6 Next-Generation Sequencing

The next-generation sequencing (NGS) enables a genome to be sequenced within hours to days through massive parallel sequencing approach. NGS approaches simultaneously generate sequence information at tens to hundreds of billions of base pair positions. Because of the substantial time and cost savings of NGS, compared with traditional sequencing methods, NGS has ushered in an era of genome-level sequencing projects that were previously unimaginable (Wheeler et al. 2008).

Roche Applied Sciences introduced and commercialized the first next-generation sequencer to become commercially available in 2004, followed by Solexa 1G Genetic Analyzer from Illumina in 2006, and the SOLiD (Supported Oligonucleotide Ligation and Detection) System from Applied Biosystems in 2007. Currently many platforms of NGS are available for sequencing. Recently next-next generation sequencing (also called as third-generation sequencing) based on the single-molecule sequencing approaches is available. In plant virology, NGS methods are used for the diagnosis and identification of unknown viruses infecting plants.

Creation of a library of DNA templates is the starting point for NGS, and a range of strategies may be employed in library creation. In general, the DNA template library is generated by first breaking the genomic DNA into homogenously sized fragments that are then ligated to adapter molecules on both ends. These adapter molecules facilitate the immobilization of DNA templates onto glass slides or beads, which produces a spatial distribution of DNA templates that allows for each of them to be imaged independently. In some cases, the immobilized DNA templates are amplified with universal primer sequences contained within the adapters, which transforms each DNA template into a cluster of identical DNA molecules. In other cases, the signal intensity generated by a single DNA template molecule is adequate for detection, and no DNA template amplification is required.

Upon generation of a library of DNA templates, most NGS approaches use dye-labelled nucleotides that are specifically incorporated during cycle sequencing and are subsequently imaged. After the imaging step of each cycle, the labelled nucleotides are modified to remove the dye molecules and to allow progression of the next cycle of incorporation and imaging. Although the chemistry and experimental design vary among platforms, every NGS platform requires the sequential imaging of fluorescent signals at fixed locations to facilitate and assemble sequence reads from the

DNA templates or clusters. A brief chemistry used by different platforms is described in Chapter 41.

The raw data generated by NGS consist of a very large number of sequences reads, the lengths and configuration of which are dictated by the platform and experimental design used for the sequencing. These sequence reads must be aligned to a known reference sequence to determine the variability. In some cases, it may be beneficial to first assemble the sequence reads with one another in the absence of the reference sequence, referred to as *de novo* assembly. In the case of structural variants, gene families with a high degree of homology, or regions rich in repetitive elements, sequence reads may align with multiple genomic locations in the reference sequence and as a result may be excluded from subsequent analysis. With *de novo* assembly, one makes attempts to harness the power of the overlapping sequence reads to place these otherwise multiply-mapping reads into their correct genomic position. Once the sequences have been assembled and aligned, one may generate a list of positions at which the experimental sample differs from the reference, and rapidly evolving bioinformatic tools can annotate sequence variants based upon characteristics such as the type of variant, the depth of coverage or quality scores of a variant, the degree of conservation of that reference position across other organisms and the predicted effect of the variant on its protein product (Kumar 2012).

Even when sequencing is performed by the most experienced sequencing centres and bioinformaticians, errors are inescapably introduced at multiple steps of the NGS workflow, leading to an errant base call rate as high as 1%. Potential sources of error are widespread and may include the limited fidelity of polymerase enzymes during the whole-genome amplification or cluster amplification processes, the incorporation of degraded dye-labelled nucleotides or the misincorporation of dye-labelled nucleotides during cycle sequencing, the presence of optical interference that precludes proper identification of bases, and the false-positive and false-negative variant identification that may result from erroneously mapped sequence reads. In response to this problem, there is a rapid emergence of both experimental and bioinformatic approaches that promise to better identify erroneously called variants. At this point, validation of NGS findings by Sanger sequencing remains the gold standard.

Although there exist an ever-growing number of NGS platforms and techniques, they each share a central schema which includes preparation of a library of DNA templates, sequencing and imaging of the template library, and finally analysis of the data. A general example of this NGS process is depicted in Fig. 44.4.

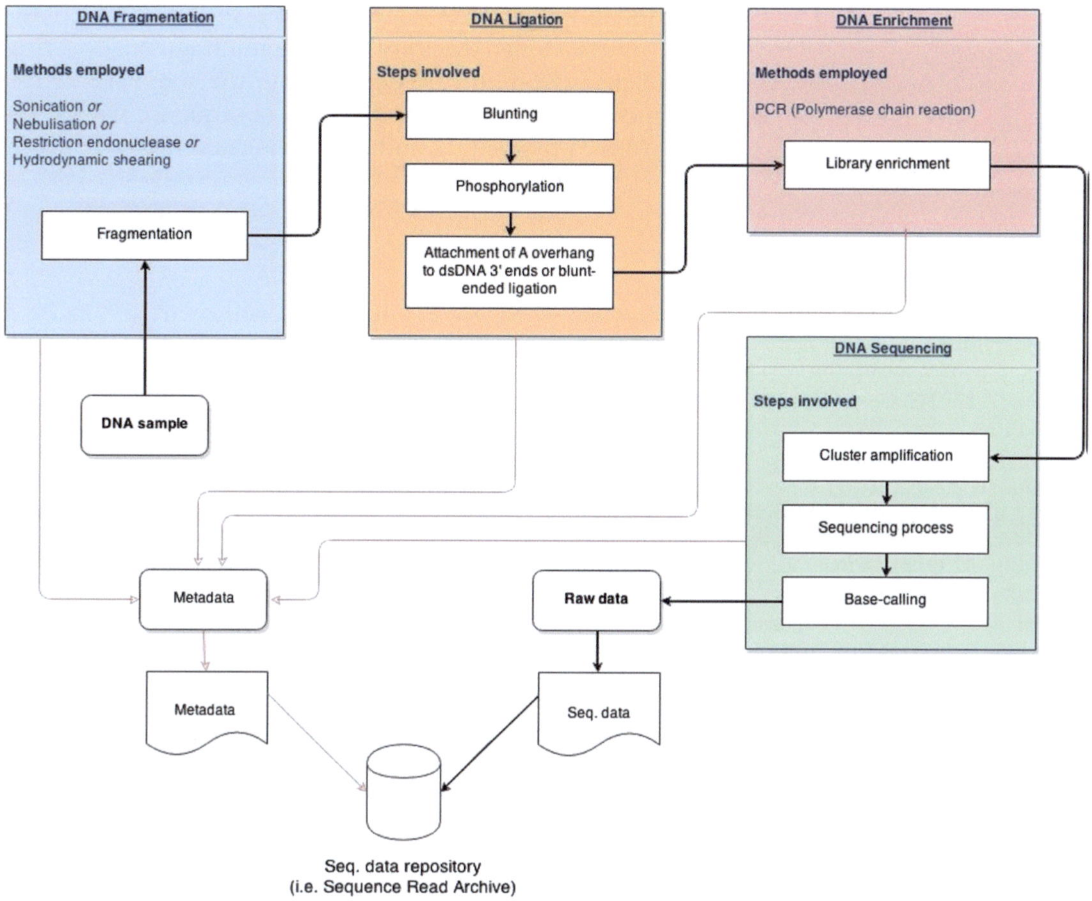

Fig. 44.4 Different steps in next-generation sequencing protocol. The sequencing workflow is shown by the black arrows; red arrows depict the metadata that should be captured from these sequencing workflow steps. (Reproduced from Alnasir and Shanahan (2015))

44.7 Notes

1. Most of the current Sanger sequencing is done through automated sequencing facility (also known as dye-terminator sequencing) that utilizes labelling of the chain terminator ddNTPs, which permits sequencing in a single reaction. Each of the four dideoxynucleotide chain terminators is labelled with different fluorescent dyes (ddA Green, ddT Red, ddG Yellow and ddC Blue), each of which with different wavelengths of fluorescence and emission. The fragment stopping at the base position is detected on the gel by a powerful laser beam (Brown 2010).
2. Both manual and automated DNA sequencing methods suffer from problems that can give rise to errors in the sequence. These are often associated with the presence of certain combinations of bases in the DNA. Some of these potential errors can

be identified and minimized by altering the reaction conditions, by sequencing a different overlapping fragment covering the problem region and by determining the sequence of the complementary strand. A good complete sequence will therefore be derived by reading and assembling several overlapping sequences in each direction.

References

Alnasir J, Shanahan HP (2015) Investigation into the annotation of protocol sequencing steps in the sequence read archive. GigaScience 4:23

Brown TA (2010) Gene cloning and DNA analysis: an introduction. Wiley, Chichester, West Sussex, UK

El-Bondkly AMA (2014) Biotechnology and biology of Trichoderma. In: Gupta V, Schmoll M, Herrera-Estrella A, Upadhyay R, Druzhinina I, Tuohy M (eds) Biotechnology and biology of Trichoderma. Elsevier, pp 377–392

Kumar BR (2012) DNA representation. In: Munshi A (ed) DNA sequencing—methods and applications. In Tech Publishers, Rijeka, Croatia, pp 3–14

Maxum AM, Gilbert W (1977) A new method for sequencing DNA. Proc Natl Acad Sci U S A 74:560–564

Sanger F, Nicklen S, Coulson AR (1977) DNA sequencing with chain terminating inhibitors. Proc Natl Acad Sci U S A 74:5463–5467

Smith LM, Fung S, Hunkapiller MW, Hunkapiller TJ, Hood LE (1985) The synthesis of oligonucleotides containing an aliphatic amino group at the 5' terminus: synthesis of fluorescent DNA primers for use in DNA sequence analysis. Nucleic Acids Res 13:2399–2412

Smith LM, Sanders JZ, Kaiser RJ et al (1986) Fluorescence detection in automated DNA sequence analysis. Nature 321(6071):674–679

Wheeler DA, Mothberg S, Rothberg JM (2008) The complete genome of an individual by massively parallel DNA sequencing. Nature 452:872–876

Chapter 45

Sequence Analysis and Phylogenetic Studies

Abstract

After determining partial or complete genome sequence of a test virus, nucleotide sequences may be submitted to the international sequence databases. The sequence data is then used to perform the following major analysis: (1) to determine the exact identity of the virus species, (2) determination of open reading frames and genome organization, (3) identification of functionally important conserved motifs and domains, (4) determination of percent sequence identity and phylogenetic relationships with viruses of the same species or different genera. For performing these analyses, several softwares available in the public domain can be used.

Key words Virus sequence, GenBank, Blast, Identification of the virus, Percent identity, Similarity, Phylogeny, Sequence comparison, Plant virus database

45.1 Introduction

Once the nucleotide sequence (either partial or complete) of a test virus is determined by experimental techniques (e.g. Sanger sequencing or next-generation sequencing), the sequence may be deposited in public domain nucleotide sequence database such as EMBL (European Molecular Biology Laboratory) (http://www.ebi.ac.uk/), GenBank at NCBI (National Centre for Biotechnology Information) (http://www.ncbi.nlm.nih.gov) or DDBJ (DNA Databank of Japan) (http://www.ddbj.nig.ac.jp) for public use. Each of these databases collects the sequence data, and updated database entries are exchanged between the databases on a daily basis. Whenever a new sequence is submitted, within two working days, the submitter would be provided a unique accession number that will never change. The sequences can be submitted to any of nucleotide sequence databases (e.g. DDBJ, EMBL and GenBank) depending on the convenience (Benson et al. 1999). There are two ways of submitting DNA sequences to GenBank database; these include (1) Bankit (www based submission tool) recommended for simple submissions and (2) sequin—used for more control over annotating the entry, segmented records or very long entries. Similarly, to submit sequences to EMBL, the WEBIN (www) based tool is used. The nucleotide sequences submitted to the databases

Alangar Ishwara Bhat and Govind Pratap Rao, *Characterization of Plant Viruses*, Springer Protocols Handbooks, https://doi.org/10.1007/978-1-0716-0334-5_45, © Springer Science+Business Media, LLC, part of Springer Nature 2020

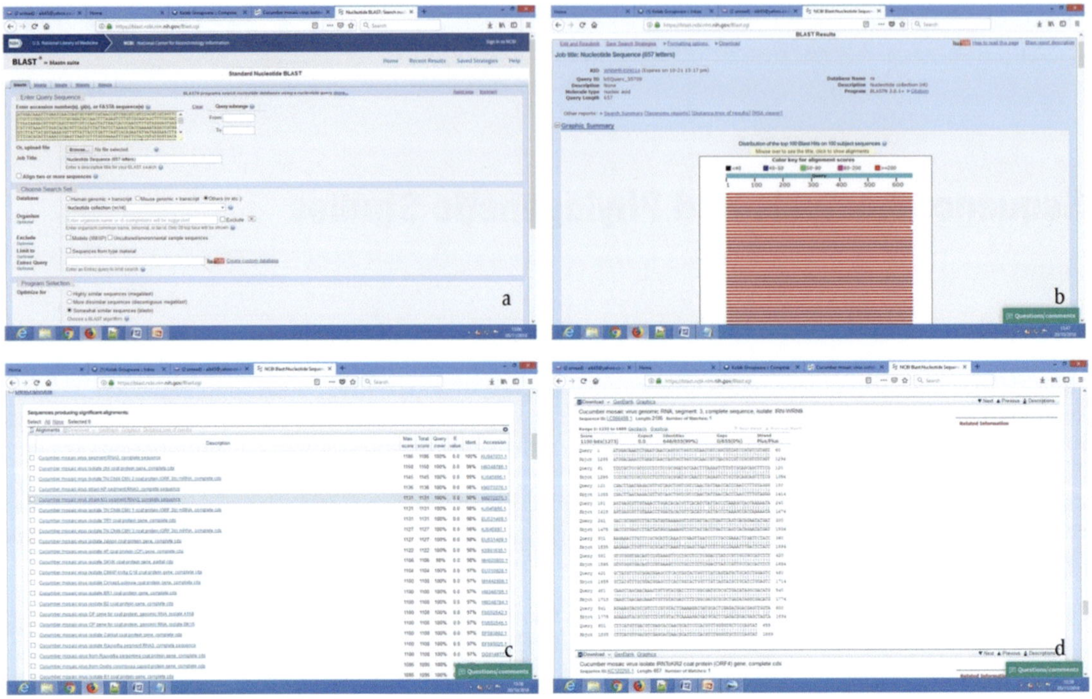

Fig. 45.1 Screen shot of a BLAST input and output. (**a**) Input page, (**b**) graphic summary, (**c**) Descriptions-displays Identity—Percentage of exact sequence match, its accession number, Expect (*E*-value or *p*-value) probability or expected frequency value, denotes statistical significance (low values indicate greater statistical significance), (**d**) Alignments—Provides sequence alignment of query sequence entered by the user (input) with subject-sequence in the database which is being compared to

should be annotated. The submitter should provide sufficient information such as nucleotide and protein sequences, coding region, gene, mRNA features and organism from which sequences were determined, its source, location, date of collection, name of the collector and so on to make the sequences useful to others.

45.2 Identification of Similar Sequences

In order to know the similarity of the new sequence with the available sequences deposited in the database, similarity search is done. BLAST (basic local alignment search tool) program is the most popular tool used for determining the similarity of the query sequence to the subject (target) sequence (Altschul et al. 1997). The program compares query with each of the subject sequences and computes the score for each alignment, and the query-target pairs with best scores are then reported to the user (Fig. 45.1). The scores reflect the degree of similarity between the query and the sequence being compared. The method involves copying the query sequence in the box provided in BLAST interface and then

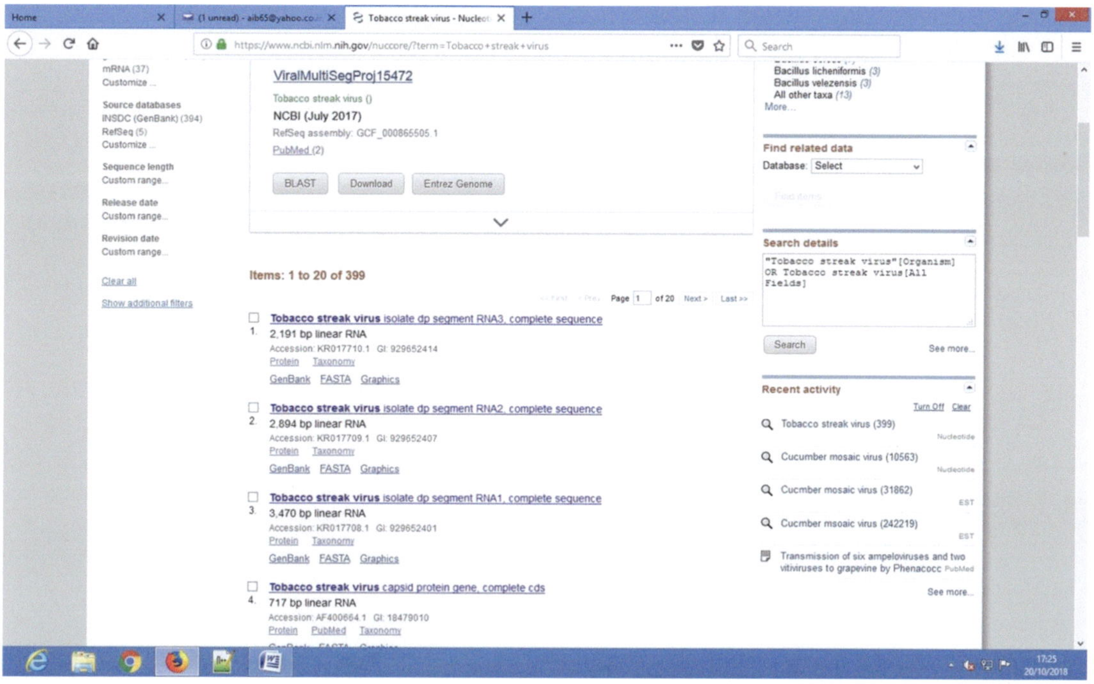

Fig. 45.2 Sample screen shot of retrieval of 'tobacco streak virus' sequences through Entrez

executing the program. The result showing percent similarity of query sequence (along with bit score, identity, query coverage, etc.) with each of the target sequences is presented as an output within a short span, depending upon the server load (Fig. 45.1).

45.3 Sequence Retrieval from Databases

Mainly three data retrieval tools [Entrez, Sequence retrieval system (SRS) and DBGET] are available. These tools allow text searching of databases and provide links to relevant information for entries that match search criteria. For example, if you are interested to retrieve sequences of 'tobacco streak virus (TSV)', use 'tobacco streak virus' as text search at NCBI Entrez. This would provide list of all of the TSV sequences available in the database (Fig. 45.2). Users can download and save individual sequences by clicking on the interested sequence from the list. Sequence can also be downloaded in FASTA format so that they can be used for multiple sequence alignment (MSA) or any other analysis.

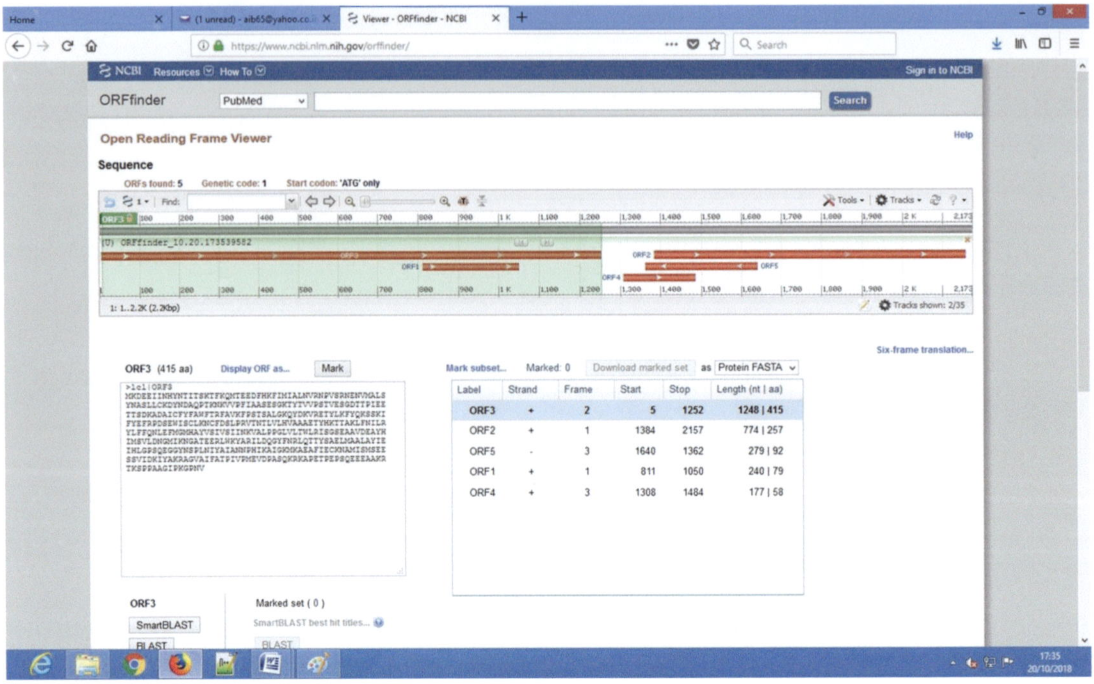

Fig. 45.3 Screen shot of ORF finder of National Centre for Biotechnology Information (https:/www.ncbi.nlm.nih.gov/orffinder/). It shows location and number ORFs in all six possible reading frames along with corresponding amino acid sequence. The amino acid sequence can be directly subjected to blast search to confirm identity of the sequence

45.4 Detecting Open Reading Frame (ORF)

It is considered as the longest reading frame uninterrupted by the stop codon (TGA, TAA or TAG). Generally the initial codon is that of methionine (ATG). Since it is also a very common amino acid, it is necessary to use additional techniques to detect 5′ un-translated sequence end or recognition of flanking sequence. There are several packages that can be used to find out ORF and to translate DNA sequence to protein such as NCBI ORF finder (https://www.ncbi.nlm.nih.gov/orffinder/), ORF finder-Bioinformatics (http://www.bioinformatics.org/sms2/orf_find.html), ORF finder-GenScript (https://www.genscript.com/sms2/orf_find.html) and SEQAID (Peltola et al. 1984) (Fig. 45.3).

45.5 Translation of Nucleotide Sequence to Protein Sequence

Nucleotide sequence can be converted into anti-sense sequence (reverse complement) or *vice versa* using different software such as Reverse Complement-Bioinformatics (https://www.bioinformatics.org/sms/rev_comp.html). Similarly, to translate a nucleotide sequence into protein, several software, such as Expasy (https://web.expasy.org/translate/), DNA to protein translation (http://insilico.ehu.es/translate/) and translate bioinformatics (www.bioinformatics.org/sms2/translate.html), are available.

45.6 Identification of Conserved Motifs and Domains

To understand the functional roles of proteins, one of the prime objectives in sequence analysis is the identification of structural motifs and domains. In a protein sequence, a particular arrangement of secondary structure or amino acids that could be present in other similar proteins also is called as a motif. Also, if that particular arrangement is somehow related to particular function of protein (i.e. DNA or protein binding, catalytic, etc.), then it is called a domain. For example, the leucine zipper motif is usually found as part of a dimerization domain in many transcription factors. Motif Scan (https://myhits.isb-sib.ch/cgi-bin/motif_scan) and MOTIF Search (https://www.genome.jp/tools/motif/) are one of the most popular web-based tools for finding all known motifs that occur in sequences. PROSITE database (https://prosite.expasy.org/) comprises information of protein domains, families and functional sites as well as associated patterns and profiles. Another important database, ProDom is a comprehensive set of protein domain families that are automatically generated from UniProt Knowledge database.

45.7 Determination of Similarity Between Sequences

Determination of percent identity of query sequence with a subject sequence will help to identify the gene and its function. If two sequences are sufficiently similar, almost invariably they have similar biological function and will be descended from a common ancestor. Similarity is measured through identity, similarity and homology. Identity represents the number of common bases between two sequences at aligned positions. In similarity, a number of residues will be replaced by ones of similar physico-chemical properties. When two sequences are evolutionarily related and stem from a common ancestor, they are called homologous. Several bioinformatics software tools such as clustal omega (https://www.ebi.ac.uk/Tools/msa/clustalo/), Bioedit (http://www.mbio.ncsu.edu/BioEdit/bioedit.html), etc. are available to determine the similarity between two or multiple sequences.

45.8 Multiple Sequence Alignment

Sequence alignment is an approach of arranging the DNA, RNA or protein sequences to infer similar regions that may reflect functional, structural or evolutionary relationships between them. Aligning two or more sequences together helps in finding out more about the sequences. Two of the most popular programs available for pairwise alignments include FastA (https://www.ebi.ac.uk/Tools/sss/fasta/nucleotide.html) and BLAST (https://blast.ncbi.nlm.nih.gov/Blast.cgi) (Corpet 1988). Pairwise alignment tool allows seeking for regions between two sequences that may be of widely differing lengths. Multiple sequence alignment (MSA) is the process of aligning more than two sequences with each other so as to bring as many similar sequence characters (nucleotides or amino acids) into register as possible. All phylogenetic methods need multiple sequence alignment to start with. The correctness of this alignment determines the correctness of resulting phylogenetic tree. Clustal W and Clustal X are generally the most widely used programs for obtaining MSA (Thompson et al. 1994, 1997). Other programs include MULTALIGN (http://multalin.toulouse.inra.fr/multalin/), PILEUP, etc. It involves progressive alignment using neighbour joining method for the generation of guide tree. It is rapid enough to align hundreds of sequences. MSA may be used to find conserved/diverged regions, unique sequence (s) present in a single sequence and possible evolutionary relationships among the sequences. It is preferable to use amino acid sequences when inferring phylogenic relationship/building trees or seeking for structural homologies. If you have a DNA sequence, then it can be translated and used. The general procedure for performing a MSA comprises of the following steps: (1) Searching for similar sequences in databases to identify all potential homologues (can be done by the BLAST program), (2) Making an input file set (collection of similar sequences under investigation in FASTA format), (3) Computing the alignments with appropriate software and (4) Checking, analysing and editing the alignment with any competitive alignment viewer (Fig. 45.4) (Felsenstein 1985; Higgins and Sharp 1988; Thompson et al. 1994, 1997; Hall 1999).

45.9 Phylogenetic Analysis

Phylogenetic analysis involves estimating the evolutionary past based on the comparison of DNA or protein sequences. It is usually depicted as branching (tree-like) diagrams, which represent a sort of pedigree of the inherited relationships among organisms (Fig. 45.5). A phylogenetic tree, also called evolutionary tree, is a statement about the evolutionary relationship between a set of

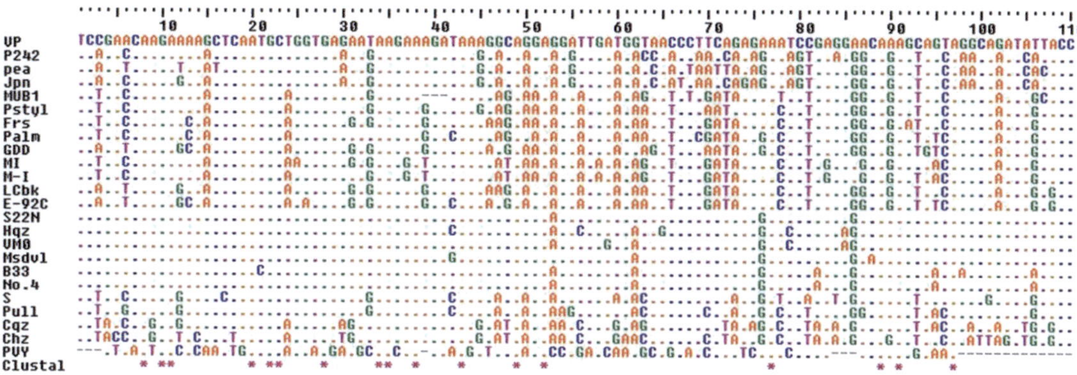

Fig. 45.4 Example of a Clustal X output of multiple sequence alignment of bean common mosaic virus isolates

homologous characters of one or several organisms. Phylogenetic trees express the patterns of relationship between organisms. It represents the evolutionary relationships among various biological species on the basis of similarities and differences in their physical or genetic characteristics. Trees help us distinguish characters shared due to common ancestry from those that converged, possible due to environmental pressures. A phylogenetic tree is composed of branches and nodes. Branches connect nodes, where a node is the point at which two or more branches diverge. A node and everything arising from it is a 'clade' (monophyletic group) where all members are derived from a unique common ancestor and have inherited a set of unique common traits from it. Root is the oldest point in the tree and a tree is rooted using an 'outgroup' (anything that is not a natural member of the group of interest).

MSA file is the input for generating phylogenetic trees. MSA builds an alignment up stepwise, starting with the most similar sequences and progressively adding the more dissimilar ones. The process begins with the construction of a 'guide tree'. This tree then determines the order in which the sequences are progressively added to build an alignment. The methods for calculating phylogenetic trees fall into two categories. These are distance-matrix, also known as clustering or algorithmic methods (e.g. UPGMA, neighbour joining, and Fitch–Margoliash), and discrete data methods, also known as tree searching methods (e.g. parsimony, maximum likelihood and Bayesian methods). PHYLIP, MEGA and PAUP are the most widely used phylogenetic packages. Bootstrapping is performed to test the reliability of the phylogenetic tree and estimates a statistics for which the underlying distribution is not known. In a phylogenetic tree, branch lengths are drawn to scale that is proportional to the amount of evolution estimated to have occurred along them (Page 1996; Ronquist and Huelsenbeck 2003).

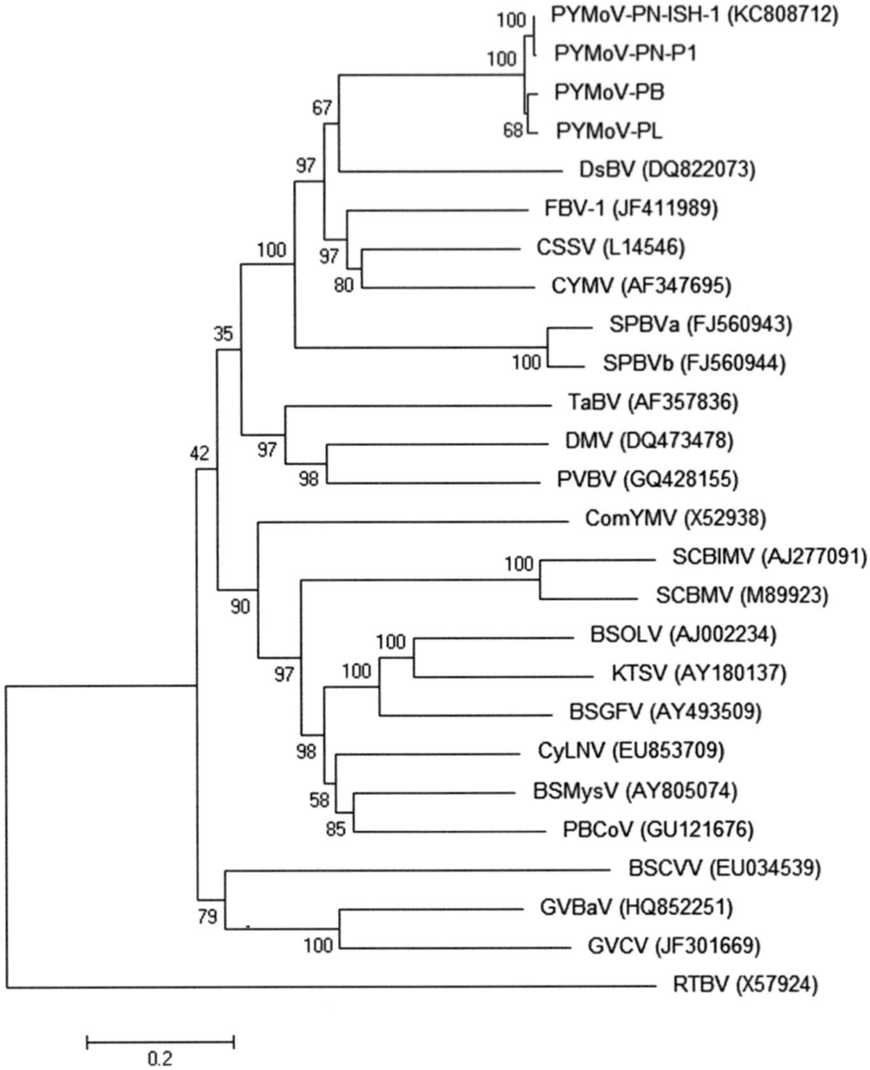

Fig. 45.5 Phylogram showing phylogenetic relationship between different isolates of badnaviruses. Rice tungro bacilliform virus (RTBV) was used as outgroup. The tree was generated by neighbour hood joining method in Clustal X. Bootstrap values are shown at the individual nodes. (Reproduced from Deeshma and Bhat 2015 with permission from Springer)

References

Altschul SF, Madden TL, Schäffer AA, Zhang J, Zhang Z, Miller W, Lipman DJ (1997) Gapped BLAST and PSI-BLAST: a new generation of protein database search programs. Nucleic Acids Res 25:3389–3402

Benson DA, Boguski MS, Lipman DJ, Ostell J, Ouellette BF, Rapp BA, Wheeler DL (1999) GenBank. Nucleic Acids Res 27:12–17

Corpet F (1988) Multiple sequence alignment with hierarchical clustering. Nucleic Acids Res 16:10881–10890

Deeshma KP, Bhat AI (2015) Complete genome sequencing of *Piper yellow mottle virus* infecting black pepper, betelvine and Indian long pepper. Virus Genes 50:172–175

Felsenstein J (1985) Confidence limits on phylogenies: an approach using the bootstrap. Evolution 39:783–791

Hall TA (1999) BioEdit: a user-friendly biological sequence alignment editor and analysis program for Windows 95/98/NT. Nucleic Acids Symp Ser 41:95–98

Higgins DG, Sharp PM (1988) CLUSTAL: a package for performing multiple sequence alignment on a microcomputer. Gene 73:237–244

Page RDM (1996) TREEVIEW: an application to display phylogenetic trees on personal computers. Comput Appl Biosci 12:357–358

Peltola H, Soderlund H, Ukkonen E (1984) SEQAID: a DNA sequence assembling program based on a mathematical model. Nucleic Acids Res 12:307–321

Ronquist F, Huelsenbeck JP (2003) MRBAYES 3: Bayesian phylogenetic inference under mixed models. Bioinformatics 19:1572–1574

Thompson JD, Higgins DG, Gibson TJ (1994) CLUSTAL W: improving sensitivity of progressive multiple sequence alignment through sequence weighting, position-specific gap penalties and weight matrix choice. Nucleic Acids Res 22:4673–4680

Thompson JD, Gibson TJ, Plewniak F, Jeanmougin F, Higgins DG (1997) The Clustal X Windows interface: flexible strategies for multiple sequence alignment aided by quality analysis tools. Nucleic Acids Res 24:4876–4882

Chapter 46

Development of Infectious Clone of Virus

Abstract

An infectious clone is a full-length DNA clone from which infectious transcripts can be obtained in vitro or in vivo with a suitable promoter. An infectious clone is the basic need to study functional genomics, replication and expression of viral proteins and in understanding host–virus interactions. Infectious clones, considered as the pools of viral genes for designing antiviral strategies, are an essential source of material for the preparation of new viral vectors. In this chapter, production of in vitro and in vivo clones of viruses is discussed.

Key words Full-length infectious clone, Full-length infectious cDNA, In vitro transcription, In vivo transcription, Bacteriophage RNA polymerase promoter, Agro-inoculation, Cauliflower mosaic virus 35S promoter

46.1 Introduction

The infectious clone has become an indispensable tool in plant virus research (Kobayashi et al. 2007). The construction of infectious full-length cDNA clones (FL-cDNAs) is still often complicated and time-consuming (Boyer and Haenni 1994). FL-cDNAs provide valuable information on the expression of viral genomes, replication and mechanism of the infection cycle. Brome mosaic virus (BMV) was the first plant virus, where infectious in vitro RNA transcripts were successfully prepared and the construction of a directly infectious cDNA clone was first reported for RNA 3 of alfalfa mosaic virus (AlMV) (Ahlquist and Janda 1984; Dore and Pinck 1988).

46.1.1 Construction of Infectious Clones

There are currently two ways of constructing infectious clones: (1) The construction of a FL-cDNA clone of a viral genome, from which an in vitro infectious transcript can be synthesized under the influence of a bacteriophage RNA polymerase promoter (T3, T7 or SP6) (2) The expression of infectious viral RNAs by in vivo transcription of cDNA containing vectors through a cauliflower mosaic virus (CaMV) 35S promoter.

Alangar Ishwara Bhat and Govind Pratap Rao, *Characterization of Plant Viruses*, Springer Protocols Handbooks, https://doi.org/10.1007/978-1-0716-0334-5_46, © Springer Science+Business Media, LLC, part of Springer Nature 2020

46.1.2 Introduction of Infectious Clones into Plants

Many technologies are available to transfer gene of interest into plant cells and mainly categorized into two classes, i.e. direct and indirect gene transfer. The direct gene transfer technique involves the use of physical equipment to transfer the gene product *viz.*, mechanical inoculation, biolistic/particle bombardment, protoplast transformation and electroporation. The indirect gene transfer technique uses the *Agrobacterium tumefaciens* bacteria as the vehicle for DNA delivery, which transfers part of its DNA (T-DNA) into the host plant genomes. Mechanical inoculation is usually done for the inoculation of in vitro run-off RNA transcripts and intact plasmid DNA of FL-cDNA clones. In this method, leaf surface is damaged with an abrasive such as celite or carborundum, which permits introduction of nucleic acid into the injured cells directly (Hull 2002; Ding et al. 2006). Agro-inoculation/agro-infection is a very efficient technique to introduce the gene of interest into the plants, because of high stability of in vivo synthesized transcripts (Annamalai and Rao 2005). *Agrobacterium*-based binary vectors have been widely used for delivering infectious full-length copies of viral genomes, inserted in the T-DNA region of the plasmids, into the plants. When the T-DNA is replaced with a cDNA clone of a virus, the virus will be transcribed, transported through nuclear pores from the nucleus to the cytoplasm, where it will get translocate, replicate and induce infection in the host plant. In vivo transcribed viral RNA from the *Agrobacterium*-delivered DNA initiates the expression of viral encoded proteins involved in replication and RNA silencing suppression, allowing the initiation of a viral cycle of replication and subsequent systemic spread of the virus throughout the plant (Leiser et al. 1992; Prufer et al. 1995). Agro-inoculation (Agro-infection) is usually performed by infiltrating leaves (or injection of stem or petioles) with *A. tumefaciens* cells carrying binary plasmids containing FL-cDNAs of virus genome components (Grimsley et al. 1986, 1987).

The agro-infiltration is simple, efficient and widely used technique to deliver the FL-cDNAs (Bendahmane et al. 2000; Voinnet et al. 2003). Agro-inoculation has been developed for a number of plant RNA viruses such as tobacco mosaic virus (TMV), potato virus X (PVX), cowpea mosaic virus (CPMV) and tobacco rattle virus (TRV) (Leiser et al. 1992; Turpen et al. 1993; Jones et al. 1999; Scholthof 1999; Ratcliff et al. 2001; Vlot et al. 2001; Liu and Lomonossoff 2002; Stephan and Maiss 2006). Importantly, agro-inoculation has also proven to be successful for several phloem-limited viruses in the family *Luteoviridae* such as beet western yellows virus (BWYV), potato leaf roll virus (PLRV) and beet mild yellowing virus (BMYV) (Leiser et al. 1992; Commandeur and Martin 1993; Prufer et al. 1995; Nurkiyanova et al. 2000; Kawchuk et al. 2002; Lee et al. 2002; Stephen and Maiss 2006). These are the RNA viruses with linear genomes which cannot be transmitted to plants by conventional mechanical inoculation, but by other means such as agro-infection.

46.1.3 Construction of FL-cDNA Under the Control of Bacteriophage Promoter (In Vitro Transcription)

Several RNA polymerase promoters have been used for in vitro transcription. For example *E. coli* Pm promoter derived from bacteriophage λ and promoters of SP6, T3 and T7 bacteriophages. Bacteriophage promoters are mainly used as they give high yield of transcripts (Dunn and Studier 1983; Melton et al. 1984). However, T7 promoter is mostly preferred over the other promoters probably due to more thoroughly studied genetics (Melton et al. 1984). To date, infectious FL-cDNA has been obtained in many viruses under the control of bacteriophage promoter, e.g. turnip crinkle virus, potato virus X (Hemenway et al. 1990), papaya ringspot virus (Chiang and Yeh 1997), TMV (Chapman 2008), soybean yellow mottle mosaic virus (Nam et al. 2009) and pepper mottle virus (Lee et al. 2011). The methodology involved in the construction of FL-cDNA under the control of bacteriophage promoter is given below.

46.1.3.1 Materials

1. As in Chapter 21 on purification of viruses.
2. As in Chapter 25 on isolation of nucleic acid from purified virus preparation.
3. As in Chapter 32 on isolation of total RNA from plants.
4. As in Chapter 35 on polymerase chain reaction.
5. As in Chapter 43 on cDNA synthesis and cloning.
6. Virus-infected plant or purified virus preparation.
7. Plasmid (transcription) vector.
8. Custom synthesized primers.
9. T7 RNA polymerase.
10. Sodium phosphate buffer (20 mM, pH 7.0).
11. Carborundum powder.
12. Healthy test plants.

46.1.3.2 Method

1. Purify the test virus from infected plant tissues (as described in Chapter 21) and extract the viral RNA (as described in Chapter 25) or isolate total RNA from infected plant (as described in Chapter 32).
2. Subject viral RNA or total RNA to cDNA synthesis using a primer hybridizing specifically to the 3′ end of the viral genome (detailed procedure provided in Chapter 43).
3. Convert the single-stranded cDNA into the double-stranded form by initiating DNA synthesis with a second primer encompassing the sequence corresponding to the nucleotides at the 5′

end of the viral RNA (detailed procedure provided in Chapter 43).

4. Clone the full-length double-stranded DNA into a plasmid vector. Transform *E. coli* cells with recombinant plasmid and select transformed colonies through antibiotic marker present in the plasmid (detailed procedure provided in Chapter 43).
5. Screen putative transformants and identify positive recombinant colony carrying full-length genome of the test virus through PCR and restriction analysis. The identified positive recombinant clone is then subjected to sequencing for confirmation of the clone.
6. Design primers to amplify the test virus genome under the control of T7 promoter. To ensure infectivity of the test virus clone, a T7 promoter should be incorporated immediately adjacent to the 5′-end of the viral genome (Boyer and Haenni 1994).
7. In order to obtain a FL-cDNA under the control of the T7 promoter, a suitable primer is designed. The primer should consist of nucleotides of the test virus at the 3′ terminus followed by nucleotides from bacteriophage T7 RNA polymerase promoter consensus sequence and six residues at the 5′ terminus to form a suitable restriction site.
8. Amplify the test virus genome with primers having T7 RNA promoter sequence (designed above) and primer with suitable restriction site at the 3′ end of the test virus using the plasmid carrying the full genome of the virus as template.
9. The amplified product from the above is again cloned in another plasmid (transcription) vector and positive recombinants identified through selection using antibiotics marker present in the plasmid.
10. Screen putative transformants and identify positive recombinant colony carrying full genome of the test virus through PCR and restriction analysis. The identified positive recombinant is then subjected to sequencing for confirmation of the clone.
11. Isolate the plasmid DNA from the above identified positive recombinant and digest with restriction enzyme to linearize the plasmid at the 3′ end of the test virus cDNA.
12. Carry out in vitro transcription by T7 RNA polymerase in the presence or absence of the cap analogue mTG (5′) ppp (5′) G as described by Nielsen and Shapiro (1986) and Holt and Beachy (1991).
13. Verify the absence of DNA template and integrity of RNA by electrophoresis in agarose gels, and estimate the concentration

of RNA by UV absorption in a spectrophotometer and ethidium bromide staining.

14. Dilute the transcription reaction with 20 mM sodium phosphate buffer, pH 7.0, and directly use for inoculation of 2-week-old test seedlings/plants. Dust the surfaces of the leaves with abrasive, carborundum and inoculate each plant mechanically corresponding to approximately 5 μg of RNA. Perform mock inoculation of control plants with phosphate buffer containing about 200 ng of test virus RNA, or 1 ng of purified test virus.
15. Place all inoculated plants under insect-proof glasshouse and observe for symptom development. Analyse the plants through ELISA, western blotting, RT-PCR, Southern and northern blotting to confirm the infection.

46.1.4 Construction of FL-cDNA Under the Control of Cauliflower Mosaic Virus 35S Promoter (In Vivo Transcription)

The second approach in construction of infectious clones is the expression of infectious viral RNAs by in vivo transcription of cDNA containing vectors through a CaMV 35S promoter. In this approach, constructs are prepared with FL-cDNA positioned downstream of a CaMV35S promoter for the initiation of transcription at the authentic 5′ end of the viral RNA in the plant nucleus (Dessens and Lomonossoff 1993). At the 3′ end nopaline synthase poly (A) signal [nos-poly (A)] (NOS) and octopine synthase (OCS) are used to permit the in vivo termination of transcription through polyadenylation (Scholthof et al. 1992). The 35S promoter would be recognized by the plant transcription machinery and produce the transcripts that are equivalent to the viral genomic RNAs. Hence viral replication would commence. The addition of the NOS terminator could efficiently prevent the synthesis of transcripts longer than the genomic size (Dessens and Lomonossoff 1993). However, Yamaya et al. (1988) and Gal-On et al. (1995) showed the infectivity of FL-cDNA clones without the NOS terminator *via* transgenic approach or by particle bombardment. Several plant RNA viruses have been exploited to generate infectious cDNA plasmids with the CaMV35S promoter in combination with a 3′ poly (A) signal, e.g. carnation mottle virus (Zhang et al. 2002), plum pox virus (Maiss et al. 1992), potato virus Y (Fakhfakh et al. 1996), citrus tristeza virus (Gowda et al. 2005), grapevine leaf roll-associated virus 2 (Kurth et al. 2012), sesbania mosaic virus (Govind et al. 2012) and potato leaf roll virus (Lee et al. 2011).

In the following, construction of FL infectious clone of a begomovirus and its betasatellite is given.

For transmission of most of the begomoviruses, it is necessary to construct dimers or partial dimers (also called bitmers) of the viral DNA components that harbour at least two origins of replication. From these constructs, unit genome component will be released by replication or recombination, but recombination is generally not common. These constructs can be transferred to *Agrobacterium* plasmids and delivered to plants by agro-infiltration for plant infectivity. The procedures of development of agro-infectious clones of chilli leaf curl virus (ChiLCuV) and associated betasatellite in binary vector pCAMBIA2300 (Cambia Labs, Canberra, Australia) are described below.

46.1.4.1 Materials

1. As in Chapter 21 on purification of viruses.
2. As in Chapter 25 on isolation of nucleic acid from purified virus preparation.
3. As in Chapter 31 on isolation of total DNA from plants.
4. As in Chapter 35 on polymerase chain reaction.
5. As in Chapter 42 on cloning PCR product.
6. Virus-infected plant or purified virus preparation.
7. Plasmid vector.
8. Custom synthesized primers.
9. *Agrobacterium tumefaciens.*
10. Binary vector pCAMBIA2300, 2301.
11. Acetosyringone: Prepare 100 mM by dissolving 0.1962 g of acetosyringone in 10 mL of ethanol, store at −20 °C.
12. 2-(*N*-morpholino) ethanesulfonic acid (MES) (0.5 M) (pH 5.6): Dissolve 9.762 g of MES in about 80 mL of water, adjust the pH 6.0 with NaOH or KOH and make up the volume to 100 mL.
13. $MgCl_2$ (100 mM): Dissolve 9.52 g of $MgCl_2$ in water and make up the volume to 1 L.
14. Infiltration media (10 mM $MgCl_2$, 10 mM MES and 200 μM acetosyringone): First prepare solution containing 10 mM $MgCl_2$ and 10 mM MES, autoclave for 15 min and add acetosyringone to a final concentration of 200 μM.
15. Syringe.
16. Gamborg B5 media composition.

Ingredients	mg/L
Macroelements	
Ammonium sulphate	134.000
Calcium chloride	113.230
Magnesium sulphate	122.090
Potassium nitrate	2500.000
Sodium phosphate monobasic	130.420

(continued)

Ingredients	mg/L
Microelements	
Boric acid	3.0000
Cobalt chloride hexahydrate	0.025
Copper sulphate pentahydrate	0.025
EDTA disodium salt dihydrate	37.300
Ferrous sulphate heptahydrate	27.800
Manganese sulphate monohydrate	10.000
Molybdic acid (sodium salt)	0.213
Potassium iodide	0.750
Zinc sulphate heptahydrate	2.000
Vitamins	
Myo-inositol	100.000
Nicotinic acid (free acid)	1.000
Pyridoxine HCl	1.000
Thiamine hydrochloride	10.000
Carbohydrate	
Sucrose	20,000.000
Distilled water (to make)	1000 mL

Adjust the pH of the medium to 5.75 ± 0.5 using 1N NaOH/HCl.

46.1.4.2 Method

Development of Infectious Construct of Chilli Leaf Curl Virus

1. Amplify and clone full-length genome of chilli leaf curl virus (ChilLCuV) in a plasmid vector. Confirm the clone by determining its nucleotide sequence (details described in Chapters 35 and 42 on polymerase chain reaction and cloning PCR product) (Fig. 46.1).
2. Construct a partial tandem repeat (1.4 mer) of the viral DNA-A of ChilLCuV in two cloning steps. Digest full-length genome with *Bam*HI and *Hind*III to release fragment of ~750 bp containing the origin of replication. This size corresponds to 0.4 mer of ChilLCuV DNA-A complete genome.
3. Separate the released fragment, 0.4 mer, by 1.5% agarose gel electrophoresis, elute and purify using gel purification kit.
4. Restrict the binary vector, pCAMBIA2300 with *Bam*HI and *Hind*III. Purify the linearized vector and ligate 0.4 mer using T4 DNA ligase.

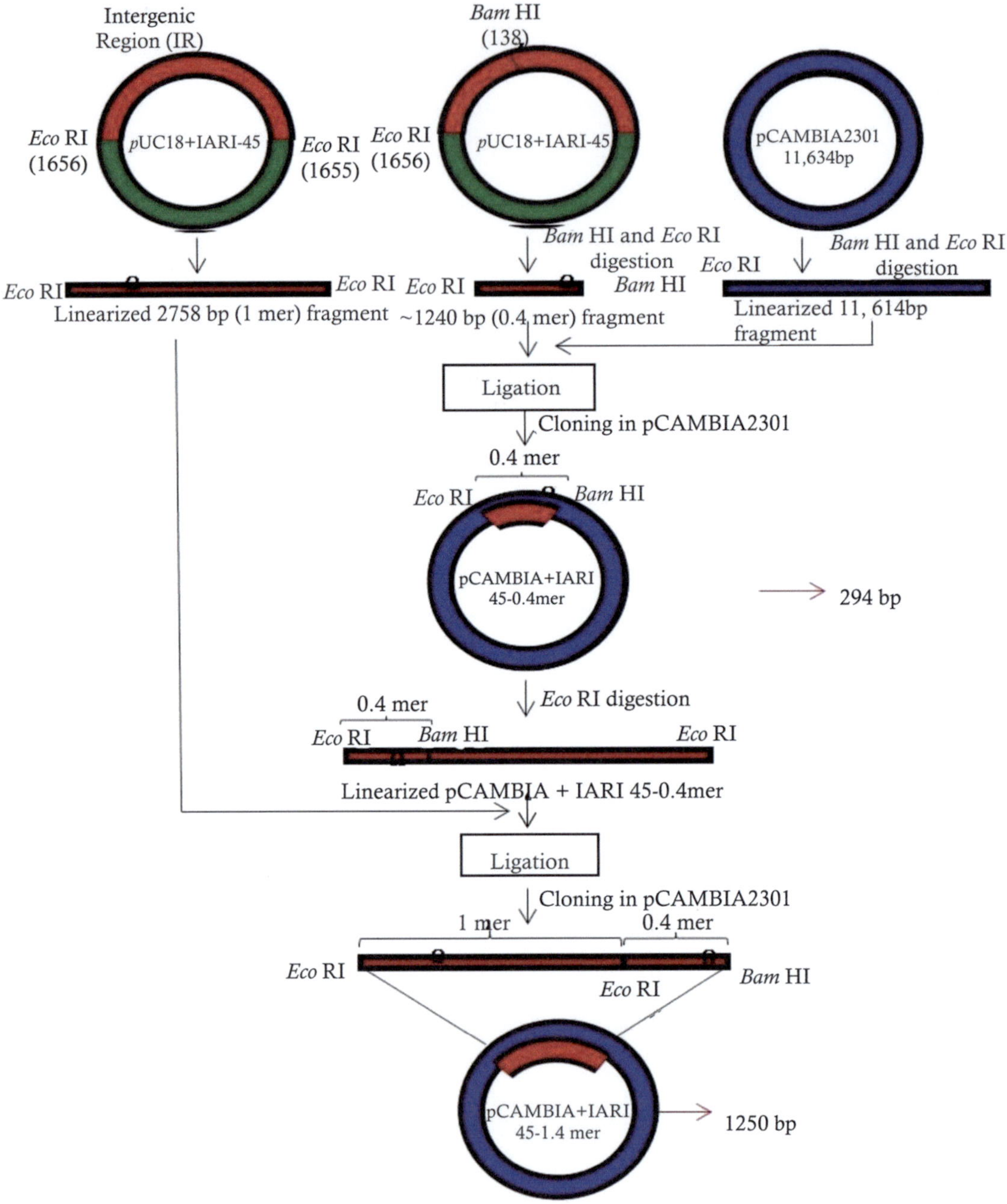

Fig 46.1 Strategy adopted for construction of the infectious clone of chilli leaf curl begomovirus. (Courtesy: Dr. R.K. Saritha, ICAR-IARI, New Delhi)

5. Transform the plasmid to *E. coli* DH5α cell and grow on Luria agar plates containing X-gal (40 μg/mL), IPTG (0.1 M) and kanamycin (50 mg/mL).
6. Screen the transformed colonies by restriction digestion of recombinant plasmid with *Bam*HI and *Hind*III. The

recombinant plasmid containing 0.4 mer genome of ChLCV in pCAMBIA 2300 is referred to as pC-Chi-0.4. This recombinant plasmid is further used for construction of 1.4 mer infectious clone of ChiLCuV DNA-A (Fig. 46.1).

7. Release complete DNA-A (1.0-mer) of ChilLCuV DNA-A cloned in the plasmid (pUC18) by restriction with *Bam*HI, and purify (Fig. 46.1).
8. Isolate and linearize the previously made recombinant plasmid pC-Chi-0.4 with *Bam*HI.
9. Ligate the linearized complete DNA-A (1.0-mer) obtained through *Bam*HI digestion to linearized pC-Chi-0.4 using T4 DNA ligase to generate the partial dimeric construct, designated as pC-Chi-1.4 (Fig. 46.1).
10. Transform the partial dimeric construct (pC-Chi-1.4) to *E. coli* DH5α cell and grow on Luria agar plates containing X-gal (40 μg/mL), IPTG (0.1 M) and kanamycin (50 mg/mL).
11. Screen the transformed colonies by colony PCR using specific primers ChLCuV FP and ChLCuV RP targeting CP gene of ChLCuV as mentioned earlier.
12. Confirm the orientation of partial dimeric construct by restriction with *Bam*HI which is expected to yield 2.7 kb fragment.
13. To test for infectivity, introduce pC-Chi-1.4 into competent cell of *Agrobacterium tumefaciens* by freeze/thaw transformation. Confirm the presence of binary vector in *A. tumefaciens* by colony PCR by using ChLCuV specific primer mentioned earlier.

Construction of Full-Length Betasatellite Associated with Begomovirus

1. Amplify and clone full-length begomovirus associated betasatellite in a plasmid vector (T/A). Confirm the clone by determining its nucleotide sequence (Fig. 46.2).
2. Construct a dimer (2.0 mer) of the betasatellite molecule in two cloning steps. In the first step, the recombinant plasmid T/A is restricted with *Bam* HI and *Kpn*I to release ~1350 bp betasatellite molecule from the recombinant plasmid. The released complete genome, 1.0 mer, is separated by 1% agarose gel electrophoresis, eluted and purified from the gel.
3. Restrict the binary vector, pCAMBIA2301 with *Bam*HI and *Kpn*I for linearization and purify the linearized vector. Ligate the complete genome, 1.0 mer using T4 DNA ligase (as described earlier) to linear pCAMBIA2301 and transform *E. coli* DH5α cell on LA plates containing X-gal (40 μg/mL), IPTG (0.1 M) and kanamycin (50 mg/mL).
4. Screen the transformed bacterial colonies by colony PCR and restriction digestion. The recombinant plasmid containing 1.0

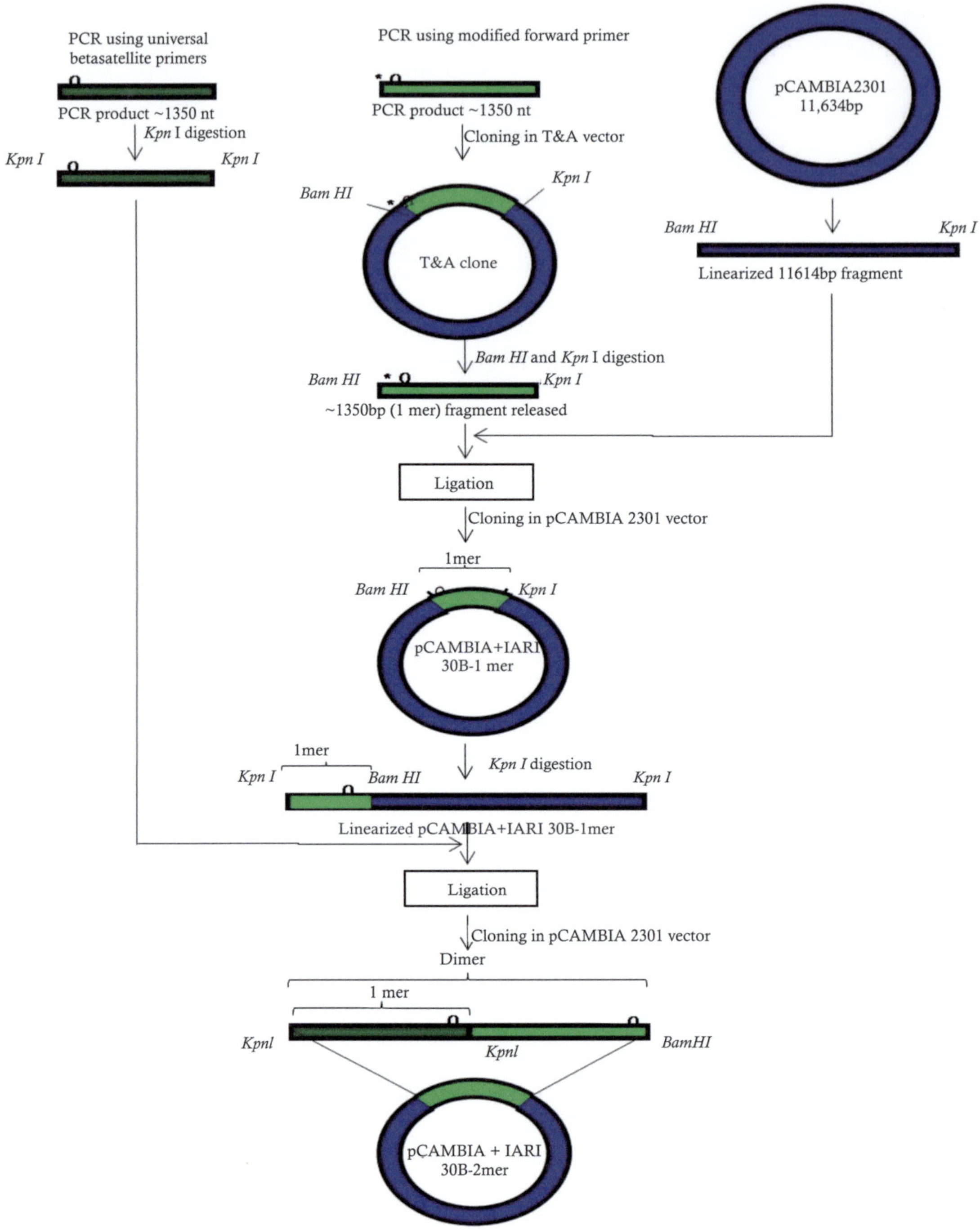

Fig. 46.2 Strategy adopted for construction of the infectious clone of chilli leaf curl begomovirus-associated betasatellite molecule. (Courtesy: Dr. R.K. Saritha, ICAR-IARI, New Delhi)

mer genome of betasatellite molecule in pCAMBIA2301 is referred to as pCAMBIA+1.0-30B (Fig. 46.2). This recombinant plasmid is further used for construction of 2.0 mer infectious clone of betasatellite molecule associated with begomovirus.

5. Amplify the second complete betasatellite molecule from total plant DNA of the begomovirus-infected plant through PCR using specific primers β01 (with *Kpn*I site) and β02 (with *Kpn*I site).
6. Restrict the PCR products with *Kpn*I and purify from the gel.
7. Linearize the previously made recombinant plasmid pCAMBIA +1.0-30B with *Kpn*I, treat with alkaline phosphatase followed by its purification.
8. Ligate the linearized second complete betasatellite DNA (1.0 mer) flanked with *Kpn*I site to linearized pCAMBIA+1.0-30B using T4 DNA ligase. Transform *E. coli* DH5α cells using ligation mixture and select the recombinant transformants by antibiotic selection.
9. Confirm the recombinant plasmid containing 2.0 mer genome of betasatellite molecule by restriction with *Bam*HI and *Kpn*I enzymes. This recombinant plasmid containing 2.0 mer genome of betasatellite molecule in pCAMBIA2301 is referred to as pCAMBIA+2.0-30B (Fig. 46.2).
10. To test for infectivity, introduce pCAMBIA+2.0-30B into competent cell of *Agrobacterium tumefaciens* by freeze-thaw transformation. Confirm the presence of binary vector in *Agrobacterium tumefaciens* by colony PCR using specific primers and restriction digestion.

Agro-Inoculation of Chilli Using Begomovirus and Betasatellite Constructs

1. Inoculate *Agrobacterium* strain containing plasmid pC-Chi-1.4 (from **step 13** under the Subheading "Development of Infectious Construct of Chilli Leaf Curl Virus") and pCAMBIA +2.0-30B (from **step 10** under the Subheading "Construction of Full-Length Betasatellite Associated with Begomovirus") separately in 5 mL LB media containing kanamycin (50 mg/L) and rifampicin (25 mg/L) and grow overnight at 28 °C with 200 rpm in a shaker incubator.
2. Add 5 mL overnight grown culture individually into fresh 50 mL media containing antibiotics, kanamycin (50 mg/L) and rifampicin (25 mg/L) and grow overnight (can also be 36 h) at 28 °C with 200 rpm.
3. Harvest bacterial cells by centrifugation at 3000 × *g* for 10 min and suspend in 5 mL of infiltration media (10 mM $MgCl_2$, 10 mM MES and 200 μM acetosyringone). The cultures are further diluted with the infiltration media (approx. 20 mL) to a

final OD_{600} value of 1.00. The resuspended culture is kept at room temperature for 3 h before use.

4. In case of agro-inoculation, harvest bacterial cells by centrifugation at 2500 × *g* for 15 min and resuspend in Gamborg B5 media.
5. Carry out agro-inoculation of chilli and *Nicotiana benthamiana* plants on 2-week-old seedlings at 4 leaf stage using pin prick method at the base of the plant using a syringe. A piece of cotton dipped in agro-culture is wrapped around the injury to ensure proper delivery of the bacterium. Maintain plants in an insect-free containment growth chamber at 28 °C.
6. In case of agro-infiltration, inoculate test plants of chilli and tobacco with agro-culture containing both virus and its betasatellite construct in 1:1 ratio. The *Agrobacterium* inocula containing infectious clones are infiltrated using a 1 mL needleless syringe to the cotyledonary leaves. A small slit of 0.1 mm size is made on the leaf using a razor blade at the site of infiltration to facilitate entry of the bacterial culture into leaf tissues. The penetration of bacterial suspension is visible as it spreads in the leaf. After agro-inoculation, keep the plants at a constant temperature (25–28 °C) under 14–16 h lighting in the insect-proof growth chamber.
7. In order to detect viral DNA components in agro-inoculated plants, subject systemically infected leaves of agro-inoculated plants to PCR amplification using virus-specific primers.

46.2 Notes

1. In some viral genomes, the synthesis of the full-length first strand cDNA appears to be a serious limiting factor, because the polymerization step is hampered by strong secondary structures on the viral RNA template. It is difficult to standardize an optimum protocol suitable for any virus, for the production of full-length cDNA clones because several variations on general scheme have been reported (Boyer and Haenni 1994).
2. There are a few limitations and pitfalls when it comes to the construction of a full-length infectious clone of a viral genome from which an in vitro infectious transcript can be produced. A critical point in the preparation and application of the infectious RNA is the transcription in vitro itself (its standardization regarding the quality and quantity of yield) and further manipulations with isolated RNA which is very sensitive to the enzyme degradation (Nagyova and Subr 2007). The possibility of producing infectious transcripts from incomplete viral cDNA clones has been reported, even though the presence of

the entire viral sequence is generally thought to be required to obtain infectious clones, but this does not ensure biological activity (Klump et al. 1990). cDNA synthesis, cloning strategies and the design of sequences bordering the viral insert strongly influence the infectivity of the viral insert (Boyer and Haenni 1994).

3. Construction of FL-cDNA under 35S promoter approach has several advantages. Firstly, the replication process can overcome detrimental effects resulting from RNA degradation, as infectivity of the clone is less dependent on RNA degradation since it presumably occurs only within cells where the RNAs are synthesized. Secondly, in vitro transcription step is not required. This is particularly important for RNA viruses for which the production of a good yield of highly infectious full-length transcripts can be problematic. This is also less expensive, because costly reagents such as the cap analogues and RNA polymerases are not required (Boyer and Haenni 1994). Further, the viral replication process and the expression of viral genes are largely independent of each other, which might be very convenient when studying the role and localization of proteins expressed by mutant viral RNAs which were unable to replicate in cells. In vivo produced viral transcripts in this way behave like messenger RNAs produced by a host RNA polymerase, still able to produce native or mutant proteins without being replicated (Van Bokoven et al. 1993). Another advantage of in vivo transcripts is the possibility to infect the plants by mechanical inoculation, particle bombardment (Gal-On et al. 1995; Fakhfakh et al. 1996) and agro infection, where viruses cannot be mechanically transmitted (Grimsley et al. 1987; Leiser et al. 1992).
4. The infectivity of transcripts is variable, in some cases reaching 100% or more in comparison with the infectivity of parental virion RNA (Hayes and Buck 1990; Hearne et al. 1990). However, there are several parameters that have dramatic influence on the infectivity of an infectious clone, namely heterogeneity, of transcript population, presence of point mutations, presence of non-viral nucleotides, effect of the cap structure, instability in bacteria and RNA polymerases.

References

Ahlquist P, Janda M (1984) cDNA cloning and *in vitro* transcription of the complete brome mosaic virus genome. Mol Cell Biol 4:2876–2882

Annamalai P, Rao ALN (2005) Replication-independent expression of genome components and capsid protein of brome mosaic virus in planta: a functional role for viral replicase in RNA packaging. Virology 338:96–111

Bendahmane A, Querci M, Kanyuka K, Baulcombe DC (2000) *Agrobacterium* transient expression system as a tool for the isolation of disease resistance genes: application to the *Rx2* locus in potato. Plant J 21:73–81

Boyer JC, Haenni AL (1994) Infectious transcripts and cDNA clones of RNA viruses. Virology 198:415–426

Chapman SN (2008) Construction of infectious clones for RNA viruses: TMV. Methods Mol Biol 451:477–490

Chiang CH, Yeh SD (1997) Infectivity assays of *in vitro* and *in vivo* transcripts of papaya ring spot virus. Bot Bull Acad Sin 38:153–163

Commandeur U, Martin R (1993) Investigations into the molecular biology of potato leaf roll luteovirus by means of agroinfection. Phytopathology 83:1426

Dessens JT, Lomonossoff GP (1993) Cauliflower mosaic virus 35S promoter controlled DNA copies of cowpea mosaic virus RNAs are infectious on plants. J Gen Virol 74:889–892

Ding XS, Schneider WL, Chaluvadi SR, Mian MA, Nelson RS (2006) Characterization of a *Brome mosaic virus* strain and its use as a vector for gene silencing in monocotyledonous hosts. Mol Plant Microbe Interact 19:1229–1239

Dore JM, Pinck L (1988) Plasmid DNA containing a copy of RNA 3 can substitute for RNA 3 in alfalfa mosaic virus RNA inocula. J Gen Virol 69:1331–1338

Dunn JJ, Studier FW (1983) Complete nucleotide-sequence of bacteriophage-T7 DNA and the locations of T7 genetic elements. J Mol Biol 166:477–535

Fakhfakh H, Valaine F, Makni M, Robaglia C (1996) Cell-free cloning and biolistic inoculation of an infectious cDNA of potato virus Y. J Gen Virol 77:519–523

Gal-On A, Meiri E, Huet H, Hua WJ, Raccah B, Gaba V (1995) Particle bombardment drastically increases the infectivity of cloned DNA of Zucchini yellow mosaic Potyvirus. J Gen Virol 76:3223–3227

Govind K, Makinen K, Savithri HS (2012) *Sesbania mosaic virus* (SeMV) infectious clone: possible mechanism of 3′ and 5′ end repair and role of polyprotein processing in viral replication. PLoS One 7(2):e31190

Gowda S, Satyanarayana T, Robertson CJ, Garnsey SM, Dawson WO (2005) Infection of citrus plants with virions generated in *Nicotiana benthamiana* plants agro-infiltrated with binary vector based *Citrus tristeza virus*. In: Hilf ME, Duran-Vila N, Rocha-Pena MA (eds) Proceedings of the 16th conference of the International Organization of Citrus Virologists. IOCV, Riverside, pp 23–33

Grimsley N, Hohn B, Hohn T, Walden R (1986) 'Agroinfection', an alternative route for viral infection of plants by using the Ti plasmid. Proc Natl Acad Sci U S A 83:3282–3286

Grimsley N, Hohn T, Davies JW, Hohn B (1987) Agrobacterium-mediated delivery of infectious maize streak virus into maize plants. Nature 325:177–179

Hayes RJ, Buck KW (1990) Infectious cucumber mosaic virus RNA transcribed in vitro from clones obtained from cDNA amplified using the polymerase chain reaction. J Gen Virol 71:2503–2508

Hearne PQ, Knorr DA, Hillman BI, Morris TJ (1990) The complete genome structure and synthesis of infectious RNA clones from Tomato bushy stunt virus. Virology 177:141–151

Hemenway C, Weiss J, O'Connell K, Tumer NE (1990) Characterization of infectious transcripts from a potato virus X cDNA clone. Virology 175:365–371

Holt CA, Beachy RN (1991) In vivo complementation of infectious transcripts from mutant tobacco mosaic virus cDNAs in transgenic plants. Virology 181:109–117

Hull R (2002) Matthew's plant virology, 4th edn. Academic Press, London

Jones L, Hamilton AJ, Voinnet O, Thomas CL, Maule AJ, Baulcombe DC (1999) RNA–DNA interactions and DNA methylation in post-transcriptional gene silencing. Plant Cell 11:2291–2301

Kawchuk L, Jaag HM, Toohey K, Martin R, Rohde W, Prufer D (2002) In planta agroinfection by Canadian and German *Potato leafroll virus* full-length cDNAs. Can J Plant Pathol 24:239–243

Klump WM, Bergmann I, Muller BC, Ameis D, Kandolf R (1990) Complete nucleotide sequence of infectious cox sakie-virus B3 cDNA: two initial 5′ uridine residues are regained during plus-strand RNA synthesis. J Virol 64:1573–1583

Kobayashi T, Antar AA, Boehme KW, Danthi P, Eby EA, Guglielmi KM, Holm GH, Johnson EM, Maginnis MS, Naik S, Skelton WB, Wetzel JD, Wilson GJ, Chappell JD, Dermody TS (2007) A plasmid-based reverse genetics system for animal double-stranded RNA viruses. Cell Host Microbe 1:147–157

Kurth EG, Peremyslov VV, Prokhnevsky AI, Kasschau KD, Miller M, Carrington JC, Dolja VV (2012) Virus-derived gene expression and RNA interference vector for grapevine. J Virol 86:6002–6009

Lee L, Palukaitis P, Gray SM (2002) Host-dependent requirement for the *Potato leaf roll virus* 17-kda protein in virus movement. Mol Plant Microbe Interact 15:1086–1094

Lee MY, Song YS, Ryu KH (2011) Development of infectious transcripts from full length and GFP-tagged cDNA clones of Pepper mottle virus and stable systemic expression of GFP in tobacco and pepper. Virus Res 155:487–494

Leiser RM, Ziegler-Graff V, Reutenauer A, Herrbach E, Lemaire O, Guilley H, Richards K, Jonard G (1992) Agroinfection as an alternative to insects for infecting plants with *Beet western yellows luteovirus*. Proc Natl Acad Sci 89:9136–9140

Liu L, Lomonossoff G (2002) Agro infection as a rapid method for propagating *Cowpea mosaic virus*-based constructs. J Virol Methods 105:343–348

Maiss E, Timpe U, Brisske-Rode A, Lesemann DE, Casper R (1992) Infectious in vivo transcripts of a Plum pox potyvirus full length cDNA clone containing the Cauliflower mosaic virus 35S RNA promoter. J Gen Virol 73:709–713

Melton DA, Krieg PA, Rebagliati MR, Maniatis T, Zinn K, Green MR (1984) Efficient *in vitro* synthesis of biologically active RNA and RNA hybridization probes from plasmids containing a bacteriophage SP6 promoter. Nucleic Acids Res 12:7035–7056

Nagyova A, Subr Z (2007) Infectious full-length clones of plant viruses and their use for construction of viral vectors. Acta Virol 51:223–237

Nam M, Kim SM, Domier LL, Koh S, Moon JK, Choi HS, Kim HG, Moon JS, Lee SH (2009) Nucleotide sequence and genome organization of a newly identified member of the genus *Carmovirus*, Soybean yellow mottle mosaic virus from soybean. Arch Virol 154:1679–1684

Nielsen DA, Shapiro DJ (1986) Preparation of capped RNA transcripts using T7 RNA polymerase. Nucleic Acids Res 14:5936

Nurkiyanova KM, Ryabov EV, Commandeur U, Duncan GH, Canto T, Gray SM, Mayo MA, Taliansky ME (2000) Tagging *Potato leafroll virus* with the jellyfish green fluorescent protein gene. J Gen Virol 81:617–626

Prufer DC, Wipf-Scheibel C, Richards K, Guilley H, Lecoq H, Jonard G (1995) Synthesis of a full-length infectious cDNA clone of cucurbit aphid-borne yellows virus and its use in gene exchange experiments with structural proteins from other luteoviruses. Virology 214:150–158

Ratcliff F, Martin-Hernandez AM, Baulcombe DC (2001) Tobacco rattle virus as a vector for analysis of gene functions by silencing. Plant J 25:237–245

Scholthof HB (1999) Rapid delivery of foreign genes into plants by direct rub inoculation with intact plasmid DNA of a *Tomato bushy stunt virus* gene vector. J Virol 73:7823–7829

Scholthof HB, Gowda S, Wu F, Shepherd RJ (1992) The full-length transcript of a caulimovirus is a polycistronic mRNA whose genes are transactivated by the product of gene VI. J Virol 66:3131–3139

Stephan D, Maiss E (2006) Biological properties of *Beet mild yellowing virus* derived from a full-length cDNA clone. J Gen Virol 87:445–449

Turpen TH, Turpen AM, Weinzettl N, Kumagai MH, Dawson WO (1993) Transfection of whole plants from wounds inoculated with *Agrobacterium tumefaciens* containing cDNA of *Tobacco mosaic virus*. J Virol Methods 42:227–239

Van Bakoven H, Venver J, Wellinck J, Van Kammen A (1993) Protoplasts transiently expressing the 200K coding sequence of cowpea mosaic virus BRNA support replication of M-RNA. J Gen Virol 74:2233–2241

Vlot AC, Neeleman L, Linthorst HJ, Bol JF (2001) Role of the 3′-untranslated regions of alfalfa mosaic virus RNAs in the formation of a transiently expressed replicase in plants and in the assembly of virions. J Virol 75:6440–6449

Voinnet O, Rivas S, Mestre P, Baulcombe D (2003) An enhanced transient expression system in plants based on suppression of gene silencing by the p19 protein of tomato bushy stunt virus. Plant J 33:949–956

Yamaya Y, Yoshioka M, Meshi T, Okada Y, Ohno T (1988) Expression of tobacco mosaic virus RNA in transgenic plants. Mol Gen Genet 211:520–525

Zhang AP, Zhu HQ, Yu SQ, Yang L, Lou Y (2002) Cloning of complete carnation mottle virus cDNA and its infectivity. Acta Horticult 568 (568):149–154

Chapter 47

Virus Elimination by Meristem-Tip Culture

Abstract

Meristem-tip culture is one of the most widely used methods for virus elimination from infected plants and production of virus-free plants. Apical meristem culture is a proven means of clonal propagation and also for eliminating viruses from infected plants. It exhibits inherent genetic stability and tremendous growth potential, making it useful material for long-term storage of virus-free germplasm. Also, due to the small length of the meristem explant, this technique will have the potential for eliminating other pathogenic organisms in addition to viruses. Meristem culture is more effective in eliminating viruses when it is combined with thermotherapy and chemotherapy. Pre-treatment of infected plants with higher temperature and usage of antiviral agents in the meristem regeneration medium have been reported to be an effective method for the elimination of many plant viruses.

Key words Vascular bundle, Meristem culture, Regeneration, Thermotherapy, Chemotherapy, Antiviral agents, Virus testing

47.1 Introduction

Virus particles that may present in the vascular system can reach the meristematic region only by movement through cell to cell (Carrington et al. 1996). This is a slow process that might take a few days. The reasons proposed for absence of virus in meristem are their high metabolic activities, usually accompanied by elevated endogenous auxin content in shoot apices, lack of vascular system and virus inactivating system may inhibit virus multiplication (Parmessur and Saumtally 2001). Based on this finding, meristem culture has been extensively used to eliminate viruses from plants. Compounds such as Ribavirin (RBV), 5-Azacytidine (AZA) and 3-Deazauridine (DZD) have been successfully used for virus eradication in many economically important crops.

The production of virus-free plants by using meristem culture involves determination of optimal size range of meristem/shoot tips and a high rate of plant regeneration (Wang et al. 2006). Various factors such as meristem size, meristem portion, detection methods, endophytic infections and the growth period of the plant also influence meristem culture and virus elimination (Cha-um et al. 2006). Virus elimination will be more when meristem culture

Alangar Ishwara Bhat and Govind Pratap Rao, *Characterization of Plant Viruses*, Springer Protocols Handbooks,
https://doi.org/10.1007/978-1-0716-0334-5_47,

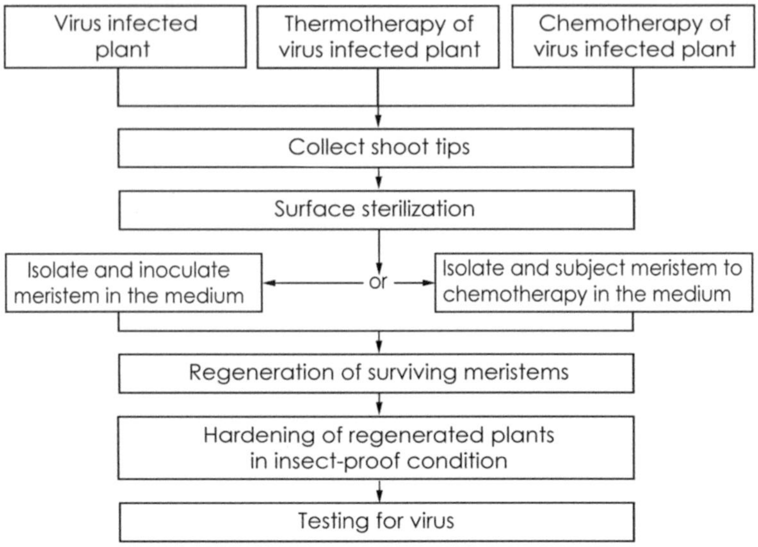

Fig. 47.1 Different stages in virus elimination by meristem-tip culture alone and combined with thermotherapy and chemotherapy

is combined with chemotherapy and/or thermotherapy (Fig. 47.1). Successful elimination of sugarcane mosaic virus and sugarcane yellow leaf virus in sugarcane (Dean 1982; Mishra et al. 2010; Ramgareeb et al. 2010), cucumber mosaic virus and dahlia mosaic virus in dahlia, potato virus X (Hossain et al. 2013) and potato virus Y (PVY) in potato (Al-Taleb et al. 2011), grapevine leaf roll-associated virus and grapevine fan leaf virus in grape (Fayek et al. 2009), cucumber mosaic virus and cymbidium mosaic virus in vanilla (Retheesh and Bhat 2010) and piper yellow mottle virus in black pepper (Sasi and Bhat 2018) were reported using meristem culture alone or combined with chemotherapy (Fig. 47.2).

47.1.1 Meristem Culture Combined with Thermotherapy

Kassanis in 1957 for the first time applied heat therapy to eliminate tobacco necrosis virus (TNV) from tulip (Parmessur and Saumtally 2001). Heat treatment of plants *in vivo* or *in vitro* reduces virus titers. It also improves virus eradication efficiently (Kassanis 1957; Mink et al. 1998). If not eliminated, higher temperatures can slow down the virus replication and infection rates, allowing new sprouts to grow (Wang et al. 2008). Thermotherapy followed by meristem culture and shoot tip grafting has enhanced virus eradication in many crops such as apple, cocoyam and garlic (Salazar et al. 1985). However, increase in temperature decreased the regeneration capacity of the meristem. Meristems obtained from virus-infected axillary buds of black raspberry grown aseptically at 4 h alternating temperature of 29 and 38 °C for 1–5 weeks tested negative for the black raspberry necrosis virus (BRNV) (Cheong et al. 2014). Thermotherapy, meristem culture and a combination of both on

Fig. 47.2 Various stages in meristem culture of vanilla. (**a**) Freshly isolated meristem with a single leaf primordium. (**b**) After 30 days of culture in induction medium. (**c–f**) Cultures 45, 60, 75 and 90 days after culturing on regeneration medium, respectively. (**g**, **h**) Fully regenerated plantlets ready for hardening. (**i**) Plants in insect-free glasshouse. (Reproduced from Retheesh and Bhat 2010 with permission from Elsevier)

infected meristem-tip explants of 0.5–1 mm in length resulted in bean yellow mosaic virus (BYMV) elimination up to 87% in gladiolus (Nezamabad et al. 2015).

47.1.2 Meristem Culture Combined with Chemotherapy

Treatment of infected plant or meristem with antiviral agents is an effective method for the elimination of many viruses (Mori 1971). Antiviral compounds can be directly sprayed on the crop or can be added along with the medium during the *in vitro* growth which inhibits virus replication (Klein and Livingston 1983). These compounds are anti-metabolites, substances that are capable of blocking the synthesis of viral nucleic acid. These compounds can be both natural and synthetic, having an antiviral effect, but none of them has a satisfactory selective effect that would enable them to be used in a particular prophylaxis and in large-scale treatment of plant viral diseases (Hansen and Lane 1985). These compounds induce antiviral resistances in plants apart from inhibiting viral replication (Nascimento et al. 2003). Compounds such as Ribavirin (RBV) (Hansen and Lane 1985; Griffiths et al. 1990; Prasada Rao et al. 1995), 3-Deazauridine (DZD) and 5-Azacytidine (AZA) (Dunbar et al. 1993) have been successfully used for virus eradication (in combination with meristem culture) in economically important crops, such as apple, peanut and *Prunus* spp. Hansen and Lane (1985) first reported eradication of apple chlorotic leaf spot virus (ACLSV) from apple shoots by ribavirin. Among three antiviral agents such as RBV, AZA and DZD, RBV showed the highest rate of PVY eradication in potato. Combination of shoot tip culture and ribavirin was used for the elimination of PDV and PNRSV from shoot tips of plum. Similarly, Hu et al. (2012) combined chemotherapy with thermotherapy (ribavirin at 25 μg/mL combined with thermotherapy at 35 °C for 40 days) for obtaining virus-free *in vitro* pear plants. Elimination of sugarcane streak mosaic virus (SCSMV) and sugarcane mosaic virus (SCMV) in sugarcane meristem treated with 10 mg of ribavirin was reported (Ramgareeb et al. 2010; Neelamathi et al. 2014).

47.2 Materials

1. Virus-infected plants.
2. Bavistin (fungicide).
3. Laminar air flow.
4. Tween 20.
5. Mercuric chloride.
6. Woody plant medium (WPM).

Stock	Constituents	Weight (g)	Stock volume (mL)	Volume/L (mL)
A	NH_4NO_3	10.00	250	10
	$Ca(NO_3)_2{\cdot}4H_2O$	13.90		
B	K_2SO_4	24.75	500	20
C	$CaCl_2$	1.812	250	10

(continued)

Stock	Constituents	Weight (g)	Stock volume (mL)	Volume/L (mL)
D	KH_2PO_4	4.25	250	10
	H_3BO_3	0.155		
	Na_2MnO_4	0.00625		
E	$MgSO_4{\cdot}7H_2O$	9.25		
	$MnSO_4{\cdot}H_2O$	0.5575	250	10
	$ZnSO_4{\cdot}7H_2O$	0.215		
	$CuSO_4{\cdot}5H_2O$	0.00625		
F	$FeSO_4{\cdot}7H_2O$	0.695	250	10
	Na_2EDTA	0.9325		
G	Thiamine HCl	0.025	250	10
	Nicotinic Acid	0.0125		
	Pyridoxine HCl	0.0125		
	Glycine	0.05		
H	Myoinositol	2.5	250	10
	Sucrose	30 g/L		
	Agar	8 g/L		
	Charcoal	2 g/L		

Sterilize by autoclaving and store at 4 °C

7. MS (Murashige and Skoog 1962) medium

Constituents	Quantity (mg/L)	Constituents	Quantity (mg/L)
(A) *Macro nutrients*		Zinc sulphate ($ZnSO_4{\cdot}7H_2O$)	8.60
Ammonium nitrate (NH_4NO_3)	1650	Sod. molybdate ($Na_2MoO_4{\cdot}2H_2O$)	0.25
Potassium nitrate (KNO_3)	1900	Copper sulphate ($CuSO_4{\cdot}5H_2O$)	0.025
Calcium chloride ($CaCl_2{\cdot}2H_2O$)	440	Cobalt chloride ($CoCl_2{\cdot}6H_2O$)	0.025
Magnesium sulphate ($MgSO_4{\cdot}7H_2O$)	370	(C) *Organic constituents*	
Pot. dihydrogen orthophosphate (KH_2PO_4)	170	Nicotinic acid	0.50

(continued)

Constituents	Quantity (mg/L)	Constituents	Quantity (mg/L)
Sod. Ethylene diaminetetraacetate ($Na_2{\cdot}EDTA{\cdot}2H_2O$)	37.3	Pyridoxine HCl	0.50
Ferrous sulphate ($FeSO_4{\cdot}7H_2O$)	27.8	Thiamine HCl	0.10
(B) *Micro nutrients*		Glycine	2.00
Potassium iodide (KI)	0.83	Inositol	100
Boric acid (H_3BO_3)	6.20	Sucrose	30,000
Manganese sulphate ($MnSO_4{\cdot}4H_2O$)	22.3	pH	5.8

Preparation and sterilization of media

- For experiments, different MS media composition are prepared by mixing the stock solutions of macro/micro elements, sucrose, vitamins and growth regulators in desired quantities.
- Adjust the pH of the medium to 5.8 before autoclaving using 0.1N NaOH or 0.1N HCl except in the experiments where effect of pH of medium on regeneration responses is to be investigated.
- Pour about 20 mL of medium in each culture vessel (culture tubes, size 25 × 150 mm or glass jars, capacity 250 mL).
- Plug the vessels with semi-permanent non-absorbent cotton plugs or polypropylene screw caps in case of glass jars.
- Autoclave at 15 $lb/in.^2$ pressure (121 °C) for 20 min. For semi-solid medium, 8 g/L agar is added to the warm medium and dissolve by continuous shaking and heating up to boiling point over a heater.
- Pour the warm medium in glass vessels, plug and autoclave as described above.
- Keep the vessels stationary in a cool dry place for solidification after autoclaving
- Add various concentrations of different plant growth regulators alone or in combinations from stock solutions to the medium in order to get specific responses (*e.g.* culture establishment from different explants, multiplication of shoots and rooting of shoots).

8. Benzyl adenine (BA) (1 mg/mL): Dissolve 10 mg BA in a 100 μL of 1N NaOH, make up the total volume to 10 mL, filter sterilize and store at 4 °C.

9. Kinetin (KN) (1 mg/mL): Dissolve 10 mg KN in a 100 μL of 1N NaOH, make up the total volume to 10 mL, filter sterilize and store at 4 °C.
10. Naphthalene acetic acid (NAA) (1 mg/mL): Dissolve 10 mg NAA in a 100 μL of 1N NaOH, make up the total volume to 10 mL, filter sterilize and store at 4 °C.
11. Tetracycline (1 g/mL): Dissolve 1 g tetracycline in 1 mL sterile distilled water, filter sterilize and store at 4 °C.
12. Spectromycin (1 g/mL): Dissolve 1 g spectromycin in 1 mL sterile distilled water, filter sterilize and store at 4 °C.
13. Streptomycin (1 g/mL): Dissolve 1 g streptomycin in 1 mL sterile distilled water, filter sterilize and store at 4 °C.
14. Sodium hypochlorite.
15. Tween 20.
16. Casein hydrolysate.
17. Tissue culture laboratory.
18. Greenhouse.
19. Ribavirin.

47.3 Method (Virus Elimination by Meristem-Tip Culture, Sasi and Bhat 2018)

The following protocol is used for the meristem-tip culture of black pepper. The same protocol may not work for all crops. Researchers may scan the literature and look for the suitable published protocols for meristem-tip culture for the crops of their interest. In case it is not available, one has to standardize the same.

1. Identify virus-infected plants from which meristems to be collected. Spray shoot tips of selected plants with the fungicide, bavistin @ 1.5% 2 days prior to explant collection.
2. Collect shoot tip explants from virus-infected plants and treat again with bavistin for 10 min, followed by treatment with water containing 1% Tween 20 for 10 min and wash under running tap water for 10 min.
3. Subject explants to surface sterilization with 0.1% mercuric chloride for 10 min and wash three times with sterile water.
4. Excise meristem of approximately 1–3 mm size from the shoot tip under a laminar air flow using a clean, sterile scalpel and inoculate into woody plant medium (WPM) containing 3 mg/L Benzyl adenine (BA) + 1 mg/L Kinetin (KN) and 1% each of tetracycline and spectromycin.

5. Incubate meristems at 28 °C with a 16-h photoperiod and allow to grow for 2–3 weeks for meristem elongation. Monitor daily and remove if any contamination is noticed.
6. Transfer extended meristems to the same medium without antibiotics till shoots are formed.
7. Transfer the shoots to the rooting medium (WPM with 3 mg/L BA + 1 mg/L NAA) and harden the rooted plantlets in the greenhouse.
8. Test hardened plants for the presence or absence of virus using PCR or RT-PCR assay along with a known infected and healthy plant. Details of PCR and RT-PCR are provided in Chapter 35.

47.4 Method (Virus Elimination by Meristem-Tip Culture Combined with Chemotherapy, Sasi and Bhat 2018)

1. Excise meristems from the shoot tips of virus-infected plants as described above (**steps 1–4** from the Subheading 47.3).
2. Inoculate the meristems in the regeneration medium (WPM with 3 mg/L BA + 1 mg/L KN + 1% tetracycline + 1% spectromycin) containing ribavirin at 10 mg/L for a period of 30 days, with two subculturing in the same medium at an interval of 15 days.
3. Transfer ribavirin-treated meristems to the same medium without ribavirin and antibiotics for proper shooting, followed by rooting as described above.
4. Harden the well-rooted plantlets in the greenhouse and test for the presence/absence of the virus by ELISA/PCR assays.

47.5 Method (Virus Elimination by Meristem-Tip Culture Combined with Thermotherapy (Ramgareeb et al. 2010; Mishra et al. 2010)

Plant material

1. Select the field grown sugarcane variety (6–12 months old), infected with virus.
2. Cut the sugarcane stalk material into single budded nodes (SBN) and incubate in a hot water bath (50 °C) containing the fungicide Bavistin @ 0.5% for 30–40 min.
3. Following hot water treatment (HWT), plant the SBN in seedling trays filled with potting soil mix.

Fig 47.3 Explant preparation of sugarcane for meristem culture; Sugarcane tops collected from field grown plants for explant preparation (**a**), 6–8 cm long segments after sterilization (**b**), dissection of meristem (**c**), inoculation of meristem explant on MS semi-solid medium (**d**), multiple shoot regeneration (**e**, **f**), rooting of the multiple shoot (**g**)

4. Incubate the trays in the glasshouse for 6–8 weeks at 30–35 °C and ambient light conditions) and water the plants twice daily.
5. Harvest the shoots developed from the nodes for meristem excision when the first node is observed (Fig. 47.3).

Explant preparation

6. Excise the apical meristem (about 6 cm long spindle segment) from the node shoot.
7. Cut the leaf roll and stem consisting of at least the first visible node from the source material.
8. Wash the segment with 1% Tween 20 for 3–5 min followed by thorough washing with tap water.
9. Immerse in 70% (v/v) ethanol for 30 s and rinse with distilled water.
10. Remove the outer leaves and the leaf roll shortened to contain the first node above which the shoot apical meristem is situated (±4 cm).
11. Finally, surface sterilize the segments with 0.1% NaOCl (sodium hypochlorite) solution for 10 min followed by several washing with distilled water.
12. Dissect the meristems of different lengths (1.0–5.0 mm) from the explants containing the apical dome and leaf primordia for inoculation (Fig. 47.3).

In vitro culture

13. Inoculate the dissected meristems on shoot induction medium (MS; 20 g/L sucrose; 10 g/L agar; 3.5 g/L activated charcoal; 1 mg/L methylene blue, pH 4.5) that contained different hormonal treatments. These included DO1 (2 mg/L BA; 1 mg/L KIN; 0.5 mg/L NAA), DO2 (0.5 mg/L BA), DO3 (2 mg/L BA) and DO4 (0.1 mg/L BA; 0.015 mg/LKIN).
14. Shift the explants to an undisturbed place on the medium in the same Petri dish after 3 days in the dark in order to reduce phenolic inhibition of shoot formation.
15. After a 1 week dark incubation period, place the meristem in the photoperiod growth room.
16. A week later, transfer them to shoot induction medium without activated charcoal.
17. Subculture the shoots fortnightly until there are ±20 shoots (±4 cm in height).
18. Growing shoots are divided in groups of two and transfer to Magenta vessels where they are further divided and placed in fresh proliferation media every fortnight for a total of 11 weeks.

Rooting and acclimatization

19. All shoots (±4 cm) are rooted in media containing half strength MS; 5 g/L sucrose; 8 g/L agar; 0.25 g/L casein hydrolysate

and pH 5.6–5.8. *In vitro* rooting is achieved in 2–3 weeks (Fig. 47.3).

20. Transfer the plants with a well-formed root system into polythene bags/tray containing potting soil (field soil, sand and fly ash, 4:1:1) and keep within a glasshouse for 2 weeks. Thereafter plants are transferred to a glasshouse chamber (±30 °C, ambient light conditions), water twice daily (2 min) and fertilize (0.2 g of N:P:K; 2:3:2) monthly.
21. Harden the plants in greenhouse for 4–6 weeks at about 25–30 °C.
22. Shift the hardened plants to net/shade house for further acclimatization.

Test for virus elimination assay

23. Extract RNA from meristem grown new micropropagated plants and check for the presence and absence of virus along with the infected mother plant used for meristem-tip culture by any appropriate method such as ELISA, RT-PCR, real-time RT-PCR or RT-LAMP.

47.6 Notes (Al-Taleb et al. 2011; Fayek et al. 2009; Mink et al. 1998; Mishra et al. 2010; Sasi and Bhat 2018)

1. The shoot tips of selected plants should be sprayed with fungicide to avoid contamination of fungi during meristem-tip culture. Fungicide to be used and dosage may vary depending on the crop.
2. The chemical used for surface sterilization of explants, its concentration and duration may vary from tissue to tissue. If information is already available on the crops you are handling you may use the same, else this needs to be standardized for each crop as excessive use (both concentration and duration) of the chemical may harm the tissue.
3. Endophytic bacterial contamination may be found in certain crops especially perennial crops. In such cases, it is necessary to select appropriate antibiotic to be used during meristem culture, its concentration and duration of treatment to eliminate endophytic bacteria.
4. Use of antibiotics may be avoided in the meristem elongation/regeneration medium if there are no contamination issues in the tissues being handled.
5. Success of virus elimination is inversely related to the size of meristem used for regeneration; lesser the size of meristem there are more chances of obtaining virus-free plants. But success of meristem-tip regeneration is directly related to the

size of meristem-tip used; more the size of meristem there are more chances of regeneration. Thus a compromise is required on the size of meristem to be used to get virus-free plants.

6. Thermotherapy and chemotherapy increase virus elimination from meristems. The temperature and duration to which the virus-infected plants should be subjected before excising the meristem will vary from one species of plant to other. Similarly, treatment with antiviral chemicals will also vary from one species to other species depending on the chemical used, its concentration and duration. Hence initial experiments with different temperatures and its duration may be required to determine the optimum condition for thermotherapy. Similarly, experiments with different antiviral chemicals, their concentration and duration of exposure need to be done to determine the right chemical, its dosage and duration to eliminate viruses during meristem culture.
7. Meristem-tip derived plants may be tested at different intervals to confirm the presence or absence of the virus in the plants. The identified virus-free plants may be protected from insects and used for further multiplication under insect-proof condition.
8. In the case of sugarcane, explants cultured on semi-solid and liquid media showed remarkable differences regarding frequency of establishment of shoot cultures. The establishment of shoot tip and meristem explants was significantly higher on semi-solid than in liquid medium.

References

Al-Taleb MM, Hassawi DS, Abu-Romman SM (2011) Production of virus free potato plants using meristem culture from cultivars grown under Jordanian environment. J Agric Environ Sci 11:467–472

Carrington JC, Kasschau KD, Makajan SK, Schaad MC (1996) Cell-to-cell and long-distance transport of viruses in plants. Plant Cell 8:1669–1681

Cha-um S, Thi-Thanh Hien N, Kirdmanee C (2006) Disease-free production of sugarcane varieties (*Saccharum officinarum* L.) using *in vitro* meristem culture. Biotechnol J 5:443–448

Cheong EJ, Jeon AR, Kang JW, Mock R, Kinard G, Li R (2014) *In vitro* elimination of *Black raspberry necrosis virus* from black raspberry (*Rubus occidentalis*). Hortic Sci 41:95–99

Dean JL (1982) Failure of *sugarcane mosaic virus* to survive in cultured sugarcane tissue. Plant Dis 66:1060–1061

Dunbar KB, Pinnow DL, Morris JB, Pittman RN (1993) Virus elimination from interspecific Arachis hybrids. Plant Dis 77:517–520

Fayek MA, Jomaa AH, Shalaby AB, Al-Dhaher AM (2009) Meristem tip culture for *in vitro* eradication of *Grapevine leaf roll associated virus-1* (GLRaV-1) and *Grapevine fan leaf virus* (GFLV) from infected flame seedless grapevine plantlets. Iniciacion Investigacion 4:1–11

Griffiths HM, Slack SA, Dodds JH (1990) Effect of chemical and heat therapy on virus concentrations *in vitro* potato plantlets. Can J Bot 68:1515–1521

Hansen AJ, Lane WD (1985) Elimination of *Apple chlorotic leafspot virus* from apple shoot cultures by ribavirin. Plant Dis 69:134–135

Hossain MA, Nasiruddin KM, Kawochar MA (2013) Effect of 6-benzyl aminopurine (BAP) on meristem culture for virus free seed production of some popular potato varieties in Bangladesh. Afr J Biotechnol 12:2406–2413

Hu GJ, Hong N, Wang LP, Hu HJ, Wang GP (2012) Efficacy of virus elimination from *in vitro*-cultured sand pear (*Pyrus pyrifolia*) by chemotherapy combined with thermotherapy. Crop Prot 37:20–25

Kassanis B (1957) Effects of changing temperature on plant virus diseases. Adv Virus Res 4:221–241

Klein RE, Livingston CH (1983) Eradication of potato viruses X and S from potato shoot tip cultures with ribavirin. Phytopathology 73:1049–1050

Mink GI, Wample R, Howell WE (1998) Heat treatment of perennial plants to eliminate phytoplasmas, viruses, and viroids while maintaining plant survival. In: Hadidi A, Khetarpal RK, Koganezawa H (eds) Plant virus disease control. APS Press, St. Paul

Mishra S, Singh D, Tiwari AK, Lal M, Rao GP (2010) Elimination of sugarcane mosaic virus and sugarcane streak mosaic virus by tissue culture. Int Sugarcane J 28:119–122

Mori K (1971) Production of virus free plants by means of meristem culture. Jpn Agric Res Q 6:1–7

Nascimento LC, Pio-Ribeiro G, Willadino L, Andrade GP (2003) Stock indexing and *Potato virus Y* elimination from potato plants cultivated in vitro. Sci Agric 60:525–530

Neelamathi D, Manuel J, George P (2014) Influence of apical meristem and chemotherapy on production of virus free sugarcane plants. Res J Recent Sci 3:305–309

Nezamabad PS, Habibi MK, Dizadji A, Kalantari S (2015) Elimination of *Bean yellow mosaic virus* through thermotherapy combined with meristem-tip culture in gladiolus corms. J Crop Prot 4:533–543

Parmessur Y, Saumtally A (2001) Elimination of *Sugarcane yellow leaf virus* and *Sugarcane bacilliform virus* by tissue culture. AMAS, Food and Agricultural Research Council, Réduit

Prasada Rao RDVJ, Pio-Ribeiro G, Pittman R, Reddy DVR, Demski JW (1995) *In vitro* culture to eliminate *Peanut stripe virus* from peanut seed. Peanut Sci 22:54–56

Ramgareeb S, Snyman SJ, Antwerpen TV, Rutherford RS (2010) Elimination of virus and rapid propagation of disease-free sugarcane (*Saccharum* spp. Cultivar NCo376) using apical meristem culture. Plant Cell Tissue Organ Cult 100:175–181

Retheesh ST, Bhat AI (2010) Simultaneous elimination of *Cucumber mosaic virus* and *Cymbidium mosaic virus* infecting *Vanilla planifolia* through meristem culture. Crop Prot 29:1214–1217

Salazar S, Fernandez R, Jarret RL, Turrialba CR (1985) Virus-free plants obtained by thermotherapy and meristem culture of white (*Xanthosoma saggitifoliu* Schott.) and purple (*X. violaceum* schott.) cocoyams. In: Proceedings of seventh symposium of the international society for tropical root crops, Gosier, Guadeloupe

Sasi S, Bhat AI (2018) *In vitro* elimination of *Piper yellow mottle virus* from infected black pepper through somatic embryogenesis and meristem-tip culture. *Crop Prot* 103:39–45

Wang Q, Liu Y, Xie Y, You M (2006) Cryotherapy of potato shoot tips for efficient elimination of *Potato leafroll virus* (PLRV) and *Potato virus Y* (PV Y). Potato Res 49:119–129

Wang Q, Cuellar WJ, Rajamaki ML, Hirata Y, Valkonen JPT (2008) Combined thermotherapy and cryotherapy for efficient virus eradication: relation of virus distribution, subcellular changes, cell survival and RNA degradation in shoot tips. Mol Plant Pathol 9:237–250

Chapter 48

Virus Elimination Through Somatic Embryogenesis

Abstract

Somatic embryogenesis is the development of plants from somatic embryos without the fusion of gametes. It can be induced by various factors such as hormones, sucrose and ethylene under in vitro conditions. Somatic embryogenesis through cyclic somatic embryogenesis is one of the best methods for large-scale micropropagation of plants and used for virus elimination from the infected plant materials. Virus elimination through somatic embryogenesis occurs due to the lack of connection between somatic embryos and the mother plant. Moreover, viruses are mainly restricted to vascular tissue, while the somatic embryos arise from the non-vascular tissues, thus increasing chances of virus elimination. Many successful reports are available for virus elimination through somatic embryogenesis in different crops such as cocoa, garlic, grapevine, sugarcane, potato and black pepper. Somatic embryogenesis combined with chemotherapy is found to be more effective in producing virus-free plants than somatic embryogenesis alone. In this chapter, different steps involved in somatic embryogenesis for the elimination of virus from infected plant are discussed.

Key words Somatic embryo, Regeneration, Virus testing, Antiviral agent, Chemotherapy

48.1 Introduction

Somatic embryogenesis is the development of haploid or diploid cells into differentiated plants without the fusion of gametes through embryo stages (Thorpe 1988). Somatic embryos are bipolar structures formed from sporophytic cells that do not have vascular connection with the maternal tissue (Ammirato 1983). Steward et al. (1958) first reported plant regeneration by somatic embryogenesis from cultured carrot cells. The somatic embryo contains both shoot and radicular meristems (Laux and Jurgens 1997). As the embryos develop, they progress through the different stages such as circular-globule shaped, heart-shaped, torpedo-shaped and cotyledonary (Pareek 2005). Somatic embryogenesis can occur directly from explant tissue cell with or without an intervening callus phase. Direct somatic embryogenesis occurs when somatic embryo arises directly from explant without callus phase while in indirect somatic embryogenesis, somatic embryo arises from proliferating callus (Pierik 1987). In vitro, somatic embryogenesis can occur from several types of explants and can

Alangar Ishwara Bhat and Govind Pratap Rao, *Characterization of Plant Viruses*, Springer Protocols Handbooks, https://doi.org/10.1007/978-1-0716-0334-5_48, © Springer Science+Business Media, LLC, part of Springer Nature 2020

be induced by various factors such as hormones, sucrose and ethylene in many different species (McKersie and Brown 1996; Karami et al. 2006).

Somatic embryogenesis is considered as a desirable and fastest mode of plant regeneration (Williams and Maheswaran 1986). Somatic embryogenesis has more advantages over organogenesis especially when embryogenic cell suspension cultures are available. Some of its distinct features are single cell origin, high proliferation capacity and little changes of development of somatic variants, and a large number of regenerates are produced (Sato et al. 1993). When somatic embryo originally derives from explants, it is termed as primary somatic embryo (PE). Superficial cells such as hypocotyls and cotyledons of primary somatic embryo are proliferated in a repetitive manner to form secondary somatic embryo (SE) and cyclic secondary somatic embryos, that is a phenomenon in which new embryos emerge from the somatic embryos. Large-scale somatic embryogenesis through cyclic somatic embryogenesis is the most appealing method of mass multiplication of plants since very large number of somatic embryo can be produced in short time (Raemakers et al. 1993, 1995; Choi and Jeong 2002). This process is reported to have some advantages such as independent nature of an explant source, high multiplication rate and long-term repeatability (Raemakers et al. 1995). By this manner, somatic embryos can be maintained for long period and in large quantities.

Somatic embryogenesis is known to be a cells-stress-response process. The stress conditions can occur from wounding (Fowler et al. 1998), pathogenesis (Jabs and Slusarenko 2000), sucrose (Karami et al. 2006), plant hormones (Kim et al. 2009), cold or salts. Under these stressful conditions, plant cells alter their metabolism, growth and development in order to acclimatize with the new environment (Dahleen and Bregitzer 2002). Somatic embryos can be induced in both Murashige and Skoog (MS) and woody plant medium (WPM) basal medium (Aboshama 2011). Selection of suitable basal medium is one of the important criteria to be considered for efficient culture of any plant species. Carbon source especially sucrose in the embryogenic induction medium is considered to have an effect on the efficiency of the embryogenic process (Levi and Sink 1990).

Somatic embryo induction usually involves the use of auxins mostly 2,4-D, with or without a low level of cytokinin (Kim et al. 2009; Criollo et al. 2014; Ajijah et al. 2016). Cyclic somatic embryogenesis have been induced in several plants from various explants such as shoot and nodal explants (Saeed and Shahzad 2015), cotyledons (Kim et al. 2009), germinating seed (Nair and Gupta 2006) and immature zygotic embryo (Singh and Chaturvedi 2009).

An important requisite of in vitro propagated plants is the maintenance of genetic stability, while using elite genotype.

Plantlets produced by direct somatic embryogenesis are said to be genetically uniform than those produced from indirect somatic embryogenesis (Rani and Raina 2000). However, possibility of variations in somatic embryo cannot be ruled out (Nookaraju and Agrawal 2012). In order to confirm the quality of the plants, it is important to establish genetic fidelity and stability of micropropagated plants (Kumar et al. 2011).

48.1.1 Virus Elimination Through Somatic Embryogenesis

Somatic embryos have a developed vascular system but this system is not connected to the explants tissue (Newton and Goussard 1990). Detailed studies on the ontogeny of somatic embryos confirmed the absence of vascular tissue connection in embryogenic callus (Newton and Goussard 1990). This character makes the use of somatic embryogenesis as a method to eliminate plant viruses, combined with or without heat shock treatments (Goussard and Wiid 1993). It is considered to be the most effective procedures for the eradication of phloem-associated viruses such as leaf roll-associated virus and fan leaf virus from grapevines (Goussard et al. 1991; Goussard and Wiid 1993). Combination of heat therapy or chemotherapy with somatic embryogenesis was more effective in eliminating viruses in grapevine, citrus (D'Onghia et al. 2001), sugarcane, cocoa (Quainoo et al. 2008), cassava (Nkaa et al. 2013) and black pepper (Sasi and Bhat 2018). Somatic embryogenesis when combined with chemotherapy produced more virus-free plants than somatic embryogenesis alone. The advantage of somatic embryogenesis over meristem-tip culture is its ability to produce multiple plantlets per explant unlike meristem-tip culture that produces only one plantlet per explant. In this chapter, virus elimination in black pepper through somatic embryogenesis alone and combined with chemotherapy is presented (Figs. 48.1 and 48.2).

48.2 Materials

1. Virus-infected plants.
2. Laminar air flow.
3. Mercuric chloride.
4. Scalpel.
5. Stereomicroscope.
6. Schenk and Hildebrandt (SH) (1972) medium:

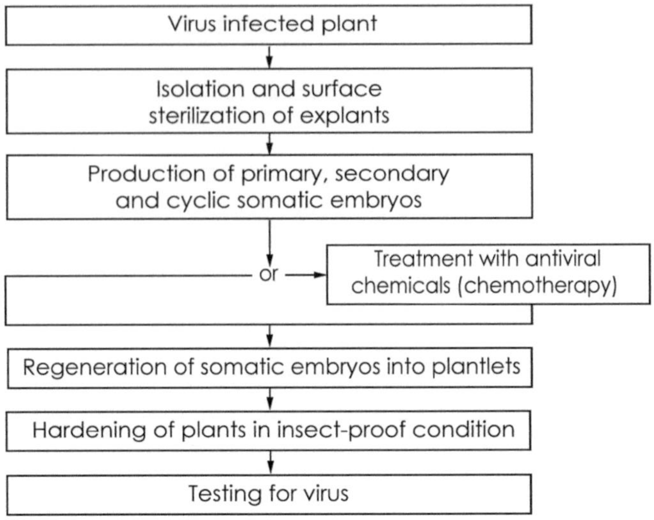

Fig. 48.1 Different stages in virus elimination by somatic embryogenesis alone and combined with chemotherapy

Stock	Constituents	Weight (g)	Stock volume (mL)	Volume/L (mL)
A	KNO_3	25		
	$MgSO_4 \cdot 7H_2O$	4.0		
	$NH_4H_2PO_4$	3.0	500	50
	$CaCl_2 \cdot 2H_2O$	2.0		
B	$MnSO_4 \cdot 2H_2O$	1.0		
	H_3BO_3	0.5		
	$ZnSO_4 \cdot 7H_2O$	0.1		
	KI	0.1	500	5
	$CuSO_4 \cdot 5H_2O$	0.02		
	Na_2MoO_4	0.01		
	$CoCl_2$	0.01		
C	$FeSO_4 \cdot 7H_2O$	1.5	500	5
	Na_2EDTA	2.0		
D	Inositol	25	250	10
E	Thiamine HCl	0.5		
	Nicotinic Acid	0.05	500	10
	Pyridoxine HCl	0.05		

(continued)

Stock	Constituents	Weight (g)	Stock volume (mL)	Volume/L (mL)
	Sucrose[a]	15–35 g/L		
	Agar	8 g/L		

Sterilize by autoclaving and store at 4 °C.
[a]SH (15), SH (30) and SH (35) media contain 15 g/L, 30 g/L and 35 g/L of sucrose, respectively

7. Woody plant medium (WPM): For preparation, please refer Chapter 47.
8. MS (Murashige and Skoog 1962) medium: For preparation, please refer Chapter 47.
9. Streptomycin (1 g/mL): Dissolve 1 g streptomycin in 1 mL sterile distilled water, filter sterilize and store at 4 °C.
10. Sodium hypochlorite.
11. Tween 20.
12. Casein hydrolysate.
13. Tissue culture laboratory.
14. Orbital shaker.
15. Greenhouse.
16. Ribavirin.

48.3 Method: Virus Elimination by Somatic Embryogenesis (Sasi and Bhat 2018)

The following protocol is used for virus elimination through somatic embryogenesis in black pepper. The same protocol may not be valid for all the crops. Researchers may scan the literature and look for the best published protocol for somatic embryogenesis for the crops of their interest.

Induction of primary, secondary and cyclic somatic embryos

1. Identify virus-infected plants based on symptoms and confirm the presence of the virus by PCR method.
2. Collect matured berries (seeds) from the above identified plants.
3. Soak the seeds in tap water for overnight and on the next day remove the outer mesocarp of the seed by slight rubbing.
4. Surface sterilize the seeds with 0.1% mercuric chloride solution for 5 min, followed by repeated washing (3–4 times) with sterile double distilled water.
5. Allow the seeds to dry on sterile filter paper in a laminar air flow for about 30 min.

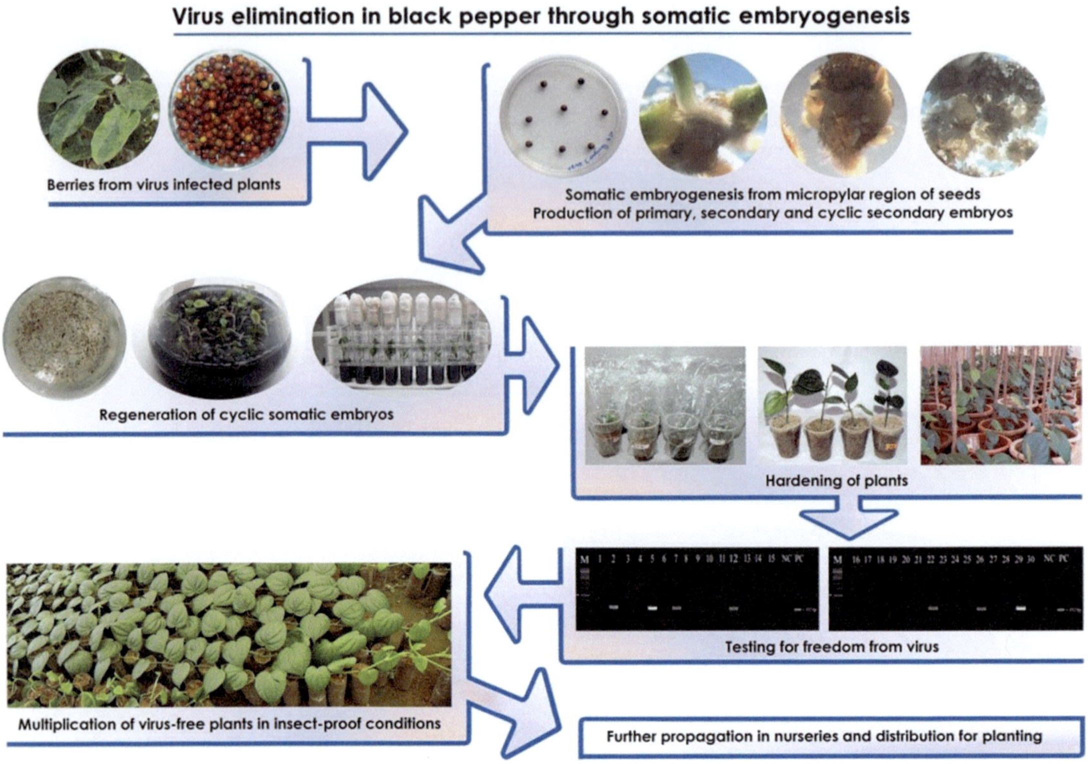

Fig. 48.2 Various stages in virus elimination through somatic embryogenesis of black pepper. (Reproduced from Bhat et al. 2018 with permission from Indian Society for Spices)

6. Dissect out the embryo along with micropylar tissue from the seeds aseptically with a sterile scalpel.
7. Culture dissected tissues on agar gelled full-strength, hormone-free SH (Schenk and Hildebrandt) medium containing 3% (w/v) sucrose (SH30) under darkness for about 90–120 days for inducing primary somatic embryos (PEs). Periodically observe and remove any contaminated tissues (Fig. 48.2)
8. Observe through stereomicroscope and inoculate the tissues showing PE on the micropylar region on full-strength, hormone-free SH medium containing 1.5% (w/v) sucrose (SH 15) and gelled with 0.8% agar and incubate in complete darkness for the production of secondary somatic embryo (SE) and cyclic SE.
9. Identify and culture PEs producing SE and cyclic SE in the same medium (SH15) gelled with 0.8% agar. Maintain cyclic SEs by regular subculturing at 30 days interval in SH15 medium.

Conversion of cyclic somatic embryos (cyclic SE) into plantlets

10. Inoculate about 100 mg of cyclic SE to hormone-free SH (liquid) medium with 3.5% sucrose (SH35) under dark for 20 days with shaking at 110 rpm in an orbital incubator shaker. Replenish the medium at every 10 days.
11. Continue shaking for another 10 days maintained under 16 h/day diffuse light.
12. Allow well-differentiated plantlets to grow under 12 h light in the same medium until they produced two to three leaves.
13. Transfer well-developed plants to WPM with 3.5% sucrose (SH35), 0.8% agar and 0.2% charcoal for another 30 days maintained under 16 h/day diffuse light for rooting.
14. Harden the well-established rooted plantlets in the greenhouse.
15. Test the hardened plants for the presence or absence of virus using PCR assay along with a known infected and a healthy plant (Fig. 48.2).

48.4 Method: Virus Elimination by Somatic Embryogenesis Combined with Chemotherapy (Sasi and Bhat 2018)

1. Use the cyclic somatic embryo (cyclic SE) produced as per the protocol described above (**steps 1–9** under the ***Induction of primary, secondary and cyclic somatic embryos***) as the starting material for this experiment.
2. Inoculate about 100 mg of cyclic somatic embryos in SH medium containing 1.5% sucrose (SH15) gelled with 0.8% agar supplemented with 10 mg/L of ribavirin.
3. Perform two subcultures at 15-day intervals in the same medium.
4. Convert ribavirin-treated cyclic somatic embryos into plantlets in liquid SH (SH 35) medium as described above and transfer well-differentiated plantlets to hormone-free WPM for proper organogenesis (**steps 10–13** above under the ***Conversion of cyclic somatic embryos (cyclic SE) into plantlets***).
5. Harden the well-established rooted plantlets in the greenhouse using sterile potting mixture.
6. Test the hardened plants for the presence or absence of virus using PCR assay along with a known infected and a healthy plant.

48.5 Method: Virus Elimination Through Somatic Embryogenesis in Sugarcane (Ramgareeb et al. 2010; Mishra et al. 2010)

Plant material

1. Select the field grown sugarcane variety (6–12 months old), infected with virus.
2. Cut the sugarcane stalk material into single budded nodes (SBN) and incubate in a hot water bath (50 °C) containing the fungicide Bavistin @ 0.5% for 30–40 min.
3. Following hot water treatment (HWT), plant the SBN in seedling trays filled with potting soil mix.
4. Incubate the trays in the glasshouse for 6–8 weeks at 30–35 °C and ambient light conditions) and water the plants twice daily.
5. Harvest the shoots developed from the nodes for meristem excision when the first node is observed.

Explant preparation

6. Excise the apical meristem (about 6 cm long spindle segment) from the node shoot.
7. Cut the leaf roll and stem consisting of at least the first visible node from the source material.
8. Wash the segment with 1% Tween 20 for 3–5 min followed by thorough washing with tap water.
9. Immerse in 70% (v/v) ethanol for 30 s and rinse with distilled water.
10. Remove the outer leaves and the leaf roll shortened to contain the first node above which the shoot apical meristem is situated (±4 cm).
11. Finally, surface sterilize the segments with 0.1% NaOCl (sodium hypochlorite) solution for 10 min followed by several washing with distilled water.
12. Dissect the meristems of different lengths (1.0–5.0 mm) from the explants containing the apical dome and leaf primordia for inoculation.

Somatic embryogenesis

13. Inoculate the surface sterilized excised meristems (1.0–2.0 mm) into culture media via indirect morphogenesis in laminar flow cabinet under aseptic environment condition.
14. The initiation media should include MS with 20 g/L sucrose; 8 g/L agar; 3.5 g/L activated charcoal; 0.5 g/L casein hydrolysate/0.2 mg/mL streptomycin, pH 5.6–5.8.

15. In the indirect somatic embryogenesis approach, incubate the meristems on MS media (agar 8.0 g/L) containing BAP and kinetin (0.5 mg/mL).
16. Subculture the meristems at biweekly intervals onto a fresh nutrient media and maintain in the dark (28 °C) for a total of 8 weeks or until the appearance of embryogenic callus.
17. Transfer the callus to MS media without charcoal or 2, 4-D and move into the photoperiod growth room (16 h light/8 h dark, 28 °C).
18. Subculture embryogenic calli fortnightly until germinated plants are approximately 3 cm in length.
19. Transfer the plants to glass vessels (half strength MS; 5 g/L sucrose; 8 g/L agar; 0.25 g/L casein hydrolysate, pH 5.6–5.8) and incubate in the photoperiod growth room for 1 month.
20. Harden the well-established rooted plantlets in the greenhouse using sterile potting mixture.
21. Test the hardened plants for the presence or absence of virus using PCR assay along with a known infected and a healthy plant.

48.6 Notes (Laux and Jurgens; 1997; Raemakers et al. 1995; Ramgareeb et al. 2010; Sasi and Bhat 2018)

1. The explants used for somatic embryogenesis vary with one plant species to another and it can be induced by various factors such as hormones, sucrose and ethylene.
2. The chemical used for surface sterilization of explants, its concentration and duration may vary from tissue to tissue. If information is already available on the crops you are handling you may use the same, else this needs to be standardized for each crop as excessive use (both concentration and duration) of the chemical may harm the tissue.
3. The cyclic somatic embryos can be regularly subcultured and used for producing somatic embryo-derived plants continuously. However, the period for which it can be used for continuous production of plants without any genetic instability may vary from one plant species to another.
4. Chemotherapy increases virus elimination from somatic embryos. However, different antiviral chemicals may have varying effect on the somatic embryo and its regeneration. Hence initial experiments with different antiviral chemicals, their concentration and duration of exposure need to be optimized to determine the right chemical, its dosage and duration to eliminate viruses during somatic embryogenesis.

5. Somatic embryo-derived plants may be tested at different intervals to confirm the presence or absence of the virus in the plants. The identified virus-free plants may be protected from insects and used for further multiplication under insect-proof condition.
6. It is important to subject somatic embryo-derived plants to morphological, physiological, cytological, molecular studies and field evaluation to assess the genetic stability and somaclonal variation.

References

Aboshama HMS (2011) Somatic embryogenesis proliferation, maturation and germination in *Cajanus cajan*. World J Agric Sci 7:86–95

Ajijah N, Hartati RS, Rubiyo R, Sukma D, Sudarsono S (2016) Effective cacao somatic embryo regeneration on kinetin supplemented DKM medium and somaclonal variation assessment using SSR marker. J Agric Sci 38:80–92

Ammirato PV (1983) Embryogenesis. In: Evans DA, Sharp WR, Ammirato PV, Yamada Y (eds) Handbook of plant cell culture, techniques for propagation and breeding. Macmillan, New York

Bhat AI, Biju CN, Srinivasan V, Ankegowda SJ, Krishnamurthy KS (2018) Current status of viral diseases affecting black pepper and cardamom. J Spices Aromat Crops 27:1–16

Choi YE, Jeong JH (2002) Dormancy induction of somatic embryos of Siberian ginseng by high sucrose concentrations enhances the conservation of hydrated artificial seeds and dehydration resistance. Plant Cell Rep 20:1112–1116

Criollo H, Perea M, Toribio M, Muñoz J (2014) Effect of the combination of NAA, kinetin and sucrose on the induction of somatic embryogenesis in lulo (*Solanum quitoense Lam.*). Agron Colomb 32:170–179

D'Onghia AM, Carimi F, De Pasquale F, Djelouah KMGP (2001) Elimination of *Citrus psorosis virus* by somatic embryogenesis from stigma and style cultures. Plant Pathol 50:266–269

Dahleen LS, Bregitzer P (2002) An improved media system for high regeneration rates from barley immature embryo-derived callus cultures of commercial cultivars. Crop Sci 42:934–938

Fowler MR, Eyre S, Scott NW, Slater A, Elliot MC (1998) The plant cell cycle in context. Mol Biotechnol 10:123–153

Goussard PG, Wiid J (1993) The use of *in vitro* somatic embryogenesis to eliminate phloem limited virus and nepoviruses from grapevines. In: Extended abstracts, vol. 11th meet ICVG, Montreux

Goussard PG, Wiid J, Kasdorf GGF (1991) The effectiveness of *in vitro* somatic embryogenesis in eliminating fan leaf virus and leaf roll-associated viruses from grapevine. S Afr J Enol Vitic 12:77–81

Jabs T, Slusarenko AJ (2000) The hypersensitive response. In: Slusarenko A, Fraser R, van Loon L (eds) Mechanisms of resistance to plant diseases. Kluwer Academic, Dordrecht

Karami O, Deljou A, Esna-Ashari M, Ostad-Ahmadi P (2006) Effect of sucrose concentrations on somatic embryogenesis in carnation (*Dianthus caryophyllus* L.). Sci Hortic 110:340–344

Kim SW, Oh JM, Liu JR (2009) Somatic embryogenesis and plant regeneration in zygotic embryo explant cultures of Rugosa rose. Plant Biotechnol Rep 3:199–203

Kumar S, Mangal M, Dhawan AK, Singh N (2011) Assessment of genetic fidelity of micropropagated plants of *Simmondsia chinensis* (Link) Schneider using RAPD and ISSR markers. Acta Physiol Plant 33:2541–2545

Laux T, Jurgens G (1997) Embryogenesis: new start in life. Plant Cell 9:989–1000

Levi A, Sink KC (1990) Differential effects of sucrose, glucose and fructose during somatic embryogenesis in asparagus. J Plant Physiol 137:184–189

McKersie BD, Brown DCW (1996) Somatic embryogenesis and artificial seeds in forage legumes. Seed Sci Res 6:109–126

Mishra S, Singh D, Tiwari AK, Lal M, Rao GP (2010) Elimination of sugarcane mosaic virus and sugarcane streak mosaic virus by tissue culture. Int Sugarcane J 28(3):119–122

Nair RR, Gupta SD (2006) High frequency plant regeneration through cyclic secondary somatic embryogenesis in black pepper (*Piper nigrum* L.). Plant Cell Rep 24:699–707

Newton DJ, Goussard PG (1990) The ontogeny of somatic embryos from *in vitro* cultured grapevine anthers. S Afr J Enol Vitic 11:70–75

Nkaa FA, Ene-Obong EE, Taylor N, Fauquet C, Mabanaso ENA (2013) Elimination of *African cassava mosaic virus* (ACMV) and *East african cassava mosaic virus* (EACMV) from cassava (*Manihot esculenta Crantz*) cv. 'Nwugo' via somatic embryogenesis. Am J Biotechnol Mol Sci. https://doi.org/10.5251/ajbms.2013.3.2.33.40

Nookaraju A, Agrawal DC (2012) Genetic homogeneity of *in vitro* raised plants of grapevine cv. Crimson Seedless revealed by ISSR and microsatellite markers. S Afr J Bot 78:302–306

Pareek LK (2005) Trends in plant tissue culture and biotechnology. Agrobios 9:334

Pierik RLM (1987) *In vitro* culture of higher plants. Martinus Nijhoff, Dordrecht

Quainoo AK, Wetten AC, Allainguillaume J (2008) The effectiveness of somatic embryogenesis in eliminating the *Cocoa swollen shoot virus* from infected cocoa trees. J Virol Methods 149:91–96

Raemakers CJJ, Bessembinder J, Staritsky G, Jacobsen E, Visser RGF (1993) Induction, germination and shoot development of somatic embryos in cassava. Plant Cell Tissue Org Cult 33:151–156

Raemakers CJJ, Jacobsen E, Visser RGF (1995) Secondary somatic embryogenesis and applications in plant breeding. Euphytica 81:93–107

Ramgareeb S, Snyman SJ, Antwerpen TV, Rutherford RS (2010) Elimination of virus and rapid propagation of disease-free sugarcane using apical meristem culture. Plant Cell Tissue Org Cult 100:175–181

Rani V, Raina SN (2000) Genetic fidelity of organized meristem derived micropropagated plants: a critical reappraisal. In Vitro Cell Dev Biol Plant 36:319–330

Saeed T, Shahzad A (2015) High frequency plant regeneration in Indian Siris via cyclic somatic embryogenesis with biochemical, histological and SEM investigations. Ind Crop Prod 76:623–637

Sasi S, Bhat AI (2018) In vitro elimination of Piper yellow mottle virus from infected black pepper through somatic embryogenesis and meristem-tip culture. Crop Prot 103:39–45

Sato S, Newell C, Kolacz K, Tredo L, Finer J, Hinchee M (1993) Stable transformation via particle bombardment in two different soybean regeneration. Plant Cell Rep 12:408–413

Schenk RU, Hildebrandt AC (1972) Medium and techniques for induction and growth of monocotyledonous and dicotyledonous plant cell cultures. Can J Bot 50:199–204

Singh M, Chaturvedi R (2009) An efficient protocol for cyclic somatic embryogenesis in Neem (*Azadirachta indica* A. Juss.). Proc Int Conf Energy Environ 3:19–21

Steward FC, Marion OM, Mears K (1958) Growth and organized development of cultured cells. II. Organization in cultures grown from freely suspended cells. Am J Bot 45:705–708

Thorpe TA (1988) *In vitro* somatic embryogenesis, ISI atlas of science. Anim Plant Sci 1:81–88

Williams EG, Maheswaran G (1986) Somatic embryogenesis: factors influencing coordinated behaviour of cells as an embryogenic group. Ann Bot 57:443–462

Chapter 49

Production of Virus-Resistant Plants Through Transgenic Approaches

Abstract

Production of transgenic plants for virus resistance is one of the success stories of genetic engineering that produce long-lasting and protected virus resistance, enabling the production of crops at commercial level. Of the various transgenes, use of translatable or non-translatable regions of the virus genome is the most successful approaches for developing virus-resistant varieties (known as pathogen-derived resistance, PDR). Of the various virus sequences, coat protein gene is the most widely used to engineer transgenic resistance. Availability of reliable regeneration systems, gene constructs in appropriate vectors, plant transformation techniques, selection of transgenic plants, characterization and evaluation of transgenic plants for resistance and commercialization of the transgenic variety are the various steps in the production and commercialization of transgenic virus-resistant plants.

Key words Transgenic resistance, Coat protein-mediated resistance, Genetic engineering, Plant transformation, Transgenic plant, Regeneration, Pathogen-derived resistance, Post-transcriptional gene silencing, Viral suppressors of RNA silencing, *Agrobacterium*-mediated genetic transformation, Biolistic

49.1 Introduction

Production of transgenic plants resistant to virus is one of the success stories of plant genetic engineering. Genetic transformation of plants using virus resistance genes from outside sources into susceptible plant species is a successful way to manage the disease. The possible genes used for virus resistance include genes coding for antimicrobial proteins/peptides, defence-related genes, genes of plant expressed antibodies (plantibodies) and pathogen-derived genes. The first successful production of transgenic plant using *Agrobacterium* as a vector was reported in the year 1980s. Transgenic technology can be used to transfer any specific genes from other background into plants (which are not possible through conventional plant breeding) to provide traits such as increased yield, insect, disease and herbicide resistance without altering their basic genetic background. This method of transferring genes

Alangar Ishwara Bhat and Govind Pratap Rao, *Characterization of Plant Viruses*, Springer Protocols Handbooks, https://doi.org/10.1007/978-1-0716-0334-5_49,

into plants from other taxa, unrelated plants, microbes and animals requires an efficient regeneration system with recombinant DNA technology.

49.1.1 Pathogen-Derived Genes for Plant Virus Resistance

Use of pathogen genes or sequences to protect plants against the attack from the same pathogen or closely related pathogen is referred as pathogen-derived resistance (PDR). This approach has huge potential in producing virus-resistant plants as plant-derived virus resistance genes are often not available. Sanford and Johnston (1985) proposed the theory of PDR which was later verified by Powel-Abel et al. (1986) by transforming tobacco plant that expressed coat protein (CP) of tobacco mosaic virus (TMV) for resistance against TMV. Since then many different viral genes and viral associated RNAs have been used as transgenes to confer resistance in plants and PDR became a reality against a range of plant viruses having positive sense ssRNA, ambisense RNA, or ssDNA (Varma et al. 2002; Dasgupta et al. 2003; Mitter et al. 2017; Pooggin 2017). Both coding and non-coding regions of viral genomes have been used for developing PDR. Coat protein (CP) gene is the most commonly used transgene for generating transgenic plants resistant to viruses belonging to different genera followed by replicase protein and movement protein genes. PDR acts either at translated transgene level (protein mediated) or at transgene transcript level (RNA mediated). The major mechanism of virus resistance in transgenic plants is through a process known as RNA interference (RNAi) (also referred to as virus-induced gene silencing, VIGS) (Watson et al. 2005).

The success of coat protein-mediated resistance (CPMR) has led to the creation of many transgenic plants of different crops expressing multiple CP genes from different viruses. Many transgenic crops engineered for virus resistance using CPMR approach were also released for commercial cultivation. They include TMV, tomato mosaic virus (ToMV) and CMV resistant tomato; zucchini yellow mosaic virus (ZYMV), CMV and watermelon mosaic virus (WMV) resistant squash; CMV, ZYMV and WMV resistant cantaloupe; potato leaf roll virus (PLRV), potato virus Y (PVY) and potato virus X (PVX) resistant potato; tomato chlorotic spot virus (TCSV), tomato spotted wilt virus (TSWV) and groundnut ringspot virus (GRSV) resistant tobacco and tobacco streak virus resistant sunflower (Fuchs and Gonsalves 1995; Fuchs et al. 1998; Pooggin 2017). Some of the transgenic lines have also been used as progenitors to create additional virus-resistant varieties through conventional breeding. Development and commercialization of transgenic papaya varieties that are resistant to PRSV in Hawaii is another success story that resulted in substantial yield increase over susceptible cultivars clearly demonstrating the ability of CPMR technology (Tripathi et al. 2007; Kung et al. 2010). Development and commercialization of transgenic plum variety, honeysweet resistant against plum pox virus (PPV) gave substantial gain to growers in five countries (Scorza et al. 2007).

49.1.2 Post-transcriptional Gene Silencing (RNA Silencing)

Post-transcriptional gene silencing (PTGS) (also called as RNA silencing or RNA interference) is the major mechanism involved in imparting virus resistance in transgenic plants. It is a specific RNA degradation mechanism that occurs in all organisms to manage unwanted excess or foreign RNA intracellularly in a homology-dependent manner. The important function in this mechanism is done by small RNAs [known as short interfering RNA (siRNA) and micro RNA (miRNA)], which act as effectors of silencing. Plants recognize double-stranded RNA (dsRNA) as foreign/aberrant and with the help of ribonuclease III-like enzyme called Dicer, dsRNA is cleaved into 21–26 nt siRNAs. Following this, one of the strands of the siRNA gets incorporated into a ribonuclease complex known as the RNA-induced silencing complex (RISC) that serves as the guide for sequence-specific degradation of homologous mRNAs. This degradation process, initiating from a cell having the elicitor RNA, spreads later within the entire plant in a systemic manner. This process is believed to have evolved as a plant defence mechanism against invading viruses. When the viral RNA is either the elicitor or target of PTGS, the degradation mechanism is known as virus-induced gene silencing (VIGS). RNA silencing is used to engineer resistance in plants against viruses. Plants expressing copy of a viral sequence in sense and/or antisense orientation can show resistance upon infection with the same or closely related viruses through this mechanism (Kjemtrup et al. 1998; Smyth 1999; Pooggin 2017).

A single copy of the transgene has the ability to induce RNA silencing; however, silencing can be enhanced by the incorporation of multiple copies of the transgene especially when arranged in inverted repeats to be able to form dsRNA. When ssRNA is used as transgene, host-encoded RNA-dependent RNA polymerase generates the dsRNA that recognize abundant transgene RNAs and copy them into dsRNA to make primary siRNAs. When inverted repeats are used as transgene, they do not depend on host RNA-dependent RNA polymerase to produce primary siRNAs. But in both cases, host RNA-dependent RNA polymerase is involved in production of secondary siRNAs to reinforce and spread silencing beyond initial site (Himber et al. 2003). Plants transformed with ssRNA transgenes give rise to low and erratic numbers of virus-resistant transgenic lines while plants transformed with inverted repeats produce high level of viral resistance. Transgenes encoding intron-spliced hairpin RNA (hpRNA) are the most efficient to confer resistance and hence, hpRNA approach is now widely used to generate virus-resistant transgenic plants. The hpRNA construct consists of a plant promoter and terminator between which an inversely repeated sequence of the target gene is inserted (with a spacer region between the repeats). The RNA transcribed from such a transgene hybridizes with itself to form a hairpin structure (Smith et al. 2000; Manamohan et al. 2013; Mitter et al. 2017).

RNAi-mediated resistance was successfully demonstrated on a large number of ssRNA (+ve sense) viruses and also in viruses with ambisense RNA genomes such as tospoviruses and viruses with ssDNA genomes such as begomoviruses. Even the fused transgene that contained portion of TSWV–N gene and turnip mosaic virus (TuMV) CP gene showed resistance to both viruses in *N. benthamiana* (Jan et al. 2000). Transgenic cassava plants harbouring AC1 gene from the African cassava mosaic virus (ACMV) gave protection against ACMV and two heterologous begomoviruses infecting cassava (Chellappan et al. 2004). Similarly, transgenic cassava plant expressing common region of ACMV DNA-A gave protection against ACMV (Vanderschuren et al. 2007).

49.1.3 Viral Suppressors of RNA Silencing (VSR)

Viruses counter the effects of RNA silencing by producing virulence factors known as suppressors of RNA silencing that suppress different steps of the RNA silencing mechanism. The VSRs vary in sequence, structure and activity from virus to virus. A single VSR may target several points in RNA silencing pathways. In addition, viruses with large genomes are known to code for many proteins to achieve this effect. Some examples include: V2 protein of tomato yellow leaf curl virus and P6 protein of cauliflower mosaic virus protein interfere dsRNA processing into vsRNAs (Glick et al. 2008). Many VSRs inhibit HEN1-mediated 2_O-methylation of vsRNA duplexes (Lozsa et al. 2008).

In view of above it is reported that production of transgenic plants using sequences of silencing suppressor is the best way to obtain virus-resistant plants. This was first demonstrated by expression of artificial microRNAs (amiRNAs) that encode the silencing suppressors of the turnip yellow mosaic virus (TYMV) and the turnip mosaic virus (TuMV) in transgenic *Arabidopsis* plants that showed resistance to both viruses. Later this was validated in other virus-host combinations also (Qu et al. 2007; Hu et al. 2011; Ntui et al. 2014).

49.2 Requirements for the Development of Virus-Resistant Transgenic Plants

The main requirements for the production of transgenic plants are: (1) availability of regeneration system, (2) a good genetic transformation protocol, (3) gene construct preparation, (4) selection of transformed plants, (5) characterization of transformants, (6) screening of transgenic plants for resistance and (7) commercialization of the transgenic plant variety.

49.2.1 Development of Reliable Tissue Culture Regeneration System

This is the first important step in developing the transformation protocol for a plant. Apical meristem, stem, cotyledon, callus obtained from different plant parts, zygotic embryos, mature somatic embryos and embryogenic clumps are reported as suitable target tissues for regeneration in different plants. In most cases, embryogenic cultures are the good target tissue for the transformation of many plant species as it is actively multiplying, possess high proliferation and high regeneration abilities and are more susceptible to *Agrobacterium* infections (Gelvin 2003). The single cell origin of somatic embryos eliminates nonchimeric transgenic plants and results in more number of transformed plantlets being regenerated (Gonzalez et al. 1998; Kothari et al. 2010).

49.2.2 Transformation Methods

Direct and indirect techniques are available for performing genetic transformation of plants. The major aim of genetic transformation is to obtain more number of stably transformed plants. Large number of methods such as biolistic bombardment, *Agrobacterium*-mediated transformation, microinjection, chemical (PEG) treatment of protoplasts and electroporation of protoplasts/intact plant cells and tissues, infiltration, silicon carbide-mediated transformation and pollen tube pathway are available for transformation. Though each method is unique, transformation using *Agrobacterium* and biolistic bombardment are the widely used methods. The advantage of biolistic bombardment is that it can be used to introduce the transgene directly into plants cells. Similarly, *Agrobacterium* has been used as the vector for genetic transformation of diverse plants belonging to both dicots and monocots (Jones et al. 2005).

49.2.2.1 Biolistic Method

In this method, high-velocity transgene DNA-coated gold microcarrier is introduced directly into intact cells or tissues using a gene gun or biolistics. Parameters such as concentration of DNA used per bombardment, distance between the target tissue and macro carrier carrying the DNA, helium pressure, osmotic treatment before transformation and explants age need to be standardized for each experiment (Sailaja et al. 2008).

49.2.2.2 Agrobacterium-Mediated Gene Transfer

Agrobacterium-mediated gene transfer was first reported by Block and co-workers in the year 1984 (Block et al. 1984). The benefits of this method over other methods used for transformation include: capability to introduce large fragments of DNA with minimal rearrangement, definite integration of transgenes, stable integration, does not need of any special equipment and integration into transcriptionally active regions of the chromosomes (Smith and Hood 1995). Many strains of *Agrobacterium* such as LBA4404, EHA101, EHA105 and AGL1; plasmids and methods developed for the transformation of different plant species are now available (Draper et al. 1988). Though, during initial stages of transformation history, monocots could not be transformed through *Agrobacterium*, currently protocols are available for the efficient transformation of monocots using embryonic tissues or embryogenic cultures with super virulent strains of *Agrobacterium* in the presence of acetosyringone (inducer of virulence genes). There are two primary pre-requisites for performing *Agrobacterium*-mediated transformation: (1) the availability of explants with high regenerative ability and (2) an efficient transformation protocol. Many factors such as plant species and its genotype, type of explants, *Agrobacterium* strain, its cell density and duration of co-cultivation, and selection strategies will affect the efficiency of transformation.

49.2.3 Preparation of Gene Constructs in Binary Vectors

Cloning transgene into T-DNA region of Ti-plasmids and its subsequent introduction into plants is a complex process as Ti-plasmids occur as low copy plasmids in *Agrobacterium* and their size is very large besides the difficulty to isolate and manipulate them in vitro. All these difficulties were later overcome due to findings of Hoekema et al. (1983) and Framond et al. (1983) who came up with simple protocols for development of gene constructs. It is based on the method where T-DNA and the *vir* region are placed in two separate plasmids referred as binary Ti vectors. The disarmed Ti-plasmids that reside in the *Agrobacterium* provide *vir* gene functions while the T-DNA within which the transgene to be cloned is provided on the other vector. Binary Ti vectors can replicate in both *E. coli* and *Agrobacterium* as it contains origin for replication of both these organisms. The antibiotic marker such as kanamycin, gentamicin, tetracycline, streptomycin and spectinomycin present in the binary vectors allows for the selection of transformants (Fig. 49.1). Once the vector construct process is over, the presence of the transgene in the binary vector can be verified by PCR, restriction digestion using enzymes with unique sites or by sequencing. Initially the construct is prepared and characterized *in E. coli*. After confirmation of the construct in *E. coli*, it is mobilized into *Agrobacterium* strain containing a *vir* helper region using any of methods such as electroporation, freeze/thaw or triparental mating. Among all, triparental mating method is the largely used method for mobilizing recombinant plasmid into *Agrobacterium* though it requires three different bacteria, different incubation temperatures and long incubation time (Ditta et al. 1980). Electroporation is another high efficient method for transformation of *Agrobacterium* while freeze-thaw method has low transformation efficiency (Wen-Jun and Forde 1989).

Some of the earliest binary vectors used for plant transformation include pBIN19 (Bevan 1984) and pBI121 (Jefferson et al. 1987). Other binary vectors used in plant transformation include pCANBIA series from the Commonwealth Scientific and Industrial Research Organization (CSIRO), Australia (Fig. 49.1). Binary vector belonging to pBIN and pCAMBIA series are usually used in the

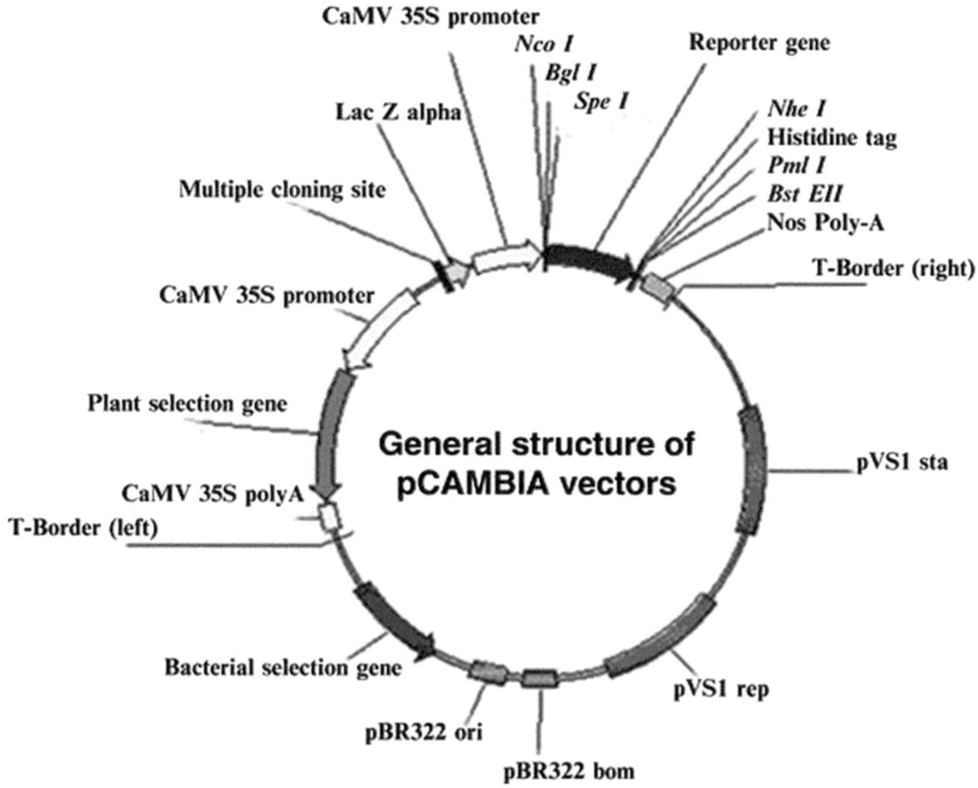

Fig. 49.1 General structure of pCAMBIA binary vector showing T-DNA region with left and right borders; plant selection gene and its promoter; reporter gene and its promoter; origin of replication and bacterial selection gene. (Source: CSIRO, Australia)

standardization of transformation experiments as it has *gus* region that helps to evaluate expression of transgene by histochemical GUS assay (Fig. 49.1). Other multipurpose and flexible binary vectors are also reported (Hellens et al. 2000; Komari et al. 2007).

49.2.4 Selection and Regeneration of Transgenic Plants

Two kinds of antibiotics are needed for the selection and regeneration of transgenic plants. The first one is used for controlling *Agrobacterium* after co-cultivation while the second one is required for the selection of transformants. In order to kill *Agrobacterium* after co-cultivation, antibiotics such as carbenicillin and cefotaxime that have minimum toxicity on most plant tissue are used. It is important to determine the optimum dosage and duration of exposure of these antibiotics required to kill *Agrobacterium* after co-cultivation for each type of explants of various plant species as very high concentration may affect regeneration of explants while low dosage may not kill *Agrobacterium* growth. In general, an antibiotic concentration ranging from 100 to 500 μg/mL is reported for the successful elimination of *Agrobacterium* from various explants belonging to different plant species (Barik et al. 2005; Li et al. 2007; Oz et al. 2009).

After co-cultivation, selection of transformants and their regeneration in the selection medium is very crucial for obtaining transgenic plants. The antibiotics or other substances to be used in the selection medium depends on the marker genes present in the T-DNA gene construct used for the transformation. The generally used selectable markers include antibiotic resistance genes, herbicide resistance genes or phospho-mannitol isomerase genes. The commonly used selective antibiotic markers are hygromycin and kanamycin. The simplicity and efficiency of kanamycin makes it as the most preferred selective agent in most transformation experiments. When kanamycin is selectable marker, the neomycin phosphotransferase II (NPT II) gene gets transferred along with the transgene into the plant genome and as a result the transformed plant becomes resistant to kanamycin (Kapaun and Cheng 1994). Kanamycin concentrations for resistant explants selection might vary from 10 to 200 μg/mL depending on plant species. In order to get more transformants, some researchers used low concentration of antibiotic in the beginning and a stepwise raise thereafter, to enable transformed tissue to express well the antibiotic resistance gene and begin cell division, thus potentially enhancing successful plant regeneration (Mondal et al. 2001; Bull et al. 2009; Retheesh and Bhat 2011). This actually reduces sudden shock to explants and increases the frequency of transformants. A few workers followed a somewhat different approach of removing selection pressure for a time to obtain a maximum number of transgenics (Cai et al. 1999). In another strategy used for papaya transformation, Kung et al. (2010) used kanamycin selection for initial 90 days and thereafter selected tissues was subcultured on the kanamycin-free medium for embryo development. The mature somatic embryos were then cultured on medium containing 50 μg/mL kanamycin for shoot development that resulted in higher transformation efficiency.

49.2.5 Screening of Transgenic Plants

49.2.5.1 GUS Assay

GUS gene is one of the normally used reporter genes for the standardization of transformation protocol for different plant species. The enzyme coded by this gene converts substrate, 5-bromo-4-chloro-3-indolyl-K-D-glucuronide into blue coloured 5-bromo-4-chloro-3-indole in the cells that display GUS activity. This technique is sensitive and can also be used score the level of expression of transgene (Jefferson et al. 1987; Jefferson and Wilson 1991).

49.2.5.2 Polymerase Chain Reaction (PCR) Assay

Polymerase chain reaction (PCR) is a quick method for primary screening of putative transgenic plants. But it is not a reliable method for the confirmation of transgenic nature of plants if DNA isolated from the transformants is contaminated with *Agrobacterium* cells harbouring Ti-plasmid. Hence PCR cannot be the method for determination of stable integration of transgenes in the plant (Shekhawat et al. 2008). However, PCR can be used for the preliminary screening of transformants as it is fast and affordable.

All PCR positive transgenic plants may be later subjected to Southern hybridization for the confirmation of stable integration of the transgene in the chromosome of the plants.

49.2.5.3 Southern Hybridization, Northern Hybridization and Western Blotting

Southern hybridization is one of the methods used for confirmation of the integration of the transgene into the genome of the transgenic plants. Besides, this test also provides information on the copy number of the transgene present in the transgenic plants. The important steps involved in the procedure are: (1) extraction of nucleic acid from transgenic plant and agarose gel electrophoresis, (2) denaturation of the DNA by treating gel with sodium hydroxide (which would separate dsDNA into ssDNA) followed by neutralization of gel with sodium chloride, (3) transfer of ssDNA from gel to membrane such as nylon or nitrocellulose membrane (Southern transfer), (4) DNA is fixed with the membrane either by using UV light which cross-links (*via* covalent bonds) the DNA to the membrane or by baking the membrane at about 80 °C for about 2 h. (5) The membrane is then incubated in pre-hybridization solution containing heterologous nucleic acids such as ssDNA from salmon sperm DNA or yeast RNA to block the remaining reactive sites on the membrane, (6) Hybridization of the membrane with labelled ssDNA probe prepared against the transgene sequences which will bind to its complementary strand, (7) removal of unhybridized excess probe by washing the membrane, (8) detection and visualization of hybridization signals using appropriate method (Southern 1975; Sambrook and Russel 2001).

The northern hybridization detects transcript of the transgene in the plants, thus confirming the expression of the transgene in the transgenic plant. In this method, total RNA isolated from transgenic plants are run on denaturing formaldehyde agarose gel. The separated RNA molecules are then transferred onto membrane by northern blotting. The blot is then hybridized with appropriately labelled DNA probe of the transgene. The important steps include: (1) isolation of total RNA from plants, (2) separation of RNA through denaturing formaldehyde agarose gel electrophoresis, (3) transfer of RNA from gel to positively charged nitrocellulose or nylon membrane (northern transfer). Like Southern blotting, transfer is usually done by capillary action that requires several hours. (4) The membrane is then incubated in pre-hybridization solution to block the remaining reactive sites of membrane using heterologous nucleic acid such as salmon sperm DNA or yeast RNA. (5) Hybridization of the membrane with labelled DNA probe against transgene which will bind to its complementary strand, (6) removal of unhybridized excess probe by washing the membrane, (7) detection (visualization) of hybridization by using appropriate detection method depending on the kind of probe used (Alwine et al. 1977). Both Southern and northern hybridization analyses have been used by many workers to confirm the integration, copy number of the transgene and production of transcripts.

If the protein-mediated resistance is expected in the transgenic plants developed in addition to Southern and northern hybridization analyses, plants also have to be subjected to western blotting to confirm the production of transgene protein. The method involves isolation of total protein from the transgenic plants and subjecting them to SDS-PAGE analysis on a 12% polyacrylamide gel. The proteins separated on the SDS-PGE gel are then transferred onto a nitrocellulose, nylon or PVDF membrane through electroblotting. The presence of the protein of interest in the membrane is detected by developing the membrane by addition of transgene-specific antibodies followed by its detection using enzyme labelled conjugate and subsequent addition of substrate for the enzyme that will produce coloured product (O'Donell et al. 1982; Sambrook and Russel; 2001).

49.2.6 Evaluation of Transgenic Plants

Evaluation of transgenic plants refers to the testing of transgenic plants for the trait for which they are produced without being a human, animal and environmental hazard. Evaluation of transgenic plants for virus resistance is done by challenge inoculation with the corresponding virus isolate. The methods of challenge inoculation are: (a) mechanical inoculation by rubbing the leaves of putative transgenic plants with sap from virus-infected plant, (b) through vector of the virus, and (c) through grafting (Praveen et al. 2005; Vassilakos et al. 2008; Jardak-Jamoussi et al. 2009).

In the following general outline for the development of transgenic plants using (1) coat protein gene of papaya ringspot virus using biolistic and (2) hairpin construct of cucumber mosaic virus using *Agrobacterium*-mediated transformation are given.

49.3 Development of Transgenic Papaya Through Coat Protein-Mediated Approach Using Biolistic

49.3.1 Preparation of Construct (Quemada et al. 1990; Fitch et al. 1990, 1992)

1. Amplify coat protein (CP) gene of papaya ringspot virus (PRSV) from total RNA isolated from PRSV-infected papaya through reverse transcription polymerase chain reaction (RT-PCR) using primers to conserved regions flanking the PRSV CP gene (details of total RNA isolation from plants and RT-PCR are provided in Chapters 32 and 35, respectively).
2. PRSV CP lacks native translation signals for CP as PRSV codes for a single polyprotein which is later processed into different functional proteins including CP. Hence, prepare a chimeric gene construct using 5′ untranslated RNA translational enhancer region and initial 16 amino acid coding sequence of the cucumber mosaic virus (CMV) CP gene fused in frame to the structural sequence of the PRSV CP including the Q/S protein cleavage site and 51 nucleotides of the non-coding

region. This is accomplished by cloning the PRSV CP structural sequence from plasmid pPRV117 (Quemada et al. 1990) in between the cauliflower mosaic virus (CaMV) 35 S double enhancer promoter–translational leader sequence and CaMV 35 S terminator of a CMV expression cassette.

3. Clone the PRSV CP expression cassette into the vector, pGA482GG that contains the npt II (neomycin phosphotransferase II) gene behind a nopaline synthase promoter and a uidA (glucuronidase [GUS]) gene behind a CaMV 35S promoter, used for kanamycin selection and colorimetric screening of transformants, respectively.

49.3.2 Transformation, Regeneration and Confirmation of Transgene Integration

1. Treat zygotic embryos obtained from immature seeds of 90–120 days old papaya fruits with 2,4-dichlorophenoxyacetic acid to be used as explant for transformation by biolistic (somatic embryos produced from seeds or hypocotyls can also be used as explants).
2. Anneal the PRSV CP plasmid construct to tungsten particles and carry out biolistic transformation through aseptic bombardment into papaya explants tissue prepared in **step 1** (Fitch et al. 1990, 1992).
3. Place the bombarded explants for 4–5 weeks on induction medium under dark without antibiotics.
4. Place the tissues on maturation medium containing 75 mg/L of kanamycin for 4 weeks under light thereafter in the same medium with higher concentration of kanamycin (150 mg/L) for 8 weeks under light.
5. Place the surviving tissues in the germination medium containing kanamycin @ 150 mg/L for 2–3 months during which resistant embryos will proliferate into green plantlets.
6. Multiply clones of resistant plantlets obtained above through micropropagation and root them in rooting media.
7. Subject leaves of surviving plantlets for GUS assay, PCR and dot-blot hybridization to test for presence of the *npt II* and PRSV CP gene.
8. Subject leaves from the transgenic plants identified based on GUS assay, PCR and dot-blot hybridization to Southern hybridization using PRSV CP gene as the probe to confirm the stable integration and to determine the copy number of the transgene.
9. Analyse the plant that tested positive in Southern hybridization for the CP gene by northern RNA blots to confirm the production of transcript of the transgene.

10. Analyse the plants that tested positive in Southern and northern hybridization by western blotting to confirm the production of transgene protein using PRSV CP antibodies.

49.3.3 Evaluation of Transgenic Papaya for Virus Resistance in the Greenhouse

1. Subject confirmed transgenic papaya plants for resistance assay using the virulent isolate of PRSV through mechanical inoculation method (details of the mechanical inoculation method are provided in Chapter 6).
2. Evaluate symptoms after 21 days of inoculation and record the levels of resistance.
3. Monitor the plants for 6 weeks and re-inoculate the plants if no symptoms are seen.
4. Observe the plants for symptoms development, if no symptoms become visible till 21 days after the second inoculation, extract sap from these plants and inoculate onto the local lesion host to check for the presence of PRSV in these plants.
5. Subject the above identified transgenic papaya plants to multiple mechanical inoculations and inoculation by grafting to confirm the resistance against PRSV. Multiple inoculation assays may be performed on new growth on transgenic papaya plants every 2–4 weeks for 10 months with PRSV-infected extracts. For inoculation by grafting, transgenic papaya seedlings should be grafted to non-transformed PRSV-infected trunks.

49.4 Development of Transgenic Plant Through RNAi Approach Using *Agrobacterium*-Mediated Transformation

49.4.1 Preparation of Hairpin (HP) Construct

1. Subject the selected sequences to dsCheck online software (http://dsCheck.RNA.jp/) (Naito et al. 2005) to select the specific region that will yield dsRNA.
2. Select a 400 bp off-target minimized region by the software mentioned above.
3. Retrieve corresponding sequences of all available isolates of that particular virus species from NCBI GenBank and study sequence polymorphism in this using the software DnaSP version 5.10 (Librado and Rozas 2009).
4. Subject the selected dsRNA sequences to siDESIGN center (http://www.dharmacon.com), online software to identify the common siRNAs present in these regions.
5. After finalizing the region to be used for HP construct, amplify both sense and antisense region of the selected target region for dsRNA through PCR using appropriate primers containing appropriate restriction sites for cloning (Fig. 49.2).

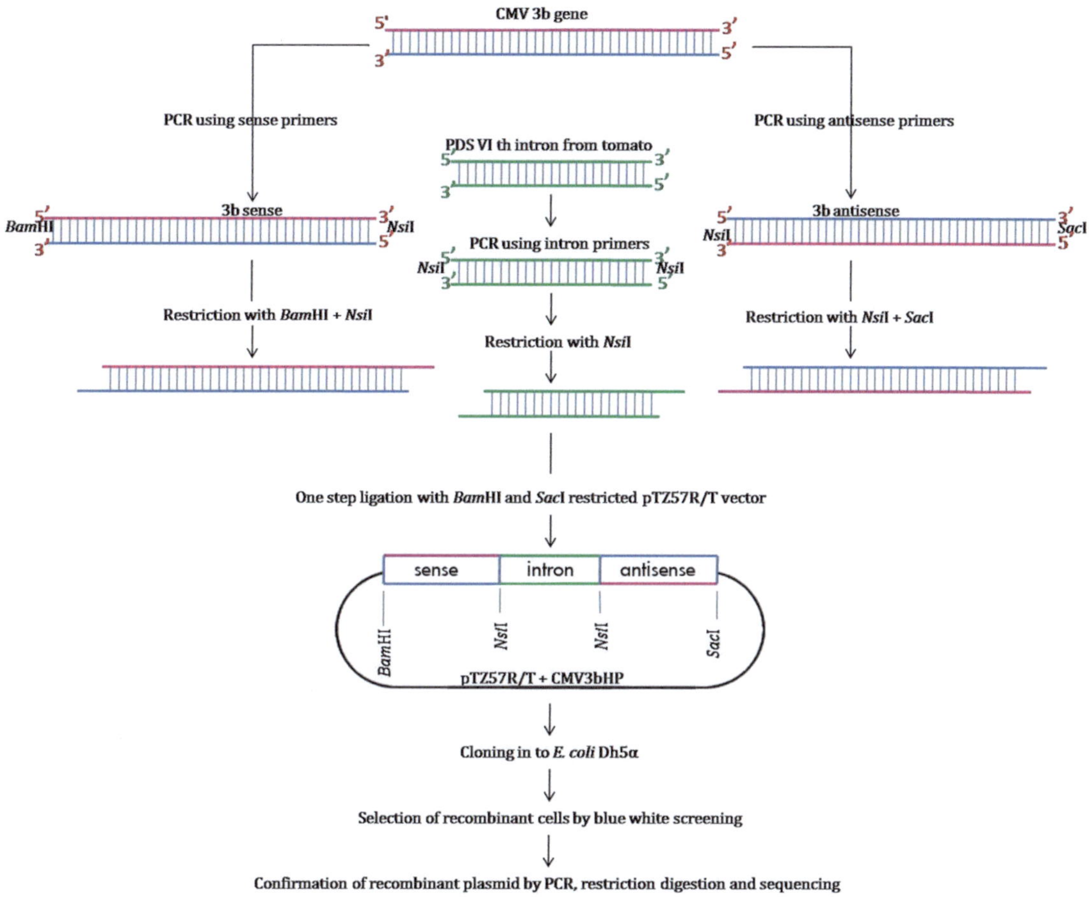

Fig. 49.2 Schematic diagram showing different steps in cucumber mosaic virus 3b hairpin construct preparation in the plasmid cloning vector pTZ57R/T. (Reproduced from Revathy and Bhat 2019 with permission from NISCAIR-CSIR, India)

6. Amplify the intron sequence to be inserted between the sense and antisense fragments flanked with appropriate restriction sites (Fig. 49.2).
7. Restrict the sense, antisense and intron PCR products with appropriate restriction enzymes for cloning.
8. Restrict the plasmid vector to be used for cloning also with same restriction enzymes.
9. The restricted sense, antisense and intron along with the double restricted plasmid vector are kept for ligation in the ratio 3:8:3:1 overnight (Fig. 49.2).
10. Transform the ligated product into *E. coli* DH5α using standard protocol as explained in Chapter 42 and identify positive recombinants through antibiotic selection.

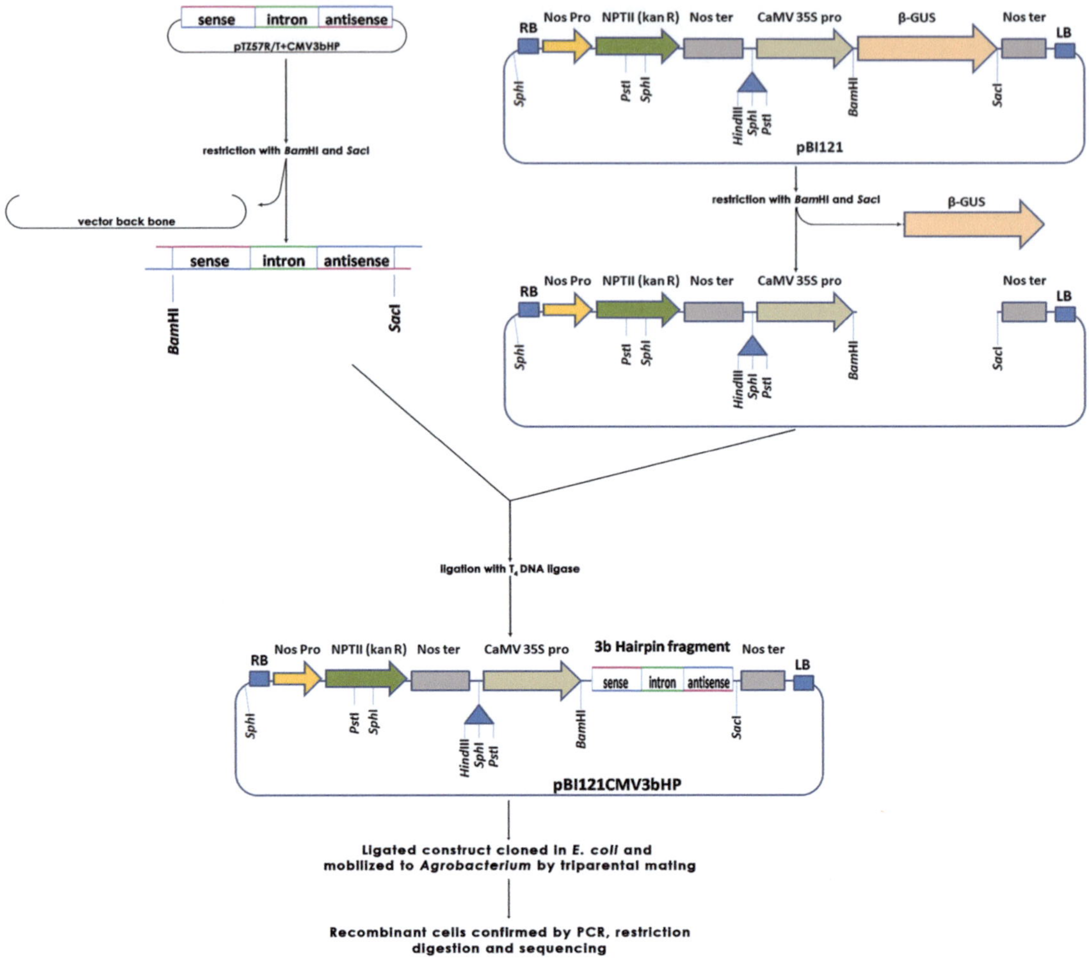

Fig. 49.3 Schematic diagram of the cucumber mosaic virus 3b hairpin construct preparation in the binary vector pBI121 and its mobilization into *Agrobacterium*. (Reproduced from Revathy and Bhat 2019 with permission from NISCAIR-CSIR, India)

11. Screen the transformants and identify positive recombinants possessing HP construct through colony PCR.
12. Isolate the plasmid from the colony PCR positive recombinant clones and confirm the presence of HP construct through PCR, restriction analysis and sequencing (Fig. 49.2).
13. Release the hairpin construct from the cloning vector through restriction digestion and ligate into a suitable binary vector such as pBI121 (Fig. 49.3).
14. The recombinant binary vector carrying HP construct is then cloned into the suitable *Agrobacterium tumefaciens* strain by triparental mating or freeze-thaw method and select transformants using appropriate antibiotics in the medium.

15. Screen transformants and identify recombinant binary plasmid through PCR, restriction enzyme and sequencing (Fig. 49.3).
16. Use confirmed clone of *Agrobacterium tumefaciens* harbouring binary vector with HP construct for transformation of plants using suitable explants.
17. A schematic diagram showing various steps involved in the HP construct preparation using CMV 3b gene in a plasmid cloning vector, its subsequent cloning into a binary vector and its transformation into *Agrobacterium* is provided in Figs. 49.2 and 49.3.

49.4.2 Agrobacterium-Mediated Transformation of Black Pepper (Nair and Gupta 2006; Jiby and Bhat 2011)

Somatic embryos produced from the micropylar region of the matured berries of black pepper are used as explant for transformation. Protocol for the production of somatic embryo is provided in Chapter 48.

1. Infect about one gram of explant (cyclic somatic embryogenic mass) with 1/5th diluted overnight grown culture ($OD_{600} = 1.5$) of *Agrobacterium tumefaciens* EHA 105 harbouring the binary plasmid pBI121 containing hairpin construct with intermittent vigorous shaking for 2 h (Fig. 49.4).
2. Pass the contents through a sterile Whatmann No. 2 filter paper and discard the filtrate containing bacteria.
3. Culture the explant on basal SH medium containing 3% sucrose for 48 h under dark.
4. Transfer the explant to plates containing selection medium (SH basal + 1.5% sucrose + 100 μg/mL cefotaxime + 25 μg/mL kanamycin) and incubate for 3–5 weeks under dark.
5. Separate responding embryogenic mass and place them in the fresh medium of same composition but with 50 μg/mL kanamycin for 4 weeks.
6. Transfer proliferated embryogenic mass to liquid SH medium containing 3% sucrose (without kanamycin) under dark with shaking at 110 rpm for 4 weeks for growth of embryos into plantlets (Fig. 49.4).
7. Allow surviving plantlets to grow for 2 weeks under light with fresh medium of same composition.
8. Transfer the plantlets to woody plant medium (WPM) containing 3% sucrose, 0.8% agar, 100 μg/mL kanamycin and 0.2% charcoal for 2 weeks (non-transformed plant if any will bleach during this incubation) (Fig. 49.4).
9. Transfer the surviving plants to fresh medium of same composition but without kanamycin for another 4 weeks.

Fig. 49.4 Steps in *Agrobacterium*-mediated genetic transformation black pepper. (**a**) Co-cultivated embryogenic mass. (**b**) Growing points under kanamycin selection. (**c**) Fully developed plantlets in liquid SH medium. (**d**) Plantlets in WPM medium. (**e**, **f**, **g**) Different stages of hardening. (Reproduced from Jiby and Bhat 2011 with permission from Springer)

10. Harden the survived plants in the greenhouse in small pots containing sterile potting mixture, cover with a polyethylene bag if needed, to maintain moisture (Fig. 49.4).
11. When plantlets attain just about 10 cm height, shift plants to bigger sized pots (18 cm diameter) containing sterile potting mixture and maintain them in the greenhouse.
12. Screen six-month-old plants for the presence of transgene by PCR, dot-blot and Southern hybridization.

13. Transgenic plants that tested positive in Southern hybridization can be subjected to RT-PCR and northern hybridization to check for the production of transcripts using npt II specific primers.

49.4.3 Testing of Transgenic Black Pepper for Viral Resistance in the Greenhouse

1. Subject transgenic plants identified from above experiment for resistance to the virulent isolate of CMV using either aphid or cleft grafting.
2. Confirm the resistance through multiple aphid inoculations and inoculation by grafting as described under transformation of papaya.

49.5 Notes

1. Research and development of transgenic crops is subject to strict biosafety rules due to their potential impact on the environment, human and animal health. The environmental risks may be in the form of effects on beneficial insects, mammals, microbes, the possibility of transgenic plant crossing with non-transgenic species and their persistence in the environment.
2. In a few transformation experiments, it was observed that antibiotic selection can reduce the recovery of potential transgenic plants. In such cases, recovery of transgenic plants can be enhanced by reducing the period under antibiotic selection (Cai et al. 1999) or by using alternative antibiotics, or by use of visual marker such as green fluorescent protein.
3. In *Agrobacterium*-mediated transformation, the marker gene and gene of interest are bordered by the T-DNA region of *Agrobacterium* while this may not be the case in biolistic-mediated transformation method (Tripathi et al. 2007).
4. In the case of biolistics, the integration event is random, and the transgene and marker genes do not always co-integrate; hence, the presence of marker gene does not necessarily indicate the integration of transgene also (Tripathi et al. 2007).
5. Coat protein expressed by the transgene within a transgenic plant may enter another invading virus infecting the plant *via* a process of hetero-encapsidation (Varma et al. 2002).
6. Sequence variability exists within different strains of the same virus species. Hence virus-resistant transgenic plant developed against one strain may not give protection against all strains of the same virus. Hence it is important to use highly conserved sequence region as transgene to ensure protection against all strains of the virus (Dasgupta et al. 2003).

7. In general, acceptance rate of transgenic crops is extremely low probably due to the complexity in the development of transgenic crops, biosafety and trade regulatory issues and its capability as a commercial product (Mitter et al. 2017).

References

Alwine JC, Kemp DJ, Stark GR (1977) Method for detection of specific RNAs in agarose gels by transfer to diazobenzyloxymethyl paper and hybridization with DNA probes. Proc Natl Acad Sci U S A 74:5350–5354

Barik DP, Mohapatra U, Chand PK (2005) Transgenic grasspea (*Lathyrus sativus* L.): factors influencing *Agrobacterium* mediated transformation and regeneration. Plant Cell Rep 24:523–531

Bevan M (1984) Binary *Agrobacterium* vectors for plant transformation. Nucleic Acids Res 12 (26):8711–8721

Block MD, Herrera EL, Vanmontagu M (1984) Expression of foreign genes in regenerated plants and in their progeny. EMBO J 3:1681–1689

Bull SE, Owiti JA, Niklaus M, Beeching JR, Gruissem W, Vanderschuren H (2009) *Agrobacterium* mediated transformation of friable embryogenic calli and regeneration of transgenic cassava. Nat Protoc 4:1845–1854

Cai W, Gonalves C, Tennant P, Fermin G, Souza M, Sarinud N, Jan FJ, Zhu HY, Gonsalves D (1999) A protocol for efficient transformation and regeneration of *Carica papaya* L. InVitro Cell Dev Biol-Plant 35:61–69

Chellappan P, Masona MV, Vanitharani R, Taylor NJ, Fauquet CM (2004) Broad spectrum resistance to ssDNA viruses associated with transgene-induced gene silencing in cassava. Plant Mol Biol 56:601–611

Dasgupta I, Malathi VG, Mukherjee SK (2003) Genetic engineering for virus resistance. Curr Sci 84:341–354

Ditta G, Stanfield S, Corbin D, Helinski DR (1980) Broad host range DNA cloning system for Gram-negative bacteria: construction of a gene bank of *Rhizobium meliloti*. Proc Natl Acad Sci U S A 77:7347–7351

Draper J, Scott R, Armitage P (1988) Plant genetic transformation and gene expression: a laboratory manual. Blackwell Scientific Publishers, Oxford

Fitch MMM, Manshardt RM, Gonsalves D, Slightom JL, Sanford JS (1990) Stable transformation of papaya via microprojectile bombardment. Plant Cell Rep 9:189–194

Fitch MMM, Manshardt RM, Gonsalves D, Slightom JL, Sanford C (1992) Virus resistant papaya derived from tissues bombarded with the coat protein gene of papaya. Biotechnology 10:1466–1472

Framond AJ, Barton KA, Chilton MD (1983) Mini Ti: a new vector strategy for plant genetic engineering. Bio/Technology 1:262–269

Fuchs M, Gonsalves D (1995) Resistance of transgenic squash Pavo ZW-20 expressing the coat protein genes of Zucchini yellow mosaic virus and Watermelon mosaic virus 2 to mixed infections by both potyviruses. BioTechnology 13:1466–1473

Fuchs M, Tricoli DM, McMaster JM, Carney KJ, Schesser M (1998) Comparative virus resistance and fruit yield of transgenic squash with single and multiple coat protein genes. Plant Dis 82:1350–1356

Gelvin SB (2003) *Agrobacterium* mediated plant transformation: the biology behind the "Gene-Jockeying" tool. Microbiol Mol Biol Rev 67:16–37

Glick E, Zrachya A, Levy Y, Mett A, Gidoni D et al (2008) Interaction with host SGS3 is required for suppression of RNA silencing by tomato yellow leaf curl virus V2 protein. Proc Natl Acad Sci U S A 105:157–161

Gonzalez AE, Schöpke C, Taylor NJ, Beachy RN, Fauquet CM (1998) Regeneration of transgenic cassava plants (*Manihot esculenta* Crantz) through *Agrobacterium* mediated transformation of embryogenic suspension cultures. Plant Cell Rep 17:827–831

Hellens RP, Mullineaux P, Klee H (2000) pGreen: a versatile and flexible binary Ti vector for Agrobacterium mediated plant transformation. Plant Mol Biol 42:819–832

Himber C, Dunoyer P, Moissiard Izenthaler C, Voinnet O (2003) Transitivity dependent and independent cell-to-cell movement of RNA silencing. EMBO J 22:4523–4533

Hoekema A, Hirsch PR, Hooykaas PJJ, Schilperoort RA (1983) A binary plant vector strategy based on separation of *vir*- and T-region of the *Agrobacterium tumefaciens* Tiplasmid. Nature 303:179–180

Hu Q, Niu Y, Zhang K, Liu Y, Zhou X (2011) Virus-derived transgenes expressing hairpin RNA give immunity to *Tobacco mosaic virus* and *Cucumber mosaic virus*. Virol J 8:41

Jan FJ, Fagoaga C, Pang SZ, Gonsalves D (2000) A single chimeric transgene derived from two distinct viruses confers multi-virus resistance in transgenic plants through homology dependent gene silencing. J Gen Virol 81:2103–2109

Jardak-Jamoussi R, Winterhagen P, Bouamama B, Dubois C, MLiki A, Wetzel T, Ghorbel A, Reustle GM (2009) Development and evaluation of a GFLV inverted repeat construct for genetic transformation of grapevine. Plant Cell Tiss Org Cult 97:187–196

Jefferson RA, Wilson KJ (1991) The GUS gene fusion system. Plant Mol Biol Rep 5:387–405

Jefferson RA, Kavanagh TA, Bevan MW (1987) GUS fusions: ß-glucuronidase as a sensitive and versatile gene fusion marker in higher plants. EMBO J 6:3901–3907

Jiby MV, Bhat AI (2011) An efficient *Agrobacterium*-mediated transformation protocol for black Pepper (*Piper nigrum* L.) using embryogenic mass as explants. J Crop Sci Biotech 14:247–254

Jones HD, Diherty A, Wu H (2005) Review of methodology and a protocol for the *Agrobacterium* mediated transformation of wheat. Plant Methods 1:5

Kapaun JA, Cheng ZM (1994) Aminoglycoside antibiotics inhibit shoot regeneration from Siberian elm leaf explants. Hortsciences 34:727–729

Kjemtrup S, Sampson KS, Peele CG, Nguyen LV, Conkling MA, Thompson WF, Robertson D (1998) Gene silencing from plant DNA carried by a geminivirus. Plant J 14:91–100

Komari T, Imayama T, Kato N, Ishida Y, Ueki J, Komari T (2007) Current status of binary vectors and super binary vectors. Plant Physiol 145:1155–1160

Kothari SL, Joshi A, Kachhwaha S, Ochoa-Alejo N (2010) Chilli peppers—a review on tissue culture and transgenesis. Biotechnol Adv 28:35–48

Kung Y, Yu T, Huang C, Wang H, Wang S, Yeh S (2010) Generation of hermaphrodite transgenic papaya lines with virus resistance *via* transformation of somatic embryos derived from adventitious roots of *in vitro* shoots. Transgenic Res 19:621–635

Li ZN, Fang F, Liu GF, Bao MZ (2007) Stable *Agrobacterium*-mediated genetic transformation of London plane tree (*Platanus acerifolia* Willd.). Plant Cell Rep 26:641–650

Librado P, Rozas J (2009) v5: software for comprehensive analysis of DNA polymorphism data. Bioinformatics 25:1451–1452

Lozsa R, Csorba T, Lakatos L, Burgyan J (2008) Inhibition of 3 modification of small RNAs in virus infected plants require spatial and temporal coexpression of small RNAs and viral silencing-suppressor proteins. Nucleic Acids Res 36:4099–4107

Manamohan M, Sharath Chandra G, Asokan R, Deepa H, Prakash MN, Krishna Kumar NK (2013) One-step DNA fragment assembly for expressing intron-containing hairpin RNA in plants for gene silencing. Anal Biochem 433:189–191

Mitter N, Worrall EA, Robinson KE, Li P, Jain RG, Taochy C, Fletcher SJ, Carroll BJ, Lu GO, Xu ZP (2017) Clay nanosheets for topical delivery of RNAi for sustained protection against plant viruses. Nat Plants 3:16207

Mondal TK, Bhattacharya A, Ahuja PS, Chand PK (2001) Transgenic tea [*Camellia sinensis* (L.) O. Kuntze cv. Kangra Jat] plants obtained by *Agrobacterium* mediated transformation of somatic embryos. Plant Cell Rep 20:712–720

Nair RR, Gupta SD (2006) High frequency plant regeneration through cyclic secondary somatic embryogenesis in black pepper (*Piper nigrum* L.). Plant Cell Rep 24:699–707

Naito Y, Yamada T, Matsumiya T, Ui-Tei K, Saigo K et al (2005) dsCheck: highly sensitive off--target search software for dsRNA–mediated RNA interference. Nucleic Acids Res 33:589–591

Ntui VO, Kynet K, Khan RS, Ohara M, Goto Y, Watanabe M, Fukami M, Nakamura I, Mil M (2014) Transgenic tobacco lines expressing defective CMV replicase-derived dsRNA are resistant to CMV-O and CMV-Y. Mol Biotechnol 56:50–63

O'Donell IJ, Shukla DD, Gough KH (1982) Electroblot immunoassay of virus infected plant sap—a powerful technique for detecting plant viruses. J Virol Methods 4:19–26

Oz MT, Eyidogan F, Yucel M, Oktem HA (2009) Optimized selection and regeneration conditions for *Agrobacterium* mediated transformation of chickpea cotyledonary nodes. Pak J Bot 41(4):2043–2054

Pooggin MM (2017) RNAi-mediated resistance to viruses: a critical assessment of methodologies. Curr Opin Virol 26:28–35

Powel-Abel P, Nelson RS, De B, Hoffman N, Rogers SG, Frayley RT, Beachy RN (1986) Delay of disease development in transgenic plants that express the tobacco virus coat protein gene. Science 232:738–743

Praveen S, Kushwaha CM, Mishra AK, Singh V, Jain RK, Varma A (2005) Engineering tomato for resistance to tomato leaf curl disease using viral rep gene sequences. Plant Cell Tiss Org Cult 83:311–318

Qu J, Ye J, Fang R (2007) Artificial microRNA-mediated virus resistance in plants. J Virol 81:6690–6699

Quemada H, L'Hostis B, Gonsalves D et al (1990) The nucleotide sequences of the 3′-terminal regions of papaya ringspot virus strains W and P. J Gen Virol 71:203–210

Retheesh ST, Bhat AI (2011) Genetic transformation and regeneration of transgenic plants from protocorm-like bodies of vanilla (*Vanilla planifolia* Andrews.) using *Agrobacterium tumefaciens.* J Plant Biochem Biotechnol 20:262–269

Revathy KA, Bhat AI (2019) Designing of siRNAs for various target genes of *Cucumber mosaic virus* subgroup IB. Indian J Biotechnol 18:119–125

Sailaja M, Tarakeswari M, Sujatha M (2008) Stable genetic transformation of castor (*Ricinus communis* L.) *via* particle gun-mediated gene transfer using embryo axes from mature seeds. Plant Cell Rep 9:1509–1519

Sambrook J, Russel DW (2001) Molecular cloning: a laboratory manual, vol I–III, 3rd edn. Cold Spring Harbor Laboratory Press, New York

Sanford JC, Johnston SA (1985) The concept of parasite-derived resistance genes from the parasite's own genome. J Theor Biol 113:395–405

Scorza R, Hily JM, Callahan A, Malinwski T, Cambra M, Capote M, Zagrai I, Damsteegt V, Briard P, Ravelonandro M (2007) Deregulation of plum pox resistant transgenic plum 'HoneySweet'. Acta Hort 738:669–673

Shekhawat UKS, Ganapathi TR, Srinivas L, Bapat VA, Rathore TS (2008) *Agrobacterium* mediated genetic transformation of embryogenic cell suspension cultures of *Santalum album* L. Plant Cell Tiss Org Cult 92:261–271

Smith RH, Hood EE (1995) *Agrobacterium tumefaciens* transformation of monocotyledons. Crop Sci 35:301–309

Smith NA, Singh SP, Wang MB, Stoutjesdijk PA, Green AG, Waterhouse PM (2000) Total silencing by intorn-spliced hairpin RNAs. Nature 407:319–320

Smyth DR (1999) Gene silencing: plants and viruses fight it out. Curr Biol 9:100–102

Southern EM (1975) Detection of specific sequences among DNA fragments separated by gel electrophoresis. J Mol Biol 98:503–517

Tripathi S, Suzuki J, Gonsalves D (2007) Development of genetically engineered resistant papaya for Papaya ringspot virus in a timely manner: a comprehensive and successful approach. Methods Mol Biol 354:197–240

Vanderschuren H, Stupak M, Futterer J, Gruissem W, Zhang P (2007) Engineering resistance to geminiviruses-review and perspectives. Plant Biotechnol J 5:207–220

Varma A, Jain RK, Bhat AI (2002) Virus resistant transgenic plants for environmentally safe management of viral diseases. Indian J Biotechnol 1:73–86

Vassilakos N, Bem F, Tzima A, Barker H, Reavy B, Karanastasi E, Robinson DJ (2008) Resistance of transgenic tobacco plants incorporating the putative 57-kDa polymerase readthrough gene of Tobacco rattle virus against rub-inoculated and nematode-transmitted virus. Transgenic Res 17:929–941

Watson JM, Fusaro AF, Wang M, Waterhouse PM (2005) RNA silencing platforms in plants. FEBS Lett 579:5982–5987

Wen-Jun S, Forde BG (1989) Efficient transformation of *Agrobacterium* spp. by high voltage electroporation. Nucleic Acid Res 17:8385

Chapter 50

Production of Virus-Resistant Plants Through CRISPR-Cas Technology

Abstract

Clustered regularly interspaced short palindromic repeats (CRISPR) are widely found in bacterial and archaeal genomes as a defence mechanism against invading viruses and plasmids. The CRISPR locus consists of segments of prokaryotic DNA with short repetitions of base sequences. Each repetition is followed by short segments of 'spacer DNA' from earlier exposed bacterial virus or plasmid. CRISPR spacers recognize and cut these exogenous genetic elements in a manner analogous to RNA interference in eukaryotic organisms. The CRISPR interference technique has enormous potential application, including altering the germline of humans, animals, other organisms and plants. The organism's genome can be cut at any desired location by delivering the Cas9 protein and guide RNAs into a cell. Thus CRISPR-Cas system represents a powerful tool in developing resistance to DNA and RNA plant viruses by editing and inserting novel traits precisely at chosen loci into plants and offers enormous potential in classical breeding. It has opened new way to get virus-resistant plants either by directly targeting and cutting the viral genome, or by modifying the host genome itself to introduce viral immunity. In this chapter, recent progress demonstrating the efficacy of the CRISPR/Cas technology against DNA and RNA plant viruses is discussed.

Key words Plant virus, CRISPR/Cas9, CRISPR/Cas13a, Genome engineering, PAM motif, Double strand break (DSB), Non-homologous end joining (NHEJ), Homology directed repair (HDR), Virus resistance

50.1 Introduction

Clustered repeats were first described in *Escherichia coli* in 1987 (Ishino et al. 1987), named initially as short regularly spaced repeats (SRSR) which was renamed as CRISPR in 2002. In 2005, it was shown that some CRISPR spacers are derived from phage DNA and plasmids indicating that spacers could have a role in adaptive immunity in bacteria. Spacers serve as a template for RNA molecules, analogous to RNA interference used by eukaryotic cells. CRISPR was first shown to work as a genome engineering/editing tool in human cell culture in 2012 (Jinek et al. 2012) and later used in a wide range of organisms including nematodes (*C. elegans*), zebrafish (*D. rerio*), fruit flies (*D. melanogaster*), plants, baker's yeast (*S. cerevisiae*), mice, monkeys and human embryos.

Alangar Ishwara Bhat and Govind Pratap Rao, *Characterization of Plant Viruses*, Springer Protocols Handbooks, https://doi.org/10.1007/978-1-0716-0334-5_50, © Springer Science+Business Media, LLC, part of Springer Nature 2020

CRISPR has been further advanced to make specific transcription factors that activate or silence the specific genes of interest (Doudna and Charpentie 2014).

50.1.1 Mechanism of CRISPR/Cas

The *cas* or *CRISPR*-associated genes are found to be associated with CRISPR repeats which encode putative nuclease or helicase proteins for the cutting and unwinding of DNA. Bacteria respond to an invading phage by transcribing spacers and palindromic DNA into a long RNA molecule, which is further cut into pieces called CRISPR RNAs by transacting CRISPR RNA and Cas9. It has two active cutting sites (HNH and RuvC), for each strand of the DNA's double helix. Doudna and Charpentier (2014) fused tracrRNA and spacer RNA into a chimeric RNA called 'single-guide RNA (sg RNA or gRNA)' molecule which could find and cut the correct DNA targets with Cas9. Their study finally concluded that these synthetic guide RNAs could be used successfully for the gene editing purpose (Doudna and Charpentie 2014). The sgRNA-Cas9 complex requires a protospacer-adjacent motif (PAM) site at the 3′ end of the target 20-bp sequence matching the protospacer to cleave and introduce double strand DNA breaks (DSBs) (Fig. 50.1). The widely used *Streptococcus pyogenes* Cas9 recognizes a 5′-NGG-3′ PAM sequence, whereas Cas9 proteins from other species recognize and bind to different PAM sequences. Recognition of the PAM by the Cas9 nuclease is thought to destabilize the adjacent sequence and allowing interrogation of the sequence by the sgRNA, thus resulting in RNA-DNA pairing.

Cas9 protein with the help of the crRNA identifies the correct sequence in DNA of the host cell and creates a single or double strand break in the DNA chain (Fig. 50.1). A single strand breaks in the host DNA can trigger the homology directed repair (HDR), which is less error prone compared to non-homologous end joining (NHEJ) that typically follows a double strand break. The NHEJ pathway generates insertions and deletions during double-stranded break (DSB) repair. HDR can seal the DSB in an error-free manner in the presence of a DNA template with homology to the sequences flanking with DSB location. In most of the cells, both of these repair pathways are supposed to be active; however, the HDR pathway is reported as less efficient than the NHEJ.

50.1.2 Application in Plants

A schematic diagram showing various steps involved in genome editing in plants using CRISPR/Cas system is provided in Fig. 50.2. CRISPR/cas system generates stable and heritable mutants that can easily segregate from cas9/sgRNA construct that results in the development of transgene-free plants. It is a new plant breeding technique that is faster than the traditional method. Plants produced through CRISPR/cas are genetically edited (GE) crops hence probably socially acceptable. It can be used to get pest resistance, enhanced nutritional value, climate

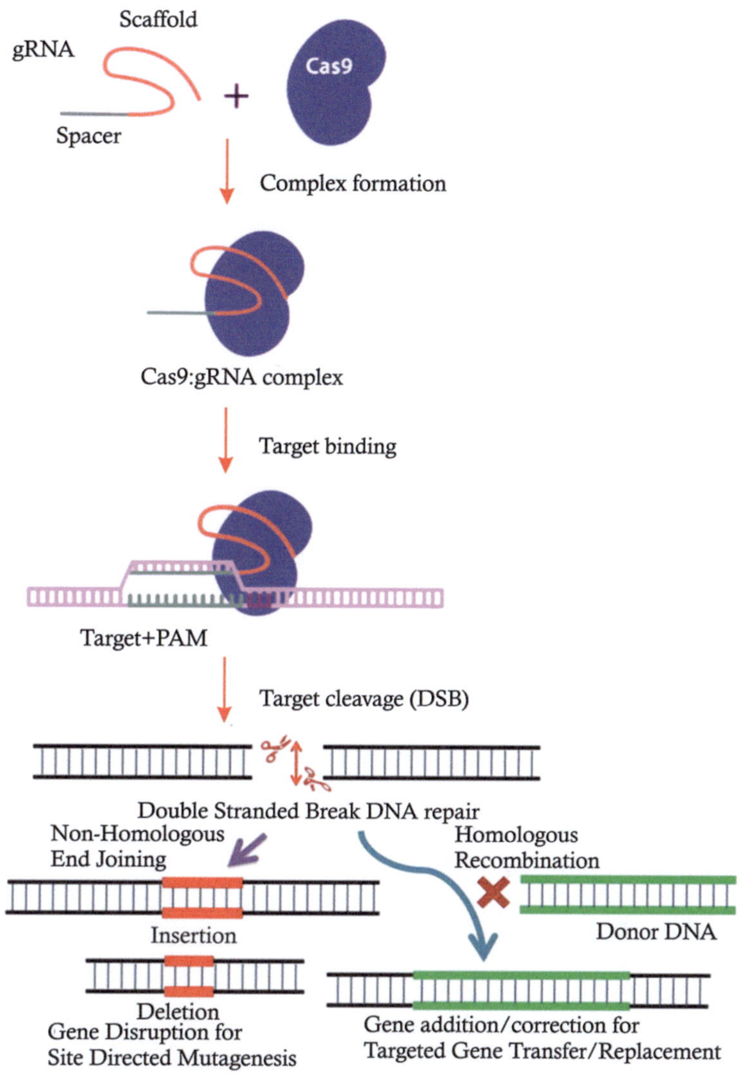

Fig. 50.1 Schematic diagram showing steps involved in CRISPR/Cas9 based genome editing

resilient crops, increased productivity, etc. Multiplex genome editing using CRISPR/Cas 9 based binary vector and a gRNA module vector is possible in plants. Plant viruses (cabbage leaf curl virus, tobacco rattle virus) can be used as vectors to deliver cas 9/sgRNA to plants. Recombinant viruses carrying cas 9/sgRNA can systemically spread throughout the plant system editing the genome in all cells of the plant.

50.1.3 Virus Resístanse in Plants via CRISPR/Cas

CRISPR/Cas9 and CRISPR–Cas13a system have received vital attention because of their efficiency, simplicity and reproducibility. Recent researches claimed CRISPR/Cas9/Cas13a as genomic

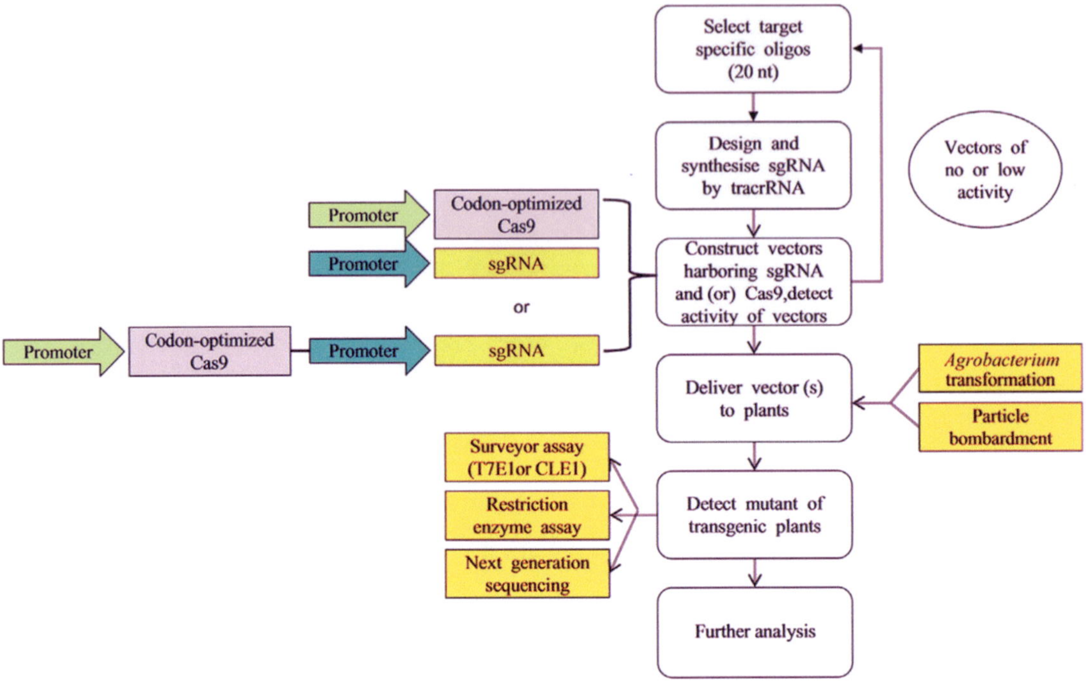

Fig. 50.2 Schematic diagram showing various steps involved in genome editing in plants using CRISPR/Cas system. (Reproduced from Liu et al. 2017 with permission from Elsevier)

tool to engineer virus resistance in plants. The viral immunity can be introduced either by directly targeting and cleaving the viral genome, or by modifying the host plant genome (Khan et al. 2018). CRISPR/Cas system is consistently reported to reduce the virus accumulation and provided reliable approaches in management of both the DNA and RNA viruses (Table 50.1). Most of the studies involving CRISPR-edited plants for virus resistance were targeted against geminivirus, cucumber mosaic virus, papaya ringspot virus and rice tungro spherical virus (Ali et al. 2015, 2016; Baltes et al. 2015; Ji et al. 2015) (Table 50.1). The maintenance of a transgene for Cas9 and sgRNA in the genome of the crop plants is considered to be the best novel genetic engineering approach for virus disease resistance.

A second strategy to develop plant viral disease resistance requires modification of plant genes that will introduce virus resistance traits through CRISPR/Cas9 tool and to release non-transgenic mutants in the field (Chandrasekaran et al. 2016; Pyott et al. 2016). RNA viruses require plant host factors, eIF4E, eIF (iso) 4E and eIF4G for maintenance of their life cycle (Sanfacon 2015). Chandrasekaran et al. (2016) developed potyviruses-resistant cucumber plants by mutating independently two different sites of the host susceptibility gene *eIF4E* with CRISPR/Cas9 technology. Non-transgenic Cucumis *eif4e* mutant plants are

Table 50.1
Successful CRISPR/Cas9 applications for virus resistance in plants

Plant	Virus	Target gene	References
Nicotiana benthamiana, *Arabidopsis thaliana*	Bean yellow dwarf virus, beet severe curly top virus	CP, Rep, IR (RCA mechanism)	Ji et al. (2015)
Nicotiana benthamiana	Beet severe curly top virus	LIR and Rep/RepA (RCA mechanism)	Baltes et al. (2015)
N. benthamiana	Tomato yellow leaf curl virus, beet curly top virus, merremia mosaic virus	CP, Rep and IR (RCA mechanism)	Ali et al. (2015)
N. benthamiana, *A. thaliana*	Cucumber mosaic virus, tobacco mosaic virus	ORF1, 2, 3, CP and 3′UTR (Replication mechanism)	Zhang et al. (2018)
Cucumis sativus	Cucumber vein yellowing virus, zucchini yellow mosaic virus, papaya ring spot virus-W	eIF4E (host factor for RNA viruses translation)	Chandrasekaran et al. (2016)
Oryza sativa L. *japonica*	Rice tungro spherical virus	eIF4G (host factor for RNA viruses translation)	Macovei et al. (2018)
A thaliana	Turnip mosaic virus	*elF(iso)4E*	Pyott et al. (2016)
N. benthamiana, *A. thaliana*	Cucumber mosaic virus and tobacco mosaic virus	Virus genome	Zhang et al. (2018)
A. thaliana	Turnip mosaic virus	Virus genome	Aman et al. (2018)

CP coat protein, *Rep* replication association protein, *IR* intergenic region, *RCA* rolling-circle amplification, *LIR* long intergenic region, *GFP1* green fluorescent protein 1, *GFP2* green fluorescent protein 2, *HC-Pro* helper component proteinase silencing suppressor, *ORF* open reading frame, *UTR* untranslated terminal repeat, *eIF4E* eukaryotic translation initiation factor 4E, *eIF4G* eukaryotic translation initiation factor 4G

produced by segregation of the CRISPR/Cas9 through backcrossing, which are safe for human consumption and for release into the environment. When challenged with cucumber vein yellowing virus (CVYV), zucchini yellow mosaic virus (ZYMV) and papaya ringspot virus-W (PRSV-W), homozygous *eif4e* mutants showed immunity against the virus; however, the non-mutant and heterozygous knockout plants did not express any resistance against these viruses. Similar strategy was used by Macovei et al. (2018) to develop virus resistance in rice against rice tungro spherical virus (RTSV) (Table 50.1).

50.1.3.1 DNA Virus Resístanse via *CRISPR/Cas9*

Geminiviridae is a large family of plant viruses causing worldwide crop losses to potato, tomato, cotton, cassava, cucurbits and legumes. The virus genome is replicated through rolling-circle amplification *via* a dsDNA replicative form, or by recombination-mediated replication (Hanley-Bowdoin et al. 2013). The most important genus of geminiviruses is *Begomovirus* and their genome is organized as monopartite or bipartite components (Fondong 2013). Utilization of CRISPR/Cas system application focusing on resistance to geminiviruses was reported with beet severe curly top virus and bean yellow dwarf virus in *Arabidopsis* and *Nicotiana benthamiana*, respectively (Baltes et al. 2015; Ji et al. 2015). Other reports employed a CRISPR/Cas9 approach for achieving resistance to begomoviruses in the host cell nucleus to target and cleave the virus genome during replication are listed in Table 50.1. Ali et al. (2015) developed sgRNA molecules using different regions of tomato yellow leaf curl virus (TYLCV) delivered *via* a tobacco rattle virus (TRV) vector into *N. benthamiana* plants stably overexpressing the Cas9 endonuclease. They found that sgRNAs were able to interfere with TYLCV genome sequences and reported a more significant reduction of viral replication and accumulation. The same CRISPR/Cas9 system was tested for targeting simultaneously against the monopartite and bipartite geminiviruses that share the same stem-loop sequence in the IR. The results demonstrated that mixed infection immunity can be developed *via* a single sgRNA specific for conserved sequences of multiple viral strains.

50.1.3.2 RNA Virus Resístanse via *CRISPR/Cas9*

Protection against RNA plant viruses is more difficult to achieve, because the SpCas9 from *Streptococcus pyogenes* only recognizes dsDNA. However, later the characterization of new nucleases have been identified as FnCas9 from *Francisella novicida* and/or Lwa Cas13a from *Leptotrichia wadei* that can bind to and cut RNA. The first successful attempt in demonstrating resistance to RNA viruses (cucumber mosaic virus, tobacco mosaic virus) was established by expressing FnCas9 and RNA-targeting sgRNAs specific in *N. benthamiana* and *Arabidopsis* plants, respectively (Zhang et al. 2018). Transgenic plants showed 40–80% reduction in CMV and TMV accumulation as compared with the control plants. The virus resistance obtained by expressing the sgRNA-FnCas9 system was found to be quite stable and active up to T6 generation. Zhang et al. (2018) observed that the endonuclease activity of FnCas9 was not essentially required for interference with the CMV genome, but FnCas9's role in RNA-binding activity was found to be essential. This application of FnCas9 can be recognized as a CRISPR interference (CRISPRi) tool, similar to catalytically inactive SpCas9 proteins programmed to mitigate gene expression (Larson et al. 2013). The use of a catalytically inactive variant of FnCas9 has the advantage of limiting the onset of mutated viral variants capable of escaping CRISPR/Cas9. Aman et al. (2018) also employed

RNA-guided ribonuclease to manipulate the turnip mosaic virus (TuMV) RNA genome using Cas13a. They targeted four different genomic regions of the virus: two targets in the green fluorescent protein (GFP) region, one in the coat protein (CP) and the one in the helper component proteinase silencing suppressor (HC-Pro). The most efficient virus interference was observed with CRISPR RNA editing HC-Pro and GFP2 genes, which reduced the replication and spread of TuMV.

50.2 General Outline of CRISPR/Cas9 Genome Editing in Plants

CRISPR/Cas9 genome editing involves simple steps of designing and cloning, with the use of Cas9 and different guide RNAs (gRNAs) targeting multiple sites in the genome. Besides the primary CRISPR-Cas9 module, several modified Cas9 cassettes (e.g. Nmcas9, Sacas9, Stcas9) have been utilized in crop plants for improving target specificity and reducing off-target cleavage. Further, the availability of Cas9 enzymes from other bacterial species has been made available to enhance specificity and efficiency of gene editing methodologies. Plant-specific RNA polymerase III promoters [AtU6 (*Arabidopsis*); OsU6 or OsU3 (rice); TaU6 (wheat)] are used to express Cas9 and gRNA in plant systems. There are several commercially available vectors for expressing Cas9 variants and gRNAs in plant systems.

50.3 The CRISPR Cleavage Methodology Involves the Following Steps

1. Identify different spacer sequence which is followed by a PAM sequence in the target gene. CRISPR Design (http://crispr.mit.edu/) and DNA2.0 CRISPR gRNA design https://www.dna20.com/eCommerce/cas9/input).
2. Off-target analysis of the spacer sequence in different plant genome sequence database using appropriate bioinformatics tools.
3. Synthesis of the single guide RNA (sgRNA) for all the selected spacer sequences.
4. *In vitro* efficacy analysis of all the spacer sequence and choose the spacers which has clearly shown the cleavage activity.
5. Synthesis of both the strands of 20 nt spacer oligos along with additional nucleotides (if required for cloning) and anneal them to make a double-stranded spacer sequence.
6. Cloning the double-stranded spacer under U6/U3 promoters and terminators into a suitable binary vector which also has gRNA scaffold and *Cas9* gene.

7. Transform bacterial host with the construct prepared above.
8. Agro-inoculation of gRNA-Cas9 construct followed by selection and regeneration of transformed plants.
9. Confirm the mutation in the target region of transformed plants by sequencing.
10. Subject the edited plants to challenge inoculation by the virus through mechanical, insect or graft inoculation methods.
11. Confirm the resistance (if found) by repeated multiple inoculation with the virus and perform quantitative (q) PCR to understand the efficacy of the construct to inhibit virus replication.
12. Screen and validate the edited crops through either restriction enzyme assay or surveyor assay (T7E1 or CLE1) or next-generation sequencing.
13. Subject the edited plants to back cross for the selection of transgene-free (cas9 and sgRNA) plant.

50.4 Notes

1. Editing *eIF4E* and *eIF(iso)4E* host factors which rely on loss-of-function mutations in host *eIFs* may provide broad-spectrum resistance against large number of RNA viruses (Chandrasekaran et al. 2016).
2. Cas9 proteins isolated from a few bacterial species have unique and longer PAM sequences that can aid in increasing on-target specificity and will minimize off-target cleavage.
3. The CRISPR/Cas9 system can be used to develop 'non-transgenic' virus-resistant varieties (Woo et al. 2015).
4. CRISPR/Cas9 can be introduced as transgenes to create the genome editing, and then progeny plants can be selected with desired edits without the Cas9 transgene through segregation (Kanchiswamy 2016). Alternatively, the Cas9 protein and gRNA may be introduced as ribonucleoprotein complex directly into cells that would result in gene editing. The resulting plants would be genetically improved and indistinguishable from plants carrying naturally occurring alleles or identified through random mutagenesis (Voytas and Gao 2014). These approaches can be applied to diverse crops against several plant viruses with RNA or DNA genomes.
5. CRISPR/Cas can target a single or multiple genetic locus in one or many viruses, and simplicity and robustness of this technology make it possible to respond to newly emerging virus strains by deploying appropriate sgRNA transgenes.

6. CRISPR/Cas9 system has disadvantage in generation of off-target cleavage sites as a result of complexing of the gRNA with mismatched complementary target DNA within the genome.
7. The CRISPR/Cas system may select for synonymous or neutral non-synonymous mutations in targeted coding sequences that would enable the virus to escape cleavage. It is also possible that CRISPR/Cas-resistant mutations arise within the targeted conserved non-coding sequences. CRISPR/Cas system is mutagenic by nature and thus acceleration of virus evolution can be expected.
8. Both Cas9 and sgRNAs are consistently expressed in the cells while controlling viruses using the CRISPR-Cas9 system. Recruiting Cas9 to viral DNAs generally depends on the presence and abundance of sgRNAs (Ali et al. 2015).

References

Ali Z, Abulfaraj A, Idris A, Ali S, Tashkandi M, Mahfouz MM (2015) CRISPR/Cas9-mediated viral interference in plants. Genome Biol 16:238

Ali Z, Ali S, Tashkandi M, Zaidi SS, Mahfouz MM (2016) CRISPR/Cas9-mediated immunity to geminiviruses: differential interference and evasion. Sci Rep 6(26):9–12

Aman R, Ali Z, Butt H, Mahas A, Aljedaani F, Zuhaib Khan M, Ding S, Mahfouz M (2018) RNA virus interference via CRISPR/Cas13a system in plants. Genome Biol 19:1

Baltes NJ, Hummel AW, Konecna E, Cegan R, Bruns AN, Bisaro DM (2015) Conferring resistance to geminiviruses with the CRISPR–Cas prokaryotic immune system. Nat Plants 1:15145

Chandrasekaran J, Brumin M, Wolf D, Leibman D, Klap C, Pearlsman M (2016) Development of broad virus resistance in non-transgenic cucumber using CRISPR/Cas9 technology. Mol Plant Pathol 17:1140–1153

Doudna JA, Charpentie E (2014) The new frontier of genome engineering with CRISPR-Cas9. Science 346(6213):1258096

Fondong VN (2013) Geminivirus protein structure and function. Mol Plant Pathol 14:635–649

Hanley-Bowdoin L, Bejarano ER, Robertson D, Mansoor S (2013) Geminiviruses: masters at redirecting and reprogramming plant processes. Nat Rev Microbiol 11:777–788

Ishino Y, Shinagawa H, Makino K, Amemura M, Nakata A (1987) Nucleotide sequence of the IAP gene, responsible for alkaline phosphatase isozyme conversion in *Escherichia coli*, and identification of the gene product. J Bacteriol 169:5429–5433

Ji X, Zhang H, Zhang Y, Wang Y, Gao C (2015) Establishing a CRISPR-Cas-like immune system conferring DNA virus resistance in plants. Nat Plants 1:15144

Jinek M, Chylinski K, Fonfara I, Hauer M, Doudna JA, Charpentier E (2012) A programmable dual-RNA–guided DNA endonuclease in adaptive bacterial immunity. Science 337:816–821

Kanchiswamy CN (2016) DNA-free genome editing methods for targeted crop improvement. Plant Cell Rep 35:1469–1474

Khan MZ, Amin A, Hameed A, Mansoor S (2018) CRISPR–Cas13a: prospects for plant virus resistance. Cell 36(12):1207–1210

Larson MH, Gilbert LA, Wang X, Lim WA, Weissman JS, Qi LS (2013) CRISPR interference (CRISPRi) for sequence-specific control of gene expression. Nat Protoc 8(11):2180–2196

Liu X, Wu S, Xu J, Sui C, Wei J (2017) Application of CRISPR/Cas9 in plant biology. Acta Pharm Sin 7:292–302

Macovei A, Sevilla NR, Cantos C, Jonson GB, Slamet-Loedin I, Čermák T, Voytas DF, Choi IR, Chadha-Mohanty P (2018) Novel alleles of rice eIF4G generated by CRISPR/Cas9-targeted mutagenesis confer resistance to Rice tungro spherical virus. Plant Biotechnol J 16 (11):1918–1927

Pyott DE, Sheehan E, Molnar A (2016) Engineering of CRISPR/Cas9-mediated potyvirus resistance in transgene-free *Arabidopsis* plants. Mol Plant Pathol 17:1276–1288

Sanfacon H (2015) Plant translation factors and virus resistance. Viruses 7:3392–3419

Voytas DF, Gao C (2014) Precision genome engineering and agriculture: opportunities and regulatory challenges. PLoS Biol 12:e1001877

Woo JW, Kim J, Kwon SI, Corvalan C, Cho SW, Kim H et al (2015) DNA-free genome editing in plants with preassembled CRISPR-Cas9 ribonucleoproteins. Nat Biotechnol 33:1162–1164

Zhang T, Zheng Q, Yi X, An H, Zhao Y, Ma S et al (2018) Establishing RNA virus resistance in plants by harnessing CRISPR immune system. Plant Biotechnol J 16(8):1415–1423

Appendix: Common Conversions, Information Sources and Software of Nucleic Acids and Proteins

Weight conversion

$1\ \mu g = 10^{-6}$ g

$1\ ng = 10^{-9}$ g

$1\ pg = 10^{-12}$ g

$1\ fg = 10^{-15}$ g

Spectrophotometric conversion

1 A_{260} unit of double-stranded DNA = 50 μg/mL

1 A_{260} unit of single-stranded DNA = 33 μg/mL

1 A_{260} unit of single-stranded RNA = 40 μg/mL

DNA molar conversions

1 μg of 1000 bp DNA = 1.52 pmole (3.03 pmoles of ends)

1 pmole of 1000 bp DNA = 0.66 μg

Protein molar conversion

100 pmoles of 100,000 Da protein = 10 μg

100 pmoles of 50,000 Da protein = 5 μg

100 pmoles of 10,000 Da protein = 1 μg

Protein/DNA conversion

1 kb of DNA = 330 amino acids of coding capacity – 3.7×10^4 Da protein

10,000 Da protein = 270 bp DNA

50,000 Da protein = 1.35 kb DNA

100,000 Da protein = 2.7 kb DNA

Alangar Ishwara Bhat and Govind Pratap Rao, *Characterization of Plant Viruses*, Springer Protocols Handbooks, https://doi.org/10.1007/978-1-0716-0334-5, © Springer Science+Business Media, LLC, part of Springer Nature 2020

DNA Data

Average weight of DNA base pair (sodium salt) = 650 Da

MW of a double-stranded DNA molecule = (# of base pair) × (650 Da/base pair)

Moles of ends of a double-stranded DNA molecule = 2 × (grams of DNA)/(MW in Daltons)

Moles of ends generated by restriction endonuclease cleavage:

(a) Circular DNA molecule: 2 × (moles of DNA) × (number of sites)

(b) Linear DNA molecule: 2 × (moles of DNA) × (number of sites) + 2 × (moles of DNA)

Picomole ends per microgram of double-stranded linear DNA

$(2 \times 10^6)/(660 \times \text{number of bases}) = \text{pmole ends}/\mu g$

Exact molecular weight of an oligonucleotide

$[(A \times 312.2) + (G \times 328.2) + (C \times 288.2) + (T \times 303.2) - 61.0] = MW(g/mol)$ of specific oligonucleotide

Amino acid abbreviations and molecular weights

Amino acid	Three-letter Abbrev.	One-letter Symbol	MW
Alanine	Ala	A	80
Arginine	Arg	R	174
Asparagine	Asn	N	132
Aspartic acid	Asp	D	133
Asparagine/aspartic acid	Asx	B	–
Cysteine	Cys	C	121
Glutamine	Gln	Q	146
Glutamic acid	Glu	E	147
Glutamine/glutamic acid	Glx	Z	–
Glycine	Gly	G	75
Histidine	His	H	155
Isoleucine	Ile	I	131
Leucine	Leu	L	131
Lysine	Lys	K	146
Methionine	Met	M	149

(continued)

Amino acid	Three-letter Abbrev.	One-letter Symbol	MW
Phenylalanine	Phe	F	165
Proline	Pro	P	115
Serine	Ser	S	105
Threonine	Thr	T	119
Tryptophan	Trp	W	204
Tyrosine	Tyr	Y	181
Valine	Val	V	117

Important DNA and Protein Information Sources and Software

- GenBank (http://www.ncbi.nlm.nih.gov)
- EMBL/EBI (http://www.ebi.ac.uk)
- DNA Data Bank of Japan (DDBJ) (http://www.ddbj.nig.ac.jp/)
- Protein Information Resource (PIR) (http://pir.georgetown.edu/)—Comprehensive, annotated, non-redundant protein sequence database
- VIDA (http://www.biochem.ucl.ac.uk/bsm/virus_database/VIDA.html)—Homologous viral protein families
- BLOCKS (http://blocks.fhcrc.org/)—Multiple alignments of conserved regions of protein families
- CDD (http://www.ncbi.nlm.nih.gov/Structure/cdd/cdd.shtml)—Alignment models for conserved protein domains
- PIR-ALN (http://pir.georgetown.edu/pirwww/dbinfo/piraln.html)—Protein sequence alignments
- International Committee on Taxonomy of Viruses (ICTV) (https://talk.ictvonline.org/)
- Plant virus database (http://www.dpvweb.net/)
- Indian Plant virus database (http://14.139.189.27/virusdb/)

Software for Sequence Analysis

DNA sequence analysis software

- AcaClone pDRAW32 (http://www.acaclone.com): DNA cloning, analysis and visualization software. freeware
- BioTools (http://www.biotools.com): DNA, protein sequence analysis, and chromatogram analysis
- Oligonucleotide Properties Calculator (Oligocalc): (http://biotools.nubic.northwestern.edu/OligoCalc.html

- BLAST (http://www.ncbi.nlm.nih.gov/BLAST/)—Pairwise sequence alignment
- Multiple sequence alignment: (https://www.ebi.ac.uk/Tools/msa/clustalo/)
- DNASIS (http://www.oligo.net/dnasis.htm)—MacOS and Windows-based program for gene analysis
- Sequence similarity searching against protein databases: (http://www.ebi.ac.uk/fasta33/)
- Pairwise sequence analysis: (https://www.ebi.ac.uk/Tools/psa/)
- *GENSCAN* (http://genes.mit.edu/GENSCAN.html)—Predicts complete gene structures in genomic sequences
- *Lasergene* (http://www.dnastar.com)—sequencing, primer design, sequence alignment, databases and database searching, protein analysis and restriction map analysis
- NIH-Repositories of Sequence Analysis Software (http://molbio.info.nih.gov/ *molbio/software.htm*)—NIH (USA)
- GCG: The Wisconsin Package of sequence analysis programs
- Biological sequence alignment editor (Bioedit): http://www.mbio.ncsu.edu/BioEdit/bioedit.html Seqaid
- Translation of a nucleotide (DNA/RNA) sequence to a protein sequence:

  https://web.expasy.org/translate/

  http://insilico.ehu.es/translate/

  https://www.bioinformatics.org/sms2/translate.html

  https://www.ebi.ac.uk/Tools/st/

  https://www.ncbi.nlm.nih.gov/Class/NAWBIS/Modules/DNA/dna21a.html
- Open Reading Frame Finder

  https://www.ncbi.nlm.nih.gov/orffinder/

  https://www.bioinformatics.org/sms2/orf_find.html
- Phylogenetic analysis software:

  Bayesian Analysis of Trees With Internal Node Generation (BATWING)

  Progressive multiple sequence alignment (ClustalW)

  Molecular Evolutionary Genetics Analysis (MEGA)

  Phylogenetic analysis using parsimony (and other methods) (PAUP)

  Phylogenetic inference package (PHYLIP)

Molecular modeling software

- Rasmol (http://www.umass.edu/microbio/rasmol/)—RasMol is a program for molecular graphics visualization tool for showing the structure of DNA, proteins and smaller molecules.
- Swiss PDB Viewer (http://us.expasy.org/spdbv/)—It is tightly linked to Swiss model, an automated homology modeling server.

Primer design and other software

- Amplify (http://engels.genetics.wisc.edu/amplify)—Software for PCR—Ver.2.52b
- CODEHOP (http://blocks.fhcrc.org/blocks/codehop.html)—PCR primers designed from protein multiple sequence alignments
- NetPrimer (http://www.PremierBiosoft.com)—Most comprehensive free primer analysis program on the web including real-time PCR, LAMP and Microarray
- PCR Rare (http://bioinformatics.weizmann.ac.it/software/PC-Rare)—PC-Rare is a very powerful (but user friendly) software that allows the choice of specific PCR primers.
- Primer 3 (http://www-genome.wi.mit.edu/cgi-bin/primer/primer3_www.cgi)—Pick primers from a DNA sequence
- Primer finder (http://eatworms.swmed.edu/~tim/primerfinder/)—Tool to design oligonucleotides suitable for PCR within any sequence
- Primer Premier 5 (http://www.PremierBiosoft.com/)—Primer design program X-Primer (http://alces.umn.edu/pv/pub/ComBin)—for designing PCR primers—the Virtual Genome Center (USA)
- LAMP primer designer tool (Primer explorer V5): (https://primerexplorer.jp/e/)
- RPA primer design: https://www.twistdx.co.uk/en/rpa/using-pcr-primers
- List of Bioinformatics Software Tools for Next-Generation Sequencing: (https://bioinformaticsonline.com/pages/view/26617/list-of-bioinformatics-software-tools-for-next-generation-sequencing)
- Selection of specific region that will yield dsRNA (dsRNA check center): (http://dsCheck.RNA.jp/)
- To design siRNA (siDESIGN center): (http://www.dharmacon.com)

Software for CRISPR/cas9 experiments

Tool	Web address	Main additional features
sgRNA Designer	http://www.broadinstitute.org/rnai/public/analysis-tools/sgrna-design	Efficacy prediction
CRISPRscan	http://www.crisprscan.org/	Efficacy prediction, off-target prediction
WU-CRISPR	http://crispr.wustl.edu/	Design for multiple purposes
E-CRISP	E-CRISP http://www.e-crisp.org/E-CRISP/	Off-target prediction
CRISPR Design	http://crispr.mit.edu/	SNP matching to target sites
Cas9 Design	http://cas9.cbi.pku.edu.cn/index.jsp	Focus on plant species and off-target prediction
CRISPR-P	http://cbi.hzau.edu.cn/cgi-bin/CRISPR	Off-target predictions
CRISPRdirect	http://crispr.dbcls.jp/ Off-target predictions	Off-target prediction
Cas9 Online Designer	http://cas9.wicp.net/	Efficacy prediction

Glossary

Accession number	An identifier supplied by the curators of the major biological databases upon submission of a novel entry that uniquely identifies that sequence (or other) entry.
Acquisition feeding time	The feeding time during which a vector feeds on an infected plant to acquire a virus for subsequent transmission (e.g. to become viruliferous).
Adapter or a linker	It is a short, chemically synthesized, single-stranded or double-stranded oligonucleotide that can be ligated to the ends of other DNA or RNA molecules.
Adenine	A purine base found in DNA and RNA.
Adjuvant	Substance injected with antigens, which non-specifically enhances the immune response to that antigen. Thus, antibody production or the reaction of well-mediated immunity is more vigorous than would be the case were the antigen injected without adjuvant.
Aetiology (etiology)	The science of the causes of diseases; the study of the causal factor, its nature and relations with the host.
Agar	A gelatin-like material obtained from seaweed and used to prepare culture media on which microorganisms are grown and studied.

Alangar Ishwara Bhat and Govind Pratap Rao, *Characterization of Plant Viruses*, Springer Protocols Handbooks,
https://doi.org/10.1007/978-1-0716-0334-5,

Agglutin	An antibody that causes a particular antigen to clump and settle out of suspension.
Alate	Winged form in the life cycle of certain insects (e.g. aphids) (see *apterous*).
Algorithm	Series of steps defining a procedure or formula for solving a problem, that can be coded into a programming language and executed. Bioinformatics algorithms typically are used to process, store, analyse, visualize and make predictions from biological data.
Alignment	Explicit mapping of characters of a sequence to characters of one or more other sequence(s).
Aliquot	An exact fractional sample or portion of a whole (used especially of solutions).
Allele	A given form of a gene that occupies a specific position or locus on a chromosome. Variant forms of genes occurring at the same locus are said to be alleles of one another.
Analogy	Reasoning by which the function of a novel gene or protein sequence may be deduced from comparisons with other gene or protein sequences of known function. Identifying analogous or homologous genes via similarity searching and alignment is one of the chief uses of bioinformatics. (See also alignment, similarity search)
Annealing	Refers to the pairing of complementary single strands of DNA to form a double helix.
Annotation	A combination of comments, notations, references and citations, either in free format or utilizing a controlled vocabulary, that together describe all the experimental and inferred information about a gene or protein. Annotations can also be applied to the description of other biological systems. Batch, automated annotation of bulk biological sequence is one of the key uses of bioinformatics tools.

Antibiotic	A substance produced by a microorganism and able to inhibit the growth of other microorganisms, or to destroy them.
Antibody	A substance that is produced in response to injection of a foreign subsistence (antigen) into an animal body, and those reacts specifically with the foreign substance. Antibodies are modified serum globulins.
Anticodon	The triplet of contiguous bases on tRNA that binds to the codon sequence of nucleotides on mRNA. Example: GGG codes for glycine.
Antigen binding site	Area of an antibody molecule that binds to antigen.
Antigen	A substance that, when injected into an animal body, stimulates the production of a substance (antibody) antagonistic to the substance injected.
Antisense	DNA or RNA composed of the complementary sequence to the target DNA/RNA. Also used to describe a therapeutic strategy that uses antisense DNA or RNA sequences to target specific gene DNA sequences or mRNA implicated in disease, in order to bind and physically inhibit their expression by physically blocking them.
Antiserum titre	The highest dilution of an antiserum that will react with its homologous virus (see *homologous* and *heterologous* reaction).
Antiserum	Serum that contains antibodies.
Apterous	Wingless stage in the life cycle of certain insects (e.g. aphids) (see *alate*).
Attenuation	Lessening of the capacity of a parasitism organism or virus to cause disease.
Autoradiography	It detects radioactively labelled molecules by their effect in creating an image on photographic film.

Avidin	A protein that has a high affinity for biotin and is used in the detection system for biotinylated probes.
Avirulent	Lacking virulence.
Base pair	A pair of nitrogenous bases (a purine and a pyrimidine), held together by hydrogen bonds, that form the core of DNA and RNA, i.e. the A:T, G:C and A:U interactions. Adenine (A), cytosine (C), guanine (G), thymine (T) and uracil (U) or their post-replicationally or post-transcriptionally modified derivatives
Bio-assay	Quantitative estimation of biologically active substances by the extent of their actions under standardized conditions on living organisms.
Biotin	A molecule that can be incorporated into dUTP and used as a non-radioactive label for a DNA probe.
Blot	To transfer DNA, RNA, or proteins, usually from an electrophoresis gel to an immobilizing matrix such as nitrocellulose or nylon membranes.
Blunt-end (ligation)	The joining of DNA fragments that contain no overhang at either end and consequently no DNA bases available for hybridization (cf. sticky-end ligation).
Buffer	A solution which resists changes in pH when acid or alkali is added to it.
Capsid	The protein shell of a virus particle.
Capsomere	The morphological subunits seen on the surface of the virus particle (virion) in the electron microscope. A capsomere is built up from varying numbers of protein subunits (polypeptide chains).
cDNA (complementary DNA)	A DNA strand copied from mRNA using reverse transcriptase. A cDNA library represents all of the expressed DNA in a cell.

cDNA library	A set of DNA fragments prepared from the total mRNA obtained from a selected cell, tissue or organism.
Centrifugation	Technique used for the separation of particles from a solution according to their size, shape, density, viscosity of the medium and rotor speed. The particles are suspended in a liquid medium and placed in a centrifuge tube. The tube is then placed in a rotor and spun at a defined speed.
Centrifuge	An equipment that puts an object in rotation around a fixed axis (spins it in a circle), applying a force perpendicular to the axis of spin (outward) that can be very strong. During centrifugation, radial acceleration causes denser particles to settle to the bottom of the tube, while low-density substances rise to the top.
Chemiluminescence	It is the production of light from a chemical reaction. Two chemicals react to form an excited (high-energy) intermediate, which breaks down releasing some of its energy as photons of light. The emitted light can be captured on an X-ray film that can be subjected to autoradiography to visualize the results. It is used in the non-radioactive detection of nucleic acid in Southern, northern and dot-blot hybridization.
Chimeric clone	A cloning artefact created by a foreign gene being inserted into a vector in an incorrect orientation resulting in the expression of a protein consisting of a fusion of two different gene products.
Chlorosis	Absence, partial or complete, of normal green colour.
Chromosome	The structure in the cell nucleus that contains all of the cellular DNA together with a number of proteins that compact and package the DNA.

Circulative virus	A virus which is transmitted by an insect in a persistent manner and which circulates from the insect's digestive tract, through the haemolymph to the salivary glands, before being transmitted in the saliva as the insect feeds.
Clone	A population of genetically identical cells or DNA molecules.
Cloning vector	A plasmid or phage that is used to 'carry' inserted foreign DNA for the purposes of producing more material or a protein product.
Cloning	The formation of clones or exact genetic replicas.
Cluster	The grouping of similar objects in a multidimensional space. Clustering is used for constructing new features which are abstractions of the existing features of those objects.
Clustered regularly interspaced short palindromic repeats (CRISPR)	This technique is a powerful tool to develop resistance to DNA and RNA plant viruses by editing and inserting novel traits precisely at chosen loci into plants and offers enormous potential in classical breeding.
Coat protein-mediated resistance (CP-MR)	Refers to the resistance of transgenic plants that produce CP to the virus from which the CP gene is derived.
Coding regions (CDS)	The portion of a genomic sequence bounded by start and stop codons that identifies the sequence of the protein being coded for by a particular gene.
Codon	The three letter grouping in coding RNA sequence that selects the specific amino acid for incorporation into the protein.
Colony (of bacteria and yeasts)	A mass of individuals, generally of one species, living together; (of mycelial fungi) a group of hyphae (frequently with spores)
Competent cells	It is the ability of a cell to alter its genetics by taking up extracellular DNA from its environment in the process called transformation.

Conformation	The precise three-dimensional arrangement of atoms and bonds in a molecule describing its geometry and hence its molecular function.
Conjugate	The product of joining two or more dissimilar molecules by covalent bonds.
Consensus sequence	A single sequence delineated from an alignment of multiple constituent sequences that represents a 'best fit' for all those sequences. A 'voting' or other selection procedure is used to determine which residue (nucleotide or amino acid) is placed at a given position in the event that not all of the constituent sequences have the identical residue at that position.
Contig	A contig is a set of overlapping DNA segments that together represent a consensus region of DNA.
Cross protection	The phenomenon in which plant tissues infected with one strain of a virus are protected from infection by other strains of the same virus.
Cross-reacting antigen	Antigen capable of combining with antibody produced in response to a different antigen. May cross-react due to sharing of determinants by the two antigens or because the antigenic determinants of each, although not identical, are closely inoculated stereochemically to combine with antibody against one of them.
Cultivars	A variety of a cultivated plant.
Culture	A growth of an organism for the purpose of experiment, especially on laboratory media (culture media); often used in the sense of isolate or strain.
Database	Any file system by which data gets stored following a logical process.
ddNTP (dideoxy nucleoside triphosphate)	An individual DNA or RNA base that cannot be extended in the 3′ direction.

Degeneracy	Degeneracy in the genetic code refers to the lack of an effect of any change in the third base of the codon on the amino acid that is represented.
Deletion	A chromosomal alteration in which a portion of the chromosome or the underlying DNA is lost.
Denaturation of protein	Describes its conversion from the physiological conformation to some other (inactive) conformation.
Denaturation of nucleic acid	Conversion of DNA or RNA from the double-stranded to the single-stranded state; separation of the strands is most often accomplished by heating.
Dendrogram	A graphical procedure for representing the output of a hierarchical clustering method. A dendrogram is strictly defined as a binary tree with a distinguished root, which has all the data items at its leaves. Conventionally, all the leaves are shown at the same level of the drawing. The ordering of the leaves is arbitrary, as is their horizontal position. The heights of the internal nodes may be arbitrary, or may be related to the metric information used to form the clustering.
Density gradient centrifugation	A centrifugation procedure in which partially purified virus is further clarified by movement through a gradient. Contaminating components may be separated from the virus particles by velocity centrifugation, usually in a low to high sucrose gradient, which separates components according to their differing sedimentation coefficients, or alternatively, by isopycnic centrifugation usually in cesium chloride or cesium sulphate gradients, which separates the components according to their differing buoyant densities.

Differential centrifugation	Cycles of low and high speed clarification and sedimentation used in the purification of a virus.
Dilution end-point	The extent to which sap from virus-infected plants can be undiluted with the water before its infectivity is lost.
Dimer	A composite molecule formed by the binding of two molecules (see homo- and heterodimers).
Disease	Harmful deviation from normal functioning of physiological growth.
Disorder	A harmful nonpathogenic deviation from normal growth.
DNA (deoxyribonucleic acid)	The chemical that forms the basis of the genetic material in virtually all organisms. DNA is composed of the four nitrogenous bases, adenine, cytosine, guanine and thymine, which are covalently bonded to a backbone of deoxyribose-phosphate to form a DNA strand. Two complementary strands (where all Gs pair with Cs and As with Ts) form a double-helical structure which is held together by hydrogen-bonding between the cognate bases.
DNA chips	It allows the simultaneous detection of thousands of cDNAs probes arrayed on a small surface of a microscope slide or chip approximately of 1 cm^2 in size.
DNA polymerase	One of several enzymes that synthesizes a new DNA strand complementary to a template strand by adding nucleotides one at a time to a 3′-OH end.
DNA sequencing	It is the process of determining the order of nucleotides in a DNA fragment.
DNAase	It is an enzyme that attacks bonds in DNA.
Domain (protein):	A region of special biological interest within a single protein sequence.

Double antibody sandwich	A method in enzyme-linked immunosorbent assays (ELISA) in which the reactants are added to the test plate in the order of antibody, virus and antibody-enzyme complex.
Electro blotting	The electrophoretic transfer of macromolecules (DNA, RNA or protein) from a gel in which they have been separated onto a support matrix such as a nitrocellulose sheet. The procedure is an alternative to the capillary transfer used in techniques such as Southern and northern blotting.
Electrophoresis	A technique for separating different types of molecules based on their patterns of movement in an electrical field.
Enation	An abnormal outgrowth (often on the leaf) caused by an increase in cell numbers (hyperplasia).
Encapsidation	The enclosure of a virus's nucleic acid genome within a protein shell.
End labelling	It describes the addition of a radioactively labelled group to one end (5′ or 3′) of a DNA strand.
Endonucleases	These enzymes cleave bonds within a nucleic acid chain; they may be specific for RNA or for single-stranded or double-stranded DNA.
Enveloped virus	Plant viruses of the reovirus and rhabdovirus groups which have an outer lipid-protein membrane surrounding the protein shell of the virus.
Enzyme-linked immunosorbent assay (ELISA):	A serological test in which the sensitivity of the antibody-antigen reaction is increased by attaching an enzyme to one of the two reactants (see *double antibody sandwich*).
Epidemiology	The study of factors affecting the outbreak and spread of infectious diseases.

Exon	The region of DNA within a gene that codes for a polypeptide chain or domain. Typically a mature protein is composed of several domains coded by different exons within a single gene.
Exonuclease	It cleaves nucleotides one at a time from the end of a polynucleotide chain; they may be specific for either 5′ or 3′ end of DNA or RNA.
Expression (gene or protein)	A measure of the presence, amount and time-course of one or more gene products in a particular cell or tissue. Expression studies are typically performed at the RNA (mRNA) or protein level.
Expression vector	A cloning vector that is engineered to allow the expression of protein from a cDNA. The expression vector provides an appropriate promoter and restriction sites that allow insertion of cDNA.
F $(ab')_2$	Fragment obtained by pepsin hydrolysis of an immunoglobulin molecule. The F $(ab')_2$ fragment consists of two fab fragments joined by disulphide bonds.
Fc fragment	The non-antigen-binding fragment of an immunoglobulin molecule produced after digestion with a papain comprising two heavy chain fragments.
Feeder cells	They supply growth factors required for the growth of hybridoma cells after fusion. Growth factors secreting cells such as peritoneal macrophages, splenocytes, thymocytes and myeloma cells of mice could be used as feeder cells.
Frame shift	A deletion, substitution or duplication of one or more bases that causes the reading frame of a structural gene to shift from the normal series of triplets.

Freeze-drying	Preservation of living microorganisms, etc. by removing water under vacuum while tissue remains in frozen state (lyophilization).
Freund's adjuvant	The mixture of mineral oil and lanoline that enhances immune responses when emulsified with antigen for immunization.
Gel double-diffusion	A serological test in which the antibody and antigen reactants diffuse towards each other in gel and react to form a visible precipitation line.
Gel	The inert matrix used for the electrophoretic separation of nucleic acids or proteins.
GenBank	Data bank of genetic sequences operated by a Division of the National Institutes of Health.
Gene chips (also gene arrays)	The covalent attachment of oligonucleotides or cDNA directly onto a small glass or silicon chip in organized arrays. Over 50,000 different DNA fragments can be presented on a single chip providing a high-throughput parallel method of probing gene expression, genotype or gene function.
Gene expression	The conversion of information from gene to protein *via* transcription and translation.
Gene families	Subsets of genes containing homologous sequences which usually correlate with a common function.
Gene library	A collection of cloned DNA fragments created by restriction endonuclease digestion that represent part or all of an organism's genome.
Gene product	The product, either RNA or protein, that results from expression of a gene. The amount of gene product reflects the activity of the gene.
Gene	A segment of DNA on a chromosome that encodes a protein and all the regulatory sequences (promoter) required to control expression of that protein.

Genetic code	The mapping of all possible codons into the 20 amino acids including the start and stop codons.
Genetic engineering (recombinant DNA technology)	The procedures used to isolate, splice and manipulate DNA outside the cell. Genetic engineering allows a recombinantly engineered DNA segment to be introduced into a foreign cell or organism, and be able to replicate and function normally.
Genetic marker	Any gene that can be readily recognized by its phenotypic effect, and which can be used as a marker for a cell, chromosome, or individual carrying that gene.
Genome	The complete genetic content of an organism.
Genomics	Use of complete genomes to explain and interpret biological phenomena.
Glasshouse (Greenhouse)	A glasshouse is a greenhouse which is used for growing plants.
Global alignment	The alignment of two nucleic acid or protein sequences over their entire length.
GUS gene	It is one of the normally used reporter gene for the standardization of transformation protocol for different plant species.
Hairpin	A double-helical region in a single DNA or RNA strand formed by the hydrogen-bonding between adjacent inverse complementary sequences to form a hairpin-shaped structure.
Heterodimer	Protein composed of two different chains or subunits.
Heteroduplex	Hybrid structure formed by the annealing of two DNA strands (or an RNA and DNA) that have sufficient complementarity in their sequence to allow hydrogen-bonding.
Heterologous reaction	A serological reaction in which an antiserum is reacted against an antigen other than the one used in its preparation.

Homologous reaction	A serological reaction in which an antiserum is reacted against the antigen used for its preparation.
Homology	Relationship by evolutionary descent from a common ancestral precursor.
Host	A plant that is invaded by a parasite and from which the parasite obtains its nutrients.
Host-range	The various kinds of host plants that may be attacked by a parasite.
Hybridoma	A hybrid animal cell produced by the fusion of a spleen cell and a cancer cell and able to multiply and to produce monoclonal antibodies.
Identity	The extent to which two (nucleotide or amino acid) sequences are invariant.
IgG	The major immunoglobulin class in the serum of man.
Immune response	The ability of an animal to produce antibodies as a result of antigens, such as proteins, entering its body either by infection with a pathogenic agent, or by artificial injection. The ability of the antigen to induce this response is referred to as *immunogenicity*, and the substances that are capable of inducing the response are called *immunogens*.
Immune	Exempt from infection.
Immunodiffusion	A serological procedure in which the antigen-antibody reaction is carried out by allowing the reactants to diffuse in gel.
Immunosorbent electron microscopy (ISEM) (syn. Electron microscope serology)	Techniques involving the visualization of the antibody-antigen reaction in the electron microscope (E.M.). These include: (1) *Trapping*: A procedure in which the E.M. grid is first coated with antiserum (referred to as an antibody-coated grid, ACG), which then attracts virus particles from a virus preparation placed on it; (2) *Decoration*: In this technique, virus particles are attached to the E.M. grid and

	then the antiserum is added. Homologous antibodies will react with the particles to coat or '*decorate*' them.
In vitro	In culture, outside the host.
In vivo	In the host.
Inclusion body	Virus-induced structures that may occur in the cytoplasm or nucleus of infected plants.
Indexing	A procedure for demonstrating the presence of virus infection in a plant.
Indicator plant	One which reacts to certain viruses or environmental factors with specific symptoms, used for identification of the viruses or the environmental factors.
Infectious disease	A disease that is caused by a pathogen which can spread from a diseased to a healthy plant.
Infectivity assay	A bioassay using mechanical sap-transmission to quantitatively determine the amount of infectious virus.
Inoculate	To introduce a microorganism or virus into an organism or into a culture medium.
Inoculation feeding period	The length of time a vector feeds on a test plant during transmission experiments.
Inoculation threshold period	The minimum feeding period a vector needs on a test plant to transmit a virus.
Insert	The piece of foreign DNA introduced into a vector molecule.
Instar	A growth phase between moults in an insect's life cycle.
Intercellular	Between cells.
Intracellular	Within a cell.
Introns	Nucleotide sequences found in the structural genes of eukaryotes that are non-coding and interrupt the sequences containing information that codes for polypeptide chains.
Isolate	A single spore or culture and the subcultures derived from it. Also, used to indicate collections of a pathogen made from different places or times.

Isolate	A virus that has been obtained from an infected plant.
Isometric	Used to describe virus particles that are approximately spherical in shape.
kb	It is an abbreviation for 1000 base pairs of DNA or 1000 bases of RNA.
Latent infection (Latency)	Infection in a plant without visual symptoms.
Latent period	Time between infection and appearance of disease symptoms or period after acquisition of virus by vector before it becomes infective (viruliferous).
Latent virus	A virus that does not induce symptom development in its host.
Lesion	A localized area of diseased or disordered tissue.
Ligand	Any small molecule that binds to a protein or receptor; the cognate partner of many cellular proteins, enzymes and receptors.
Ligation	Formation of a phosphodiester bond to link two adjacent bases separated by a nick in one strand of a double helix of DNA (The term can also be applied to blunt-end ligation and to joining of RNA).
Local alignment	The alignment of some portion of two nucleic acid or protein sequences.
Local lesion	A localized spot produced on a leaf upon mechanical inoculation with a virus.
Longevity *in vitro*	The storage time after which a virus in a crude sap preparation loses its infectivity. Usually determined at 0 or 20 °C.
Loop-mediated isothermal amplification (LAMP)	It is a molecular detection tool, which can amplify nucleic acid with high specificity, sensitivity and speed under isothermal conditions.
Lyophilization (syn. Freeze-drying)	A technique by which water is removed under vacuum while the preparation or tissue is frozen; used to preserve viruses or antisera.

Mechanical inoculation	Inoculation of a plant with a virus through transfer of sap from a virus-infected plant to a healthy plant. May also occur in the field when virus is transmitted from one plant to another by leaves rubbing or root contact.
Medium (culture medium)	A substance or solution for the culture of microorganisms.
Melting (of DNA)	The denaturation of double-stranded DNA into two single strands by the application of heat. (Denaturation breaks the hydrogen bonds holding the double-stranded DNA together.)
Melting temperature (T_m)	It is the midpoint of the temperature range over which DNA is denatured.
Meristem-tip culture	The meristem dome of cells and one or two pairs of primordial leaves (0.5–1 mm in diameter), which comprises the explants removed from a bud and grown in tissue culture to produce a virus-free plant.
Messenger RNA (mRNA)	The complementary RNA copy of DNA formed from a single-stranded DNA template during transcription that migrates from the nucleus to the cytoplasm where it is processed into a sequence carrying the information to code for a polypeptide domain.
Microarray	A 2D array, typically on a glass, filter or silicon wafer, upon which genes or gene fragments are deposited or synthesized in a predetermined spatial order allowing them to be made available as probes in a high-throughput, parallel manner.
Monoclonal antibodies (MAb)	Identical immunoglobulins targeted to a single epitope, generated from a single B-cell clone. These antibodies recognize and bind to a single epitope in an antigen.
Mosaic	Patchy variation of normal green colour, symptomatic of virus diseases.

Motif	A short conserved region in a protein sequence. Motifs are frequently highly conserved parts of domains.
Mottle	Arrangement of spots or confluent blotches of colour, often symptomatic of virus diseases.
Multicomponent virus	A virus whose genome is divided into two or more parts, each part being separately encapsidated. Hence two or more components are needed to initiate an infection.
Multiple (sequence) alignment	A multiple sequence alignment is a sequence alignment of three or more biological sequences, generally protein, DNA or RNA. In many cases, the input set of query sequences are assumed to have an evolutionary relationship by which they share a linkage and are descended from a common ancestor.
Multiplex sequencing	Approach to high-throughput sequencing that uses several pooled DNA samples run through gels simultaneously and then separated and analysed.
Mutant	An organism that shows one or more discrete heritable differences from standard type.
Naked DNA	Pure, isolated DNA devoid of any proteins that may bind to it.
Necrosis	Death of plant cells, especially resulting in darkening of the tissues.
Nested PCR	The second round amplification of an already PCR-amplified sequence using a new pair of primers which are internal to the original primers. Typically done when a single PCR reaction generates insufficient amounts of product.
Next-generation sequencing (NGS)	Enables a genome to be sequenced within hours to days through massive parallel sequencing approach.

Nitrocellulose	A nitrated derivative of cellulose that is made into membrane filters of defined porosity. These filters have a variety of uses in molecular biology in which the nucleic acids are transferred from an agarose gel to a nitrocellulose filter.
Non-persistent virus	The virus which remains infective within its insect vector for only a short period (few seconds to minute).
Northern blotting	Technique for transferring RNA from an agarose gel to a nitrocellulose filter on which it can be hybridized to a complementary DNA.
Nucleotide	A nucleic acid unit composed of a five carbon sugar joined to a phosphate group and a nitrogen base.
Open reading frame (ORF)	Any stretch of DNA that potentially encodes a protein. Open reading frames start with a start codon and end with a termination codon. No termination codons may be present internally. The identification of an ORF is the first indication that a segment of DNA may be part of a functional gene.
Pathogen derived resistance (PDR)	The use of pathogen genes or sequences to protect plants against the attack from the same pathogen or closely related pathogen.
Pathogen	A parasite able to cause disease in a particular host or range of hosts.
Persistent transmission	A type of insect transmission in which the virus is acquired by the vector only after a long acquisition feeding period, and in which there may be a latent period following the acquisition feed, before the vector can transmit the virus. The vector remains viruliferous for a long period.
Persistent virus	A virus which remains infective within its insect vector for a long period.
Phloem	Food-conducting tissue, consisting of sieve tubes, companion cells, phloem parenchyma and fibres.

Phylogenetic analysis	Estimating the evolutionary past based on the comparison of DNA or protein sequences. It is usually depicted as branching (tree-like) diagrams, which represent a sort of pedigree of the inherited relationships among organisms.
Plasmid	An autonomous self-replicating extrachromosomal circular DNA.
Polymerase chain reaction (PCR)	Technique used to amplify or generate large amounts of replica DNA of a segment of any DNA whose 'flanking' sequences are known.
Polyphagus	An insect such as an aphid that feeds on various secondary host species.
Precipitation (syn. precipitin) reaction	A visible precipitation reaction that occurs when antibodies and antigens react to form an insoluble lattice.
Primary host	The plant on which the sexual forms of an aphid mate and lay eggs to overwinter.
Primary sequence (protein)	The linear sequence of a polypeptide or protein.
Primer	A short sequence (often of RNA) that is paired with one strand of DNA and provides a free 3′-OH end at which a DNA polymerase starts synthesis of a deoxyribonucleotide chain.
Probe	A defined nucleic acid sequence that has been labelled with a radioactive isotope or fluorescent dye. It is used to identify specific DNA and RNA molecules that have the complementary sequence. Also, used to perform hybridization to detect a specific gene or transcript.
Promoter	A region of DNA involved in binding of RNA polymerase to initiate transcription.
Propagative virus	A virus that multiplies within its insect vector.
Protein subunit	A small protein molecule that is the structural and chemical unit of the protein coat of a virus.

Proteome	The entire protein complement of a given organism.
Proteomics	The study of the proteome. Typically, the cataloguing of all the expressed proteins in a particular cell or tissue type, obtained by identifying the proteins from cell extracts using a combination of 2D gel electrophoresis and mass spectrometry. The large-scale analysis of the protein composition and function. (cf genomics)
PSI-BLAST (Position-Specific Iterative BLAST)	An iterative search using the BLAST algorithm. A profile is built after the initial search, which is then used in subsequent searches. The process may be repeated, if desired with new sequences found in each cycle used to refine the profile.
Purification	The isolation and concentration of virus particles in a pure infective form, free from cell components.
Pyrosequencing	A method of DNA sequencing based on the 'sequencing by synthesis' principle, in which the sequencing is performed by detecting the nucleotide incorporated by a DNA polymerase.
Reading frame	A sequence of codons beginning with an initiation codon and ending with a termination codon, typically of at least 150 bases (50 amino acids) coding for a polypeptide or protein chain (see ORF).
Real-time PCR (or quantitative PCR)	Technique that couples amplification of a target DNA sequence with quantification of the concentration of that DNA species in the reaction.
Recombinant DNA (rDNA)	DNA molecules resulting from the fusion of DNA from different sources. The technology employed for splicing DNA from different sources and for amplifying the resultant heterogenous DNA.

Recombinase polymerase amplification	An isothermal amplification method that is carried out at a constant single temperature used for detection of DNA without PCR machine. It has high specificity and sensitivity and does not require denaturing of the template DNA by heating. Denaturing is done by an enzymatic activity that will help in annealing of primers to the complementary sequences in the target region.
Reporter gene	It is a coding unit whose product is easily assayed (such as chloramphenicol acetyl transferase, GUS); it may be connected to any promoter of interest so that expression of the gene can be used to assay promoter function.
Resistance	The power of an organism to overcome, completely or in some degree, the effect of a pathogen or other damaging factor.
Restriction enzyme (restriction endonuclease)	A type of enzyme that recognizes specific DNA sequences (usually palindromic sequences 4, 6, 8 or 16 base pairs in length) and produces cuts on both strands of DNA containing those sequences only. The 'molecular scissors' of rDNA technology.
Reverse transcriptase-PCR (RT-PCR)	Procedure in which PCR amplification is carried out on DNA that is first generated by the conversion of mRNA to cDNA using reverse transcriptase.
Reverse transcription	Synthesis of DNA on a template of RNA; accomplished by reverse transcriptase enzyme.
Ribonucleic acid (RNA)	A category of nucleic acids in which the component sugar is ribose and consisting of the four nucleotides, thymidine, uracil, guanine and adenine. The three types of RNA are messenger RNA (mRNA), transfer RNA (tRNA) and ribosomal RNA (rRNA).
RNA silencing (or post-transcriptional gene silencing, PTGS or RNA interference)	It is the major mechanism involved in imparting virus resistance in transgenic plants.
RNase	An enzyme whose substrate is RNA.

Rolling circle amplification (RCA)	A method of isothermal amplification of circular DNA molecules. RCA assay involves DNA amplification using a DNA ϕ polymerase with strand displacement activity to extend a single or multiple primers annealed to a circular DNA template.
Sedimentation coefficient	The rate of sedimentation of a virus per unit centrifugal field measured in *Svedberg units* (*S*).
Semi-persistent transmission	Virus transmission by an insect vector that is intermediate between *non-persistent* and *persistent transmission*.
Sense strand	The strand of double-stranded DNA that acts as the template strand for RNA synthesis. Typically only one gene product is produced per gene, reading from the sense strand only. (Some viruses have open reading frames in both the sense and the antisense strands.)
Sequence	An ordered succession of characters, units or symbols, *i.e.* a string.
Serology	A method using the specificity of the antigen-antibody reaction for the detection and identification of antigenic substances and the organisms that carry them.
Serum	Fluid expressed from a blood clot as it contracts after coagulation of the blood. The essential difference between plasma and serum is that the latter does not contain fibrinogen.
SI nuclease	An enzyme that specifically degrades unpaired (single-stranded) sequences of DNA.
Similarity (homology) search	Method used for the prediction of structure and function of newly sequenced DNA based on the detection of significant extended sequence similarity to a protein of known structure, or of a sequence pattern characteristic of a protein family.

Somatic embryo	Somatic embryos are formed from plant cells that are not normally involved in the development of *embryos*, *i.e.* ordinary plant tissue.
Somatic embryogenesis	Development of plants from somatic embryos without the fusion of gametes through embryo stages.
Southern blotting	The procedure for transferring denatured DNA from an agarose gel to a nitrocellulose filter where it can be hybridized with a complementary nucleic acid.
Southern hybridization	One of the methods used for confirmation of the integration of the transgene into the genome of the transgenic plants.
Sterilization	The elimination of pathogens and other living organisms from soil, containers, etc., by means of heat or chemicals.
Structural gene	Gene which encodes a structural protein (for example, coat protein in a virus).
Stylet-borne	A virus borne on the stylet of its vector; a non-circulative virus.
Subculture	A culture derived from another one.
Substitution	The presence of a non-identical amino acid at a given position in an alignment. If the aligned residues have similar physico-chemical properties, the substitution is said to be 'conservative'.
Symptom	The external and internal reactions or alterations of a plant as a result of a disease.
Symptomless carrier	A plant which, although infected with a pathogen (usually a virus), produces no obvious symptoms.
Template nucleic acid	A template is a single-stranded DNA or RNA polymer that is used to direct synthesis of another polymer such as DNA, RNA or protein. DNA polymerases use template DNA by covalently linking deoxyribonucleoside 5′-triphosphates that base pair with template DNA to form a new, complementary DNA strand.

Termination codon	One of three triplet sequences. DAG (amber), UAA (ochre) or UGA that cause termination of protein synthesis; they are also called nonsense codons.
Thermal inactivation point	The lowest temperature at which heating for a limited period (usually 10 min) is sufficient to cause a virus to lose its infectivity or an enzyme its activity.
Thermoscript RT	A reverse transcriptase enzyme which can withstand temperature up to 65 °C and thus suitable for RT-LAMP assay.
Titre	In serological reactions, a relative measure of the amount of antibody in an antiserum per unit volume of original serum. The antibody is serially diluted and antigen is added. Serum titre is indicated as the reciprocal of the highest serum dilution producing a discernible antigen-antibody reaction.
Transcapsidation	The encapsidation of the nucleic acid of one virus strain with the protein of another, during simultaneous infection and replication of two strains.
Transcription factors	A group of regulatory proteins that are required for transcription in eukaryotes. Transcription factors bind to the promoter region of a gene and facilitate transcription by RNA polymerase.
Transcription	Synthesis of RNA on a DNA template.
Transfer RNA (tRNA)	A small RNA molecule that recognizes a specific amino acid, transports it to a specific codon in the mRNA, and positions it properly in the nascent polypeptide chain.
Transformation of bacteria	Acquisition of new genetic markers by a bacterium by incorporation of added DNA.
Transgenic plants	Created by introducing new DNA sequences into the host DNA either through biolistic or through *Agrobacterium*-mediated delivery.

Translation	The process of converting RNA to protein by the assembly of a polypeptide chain from an mRNA molecule at the ribosome.
Transmission	The transfer or spread of a virus or other pathogen from one plant to another.
Transovarial transmission	When virus is transmitted through the eggs of the infected vector to its progeny.
Transstadial blockage	When virus is retained through the moult of its insect vector.
Uracil	Nitrogenous pyrimidine base found in RNA but not DNA.
Vector	Any agent that transfers material (typically DNA) from one host to another. Typically DNA vectors are autonomous DNA elements (such as plasmids) that can be manipulated and integrated into a host's DNA or recombinant viruses.
Vector	An organism able to transmit a virus.
Virion	The complete virus particle consisting of ribonucleic acid and protein shell.
Viroid	A pathogenic agent consisting of ribonucleic acid of low molecular weight without a protein coat.
Virulent	Strongly pathogenic.
Viruliferous	A vector that carries or contains virus.
Virus	A submicroscopic obligate parasite consisting of nucleic acid and protein.
Western blot	A procedure for transfer of proteins after separation on a polyacrylamide gel to a suitable immobilizing matrix such as a nitrocellulose membrane. The proteins attached to the support matrix can then be probed with a specific antibody to identify a particular protein species.
Wild type	Form of a gene or allele that is considered the 'standard' or most common.

MIX
Papier aus verantwortungsvollen Quellen
Paper from responsible sources
FSC® C105338

If you have any concerns about our products,
you can contact us on
ProductSafety@springernature.com

In case Publisher is established outside the EU,
the EU authorized representative is:
Springer Nature Customer Service Center GmbH
Europaplatz 3, 69115 Heidelberg, Germany

Printed by Libri Plureos GmbH
in Hamburg, Germany